Sustainable Civil Infrastructures

About this Series

Sustainable Infrastructure impacts our well-being and day-to-day lives. The infrastructures we are building today will shape our lives tomorrow. The complex and diverse nature of the impacts due to weather extremes on transportation and civil infrastructures can be seen in our roadways, bridges, and buildings. Extreme summer temperatures, droughts, flash floods, and rising numbers of freeze-thaw cycles pose challenges for civil infrastructure and can endanger public safety. We constantly hear how civil infrastructures need constant attention, preservation, and upgrading. Such improvements and developments would obviously benefit from our desired book series that provide sustainable engineering materials and designs. The economic impact is huge and much research has been conducted worldwide. The future holds many opportunities, not only for researchers in a given country, but also for the worldwide field engineers who apply and implement these technologies. We believe that no approach can succeed if it does not unite the efforts of various engineering disciplines from all over the world under one umbrella to offer a beacon of modern solutions to the global infrastructure. Experts from the various engineering disciplines around the globe will participate in this series, including: Geotechnical, Geological, Geoscience, Petroleum, Structural, Transportation, Bridge, Infrastructure, Energy, Architectural, Chemical and Materials, and other related Engineering disciplines.

More information about this series at http://www.springer.com/series/15140

D.N. Singh · Abdullah Galaa
Editors

Contemporary Issues in Geoenvironmental Engineering

Proceedings of the 1st GeoMEast International Congress and Exhibition, Egypt 2017 on Sustainable Civil Infrastructures

Springer

Editors
D.N. Singh
Indian Institute of Technology Bombay
Bombay
India

Abdullah Galaa
Housing and Building National Research Center
Giza
Egypt

ISSN 2366-3405 ISSN 2366-3413 (electronic)
Sustainable Civil Infrastructures
ISBN 978-3-319-61611-7 ISBN 978-3-319-61612-4 (eBook)
DOI 10.1007/978-3-319-61612-4

Library of Congress Control Number: 2017946433

Printed on acid-free paper

This Springer imprint is published by Springer Nature
The registered company is Springer International Publishing AG
The registered company address is: Gewerbestrasse 11, 6330 Cham, Switzerland

Preface

Toward building sustainable and longer civil infrastructures, the engineering community around the globe continues undertaking research and development to improve existing design, modeling, and analytical capability. Such initiatives are also the core mission of the Soil-Structure Interaction Group in Egypt (SSIGE) to contribute to the ongoing research toward sustainable infrastructure. This conference series “GeoMEast International Congress and Exhibition” is one of these initiatives.

Ancient peoples built their structures to withstand the test of time. If we think in the same way, our current projects will be a heritage for future generations. In this context, an urgent need has quickly motivated the SSIGE and its friends around the globe to start a new congress series that can bring together researchers and practitioners to pursue “Sustainable Civil Infrastructures.” The GeoMEast 2017 is a unique forum in the Middle East and Africa that transfers from the innovation in research into the practical wisdom to serve directly the practitioners of the industry.

More than eight hundred abstracts were received for the first edition of this conference series “GeoMEast 2017” in response to the Call for Papers. The abstracts were reviewed by the Organizing and Scientific Committees. All papers were reviewed following the same procedure and at the same technical standards of practice of the TRB, ASCE, ICE, ISSMGE, IGS, IAEG, DFI, ISAP, ISCP, ITA, ISHMII, PDCA, IUGS, ICC, and other professional organizations who have supported the technical program of the GeoMEast 2017. All papers received a minimum of two full reviews coordinated by various tracks chairs and supervised by the volumes editors through the Editorial Manager of the SUCI “Sustainable Civil Infrastructure” book series. As a result, 15 volumes have been formed of the final +320 accepted papers. The authors of the accepted papers have addressed all the comments of the reviewers to the satisfaction of the tracks chairs, the volumes editors, and the proceedings editor. It is hoped that readers of this proceedings of the GeoMEast 2017 will be stimulated and inspired by the wide range of papers written by a distinguished group of national and international authors.

Publication of this quality of technical papers would not have been possible without the dedication and professionalism of the anonymous papers reviewers. The names of these reviewers appear in the acknowledgment that follows. For any additional reviewers whose names were inadvertently missed, we offer our sincere apologies.

We are thankful to Dr. Hany Farouk Shehata, Dr. Nabil Khelifi, Dr. Khalid M. ElZahaby, Dr. Mohamed F. Shehata, and to all the distinguished volumes editors of the proceedings of the GeoMEast 2017. Appreciation is extended to the authors and track chairs for their significant contributions. Thanks are also extended to Springer for their coordination and enthusiastic support to this conference. The editors acknowledge the assistance of Ms. Janet Sterritt-Brunner, Mr. Arulmurugan Venkatasalam in the final production of the 15 edited volumes "Proceedings of GeoMEast 2017".

Contents

The Influence of Filler Content on the Compactness of the Bituminous Mixture

Sami Yaich[1(✉)] and Kaouther Machta[2]

[1] Maitre Technologue, Institut Supérieur des Etudes Technologiques Sfax, Sfax, Tunisia
Sami.yaich@gmail.com
[2] Maitre Technologue, Institut Supérieur des Etudes Technologiques Rades, Radès, Tunisia
Kaouther.machta@planet.tn

Abstract. The study of asphalt concrete mix design consists of a series of compromises. The behavior of the mixture oscillates between sensitivity to permanent strain and fatigue cracking and the stripping or peeling away aggregates. In a mix design study, we investigate the proportion of the various components able to reach and ensure, during the achieved pavement life, the maintenance of a satisfactory level of performance. In this paper the influence of filler content for different bitumen content on the (density) compaction of the Tunisian asphalt mix is studied based on the Marshall test in accordance with the AASHTO standard.

Keywords: Mix design · Filler · Bitumen · Compactness · Marshall flow and stability

1 Introduction

Design methods for bituminous mixes have been developed in order to satisfy the latest requirements issued by road builders and engineering companies. Rising traffic loads, along with the integration of safety, comfort, durability, maintenance and user nuisance considerations, under given climatic conditions and within a defined technical context (design and dimensioning of pavements layers) have incited increasingly-complex material design approaches [4, 8].

In Tunisia, the mix design of bituminous mixtures for coating layers is based on the French standard which defines levels of study for the formulation of asphalt taking into account the classification of the road in the road network [1]. For this test made are Duriez, press shear Gyrator and the ornerier for heavy traffic roads [7]. The Marshall test is no longer considered in these studies in the French specifications. All laboratories in our country are equipped with the Duriez and Marshall testing equipment, only one has the shear gyratory compactor (SGC). For highways, where traffic is important, mix design is performed in a French laboratory. In this work, we seek to review the mix design process of asphalt concrete based on the Marshall test with AHTOO standard and to investigate the influence of the fines content on compactness and stability of the mixture [2].

D.N. Singh and A. Galaa (eds.), *Contemporary Issues in Geoenvironmental Engineering*, Sustainable Civil Infrastructures,
DOI 10.1007/978-3-319-61612-4_1

2 Fundamentals of Asphalt Paving Mix Design

The overall objective of the design of asphalt paving mixes is to determine a cost-effective blend and gradation of aggregate and asphalt that yields a mix having [3, 6]:

- sufficient Asphalt to ensure a durable pavement
- sufficient mix stability to satisfy the demands of traffic without distortion or displacements
- sufficient voids in the total compacted mix to allow for a slight amount of additional compaction under traffic loading and slight amount of asphalt expansion due to temperature increases without flushing, blending, or loss of stability
- a maximum void content to limit the permeability of harmful air and moisture in the mix
- sufficient Workability to permit efficient placement of the mix without segregation and without sacrificing stability and performance
- proper aggregate texture and hardness to provide sufficient skid resistance in unfavorable weather conditions.

Surface course mixes must be designed to have sufficient stability and durability to both carry the anticipated traffic loads and to withstand the detrimental effects of air, water, and temperature changes.

3 Materials Specifications

The study of the composition formula of asphalt is made with the following materials:

- Gravel 0 / 14 from “FAIEDH” career
- Bitumen 35/50 from Oil Libya asphalt plant in Sfax
- The Lime from the cement plant in Gabes

3.1 Sieve Analysis

Grading curves are established for the three fractions: 0/4, 4/8, and 8/14.

The aggregates were mixed in some proportions, but the size analysis of the mixture led to an erratic that we adjusted with a blended aggregate gradation that approaches the median of the specification limits (Fig. 1).

3.2 Physical Characteristics of Materials

Tests are run to characterize the aggregate. These tests determine gradation and size, voids in mineral aggregates MVA and effective specific gravity of aggregate MVR.

Gradation has a profound effect on material performance. The best gradation is one that produces the maximum density. This would involve a particle arrangement where smaller particles are packed between the larger particles, which reduces the void space between particles.

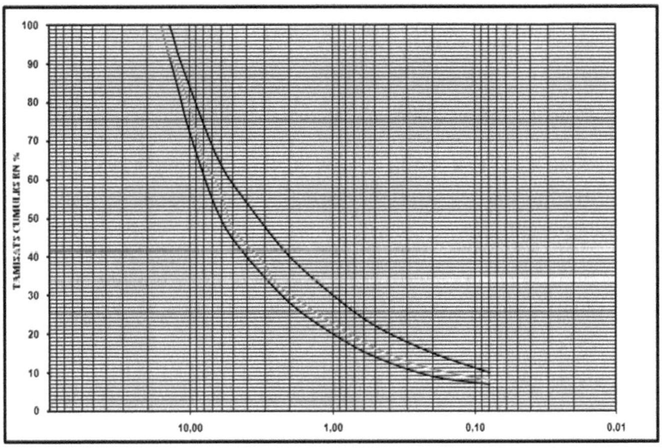

Fig. 1. Sieve analysis curve 0/14

Table 1. Physical properties of materials

Granulometric fraction	MVR (g/cm^3)	VMA (g/cm^3)
Sand 0/2	2,84	1,60
Gravel 2 /4	2,75	1,50
Gravel 4/8	2,67	1,48
Gravel 8/14	2,64	1,55

3.3 Mechanical Tests

Aggregate abrasion characteristics are important because the constituent aggregate in HMA must resist crushing, degradation and disintegration in order to produce a high quality HMA. Choc resistance of used aggregate is LA = 15.

Friction resistance can be considered as the measure of the resistance of a pavement surface to vehicles sliding and skidding, and it is directly related to the forces developed due to the tire–pavement interaction.

Wet friction resistance MDE = 12.5.
Dry friction resistance MDS = 9.66.

4 Mix Specifications

The mixture obtained is a hot mix asphalt (HMA) continuous granularity, consisting of gravel, filler, binder and granularity 0/14. Different formulas to study.

Binder content are usually expressed in terms of percentage of asphalt binder by the total weight of HMA. This ratio has a profound effect on the Hot mix asphalt pavement performance. Because of the wide differences in aggregate specific gravity, the proportion of asphalt binder expressed as a percentage of total weight can vary widely even though the volume of asphalt binder as a percentage of total volume remains quite constant [3].

$$\text{Binder content (\%)} = [(\text{weight of binder})/(\text{aggregate dry weight})] \times 100$$

The percentage of binder is connected with the richness modulus concept characterized somewhat by the thickness of the asphalt film surrounding the aggregate given by the following relationship:

Binder content:

$$\mathrm{TL} = \mathrm{K}.\alpha.\sqrt[5]{\Sigma} \qquad (1)$$

α = correction factor to take into account the actual density of aggregates (MVR)

$$\alpha = \frac{2.65}{\mathrm{MVRg}} \qquad (2)$$

K = richness modulus

It is a characteristic of the film thickness of asphalt that coats the grains.

Σ = specific area of aggregates

The specific surface of aggregates expressed in m2/kg, that is to say, the developed area would have aggregates assimilated to sphere shape.

The surface is calculated by cutting the curve in small size fractions which are assigned a specific area average.

$$\sum = \mathbf{0.25G + 2.3S + 12s + 135f} \qquad (3)$$

G = weight proportion of the top to 6.3 mm.
S = weight proportion of components between 6.3 and 0.315 mm.
s = weight proportion of components between 0.315 and 0.08 mm.
f = proportion of items below 0.08 mm.

$$\mathrm{K} = \frac{TL}{\alpha \times \sqrt[5]{\Sigma}} \qquad (4)$$

Bulk specific gravity of the coated:

$$\mathrm{MVR} = \frac{100 + TL}{[(\%\mathrm{Gi/MVRgi}) + (\mathrm{TL/MVLiant})]} \qquad (5)$$

In determining the asphalt content design for a particular blend or gradation of aggregates and for a different filler's content, a series of test specimen (48) are prepared for a range of asphalt contents.

Table 2. Asphalt content design for a particular blend or gradation of aggregates

Specimen	Aggregates weight (g)	Bitumen content (%)	Filler content (%)	K	MVR (g/cm^3)	Specimen weight (g)	Asphalt mix weight (g)
1	3787	5	0	3.088	2.520	3787	3976,35
2	3690		4		2.533	3837,6	4029,6
3	3635		6		2.538	3853,1	4046,1
4	4005		8		2.544	4325,4	4541,4
5	3755	5,10	0	3.150	2.517	3755	3947
6	4150		4		2.529	4316	4536
7	4685		6		2.535	4966,1	5219,3
8	4395		8		2.541	4746,6	4988,6
9	3687	5,20	0	3.212	2.514	3687	3879
10	4880		4		2.526	4035,2	4244,2
11	4890		6		2.532	4123,4	4337,8
12	4065		8		2.537	4390,2	4618,5
13	3811	5,30	0	3.273	2.510	3811	4012
14	4895		4		2.522	5090,8	5360,6
15	4230		6		2.528	4483,8	4721,4
16	3605		8		2.534	3893,4	4099,7

5 The Marshall Method

The Marshall method uses standard test specimen of 64 mm height and 102 mm diameter. These are prepared using a specified procedure for heating, mixing, and compacting the asphalt-aggregate mixture. The two principal features of the Marshall method of mix design are the density-voids analysis and the stability-flow test of compacted test specimens. The stability of the test specimen is the maximum load resistance in Newton that the standard test specimen developed at 60° when tested. The flow value is the total movement or stain, in units of 0,25 mm occurring in the specimen between no load and the point of maximum load during the stability test [5] (Fig. 2).

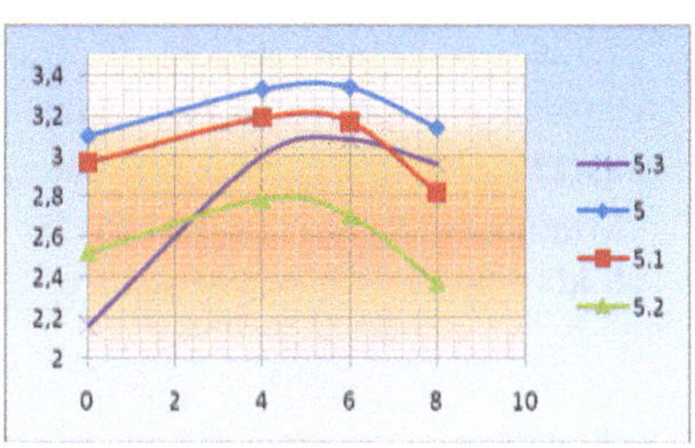

Fig. 2. Influence of filler on determination of flow

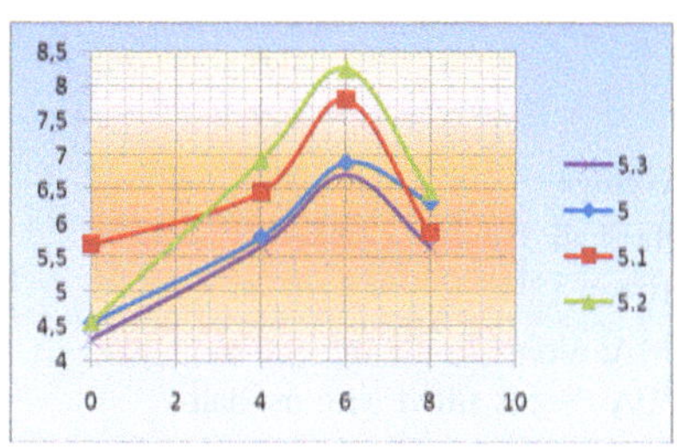

Fig. 3. Influence of filler on determination of stability

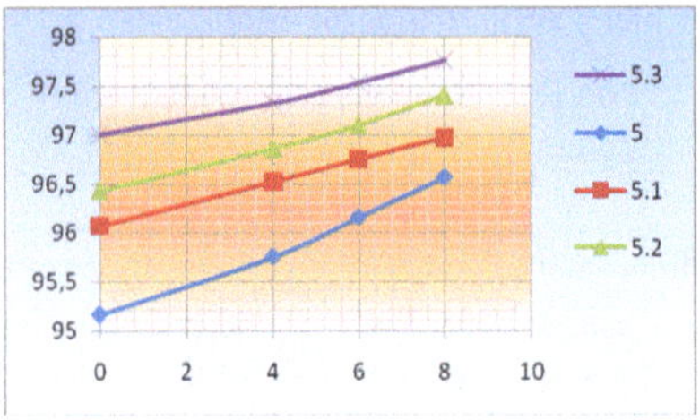

Fig. 4. Influence of filler on determination of compact

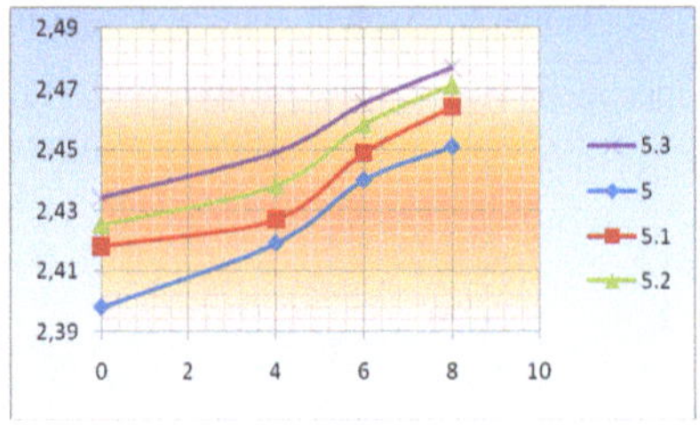

Fig. 5. Influence of filler on density

The stability increases with the filler content percentage to a 6% optimum of the filler content and then decreases. With no filler added, the specimen with 5.1% of binder content gives the best stability (Fig. 3).

The density and the compactness of the mix grow with the filler content and the binder content (Fig. 4).

The admixture of filler in the mix design requires more binder content; the specimen with 5.2% of binder content gives the best stability. When the binder content reaches 5.3%, we note a decrease of the stability. In fact, the mix becomes soft (Fig. 3).

The compacted paving mixture with 5,2% bitumen content and 6% filler content is analyzed in the table below. Specific data for the relevant mixture are (Fig. 5) (Tables 1, 2 and 3):

Maximum specific gravity of asphalt mix: 2,526
Aggregate specific gravity: 2,717
Specific gravity of Bitumen: 1, 03
Number of blows: 75 on both sides

Table 3. Marshall test results

Sample	Thickness (mm)	Volume (cc)	Bulk specific gravity (g/cc)	Air content (%)	Stability (kN)	Flow (units)	VMA (%)	VFA (%)
1	62,1	487,2	2,453	3,12	8,751	7,60	14,02	77,70
2	61,9	490,5	2,446	3,40	7,729	11,48	14,25	76,10
3	62,2	497,7	2,432	3,95	9,988	8,40	14,74	73,20
Average				**3.49**	**8,822**	**9,16**	**14,34**	**75,66**
Marshall criteria				**3–5**	**Minimum 8 kN**	**8–14**	**13–15**	**65–75**

VMA: Voids in the mineral aggregate.
VFA: Voids filled with asphalt.

6 Conclusion and Perspectives

Incorporation of filler in the composition of job mix design formula, to some extent, improves the mechanical strength of the mixing from the point of view Marshall Stability and flow. This is also confirmed by the compactness of the mixture. This adding leads to the increase of the binder content but improves the quality and the durability of the hot mix asphalt.

It could be interesting to replicate this study by considering other quality mineral filler such as limestone, other aggregate sources and other gradations.

Finally, we hope to acquire shear gyratory compactors in order to characterize and appreciate the influential parameters of asphalt concrete in their composition study.

References

1. Delorme, J.-L., de la Roche, C., Wendiling, L.: LCP Bituminous Mixture Design Guide under the supervision
2. Mix design methods for Asphalt concrete and others Hot-Mix Types Asphalt Institute Manual Series N°2 six edition
3. Hot mix Asphalt pavement design guide Penysylvania Asphalt Pavement Association PAPA
4. Pavement design guide Manuel Notice Thomas R. Bohuslav, P.E., Director, Construction Division
5. Marshall Mix design Pavement interactive. http://pavementinteractive.org
6. Hot mix design fundamentals. http://training.ce.washington.edu/PGI/Modules/05_mix_design/05-2_body.htm
7. Brosseaud, Y.: Senior research in LCPC Nantes – France French bituminous mix design versus pavement design: a good tool for innovations in accordance with the European references
8. Impact of Construction Variability on Pavement Performance Research Division Nevada Department of Transportation Research Report No. 13DN-2. University of Nevada, December 2005

Thin Sections Images Processing Technique for the Porosity Estimation in Carbonate Rocks

N.D. Nurgalieva[1] and N.G. Nurgalieva[2(✉)]

[1] Department of General and Applied Physics,
The Moscow Institute of Physics and Technology, Dolgoprudny, Russia
[2] Institute of Geology and Petroleum Technologies,
Kazan (Volga Region) Federal University,
Kazan 420008, Republic of Tatarstan, Russian Federation
nurgal07@mail.ru

Abstract. In present paper we used program Cluster Image which was created in Java to process thin section image of carbonate rock to estimate its porosity on image of any format and with a strong color contrast between the mineral part and pores in thin sections under polarized light. For the experiment the images of thin sections of carbonate rocks of the Carboniferous age were used. Cluster Image does picture clustering with parameters given by user. After opening the program, the picture should be downloaded and parameters should be chosen. A thin section photo in polarized light can be downloaded in any format; also a folder or an URL address containing pictures can be chosen as a material for clustering. It is necessary to specify the number of clusters. To process clustering ISODATA algorithm is preferred because it is iterative and accurate. It is necessary to specify the number of clusters, the percentage of their convergence and the minimum size of one cluster (in pixels). Each pixel's color can be represented as vector of three components in RGB basis. As a result, the picture is a set of vectors which have to be divided into separate groups according to their coordinates. The total number of groups is given by the number of clusters, while the convergence specifies the accuracy rate within the group and bounds the number of algorithm iterations. The program creates a completely new image in which pixels of a particular group are all colored in average color of the group. Since the pores in the photo are black, the program can recognize them as a separate group. Digital estimation of porosity was made for cores from two wells in comparison with liquid injection method of porosity measuring. The features of digital porosity were explained by porosity genesis.

1 Introduction

Digital estimation of porosity of carbonate rocks includes cluster data using either a single pass or an iterative (ISODATA) clustering algorithm; saves the results for display as a thematic map. Cluster statistics can also be saved as class statistics. Use of clustering followed by ECHO spectral/spatial classification provides an effective multivariate scene segmentation scheme (Nurgalieva et al. 2014; 2015; 2016).

D.N. Singh and A. Galaa (eds.), *Contemporary Issues in Geoenvironmental Engineering*, Sustainable Civil Infrastructures,
DOI 10.1007/978-3-319-61612-4_2

To process clustering ISODATA algorithm is preferred because it is iterative and more accurate. It is necessary to specify the number of clusters, the percentage of their convergence and the minimum size of one cluster (in pixels) (Nurgalieva et al. 2016).

Elementary operation of the program can be described as following. Pixels are divided into separate groups (or the image is divided into sections) on the basis of similarity of color and painted in one of the bright spectral colors. The total number of colors is given by the number of clusters. The convergence specifies accuracy rate within the range of spectral colors. The minimum size of the cluster allows adjusting the total number of pixels of the same color spectrum, without making it too small (to expand the range of pixels' location). Since the pores in the photo are black, the program can recognize them as a separate group.

The output text is the final clustering statistics, which reports the number of pixels that correspond to each spectral color, as well as the number of pixels which in one way or another are not included in the classification of the percentage of convergence.

Porosity is determined by the ratio of pore area to the total area of the image, i.e. the ratio of the number of pixels corresponding to pores to the total number of pixels.

However, it should be noted that the pixel portion corresponding to pores cannot fall into the classification and remain unrecorded. Therefore, to calculate the total number of pixels of the third cluster we take the estimated percentage of unclassified pixels (average value of the porosity of studied rocks according to geophysics ~ 0.1).

In present paper we used program Cluster Image which was created in Java especially to process thin section of carbonate rock with to estimate its porosity on image of any format and with a strong color contrast between the mineral part and pores in thin sections under polarized light.

Carbonate rocks are characterized by voids of different genesis (Mazzullo 2004; Nurgalieva 2012; Nurgalieva et al. 2016).

Porosity of carbonate rocks is composed of primary and mostly secondary pores because the primary porosity reduced by cementation and compaction during post-depositional burial. Tournaisian carbonate reservoirs in Southern slope of South-Tatar Arc (Republic of Tatarstan, Russia) lay beneath unconformity (Nurgalieva 2012; Nurgalieva et al. 2015). The porosity values are higher in zone associated with unconformity. Therefore, secondary porosity can be explained by meteoric eogenetic or telogenetic environment. Investigated carbonate rocks consist of mineralogically stabilized to low-magnesium calcite and can be correspond to late eogenetic or telogenetic freshwater exposure of older limestones. Particle selective pores, vugs and caverns formed in these rocks. Regularities and differences in 2D and 3D porosity distribution are caused by complicated combination of primary and secondary porosity.

In this paper we continue to study these features of Tourneisian carbonate rocks of Southern slope of South-Tatar Arc on example of new log.

2 Object of Research

The carbonate section of Tourneisian rocks was probed in interval of Kizelian horizon. In common it is a progradational sedimentary sequence bottom-up from wackestones to grainstones. A characteristic feature of the selected microfacies is the lack of clay materials in the rock composition and the content of micrite component, zoning respectively, decreases up the section. Biogenic components are remnants of the shells of foraminifera, pelecypods, bryozoans, remnants of green algae. Dominating postsedimentary processes expressed in thin sections are the leaching, calcitation, fracturing and stylolization process. Leaching occurs primarily on the aragonite and low-magnesia calcite, metasomatically replaced with calcite. The leaching processes and calcite mineralization are likely to occur from the beginning of diagenesis with regime change pH. Fractures and movement joints, non-conformal primary biogenic and sediment bioclastic structure and cross elements leaching and calcite mineralization are likely to dominate at the later stages of diagenesis and at the later stages of the species at activation of geodynamic regime. The absence of clay minerals in the rock composition indicates primary pure lime sludge, supported with benthic foraminifera, bryozoans, molluscs, green algae. This version of contribution is properties of the carbonate grade.

The lack of clay parts, a well contrast among the carbonate part (dark section) and pores (black section) in fine areas under crossed nicols let to take picture of the considered instances for 2D estimation of porosity.

3 Research Method

The clustering is based on the algorithm ISODATA, which is preferable due to the iterative properties and higher accuracy. It is ordered the number of ranges, percent of convergence and smallest cluster dimension in pixels.

The photo of thin section in cross nicols is downloaded in Cluster Image (e.g. photo A on Fig. 1). The color of each pixel can be represented as a vector in the RGB three component bases. The result is an image as a group of vectors, which may be classified into sections based on their orientation. The total number of groups is defined by the number of clusters, and the convergence determines the degree of precision in the group and specifies the number of repetitive in the approach (Nurgalieva et al. 2015).

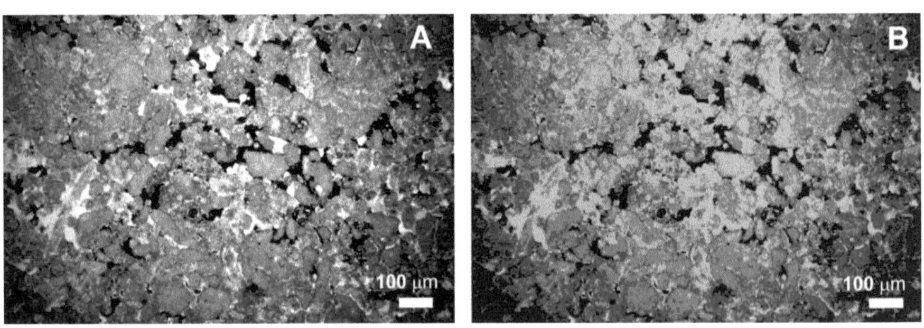

Fig. 1. The images of thin section of carbonate rock (depth 1205,8 m): A – before processing; B – after processing

The program creates a new image, wherein pixels in each range are colored the mean color (photo B on Fig. 1). With respect to that, the pores in the picture are black; the software can specify them as part of a particular range.

The image is accompanied by a diagram which shows the contribution of each cluster, e.g. porous space, group in the image (Nurgalieva et al. 2015).

4 Research Results and Discussion

Figure 2 demonstrates the variation of 2D porosity evaluation based on Cluster Image and 3D porosity evaluation according to the method of fluid saturation. Variations of 2D porosity can be considered as reflecting more or less 3D porosity. The difference

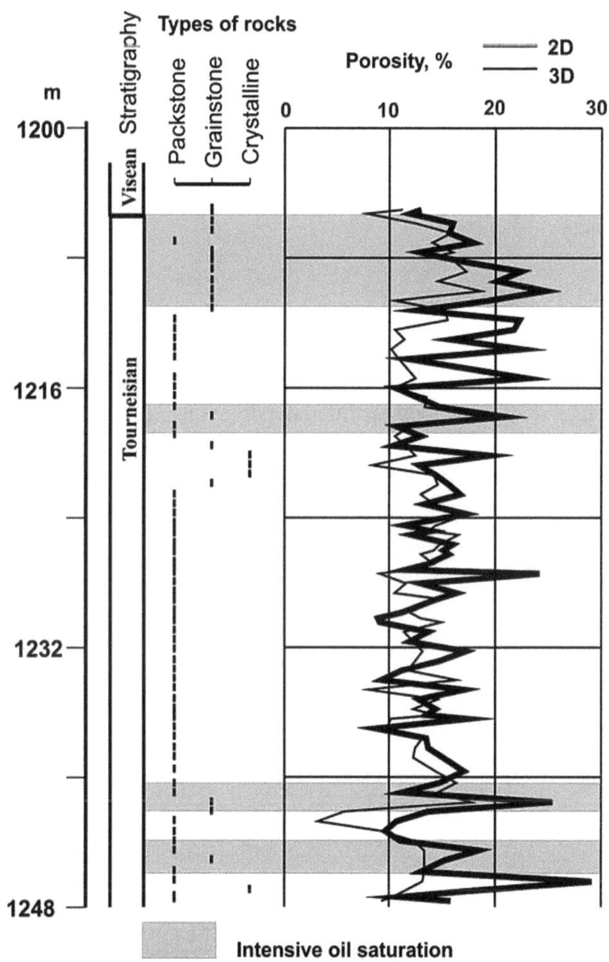

Fig. 2. The well section: comparing of porosity estimation by method of liquid saturation (3D porosity) and Cluster Image processing (2D porosity)

between two types of porosity associates both with the methodological features used according to the methods for measurement the porosity and the nature of the void space. The method of fluid saturation estimates only open porosity in large pores. Digital 2D method permits to estimate all porous space. The difference between these types of estimation depends from types of carbonate rocks (e.g., grainstones, packstones or crystalline rock). In grainstones this difference is less than in packstones and crystalline rocks. Oil saturation zones are also characterized by less deviation of 2D porosity from 3D porosity. It points on higher reservoir properties in rocks with structural type of porous space in comparing with non-structural type of this one.

The Tourneisian limestone lithogenesis was controlled by sedimentation and diagenesis due by supersaturated solutions. In the studied section, secondary porosity was probably to be formed in a multiphasic way by both models of multimineral and stabilized limestones, wherein porosity distribution in the section can be explained by meteoric eogenetic or telogenetic environment. Investigated carbonate rocks consist of mineralogically stabilized to low-magnesium calcite and can be correspond to late eogenetic or telogenetic freshwater exposure of older limestones. The erosion boundary between the Tournaisian limestone and the Visean clay is hardly expressed by the influence of supergene meteoric agents (Nurgalieva et al. 2015, 2016). Particle selective pores, vugs and caverns formed in these rocks. Regularities and differences in 2D and 3D porosity distribution are caused by complicated combination of primary and secondary porosity.

5 Conclusion

The program Cluster Image helps to differ structured and unstructured types of voids and avoid the use of unclassified estimated percentage of pixels in the process of assessing the porosity from the photos of thin sections of carbonate reservoir rocks with reliably established voids in the image.

According to the estimation of the porosity of carbonate rocks using Cluster Image it is identified specific variations of the indicated capacitive parameter of the well section, depending on the rock lithotype and the structure of voids. These variations correspond to the zonal structure of natural oil-saturated carbonate reservoir created by the particular lithogenetic history of Tournaisian rocks in the east of the Russian Plate.

Acknowledgments. The work is supported by the Russian Government Program of Competitive Growth of Kazan Federal University.

References

Nurgalieva, N.D., Nurgalieva, N.G.: Porosity estimation of carbonate rocks with Multispec processing technique. ARPN J. Eng. Appl. Sci. **1**, 20–24 (2014)

Nurgalieva, N.D., Nurgalieva, N.G.: Cluster Image processing technique for porosity estimation of carbonate rocks. ARPN JEAS **10**(4), 1668–1671 (2015)

Nurgalieva, N.D., Nurgalieva, N.G.: New digital methods of estimation of porosity of carbonate rocks. Indian J. Sci. Technol. **9**(20) (2016). doi:10.17485/ijst/2016/v9i20/93750

Nurgalieva, N.G.: Microfacies petrophysics and sequence-stratigraphic frame of carbonate reservoir rocks of Kizelovskian formation. J. Oil Ind. **3**, 38–40 (2012)

Mazzullo, S.J.: Overview of porosity evolution in carbonate reservoirs, pp. 1–19 (2004). www.searchanddis-covery.com/documents/2004/mazzullo/images/mazzullo.pdf

Landslide Study of Lands in Quarrys. Case Chouf Amar - M'sila, Algeria

Abderrazak Saadoun[1,2(✉)], Abdallah Hafsaoui[1,3], and Mohamed Fredj[1,2]

[1] Laboratory: Mineral Resources and Planning, University of Annaba, Annaba, Algeria
abdoumines23@yahoo.com, fredj_khiero@yahoo.com, ahafsaoui@yahoo.fr
[2] Faculty of Technology, Mining and Geology Department, University of Bejaia, Bejaia, Algeria
[3] Faculty of Earth Science, Mining Department, University of Annaba, BP 12, Annaba, Algeria

Abstract. Mining operations open induce risks of instability that can cause landslides and collapse at the bleachers slope. These risks may occur both during and after the operation phase. The magnitude of these risks depends on the mechanical and physical characteristics of the rock mass, the geometrical dimensions of ore bodies, their spatial arrangement and the state of the operated area. If security and technology measures are not taken into account for this purpose, the environment will be affected.

Currently, land stability can be evaluated using analytical and numerical methods. These methods determine the safety factor which represents the ratio between the forces which tend to retain a volume of rock, bounded by the free faces of the slope and a potential fracture surface, and the forces which tend to set it in motion. In this article, an application is considered. It relates to the study of land and terraces, slope in the quarry of Chouf-Amar and the objective of this work is to assess these risks by analytical methods.

The results obtained have allowed us to obtain an acceptable factor of safety and stability study of the quarry.

1 Introduction

To study the mechanical behavior of a rock mass, it is essential to know the degree of fracturing and the distribution of discontinuities in space. In-situ measurements to define the various families of discontinuities and their parameters statistically.

Recognition of rock masses allows us to understand the problems that can arise in the study of a construction project. Recognition of land properties is the link between the cause of a disaster and the remedies it is proposed to set up.

The physico-mechanical properties of rocks and soils are a major source in the understanding of their behavior, the complexity is proven. These have a great influence on the stability of slopes.

D.N. Singh and A. Galaa (eds.), *Contemporary Issues in Geoenvironmental Engineering*, Sustainable Civil Infrastructures,
DOI 10.1007/978-3-319-61612-4_3

Slope stability analysis is an important area in Geotechnical engineering. Most text and books on soil mechanics include several methods of slope stability analysis. A detailed review of equilibrium methods of slope stability analysis Arai and Tagyo (1985).

The mechanical behavior of the rock mass is a key factor in the design of structures that are running there. In order to understand, explain and model this behavior, it is necessary to know the geometric structure or more precisely the geometric distribution pattern of fractures, as well as the mechanical properties of each of the components that are the rock matrix and discontinuities.

Quite a number of techniques are available for these analyses and in this work the more widely used techniques are discussed. Extensive reviews of stability analyses have been provided by Chowdhury (1978) and by Schuster and Krizek (1978).

Slope stability problems are frequently encountered in the construction of roads, canals, dikes, dams and open pit operations. Therefore, this phenomenon is one of the major concerns of Geotechnical engineers and managers of structures and mining. In addition, some natural slopes can become unstable. A break of a slope can be catastrophic and cause loss of life and considerable damage natural.

From a stability point of view, the study of an open pit reduces the sizing slope of the career.

The estimate of the actual security in this rupture risk is a complex issue, especially in the case of limited or little known data.

According to Schroeder (2010), the different forms of rock slope instabilities depend on factors “internal” and own the massive “external” that interact with it.

1.1 Slop Stability

The slope stability analyses in open pit mine or quarry using the geological, structural, material property and hydrogeological information that has been brought together in the geotechnical model. The fundamental objective of the slope design process is to enable a safe and economic design for the pit walls at the bench, inter-ramp and overall slope scales.

The majority of rock slope instability is caused by individual discontinuities. This is because the strength of the intact rock can be high, with the result that the pre-existing discontinuities are the weakest link.

The four diagrams in Fig. (1) illustrate the mechanisms which are traditionally regarded as the four basic instability mechanisms for rock slopes. The geometry of the slip in Fig. 1(a) is a function of the geometry of the slope and the strength of the material forming the slope, but the slope instabilities in Fig. 1(b)–(d) show how the boundaries of the instability are governed by the discontinuities, giving essentially planar faces to the sliding and toppling blocks.

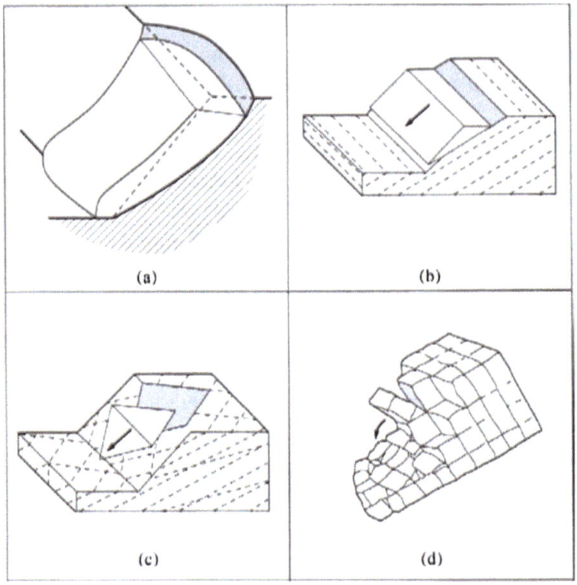

Fig. 1. The four basic mechanisms of rock slope instability: (a) circular slip; (b) plane sliding; (c) wedge sliding; and (d) toppling

The components of a bench are:

- the bench height, which is the vertical height between catch benches. Where benches are stacked this will be a multiple of the operating bench height;
- the bench width;
- the bench face angle.

The relationship is illustrated in Fig. 2.

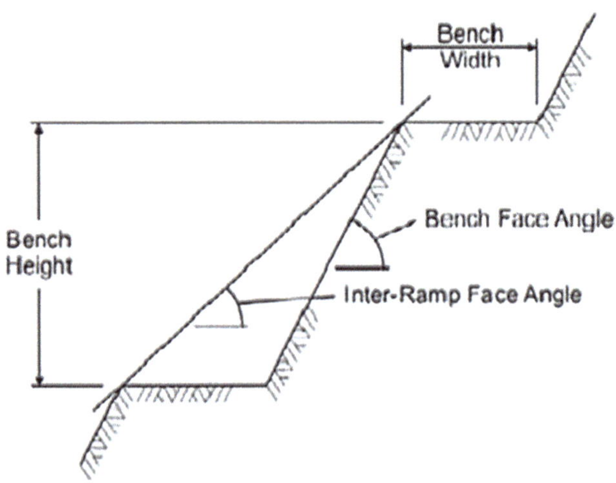

Fig. 2. Components of bench configurations

1.2 Rock Mass Analyses

Rock mass stability analyses examine the potential for inter-ramp and overall slope failures where the failure mechanisms are controlled by the strength of the rock mass. They are an essential final step in the design process, to check that there are no fatal flaws in the inter-ramp slopes selected by the kinematic analysis process and that the rock mass can sustain the proposed design over the full height of the slope.

Analysis of rock mass controlled failures commenced in the 1950s and 1960s and was based on soil mechanics experience and methodology. The analyses incorporated fundamental assumptions of scale and discontinuity density, for example that the size of rock particles in high, closely jointed rock masses were considered equivalent to an isotropic mass of soil particles. This assumption enabled slope design practitioners to adopt the Mohr-Coulomb measures of friction (Ø) and cohesion (c) to represent the strength of the rock mass. This led to the direct use of the emerging limit equilibrium slope failure analyses such as Bishop, Janbu, Morgenstern and Price and Spencer in slope design, and the embedment of Mohr-Coulomb strength parameters in the limit equilibrium stability chart procedures introduced in the 1970s and 1980s. The use of Mohr-Coulomb strength parameters was also carried over into all the continuum and discontinuum numerical modelling tools now common in pit slope design.

1.3 Empirical Methods

At an early stage of project development when data are limited and the geotechnical model has not been fully developed, empirical charts can be extremely useful for establishing preliminary slope designs, provided that their limitations are recognized.

1.4 Slope Angle Versus Slope Height Charts

A number of authors have published slope angle versus slope height charts. Well-known examples include charts published by Hoek (1970) and Sjöberg (1999). Hoek reported the investigation of many slopes and included a range of slope angles, shown in Fig. 3. Similar charts appear in Hoek and Bray (1981) and Wyllie and Mah (2004).

Hoek's approach of simply comparing slope angle with slope height from actual successful and failed slopes was used in a more recent study by Sjöberg (2000), which also classified the slopes by the characteristic rock-hardness rating. Sjöberg's data for the two cases are plotted in Fig. 4, using the notation that open symbols represent successful slopes and solid symbols represent failed slopes

Hoek's correlation of slope angle versus slope height to an approximate FoS trend is also indicated in Fig. 4. As with Hoek's findings, some slopes appear stable when a slope angle versus slope height classification would suggest failure, while others failed where stability might have been expected. Sjöberg's update of Hoek's work suggests a wider range of uncertainty.

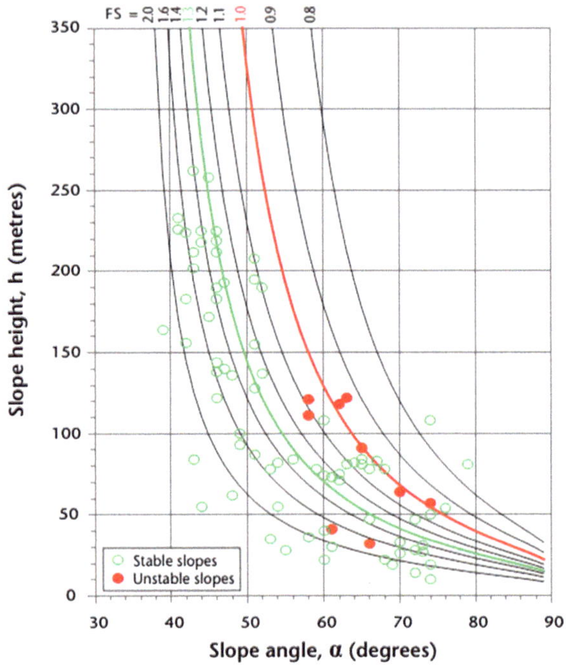

Fig. 3. Rock slope versus slope height, distinguishing between failures and non-failure

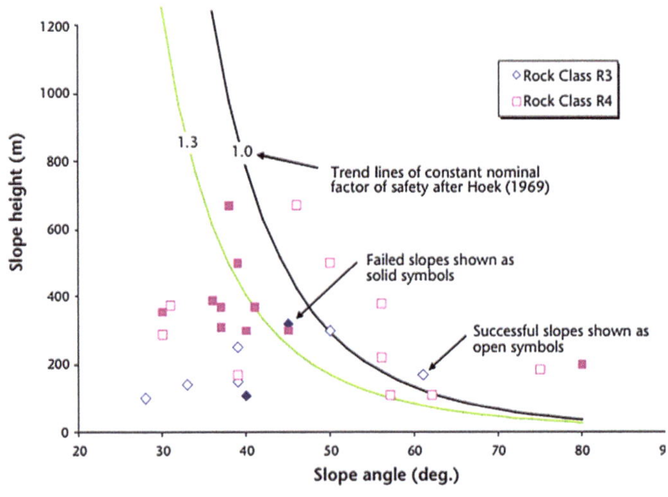

Fig. 4. Rock slope success and failure designated by rock strength

1.5 Empirical Design Charts

Design charts provide design slope angle and slope height guidelines by combining the experience gained from the known performance of slopes at various mine sites with the information provided by a classification scheme. One of the best-known and most widely used charts is that published by Haines and Terbrugge (1991), which is based on the Laubscher MRMR rock mass rating scheme. The chart is shown in Fig. 5.

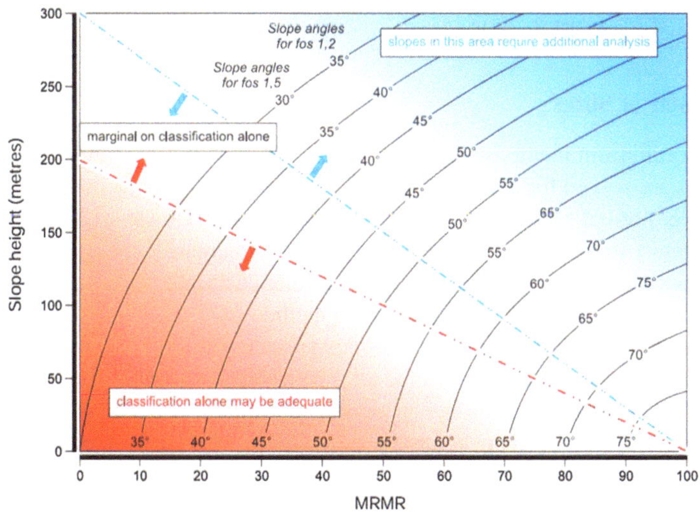

Fig. 5. Haines & Terbrugge chart for determining slope angle and slope height

The limitations of using design charts are their experiential and semi-quantitative nature. These limitations are clearly recognised in the Haines and Terbrugge chart, which indicates where it believes slope angles and slope heights can be determined solely on the basis of the MRMR rating of the rock mass, where the estimate is marginal and where additional analyses are required.

1.6 Limit Equilibrium Methods

Limit equilibrium methods use representative geometry, material and/or joint shear strength, material unit weights, groundwater and external loading/support conditions to determine slope safety factors based on a set of simplifying mechanical assumptions that are outlined below. A practical, complete and accessible description of limit equilibrium methods is in Duncan and Wright (2005). Depending on the values of the safety factor, one can assess the state of slope stability as shown in Table 1.

Table 1. Balance slope based on theoretical values of the safety factor

Safety factor S_F	State of the work
Fs < 1	Danger
Fs = 1	Limited stability
Fs ϵ]1–1,25]	Questionable security
Fs ϵ]1,25–1,40]	Satisfactory security for small-scale works but against it is a questionable safety for slopes of open pits
Fs > 1,4	Satisfactory security

1.7 Method of Slices

The limit equilibrium methods used to determine the stability of sliding planes, blocks and wedges are solved for a single free body and do not depend on the distribution of the effective normal stresses along the failure surface. However, if the mobilised Mohr-Coulomb strength of the rock mass is to be calculated, the distribution of the effective normal stresses along the candidate failure surface must be known. Solutions for this condition are usually based on the 2D method of slices, which divides the body into nslices above the candidate surface. This surface is often assumed to be circular, but may take any shape as the method of slices can readily accommodate complex slope and candidate failure surface geometries, variable rock mass conditions and external boundary loads. However, the problem is statically indeterminate as the solution has more unknowns than equations.

The forces acting on an individual slice in the method of slices are illustrated in Fig. 6.

The important choice that depends on the means that can be implemented, must be made between a method modeling the entire soil mass (finite element method) and a cinematic method, defining a fracture surface for example (limit equilibrium method). In the case of an interesting method the whole mass, the calculation directly provides the most probable rupture zone, while a method based on a curve previously defined will be repeated many times for a similar result. This choice must be made by examining the resources available, the overall behavior of the slope, but also by ensuring the availability of the calculation parameters for the model.

The limit equilibrium methods are very simple to use and well suited for a first flash estimate of slope stability. They can be applied to the study of fracture plane sliding, dihedral or failover. These methods are widely used to study the stability of Hoek and Bray (1981). They are widely used for the study of breaking through a circular shear surface.

Methods known to limit equilibrium are very suitable because it is easy to write equations that link the variables, but, except for the simplest cases, the number of unknowns is greater than the number of equations. To solve the equations, it is then necessary to introduce additional and simplifying assumptions in order to equal the number of unknowns and the number of equations.

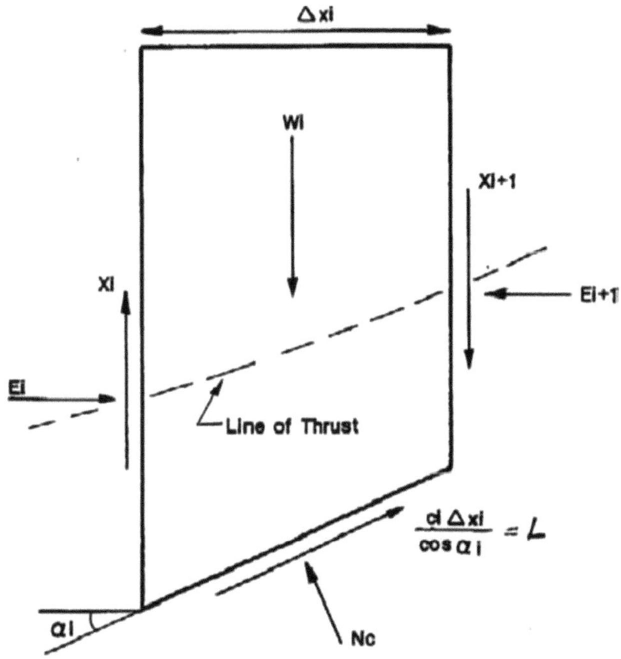

Fig. 6. Forces acting on an individual slice in the method of slices

Limit equilibrium methods This method is to consider the forces that tend to retain a certain amount of land, bounded by the free forces of the slope and a potential failure surface, and those that tend to set it in motion, Fig. 7, Seddiki (2008).

If there's a whole series of limit equilibrium procedures was developed to analyze the static slope stability. Homogeneous slopes are usually analyzed by the methods described above, simplified Bishop (1955); Janbu (1957), who assume the circular

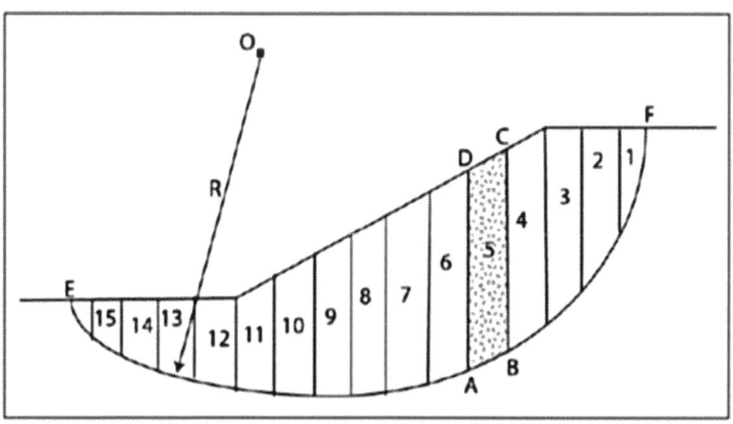

Fig. 7. Cutting a sliced slope.

failure surfaces. When soil conditions are not homogeneous, the fracture surfaces may be non-circular. In these cases, it is preferable to use methods such as Morgenstern, Price (1965).

There's also the method of Sarma (1973, 1979), where he developed a different approach to determine the safety of a slope coefficient and that checks all limit equilibrium conditions.

Since the surfaces of sliding for many slope failures have been observed to follow approximately the arc of a circle, most of the commonly used analytical techniques for calculation of slope stability involve the assumption of a circular failure arc. Most of the techniques discussed in this chapter are based upon this assumption. For composite failure surfaces, analyses have been developed by Morgenstern and Price (1965) and by Janbu (1973). A classification was proposed by the International Society of Rock Mechanics, which is as follows Table 2.

Table 2. Balance slope depending on the experimental values of the safety factor.

Fs < 1	Unstable slope
1 < Fs < 1,5	Can slide
Fs > 1,5	Generally stable

1.8 Numerical Methods

Although limit equilibrium methods of analysis such as the method of slices are simple to use and have been well adapted to slope stability problems in jointed rock masses, they cannot represent deformation and/or displacement of the failing rock mass. This deficiency has largely been filled by numerical methods of analysis, which can model many of the complex conditions found in rock slopes such as nonlinear stress–strain behaviour, anisotropy and changes in geometry. They can be used to help explain the observed physical behaviour of the rock mass and to evaluate different geological models, failure mechanisms and slope design options.

To adequately capture stress and strain gradients within the slope, it is necessary to use relatively fine discretisations. By experience, it has found that at least 30 lower-order elements (elements with constant or uniform stress) are required over the slope height of interest. Figure 8 shows results when only 10 elements per zone height are used. The figure also shows that higher-order elements, or elements employing mixed discretisation, show reasonably accurate results when only 10 elements are used in the slope heights. Finite element programs using higher-order elements probably require fewer zones than the constant-strain/constant-stress elements common in finite difference codes. If flexural toppling is involved, a minimum of four zones across the rock column is usually required.

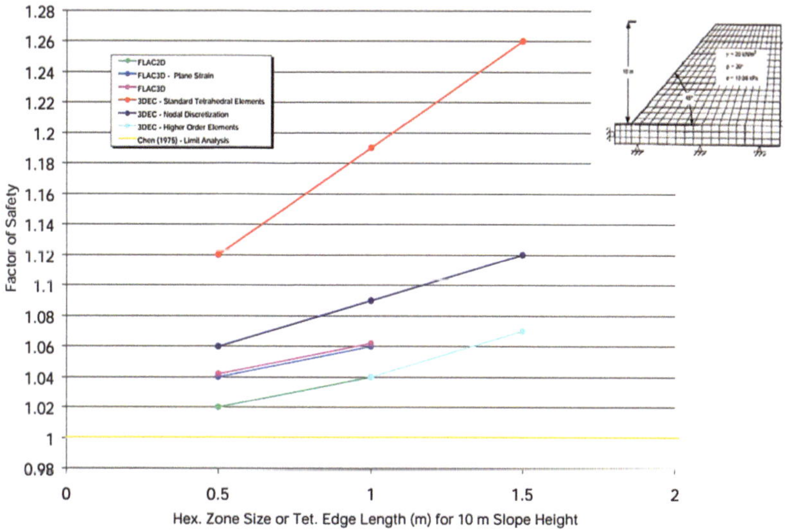

Fig. 8. Comparison of calculated FoS s for the Chen (2007) solution for different element types. (Inset: numerical mesh for homogeneous embankment.)

1.9 The Geographical and Administrative Situation of the Study Area

The Wilaya of M'sila is located in the Central East parties of the Algeria, 250 km southeast of the capital Algiers; it borders with the provinces of Bouira and Bordj Bou Arreridj to the north, east Batna, Biskra southeast, Djelfa and Medea west.

Field studies and analysis of stability of neighborhoods in the career of Chouf Amar, lead us to consider a career as a sensitive area of stability perspective. Production shots made in a repetitive manner can further decrease the stability of the slope: first because of the destabilizing forces of inertia they can generate, and secondly because they degrade the long term mechanical properties of the massif.

The Morphologically limestone deposit of the career Chouf Amar plant is part of a massive array stretched in a longitudinal direction over more than 150 km. Locally, it is in the form of two separate compartments monoclinal, with an average height difference of 250 m, stretched in one direction in a dip angle of 10° to 15° separated by a relatively deep trough and delimit two butes which constitute the southern limit, and the situation of the quarry it represent in the Fig. 9.

Basically, just above the dolomitic limestones of the Turonian, based essentially marl foundation of a 70 m thick about, consists of gray marl with intercalations of gray and nodular limestone marl, often marl gray limestone and lumachelles d 'Oysters. This layer of marl plays the role of soap to slip and fragilise the quarry stability.

The current knowledge of the structure of the deposit does not establish with certainty the position, nature and accident parameters. An additional study, more refinement is needed. Pending the results of this study,

The stability of the front is related to Geotechnical conditions (initial fracturing) and breakage induced by shots that are superimposed with previous discontinuities.

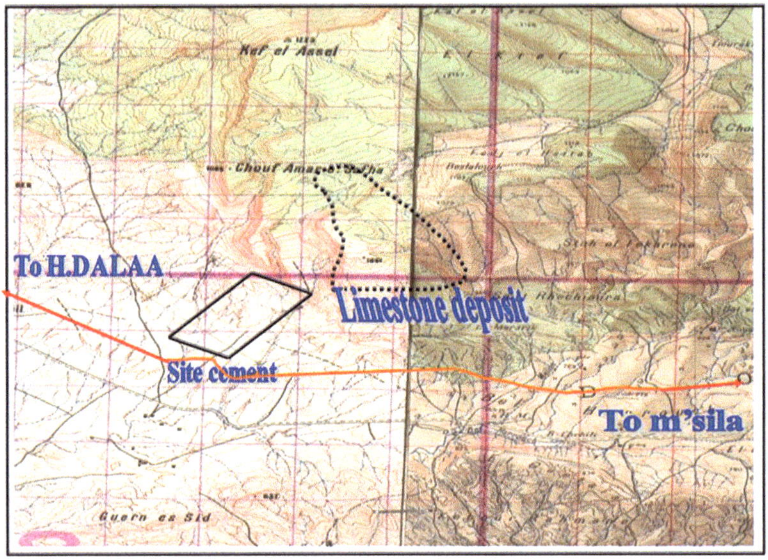

Fig. 9. Location map of limestone deposit of quarry Chouf Amar

- Shooting with the dip often leads to instabilities with conventional banks slide towards the empty and large rear effects that can potentially compromise the correct positioning of the future row of mines.
- Shooting against the dip generally limits the instabilities of the front and rear effects, but can induce rock fall risk more or less embedded in the mountain.
- Shooting across banks reveals strong irregularities forehead and a bad job of the foot.

Given their topographic position on the top parts of the reliefs, their inclination to the south, Chouf Amar deposit of limestone can not constitute a reservoir suitable to host an aquifer, and this staged by the absence of any seepage of water at the base of limestone, particularly along the cliffs, and the total water losses in the realization of prospecting surveys. The extension of the limestone to the South, deep under the cover of marl Campanian, present similar conditions and can be an aquifer that could give an equivalent flow.

In the Chouf Amar career, the bedrock is highly fractured. In addition, important structures are observed, especially on the south side where the limestones are affected by faults, and this can produce different types of slope failure bleachers. The ore is less often fractured the surrounding rock. When the flaws affect the limestone, they allow water infiltration; this latter can lead to the development of a large hydraulic pressure, unfavorable to stability. So the flaws and cracks are structures that control the possible instabilities small, medium and large scales.

1.10 Stability and Rear End Problem

The rear effects develop preferentially in rock masses with a high density of discontinuities, especially when they are communicating and openly. At this time, expansion of the gases generated by the firing is maximum and these gas enlarges the openings and discontinuities themselves behind the last firing line. It is very difficult to control these effects back in this situation, Figs. 10 and 11.

Fig. 10. Career of Chouf Amar before sliding

Fig. 11. Career of Chouf Amar after sliding

The dip of discontinuity need not be important to create problems. Stability can also be understood in terms of shot holes and complicate either the drilling or the loading of the explosive (presence blocks obstructing the hole) or operation of firing at the hole (break in the continuity of lateral or priming hole).

Numerical Simulation always provides information prior to actual mining operation to save the undesirable excavation. The jointed rock shows very complex behavior and therefore, numerical simulation of such material becomes difficult to model. The conventional practice like limit equilibrium method (LEM) does not consider the joints or any other rock mass parameters for stability analysis, therefore, it becomes imperative to use high end numerical tools. Moreover, in rocks, failure is guided through the orientation of discontinuities and depending on the frequency and number of joint sets, mode of failure may vary. In such cases, stress-strain attributes of the material involved also becomes very important. All such features are absent in the LEM, hence it was decided to go for advanced numerical simulation which can replicate the original in-situ conditions. Some studies related to dump slopes and hill slope stability have already been done with these methods, Gupte et al. (2013); Sinha et al. (2008); Umrao et al. (2011).

2 Calculation of Safety Factor

In this part, we will study the stability of the Chouf Amar career and provide a potential risk of instability represented by a safety factor determined by the limit equilibrium methods.

For the analysis illustrated in Fig. 12, the optimised non-circular failure surface search option in the RocScience program Slide has been used to calculate the factor of safety. The figure also shows the failure surfaces with factors of safety of less than 1.3. Evaluation of the limit equilibrium analysis results should include consideration of the zone of instability that develops. The width of the zone containing surfaces with FoS < 1.3, for example, represents the zone of possible slumping or cracking and should be a consideration in determining setbacks for infrastructure, and in the evaluation of potential instability of the pit walls.

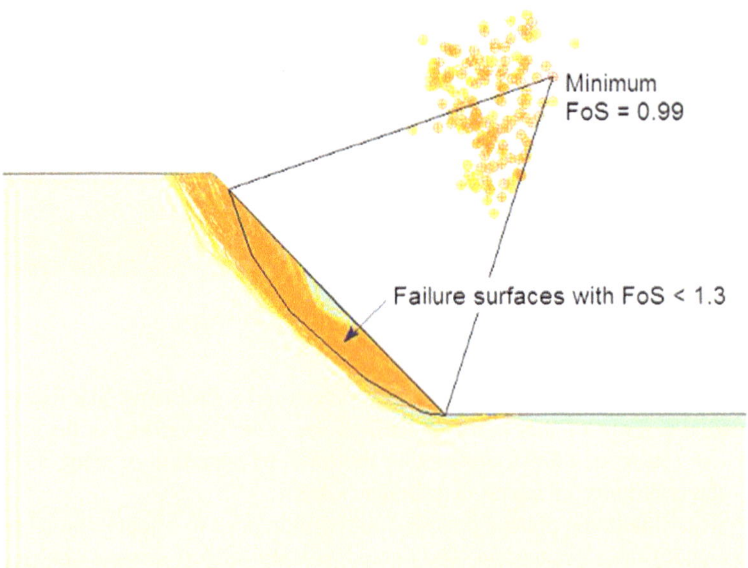

Fig. 12. Optimised non-circular search for a minimum factor of safety in the slope

An alternative method for estimating the factor of safety of the slope is the Shear Strength Reduction (SSR) method (Dawson et al. 1999; Diederichs et al. 2007).

This method was used by Lorig (1999) to investigate the influence of quarry geometry and in situ stresses.

The limit equilibrium results show no significant difference between any of the models with the exception of marginally lower values for the case where the horizontal to vertical stress ratio is 2.0. The SSR results show no change in the factor of safety for changes in the in situ stress ratio but a small increase in stability with decreasing pit bottom radius. The SSR results exhibit similar trends to those obtained by Lorig (1999), who also used the SSR technique (with different rock mass properties).

The results obtained from both limit equilibrium and shear strength reduction calculations calculated the factor of safety are presented in Table 3.

Table 3. Results of approximate factor of safety calculations for different in situ stresses and open pit geometries.

In situ stress ratio	Pit bottom radius–m	FoS–Limit Equilibrium	FoS–SSR
0.5	∞	0.99	1.04
1.0	∞	0.97	1.04
2.0	∞	0.93	1.03
1.0	100	1.0	1.25
1.0	250	0.98	1.21
1.0	500	0.98	1.15

3 Conclusions

The study of stability established by the limit equilibrium method shows that the slope of the Chouf Amar career is unstable, which checks the results obtained following the application of methods of semi-empirical classification concerning the situation of the massif. With lower values of the step height and the angle of inclination of their slope, there is increasing Sf values and thus improving the stability.

References

Arai, K., Tagyo, K.: Determination of non circular slip surface, giving the minimum factor of safety in slope stability analysis. Soils Found. (1985)

Bishop, A.W.: The use of the slip circle in the stability analysis of earth slopes. Geotechnique (1955)

Chowdhury, R.N.: Slope Analysis. Elsevier Scientific Pub., Co., Amsterdam (1978)

Gupte, S.S., Singh, R., Singh, T.N.: In-pit waste dump stability analysis using two dimensional numerical models. Min. Eng. J. (2013)

Duncan, J.M., Wright, S.G.: Soil Strength and Slope Stability. John Wiley & Sons, New Jersey (2005)

Dawson, E.M., Roth, W.H., Drescher, A.: Slope stability analysis by strength reduction. Géotechnique **49**(6), 835–840 (1999)
Diederichs, M.S., Lato, M., Hammah, R., Quinn, P.: Shear strength reduction approach for slope stability analyses. In: Proceedings First CANUS Rock Mechanics Symposium. Vancouver (2007)
Hoek, E.: Estimating the stability of excavated slopes in opencast mines. Trans. Inst. Min. Metall., Sect. A: Min. Ind. **79**, A109–A132 (1970)
Hoek, E., Bray, J.: Rock Slope Engineering, 3rd edn. Institute Mining and Metallurgy, London (1981)
Haines, A., Terbrugge, P.J.: Preliminary estimate of rock slope stability using rock mass classification systems. In: 7th Congress of International Society of Rock Mechanics, Aachen, Germany, pp. 887–892 (1991)
Janbu, N.: Stability analysis of slopes with dimensionless parameters. Harvard University Soil Mechanics Series, vol. 46 (1957)
Janbu, N.: Slope Stability Computations. In: Hirshfeld, R.C., Poulos, S.J. (ed.) Embankment Dam Engineering. John Wiley & Sons, New York (1973)
Lorig, L.: Lessons learned from slope stability studies. In: Detournay, C., Hart, R. (eds.) Proceedings of the Conference FLAC and Numerical Modeling in Geomechanics, Minneapolis, September 1999, pp. 17–21. A.A. Balkema, Rotterdam (1999)
Morgenstern, N., Price, V.E.: The analysis of the stability of general slip surfaces. Geotechnique **15**(1), 79–93 (1965)
Sarma, S.K.: Stability analysis of embankments and slopes. Geotechnique **23**(3), 423–433 (1973)
Sarma, S.K.: Stability analysis of embankments and slopes. J. Geotech. Eng. Div., ASCE **105** (GT12), 1511–1523 (1979)
Schroeder, Ch.: Study of stability of rock faces. Belgian company review of engineering geology and rock mechanics (SBGIMR) (2010)
Seddiki, A.: Analysis of slope stability under earthquake. Doctorate Memory, Mohamed Boudiaf University M'SILA (2008)
Schuster, R.L., Krizek, R.J. (eds.): Landslides - Analysis and Control (1978)
Sinha, S.C., Sarkar, K., Singh, T.N.: Geotechnical investigation of road hill slope near Bhatan tunnel along Mumbai-Pune expressway, Maharashtra. Min. Eng. J. **10**(3), 24–27 (2008)
Sjoberg, J.: Analysis of large scale rock slopes. PhD thesis (unpublished). Lulea University of Technology, Lulea (1999)
Umrao, R.K., Singh Rajesh, A.M., Singh T.N.: Stability analysis of Cut Slopes Using Continuous Slope Mass Rating and Kinematic (2011)
Wyllie, D.C., Mah, C.W.: Rock Slope Engineering, Civil and Mining, 4th edn. Spon Press, London (2004)
Klerck, P.A.: The finite element modelling of discrete fracture of brittlematerials. PhD thesis. University of Wales, Swansea, UK (2000)

Permeability Characterization of Clay Under Different Hydraulic Gradient

Olusegun Alabi(✉)

Solid Earth Physics Research Laboratory,
Department of Mathematical and Physical Sciences,
Osun State University, Osogbo, Nigeria
geosciencealabi@yahoo.com

Abstract. The study presents the results of vertical infiltration on saturated clay of three different particle of sizes 0.01, 0.001 and 0.002 mm under different hydraulic gradients. A range of hydraulic gradients were achieved by varying the clay length, L in respect to head, h in the permeameter. The aim of the study is to establish a model to determine the saturated hydraulic conductivity vis-à-vis permeability of soils as a function of time, for accurate estimate of volume of water available for plant growth in surface irrigation. The equations obtained show that permeability $K - t$ curves show that the declination of permeability will never be zero, except at an initial state or stage before the commencement of infiltration process. This equation from the hydraulic conductivity – time (K – t) curve is simple and practically useful for the determination of permeability at a particular time.

Keywords: Infiltration · Permeability · Hydraulic gradient · Volume flux · Time

1 Introduction

Spatial and temporal variations in the infiltration behavior of surface irrigated soils is the major physical constraint to achieve higher irrigation application efficiencies. To improve the hydraulic performance of surface irrigation system requires that these soil absorption rates be predictable functions of the soil properties and surface hydraulic properties. Hydraulic influence and soils properties are some of the major factors that determine the rate at which irrigation water moves into the soil (Green and Ampt 1911).

Infiltration of water into soils is an important phenomenon affecting much of agricultural production. Water is a necessary input for crop production. In most agricultural production, water is supplied to the plant from the soil water reservoir. The recharge or refilling of this reservoir occurs as water infiltrates through the soil surface and percolates into the soil profile. Localized excesses and shortage of water can decrease production in both irrigated and dry land agriculture.

One of the most important considerations in the design and management of surface irrigation is the rate at which water infiltrates into the soil. Infiltration affects the advance of the irrigation stream, the rate of flow and flow depth within the field, the recession of water from the soil surface, and the total amount of water that enters the

D.N. Singh and A. Galaa (eds.), *Contemporary Issues in Geoenvironmental Engineering*, Sustainable Civil Infrastructures,
DOI 10.1007/978-3-319-61612-4_4

soil. In addition, the water flow on the surface can change soil surface conditions, which can change infiltration.

A considerable amount of research has been conducted on the infiltration of water into soils. The basic principles of soil-water flow are well understood. However, several problems arise when trying to apply infiltration theory to field conditions. First, infiltration is greatly affected by the conditions at the soil parameters such as permeability, porosity, surface moisture content, and bulk density are changing, when assumed constant during irrigation (Clemmens 1981). It is necessary to measure all the soil physical properties prior to an irrigation, to determine how these properties will change during the subsequent irrigation, to determine how these properties distributed over the irrigation rate-time relationship from which to make meaningful management decisions (Clemmens 1982). The aim of this study is to determine a simple and more efficient equation to represent the change in separated hydraulic conductivity or permeability with time for accurate estimation of volume of water for plant.

2 Theoretical Background

Many soil physicists and hydrologists have attempted to find certain relationships between permeability and time of infiltration. One reason for searching for such relationship is that infiltration decreases with increase in time is as a result of reduction in permeability formed by clogging of pores with fines deposited by infiltration water.

Some common equations which are used to produce infiltration behavior are

$$Z = A\tau^{1/2} + C\tau \quad \text{(Philip 1957a)} \tag{1}$$

$$Z = A\tau^{B} + C\tau \quad \text{(Philip 1957b)} \tag{2}$$

$$Z = A\tau^{B} \quad \text{(Kostiakov 1937)} \tag{3}$$

Where Z is the cumulative infiltration which can be either depth or infiltrated volume per unit length, τ is the opportunity time, and A, B, C are empirical constants

The exponential decay function to represent the change in saturated hydraulic conductivity of the surface seal with time is expressed as

$$k(t) = k_f + \left(k_i - k_f\right)e^{-at} \quad \text{(Moore 1979 and Moore et al. 1980).} \tag{4}$$

Where

$k(t)$ = saturated hydraulic conductivity of the surface soil at time t

k_i = initial saturated hydraulic conductivity

k_f = final saturated hydraulic conductivity of a well-established stable surface seal and a is a constant.

'Clogging effect' is caused by infiltration with the soil body pore being blocked and permeability decreasing under seeping gradually (Moore 1979). The clogging effect is the result of internal and external factors combined action (Moore et al. 1980).

In finely ground particles such as clay and the fluid state of motion in the infiltration process, are intrinsic factor to produce clogging effect (Moore 1979; Moore et al. 1980; Huang et al. 2009).

The infiltration clogging effect is divided into four types according to the action mechanism, namely mechanical clogging, chemical clogging, biological clogging and comprehensive clogging (Huang et al. 2009). It has been observed that the influence of chemical clogging and biological clogging is small, so the principal one is mechanical clogging (Moore et al. 1980). It is found that mechanical clogging can also be divided into the following.

a. Clogging with silt refers to a phenomenon when small soil particles enter the loose medium pores and thereby reduce area of flow section and results into decrease in soil permeability.
b. Blocking, refers to a phenomenon that the particles are not easy to enter into the pore if they are of similar size of pores in loose medium, but they can still block partially the water and therefore reduce the water cross sectional area, and decrease soil permeability.
c. Occluding, refers to a phenomenon that the tiny clay will gather on the surface pore and form mud pie to seal up the pore water cross section and reduce the soil permeability.

The above three types are not mutually independent since clogging effect is the result of the combined action synthesis (Moore et al. 1980).

2.1 Darcy's Law

Theoretical studies of the movement of water in soils are all based on the Darcy theory of flow in porous media. In general, soil scientists have adapted this theory to formulate a readily usable algebraic equation for natural soil profiles (Green and Ampt 1911). Darcy's law in generalized form can be expressed as

$$v_s = -k/\eta(\frac{\partial p}{\partial s} - \gamma g cos\theta) \quad \text{(Darcy 1856)} \tag{5}$$

Where

$v_s =$ rate of flow through the unit of area in the directions
$\theta =$ angle between the line of the vertical and the direction of flows
$\eta =$ viscosity of fluid
$\gamma =$ density of fluid
$g =$ acceleration of gravity
$p =$ pressure of the fluid
$k =$ coefficient of permeability

In reference to the present study, the experimental set up used was in the form of vertical flow of water through sand in the permeameter.

Thus, Eq. (5) reduced to a simple equation

$$v_s = ki \tag{6}$$

Where
$k =$ hydraulic conductivity (ms^{-1})
$i =$ $\frac{\partial p}{\partial s}$ = one dimensional flow of water through sand (dimensionless)
$v_s =$ volume of water flowing across a area per unit time or volume flux (ms^{-1})

and by using Hubert King relation,

$$\mathrm{K} = \frac{\rho g}{\mu}\mathrm{k} \qquad \text{(Hubbert 1940)} \tag{7}$$

Where
$K =$ hydraulic conductivity (ms^{-1}),
$k =$ permeability (m^2),
$\mu =$ viscosity of the fluid (Nsm^{-2}), and
$\rho =$ density of the fluid, (kgm^{-3})

By using Eq. (7) in Eq. (6), Eq. (6) becomes

$$q = K\left(\frac{h}{L} + 1\right) \tag{8}$$

Where q is the volume flux (ms^{-1}), $\left(\frac{h}{L} + 1\right)$ is the hydraulic gradient, h is the head constant and L is the clay sample length.

Equation (6) is generally used to determine the hydraulic conductivity. This can be done by plotting v_s against i and the slope of the curve gives hydraulic conductivity of the clay permeability is obtained from the hydraulic conductivity by using Eq. (7).

The Darcy flux is defined as

$$q = Q/A \tag{9}$$

Where
$q =$ Darcy flux (ms^{-1}),
$Q =$ Volumetric flow rate (m^3s^{-1}), and
$A =$ Cross-sectional area (m^2)

3 Materials and Method

Sizeable quantities of the clay samples were brought to the laboratory in polythene bags washed and rinsed using clean water in order to remove fine organic materials that were in the process of decaying as a result of the work of soil microorganism. This action was necessary so as to avoid 'swelling' and to ascertain that no any other material was present except clay. The resulting samples were sun-dried and thereafter placed in an oven at 120 °C for an hour. After the samples were allowed to cool down. In order to obtain samples of different porosities, five different types of sieves were used to sieve the available sand samples into different grain sizes. The sizes are 0.01, 0.001 and 0.002 mm. The sieves were arranged in a stack with the coarsest on top and the finest at the base followed by the pan which is the last component of the stack. The stack is then placed on a sieve shaker. The sand samples were poured into the topmost sieve and covered with the cover of the shaker. The shaker is an electrical device that vibrates to provide energy for sieving process. After sieving, the grains retained by each sieve were tagged samples A–C.

3.1 Determination of Permeability at Different Hydraulic Gradients

The clay samples were soaked (saturated) overnight before being transferred to the transparent cylindrical tube of cross-sectional area $2.69 \times 10^{-4}\text{m}^2$. To ascertain uniform compaction throughout the samples, the screened end was blocked so as to prevent the water passage, the soaked sand was then poured into the water column up to the height h. After this, the material that was used to block the passage was removed. A continuous supply of water was fed through the sample and at height h, a hole was drilled, this enabled the height to be maintained as excess water was drained through it (Fig. 1). The water that passed through the sample for 60 s was collected with a beaker and the volume measured by measuring cylinder and steady-state flow has been maintained at a constant head. This procedure was repeated for all the samples.

By adoption of Darcy's fluid flow (Eq. 8), length of clay, L is varied in order to obtain different hydraulic gradients, i. Measurements were made at gradients of $i = 5.67$, 7.00, 9.00, 12.33 and 19.00 for each sample. The value of the volume flux, q at each hydraulic gradient was computed from the value discharge volumetric flow rate, Q, knowing that volume flux, $q = Q/A$, where $A = \pi(d/2)^2$ and d is the diameter of the cylindrical tube used, given as $1.85 \times 10^{-2}\text{m}$.

Based on the Darcy's law an instrument for infiltration of water into soils, called permeameter was designed (Fig. 1). The clay samples of different sizes were subjected to water infiltration. Seepage process was conducted for different hydraulic head within the same regular time interval to obtain the variation of soil permeability with time. The volume flux at different hydraulic head is plotted against hydraulic head. The slope of volume flux-hydraulic gradient gives permeability at various time intervals. A plot was prepared between volume flux, q and hydraulic gradient, i for samples A–C. The slope of the respective lines gave the hydraulic conductivity of the medium concerned.

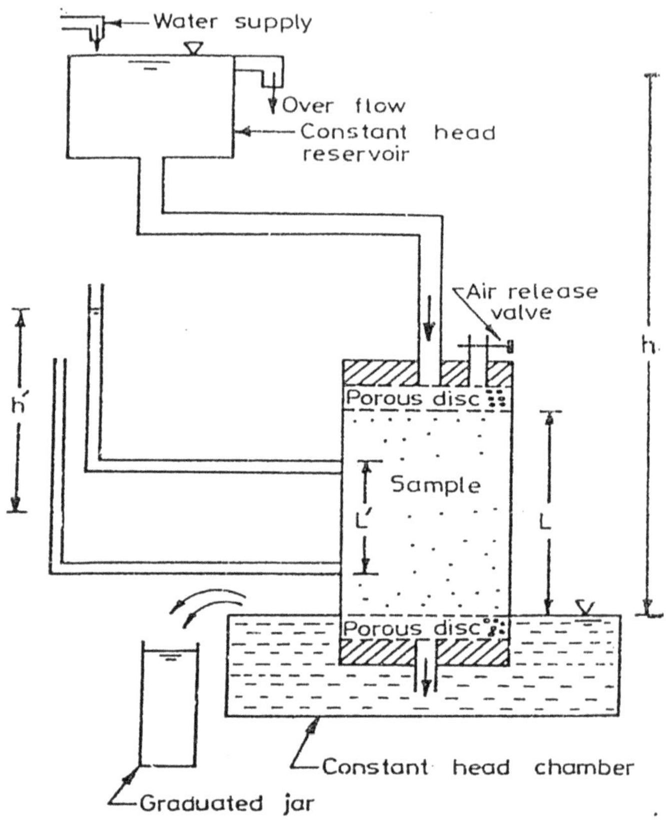

Fig. 1. Sand Model for Vertical Flow under hydraulic head (Arora 2009)

The hydraulic conductivity/permeability time curve (that is k-t curve) was obtained for three clay samples of different particles sizes. The relationship between permeability and time were obtained by best fitting curve. The analysis indicates that the permeability reduces gradually with increase in time, which proves the existing of clogging effect as the time of flowing through soil increases.

4 Results and Discussion

A range of hydraulic gradients were achieved by varying the clay length, *L* in respect to head, h in the permeameter in order to obtain different hydraulic gradients. The outflow from the permeameter for each gradient was measured by graduated measuring cylinder at interval of minutes. The volume flux, which is the rate of volume of water discharge per unit time per unit area were determined and presented in Tables 1, 2 and 3 for

Table 1. Discharge volume, volumetric flow rate and volume flux at different hydraulic head (SAMPLE A)

i	$V \times 10^{-6}$ (m^3)	$Q \times 10^{-8}$(m^3/s)	$q \times 10^{-5}$(ms^{-1})
Time 1 min			
5.67	0.30	0.05	0.41
7.00	1.10	1.80	1.50
9.00	1.60	2.70	2.20
12.33	1.30	2.20	1.80
19.00	5.50	9.20	7.40
Time 2 min			
5.67	0.80	0.67	0.54
7.00	1.80	1.50	1.20
9.00	2.80	2.33	1.90
12.33	3.10	2.60	2.10
19.00	9.25	7.70	6.30
Time 3 min			
5.67	1.00	0.56	0.45
7.00	2.60	1.44	1.20
9.00	3.60	2.10	1.70
12.33	4.60	2.50	2.00
19.00	13.50	7.50	6.10
Time 4 min			
5.67	1.00	0.42	0.34
7.00	3.50	1.50	1.20
9.00	5.00	2.10	1.70
12.33	6.30	2.60	2.10
19.00	17.00	7.10	5.80
Time 5 min			
5.67	1.60	0.53	0.43
7.00	4.20	4.00	1.10
9.00	6.50	2.20	1.80
12.33	7.90	2.60	2.10
19.00	20.75	6.90	5.60

samples A, B and C respectively. The results show that volume flux increases with hydraulic gradient. This is so because hydraulic gradient acts as a ‘driving force’ and the higher the driving force, the more the volume flux (or volume of water discharge). This is in support of Darcy’s law experiment and other previous investigations. All the samples exhibited a marked increase in volume flux as the hydraulic gradient increased from 5.67 to 19.00.

Table 2. Discharge volume, volumetric flow rate and volume flux at different hydraulic head (SAMPLE B)

i	$V \times 10^{-6}$ (m^3)	$Q \times 10^{-7}$(m^3/s)	$q \times 10^{-4}$(ms^{-1})
Time: 1 min			
5.67	4.95	0.83	0.67
7.00	8.6	1.4	1.2
9.00	12.25	2.4	1.7
12.33	19.5	3.3	2.6
19.00	59	9.8	8.0
Time: 2 min			
5.67	19	1.6	1.3
7.00	24	2.0	1.7
9.00	31	2.6	2.1
12.33	54	4.5	3.7
19.00	156.5	13	11
Time: 3 min			
5.67	13.25	7.4	0.60
7.00	17.75	0.94	0.77
9.00	22.25	1.24	1.0
12.33	37.5	2.1	1.7
19.00	103.75	5.8	4.7
Time: 4 min			
5.67	20	0.83	0.68
7.00	30.5	1.3	1.0
9.00	41	1.7	1.4
12.33	51	2.1	1.7
19.00	204.75	8.5	6.9
Time: 5 min			
5.67	35.5	1.2	0.96
7.00	42.5	1.42	1.2
9.00	49.5	1.7	1.34
12.33	87	2.9	2.4
19.00	262.75	8.8	7.1

Hydraulic conductivity for each of the samples at an interval of 1 min were obtained from the volume flux-hydraulic gradient curve (*q-i* curves) which are the gradient of the curves (Figs. 2, 3, 4, 5, 6, 7, 8, 9, 10, 11,12, 13, 14, 15 and 16). Permeabilities were determined by Hubert King's relation. Hydraulic conductivity and permeability at different time were presented in Table 4. Both hydraulic conductivity and permeability were found to be declined as the time increased. This pattern of

Table 3. Discharge volume, volumetric flow rate and volume flux at different hydraulic head (SAMPLE C)

i	V × 10^{-6} (m^3)	Q × 10^{-7}(m^3/s)	q × 10^{-4}(ms^{-1})
Time: 1 min			
5.67	15.45	2.60	2.10
7.00	24.00	4.00	3.30
9.00	35.5	5.90	4.80
12.33	37.50	6.30	5.10
19.00	57.00	9.50	7.70
Time: 2 min			
5.67	26.50	2.21	1.80
7.0	47.00	3.92	3.20
9.00	65.50	5.50	4.40
12.33	70.00	5.80	4.70
19.00	99.00	8.30	6.70
Time: 3 min			
5.67	40.00	2.22	1.80
7.00	64.00	3.56	2.90
9.00	88.50	4.92	4.00
12.33	97.75	5.40	4.40
19.00	136.0	7.56	6.10
Time: 4 min			
5.67	54.50	2.30	1.80
7.00	81.00	3.40	2.70
9.00	105.2	54.4	3.60
12.33	117.0	4.61	4.00
19.00	177.0	7.00	6.00
Time: 5 min			
5.67	63.5	2.12	1.70
7.00	96.5	3.22	2.60
9.00	135.5	4.40	3.60
12.33	142.5	4.61	3.70
19.00	206.0	7.00	5.60

declination was exhibited by all the samples. Hydraulic conductivity was found to be ranging from 0.48 × 10^{-5} to 0.36 × 10^{-5} ms^{-1}, 5.42 × 10^{-5} to 4.65 × $10^{-5}ms^{-1}$, and 3.87 × 10^{-5} to 2.66 × $10^{-5}ms^{-1}$ for samples A,B, and C respectively. This shows that irrespective of the grain size, the highest permeability for each was obtained for the first 1 min.

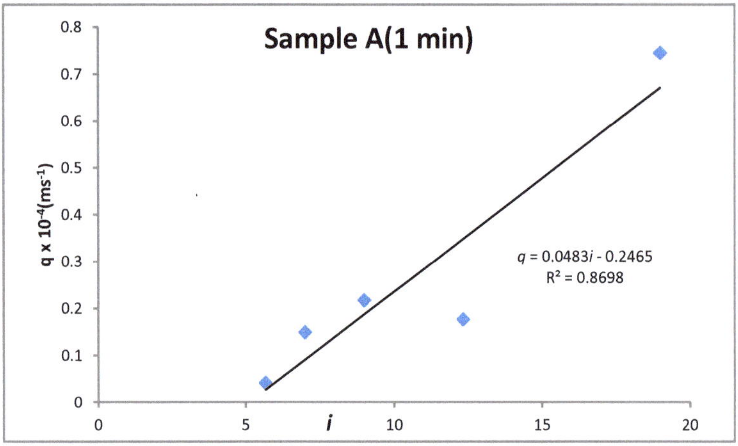

Fig. 2. Volume flux against hydraulic gradient (sample A, 1 min.)

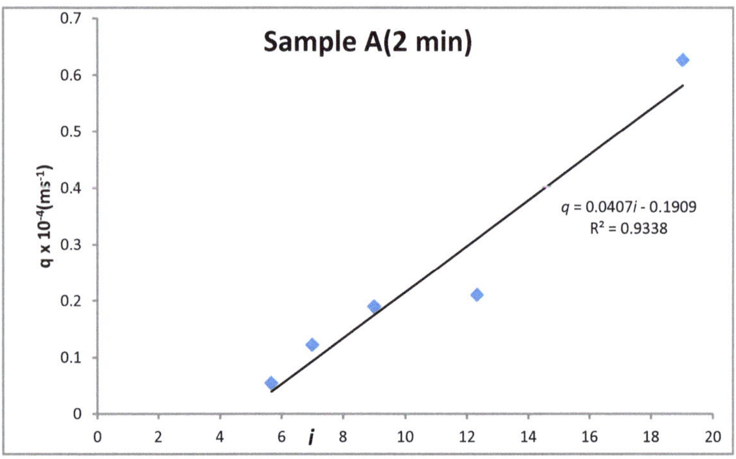

Fig. 3. Volume flux against hydraulic gradient (sample A, 2 min.)

Furthermore, by best curve fitting, the $K - i$ curve yield $K = 0.9216t^{-0.164}$, $K = 7.8053t^{-0.097}$ and $K = 9.7788t^{-0.228}$ for samples A, B and C respectively (Figs. 17, 18 and 19). These equations show that hydraulic conductivity values decline with increase in time and can only be zero whenever time, t is zero.

It is clear that hydraulic conductivity vis-à-vis permeability is strongly influenced by time. The classic law of Darcy is almost universally employed to analyze the flow of fluid through soils and sands. In the analysis and modeling of groundwater flow such as infiltration of water in soils by Darcy's law as related to irrigation, there is possibility of

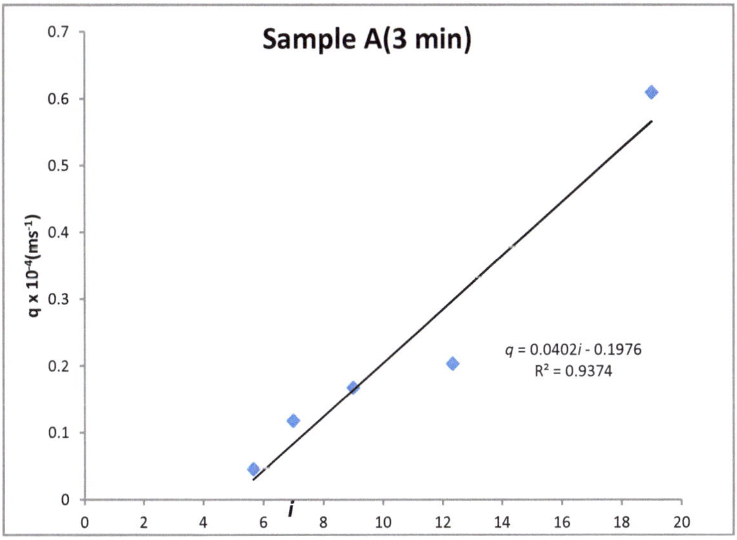

Fig. 4. Volume flux against hydraulic gradient (sample A, 3 min.)

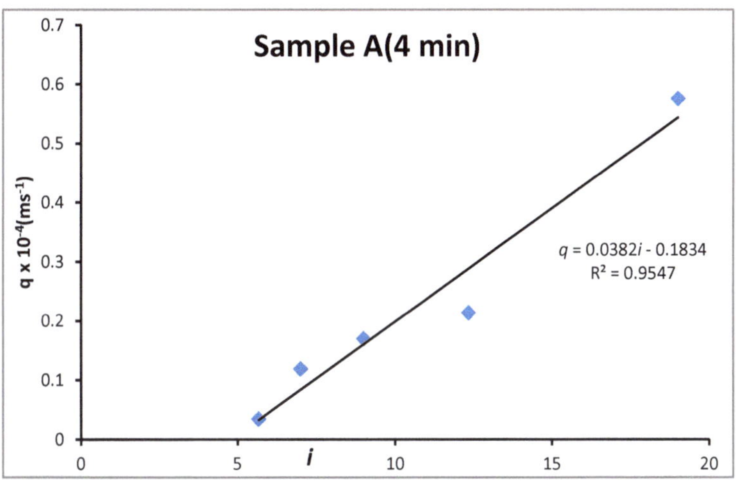

Fig. 5. Volume flux against hydraulic gradient (sample A, 4 min.)

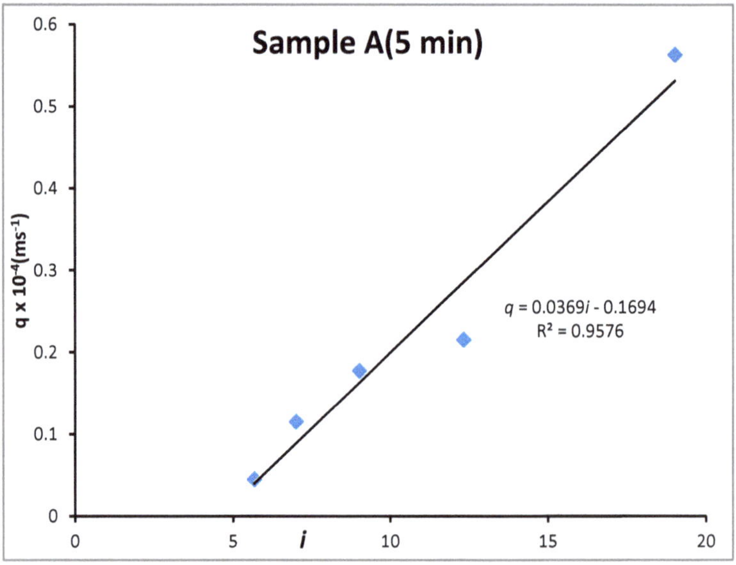

Fig. 6. Volume flux against hydraulic gradient (sample A, 5 min.)

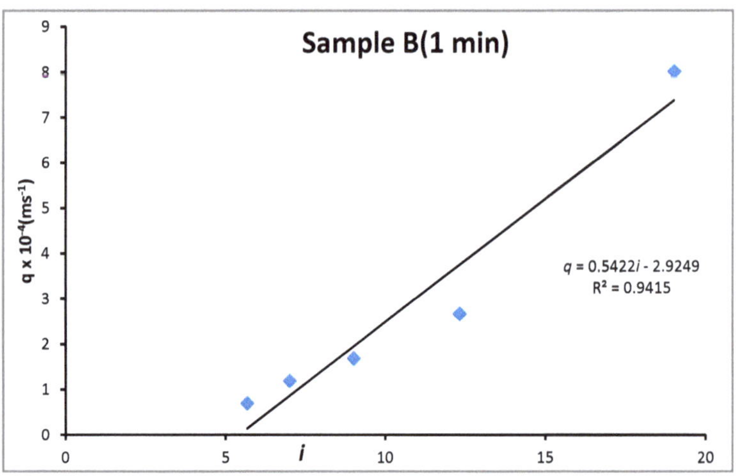

Fig. 7. Volume flux against hydraulic gradient (sample B, 1 min.)

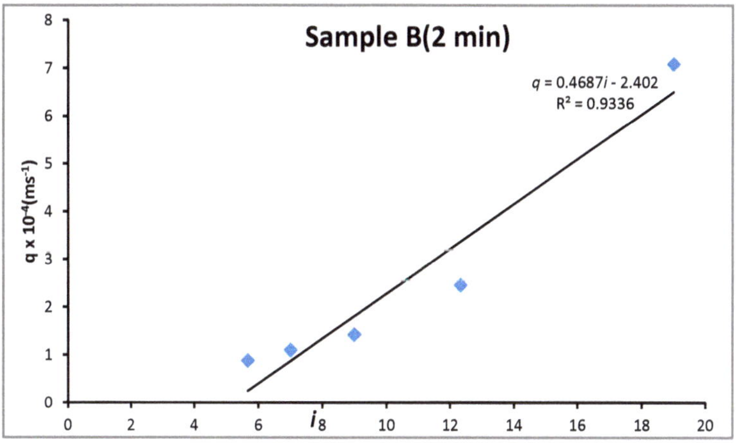

Fig. 8. Volume flux against hydraulic gradient (sample B, 2 min.)

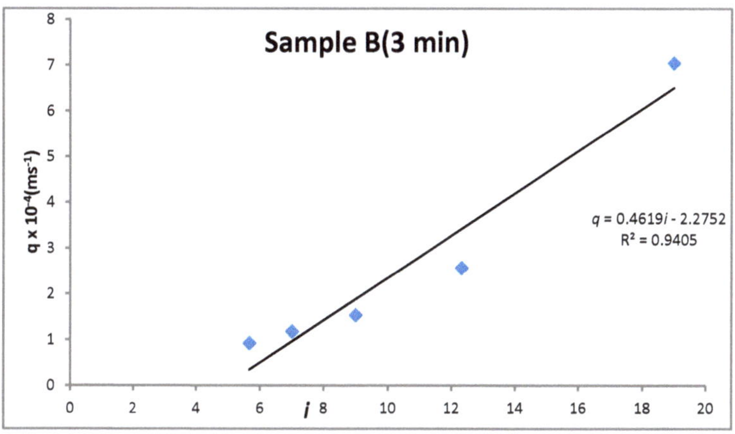

Fig. 9. Volume flux against hydraulic gradient (sample B, 3 min.)

overestimate or underestimate of volume of water supply to plant due to variability of hydraulic conductivity by Darcy's law. The understanding and introduction of declination of permeability with time is a reformed for accurate estimation of volume of water available for plants grow especially in surface irrigation. If a constant permeability of soils as proposed by Darcy's is assumed there might be either excess or shortage of water.

Moore and co-workers obtained very good fit for the relationship between the saturated hydraulic conductivity of the surface seal at time t, initial saturated hydraulic conductivity, and final saturated hydraulic conductivity (Moore 1979; Moore et al.

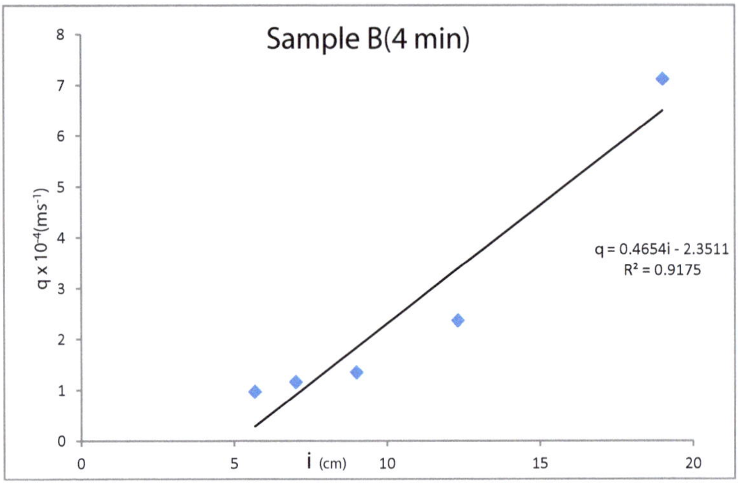

Fig. 10. Volume flux against hydraulic gradient (sample B, 4 min.)

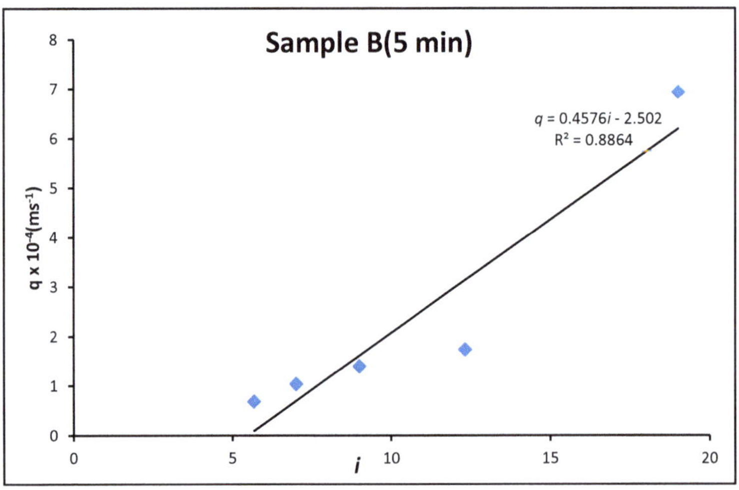

Fig. 11. Volume flux against hydraulic gradient (sample B, 5 min.)

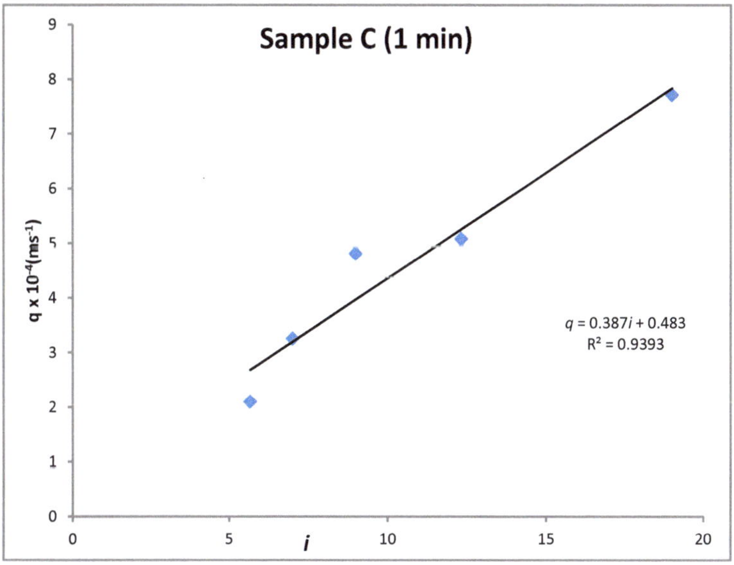

Fig. 12. Volume flux against hydraulic gradient (sample C, 1 min.)

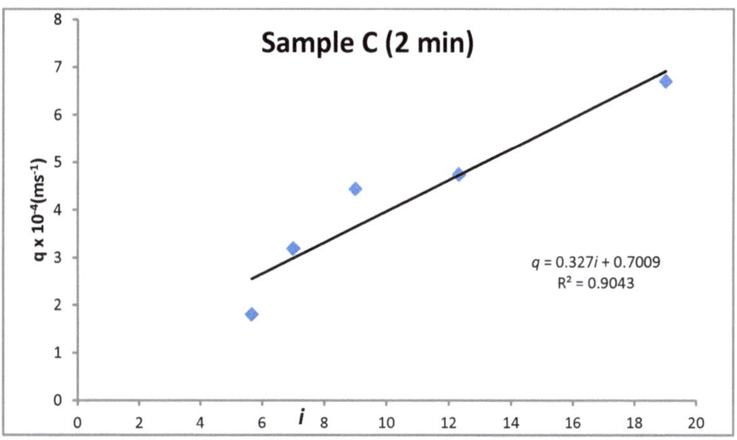

Fig. 13. Volume flux against hydraulic gradient (sample C, 2 min.)

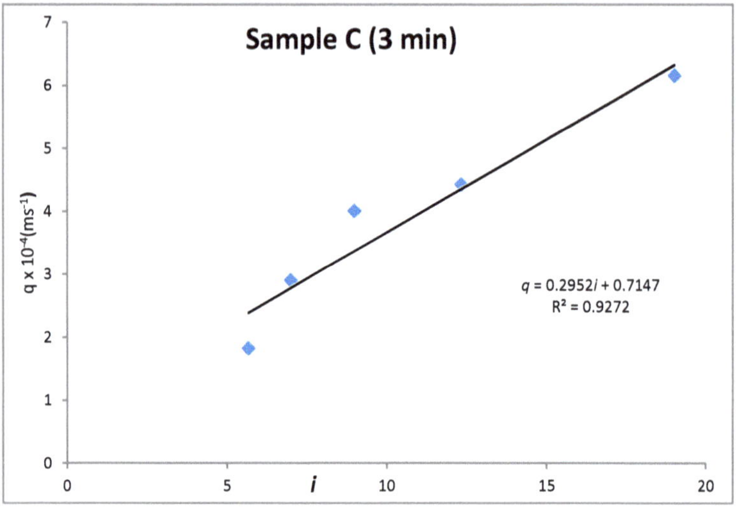

Fig. 14. Volume flux against hydraulic gradient (sample C, 3 min.)

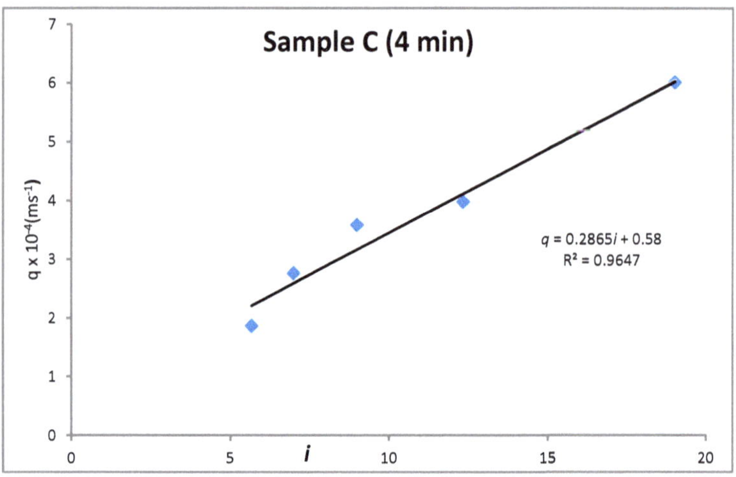

Fig. 15. Volume flux against hydraulic gradient (sample C, 4 min.)

1980). It was found that hydraulic conductivity decline exponential with time. The equations obtained show that as time increase, the value of hydraulic conductivity will never be zero. The present study result corroborates with Moore's results, because as time of infiltration and seepage is increasing with time, zero hydraulic conductivity will never occur except when time of infiltration is zero, and this is a reasonable condition.

The equation obtained by Moore requires a prior knowledge of both initial and final saturated hydraulic conductivity for computation or estimation of saturated hydraulic conductivity at a particular time. However, saturated hydraulic conductivity can be

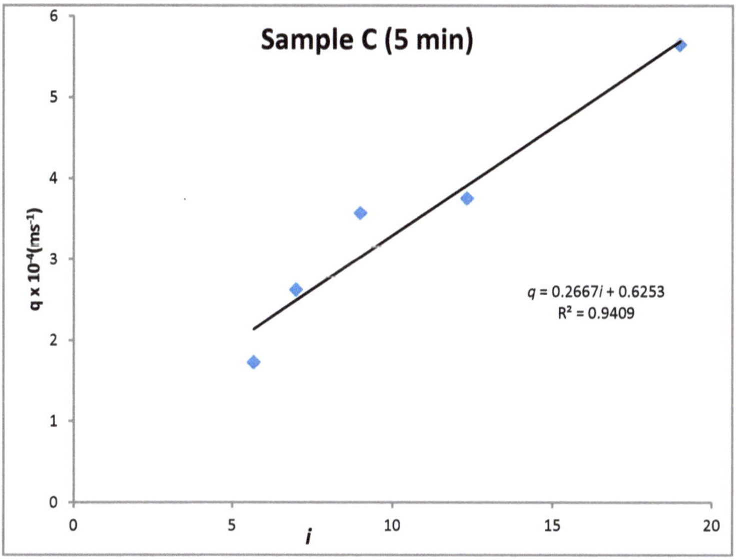

Fig. 16. Volume flux against hydraulic gradient (sample C, 5 min.)

Table 4. Hydraulic conductivity and permeability at different time

Time (s)	K $\times$ 10^{-5} (m/s)	k $\times$ 10^{-12} (m^2)
SAMPLE A		
60	0.480	0.4896
120	0.400	0.408
180	0.400	0.408
240	0.380	0.3876
300	0.360	0.3672
SAMPLE B		
60	5.42	5.53
120	4.68	4.77
180	4.61	4.70
240	4.57	4.66
300	4.65	4.10
SAMPLE C		
60	3.87	3.95
120	3.27	3.34
180	2.95	3.01
240	2.86	2.92
300	2.66	2.713

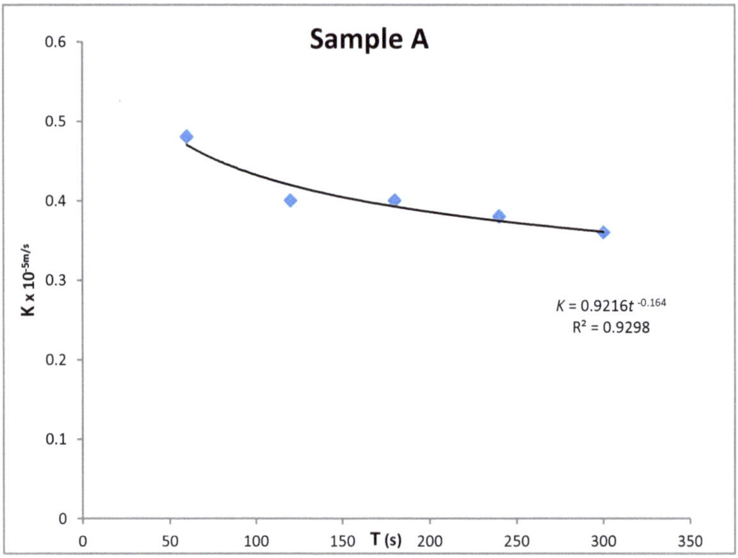

Fig. 17. Hydraulic conductivity against time (sample A)

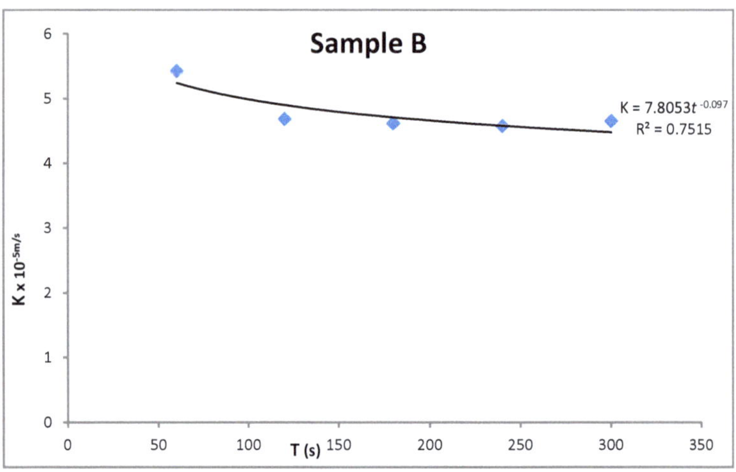

Fig. 18. Hydraulic conductivity against time (sample B)

obtained by the new proposed equation from this study without any prior information like Moore's equation. It should be noted that infiltration of water into soils is a continuous process, whereby the determination of the initial and final saturated hydraulic conductivity might not be relevant or difficult to obtain. Thus, Moore's equation may not be applicable for accurate estimation of volume of water available for plant in irrigation. Although, Moore's equation may be applicable in some other areas. In addition, the present equation or model is simple and practically useful because once

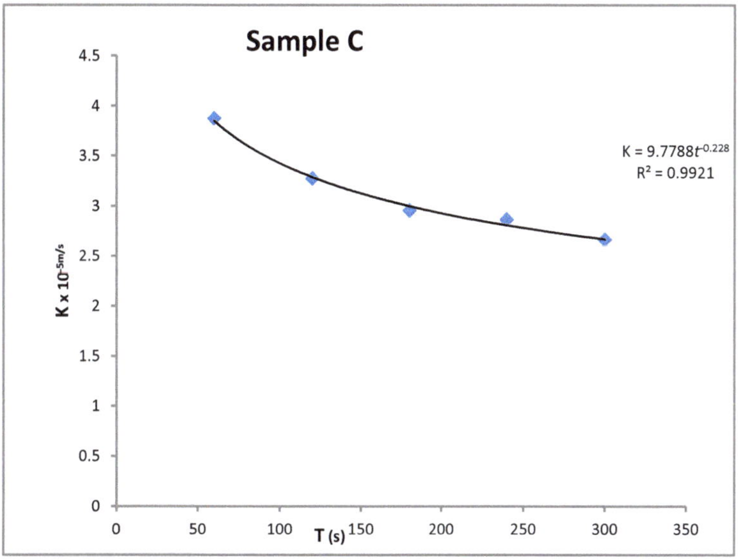

Fig. 19. Hydraulic conductivity against time (sample C)

the time is known, the hydraulic conductivity can be predicted for proper management in water irrigation, especially in surface irrigation.

5 Conclusions

A considerable amount of research has been conducted for the filtration of water into clay soils. The present study considered the laboratory test of one-dimensional vertical infiltration through clay of different particle sizes subjected to different hydraulic head. The hydraulic conductivities were determined at one minute the interval while permeabilities were obtained by Hubert King's relation. The results show that saturated hydraulic conductivity vis-à-vis permeability decline or decrease with increase in time. The equation obtained from $K - t$ curves shows that the declination of permeability will never be zero, except at an initial state or stage before the commencement of infiltration process (that is at time zero). Also the equation obtained is simple and practically useful for estimation of the permeability at a particular time, for accurate estimate of the volume of water available for plant grows in surface irrigation.

References

Arora, K.R.: Soil Mechanics and Foundation Engineering. Geotechnical 7th Engineering. Standard Publishers' Distributors, Delhi (2009)

Clemmens, A.J.: Evaluation of infiltration measurements for border irrigation. Agric. Water Management **3**(4), 251–267 (1981)

Clemmens, A.J.: Evaluating infiltration measurements for border irrigation model. Agric. Water Management **5**(2), 159–170 (1982)

Darcy, H.: Les Fountaines publiquues de la ville de Dijon. Victor Dalmont, Paris (1856)

Green, W.R., Ampt, G.A.: Studies on soil physics. The flow of air and water through soils. J. Agric. Sci. **4**, 1–26 (1911)

Huang, Z.Q., et al.: Research on infiltration clogging effect and its application prospect in anti-seepage project. Global Geol. J. **12**(2), 112–116 (2009)

Hubbert, M.K.: The theory of groundwater motion. J. Geol. **48**, 785–944 (1940)

Kostiakov, A.N.: On the dynamics of the coefficient of water percolation in soils and on the necessity for studying it from a dynamic point of view for the purposes of amelioration. Trans. 6th Comm. Intern. Soc. Soil Sci. Part A, 17–21 (1937)

Moore, I.D.: Infiltration into tillage affected soils. Unpublished Ph.D. Thesis, University of Minnesota, St. Paul, MN (1979)

Moore, I.D., et al.: Predicting infiltration and micro-relief surface storage for cultivated soils. Water Resources Centre, Bull. No. 102, University of Minnesota, St. Paul, MN (1980)

Philip, J.R.: The infiltration equation and its solution. Soil Sci. **83**, 345–357 (1957a)

Philip, J.R.: The theory of infiltration: 4, sorptivity and algebraic infiltration equations. Soil Sci. **84**, 257–264 (1957b)

Wetting and Drying Compacted Soil-Lime Mixtures

N. Maafi[1(✉)], M. Akchiche[1], and R. Sara[2]

[1] Department of Civil Engineering, University of Houari Boumedien (USTHB), Algiers, Algeria
maafinabil@ymail.com, akchiche-m@yahoo.fr
[2] Faculdade de Engenharia de Universidade do Porto, Rua Roberto Frias, s/n, 4200-465 Porto, Portugal
sara.rios@fe.up.pt

Abstract. Soil treatment technique with lime is widely applied in the world in the field of earthworks. It allows reuse of materials useless for implementation embankments. But the lack of knowledge of behavior of cycles, drying/wetting in terms of volume and water change, the mechanical performance of the treated soils, the optimal lime dosage are the origin of the limited application of this technique to embankment and the no mentioned in the GTS (technical Guide soil treatment with lime and/or hydraulic binders) of this technique. The soil studied is A-4. Lime dosages are chosen based on two objectives: the amelioration and stabilization to cycles drying/wetting of soil.

Keywords: Soil · Treatment · Lime · Cycles wetting/drying · Stabilization

1 Introduction

The soil with lime treatment is a technique widely used in the field of Earthworks (road, rail, airport infrastructure, platforms, and others.) since 20 years ago. It is now considered a classic technique to reuse unsuitable soils (generally, the fine limono clay soils) in the construction of embankments as well for minor roads for heavy infrastructure (highways, high-speed lines, airports, and others.). The goal of the use of this technique is to drying the soil, improve her manipulation and the establishment of compaction, to reach the capacity required. In addition, the soil with lime treatment to stabilize slopes and embankments. This technic have double impact, the first to improve soil properties (short term) and the second to stabilize the soil (in the medium and long term).

Lime stabilization is one of the most economical techniques for improving the engineering behavior of clayey soils (Ingles and Metcalf 1973; Little 1995; Paige-Green 2008; Lemaire et al. 2013). The addition of lime to soil causes two basic sets of reactions, one being a short-term reaction while the second is a long-term reaction (Little 1995). The immediate effect of lime addition is to cause flocculation and agglomeration of the clay particles caused by cation exchange at the surface of the soil particles. The result of this short-term reaction is mainly to improve workability and plasticity (Little 1995; Mathew and Rao 1997). The long-term reactions may

D.N. Singh and A. Galaa (eds.), *Contemporary Issues in Geoenvironmental Engineering*, Sustainable Civil Infrastructures,
DOI 10.1007/978-3-319-61612-4_5

require weeks, months or even years to complete, depending on the rate of chemical decomposition and hydration of the silicates and aluminates. This results in the formation of cementitious materials, which bind the soil particles together and improve the mechanical properties of the lime treated soil.

This present study the behavior of soil treated with lime face-to-face to wetting-drying cycles.

2 Materials

2.1 Materials

The soil samples used were collected from Khelil de Bordj Bou Arréridj (Algeria). (Tables 1, 2) and (Fig. 1) give the identification test results carried out on these soil samples and their chemical composition respectively. According to standard ***ASTM D3282-73*** the soil used in this study is classified in class **A-4**.

Table 1. Geotechnical properties of khelil soil (A-4).

Parameters	Symbols	Range of variation	Mean values
Liquid limit	WL (%)	41.203–41.831	41.525
Plastic limit	WP (%)	29.024–29.911	29.307
Plasticity index	IP (%)	12.179–12.255	12.218
Methylene blue value	MBV	1.590–1.623	1.568
Over to 2 mm	% 2 mm	90.9–92.8	91.534
Over to 0.08 mm	% <0.08 mm	65.9–68.7	67.10
Clay content	C2 μm (%)	4.274	4.274
Activity of clay	Ac	2.850–2.867	2.859
Optimum water content	Wopt (%)	22.274–22.321	22.300
Maximum dry density	γd-max	15.710–15.990	15.880

Table 2. Chemical composition of khelil soil (A-4).

Constituents	SiO_2	AlO_3	Fe_2O_3	CaO	MgO	K_2O	Na_2O	SO_3	Cl
%	31.43	4.19	1.22	32.5	1.89	0.36	0.40	0.04	0.01

Quicklime used comes from ERCO's company of Hassasna, wilaya of Saïda (Table 3).

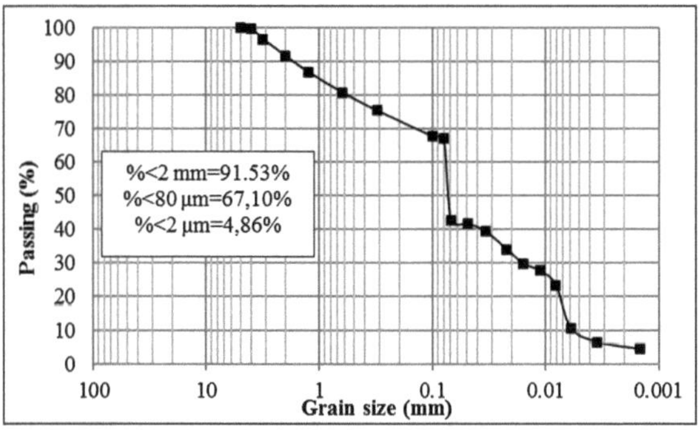

Fig. 1. Grain size distribution curve of soil A-4.

Table 3. Physico-chemical properties of Saida's Quicklime

Designation	BMSD
Chemical composition CaO	82,77
MgO	1,83
Fe_2O_3	0,87
Al_2O_3	3,27
SiO_2	1,35
SO_3	0,11
K_2O	0,151
Na_2O	0,064
Insolubles in Hcl	<1,00
Physical properties Bulk Density	731 g/l

3 Results and Discussion

In the following, we proceed to the analysis, the interpretation and the presentation of the results obtained in the study. The properties studied are: The point of fixation of lime **(PFL)**, the liquid limit **(LL)**, the plastic limit **(PL)** and plasticity index **(IP)**, Methylene blue value **(MBV)**, normal Proctor compaction **(γd, Wopn)**, Unconfined compressive strength **(UCS)**, ultrasonic pulse velocity **(km/sec)** and wetting–drying cycles.

These tests were carried out on untreated soil (control sample) and on treated soil with a mixture of various lime contents (Table 4).

3.1 The Point of Fixation of Lime (PFL)

The attachment point of the lime (PFL), proposed by Hilt and Davidson (1960), is the parameter for estimating the dosage of lime to transition between improvement and stabilization. It corresponds to the dosage of lime to the beyond which there is more

Table 4. Physico-chemical properties of the treated soil

Parameters	0%	1%	3%	5%	7%
pH	11.17	12.49	12.76	12.79	12.78
LL (%)	41.203–41.831	45.809–45.424	42.722–43.114	41.676–42.234	42.325–42.900
	41.525	45.621	41.915	41.953	42.617
PL (%)	29.024–29.911	31.991–32.430	35.036–35.616	35.142–35.739	38.138–38.762
	29.307	32.222	35.326	35.444	38.447
PI (%)	12.179–12.255	13.179–13.818	7.290–8.078	5.937–7.092	3.563–4.770
	12.218	13.399	7.589	6.509	4.170
MBV	1.590–1.623	1.484–1.567	1.132–1.245	1.070–1.150	0.817–0.934
	1.568	1.526	1.200	1.110	0.869
Wopt (%)	22.274–22.321	24.034–24.157	24.178–24.214	26.741–27.058	26.689–26.921
	22.300	24.100	24.200	26.900	26.800
γd-max	15.710–15.990	15.070–15.260	14.760–14.880	14.630–14.710	14.510–14.590
	15.880	15.140	14.830	14.680	14.560
ρd-max/Wopt	0.0711–0.0713	0.0657–0.0661	0.0656–0.0657	0.0587–0.0594	0.0590–0.0595
	0.0712	0.0659	0.0656	0.0590	0.0593

change of limit of plasticity. When the dosage of lime exceeds this dosage, the lime is adsorbed by clay minerals and becomes available for Pozzolanic reactions. The PFL is determined by the test of pH of the soil-lime mixture described in the standard **ASTM D6276-99A** that originated in the study of Eades and Grim (1960). This method allows determining the minimum dosage in lime so that the soil-lime mixture reaches a level of pH equal to 12.4. From this pH value, additional lime is supposed to be available for the development of the Pozzolanic reactions.

pH testing was performed on the soil **A-4**, and the corresponding point of fixation of lime (PFL) has been estimated at 1% (Fig. 2). Based on this measure and the performance evaluated by the Proctor curves, additions of 1%, 3%, 5% and 7% were chosen, corresponding to the minimum, intermediate and high dosages so that the A-4 soil treatment can meet the 3 objectives: (1) improvement, (2) stabilization and insensitivity to the water, (3) stabilization and resistance wetting–drying cycles.

3.2 Consistency Limits

The increase in the percentage of lime cause a significant reduction on the plasticity index, increasing the plasticity limit and decreasing the liquid limit (Fig. 3). However, the plastic limit shows a higher increase than the reduction of the liquid limit. The same

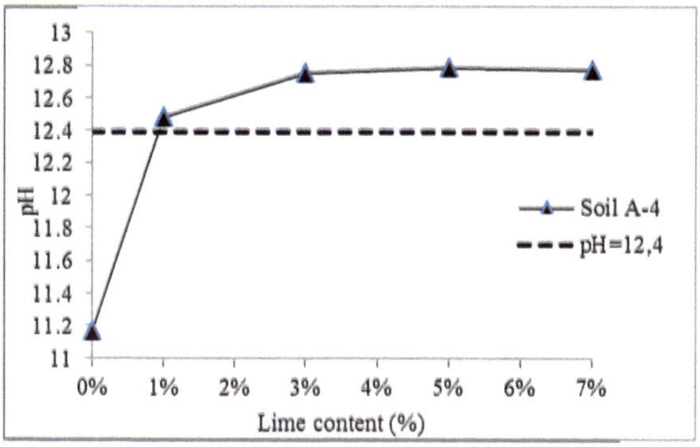

Fig. 2. The point of fixation of lime of soil A-4.

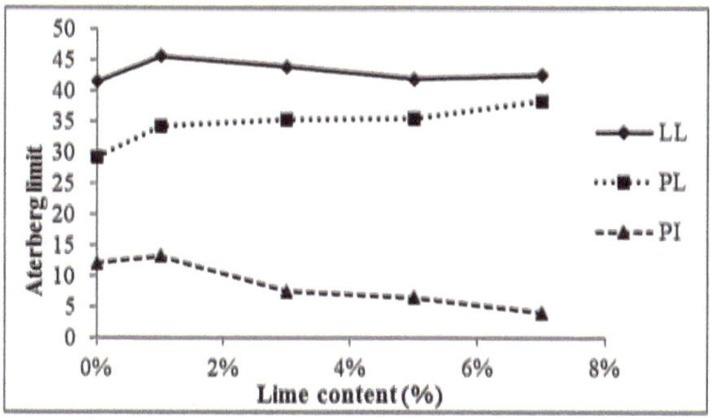

Fig. 3. Consistency limits soil A-4.

results are seen by many researchers (Afes and Didier 1999; Bengraa et al. 2004; Yucel et al. 2007).

Casagrande plasticity chart adapted to expansive soils in accordance with (Dakshanamurthy and Raman 1973 and Chen 1988) can be used to analyze their behavior after the treatment. Atterberg limits test for this soil treated with a mixture of lime indicate again that the liquid limit and plasticity index decrease significantly for treated soil (Fig. 4). It results in a reduction of the plasticity of soil, i.e. a reduction of its high swelling potential from medium to low swelling potential according to (Chen 1988) classification being therefore in the low section at for two classifications. Thus, the soil becomes less sensitive to moisture, so not very expansive and better compactable. Reduction of soil swelling improved stability with regard to deformations due to seasonal changes in the water content and, therefore, a more constant behavior of the compacted soil in what concerns the generating of fine particles.

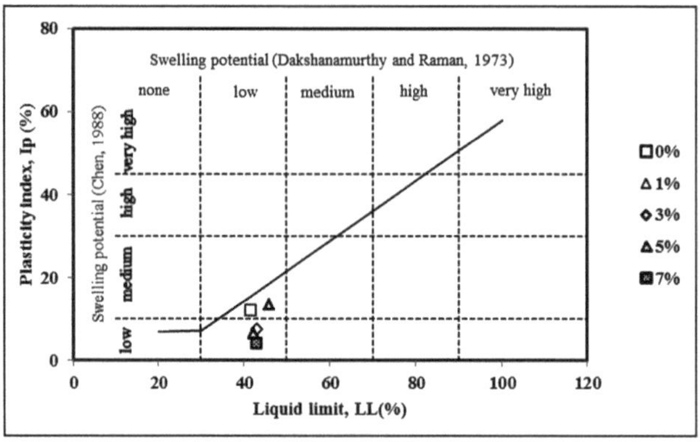

Fig. 4. Classification of soil and the treated soil according to its consistency limits.

3.3 Methylene Blue Value

The methylene blue test is considered as parameter of identification and classification of soils, (**MBV**) and it can be used to estimate the swelling potential of the soil. From (Fig. 5) it is clear that the MBV (i.e. the corresponding Sst = 21MBV) is also affected by the lime contents. In addition, Sst sound mitigation measures seems to indicate that his treatment of mixture of lime reduces its potential of swelling by changing its texture.

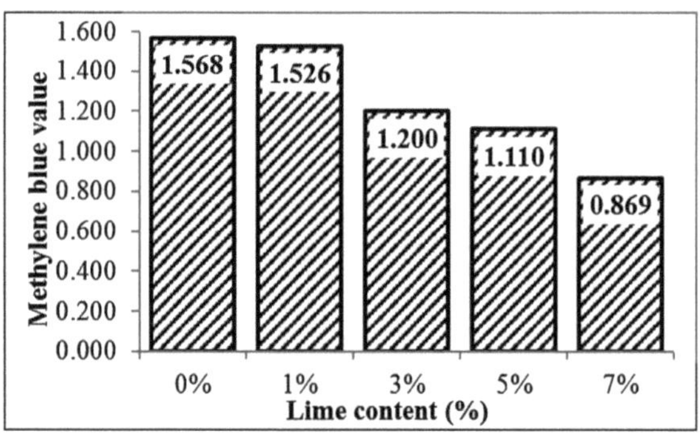

Fig. 5. Methylene blue value of the treated soil.

Results (Fig. 5) show an irreversible behavior, a linear decrease of **MBV** with the increase in the percentage of lime, as showed before for the Atterberg limits. A reduction of 45% was reached with 7% of lime.

3.4 Normal Proctor Compaction

The sensitivity to water mixtures noticed by the shape of the Proctor curves. According to (Fig. 6), the compaction curve of the soil in the natural state give an accentuated shape, which explains the great sensitivity of this soil. For lime contents of 1, 3, 5 and 7% blends, the Proctor curves are to the right and down compared to the curve of the soil control they present a flattened shape, reflecting the low sensitivity of mixtures by water, is all the more marked the soil reacts well with lime.

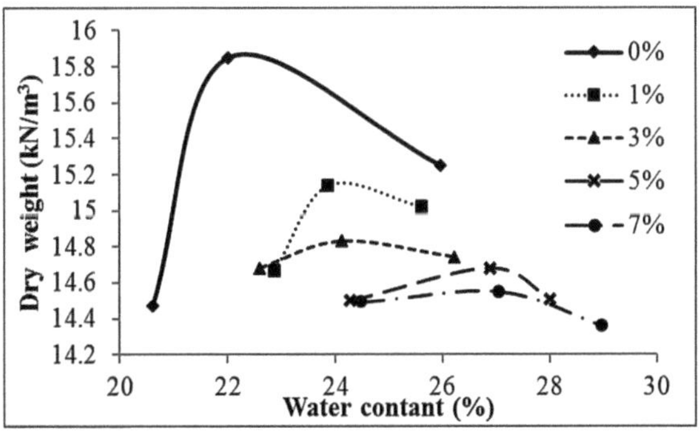

Fig. 6. Normal Proctor compaction.

The variation of the maximum dry density is represented in (Fig. 7). We can see that an increase in the percentage of lime from **1** to **7%** led to a decrease of the dry density at optimum normal proctor **(OPN)** from **15.88** to **14.56 (kN/m^3)**.

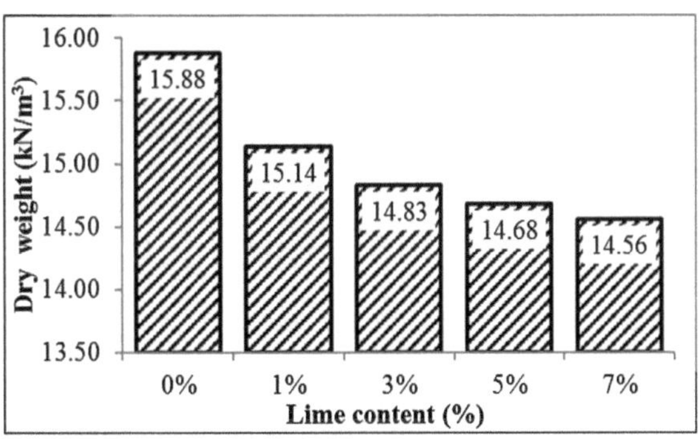

Fig. 7. The variation of the maximum dry density.

Furthermore, since quicklime was used the **OPN** (Fig. 8) water content increases with the percentage of added lime, reaching its maximum value at an **OPN 26.9%** for a percentage of **5%** lime. The treatment of soil with lime reduced the maximum value of the dry apparent density and increases the value of the optimum water content.

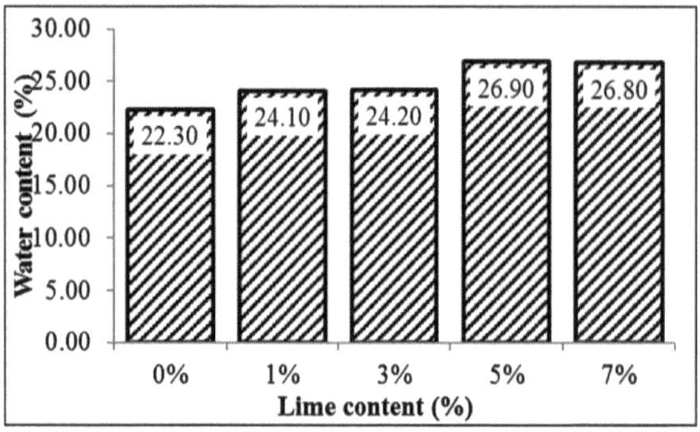

Fig. 8. The variation of the maximum water content.

The optimum water content increasing is due to the reaction of hydration of lime, and then the reason the reduction of the maximum dry density is the low specific weight of lime (Necmi et al. 2007).

After (Fig. 9) normal Proctor compaction tests are conducted on the soil treated with different content of lime in optimal conditions of Proctor. The corresponding test results are a guarantee of good repeatability of the compaction test and indicate a good reconstruction of soils under the conditions for which the massif of soil should be submitted on the spot.

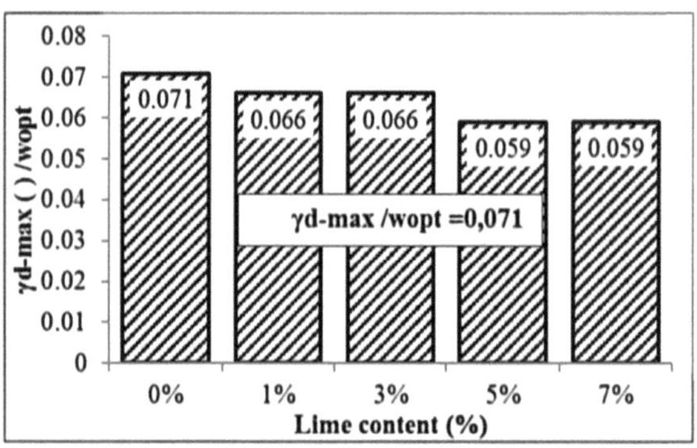

Fig. 9. Lime mixture treatment effect on the normal Proctor compaction test results.

3.5 Unconfined Compressive Strength (UCS)

Figure 10 represents the development of compressive strength based on different conditions of cure for treated and untreated mixtures. It is noticed that at 3 h after treatment, the compressive strength keeps its low value for all the mixtures.

For the period of 28-days of treatment, there is an important development with the increase in the percentage of lime, the strength values obtained for lime content of 1, 3, 5 and 7% are respectively 0.831, 0.839, 0.923 and 0.991 MPa compared with

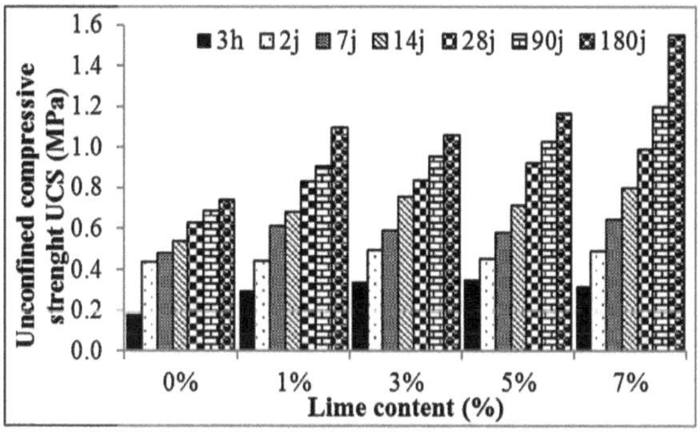

Fig. 10. The variation of the unconfined compressive strength UCS

0.628 MPa for the soil without lime. For the duration of treatment of 90 and 180 days, there is a slight increase in compressive strength to mix (1, 3 and 5) however a remarkable increase the 7% mixture. The same ability of variation is found by several authors (Kolias et al. 2005; Yi et al. 2006).

Compressive strength increases with the increase in the percentage of lime in treated blends, this increase can be explained at early ages by the addition of the fine materials to the soil, which results in better compaction. The reason for the sharp increase in the compressive strength after 28 days is justified by the reaction of hydration of lime during this period (Necmi et al. 2007). An increase in UCS with curing (from 2 to 28 days) between the soil, lime and reactions with the formation of ettringite and/or CAH and CSH. We know that lime improves the compressive strength of the soil by changing the course of hydration of the calcium silicate, which is mainly formed in the early stages of hydration.

3.6 Wave Velocity

The measurement of P waves velocities is one of the non-destructives tests methods to assess the stiffness properties of materials. P wave velocities results (Fig. 11) show

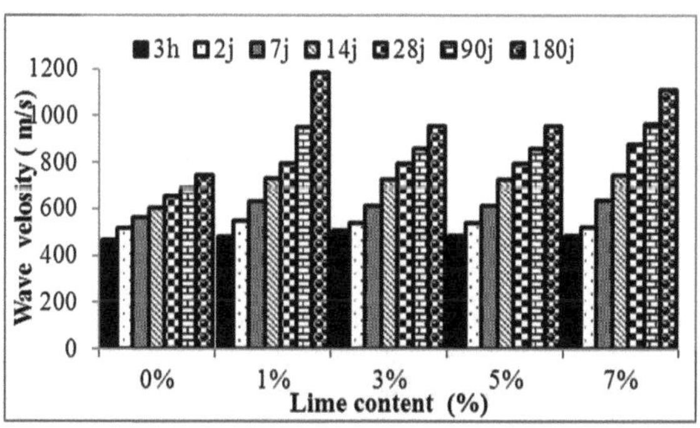

Fig. 11. The variation of the wave velocity P.

similar trends to the results of the UCS tests. For all lime content, the P waves velocities have increased with curing times. After 180 days of hardening, samples of soil with 1, 3, 5 and 7% of lime reaches maximum values of the P wave velocities on the order of 1184, 0957, 0958 and 1110 m/s 0789, 0798, which gives a ratio of improvement of 1.6, 1.3, 1.3, and 1.5 relation to the P wave velocity of the untreated soil samples, respectively. During the reactions occur between the soil and the stabilizing agent (i.e. lime).

Then, the development of a cementitious bonding between soil particles produced by the pozzolanic reaction increases the P-wave velocity measured in the stabilized soil samples. It is known that the P-wave velocity propagation increases as the compaction of the solid increases, in a soil mass, and the amount of voids filled with air and/or water decreases. Similar observations were made for lime-treated clay soil (Al-Mukhtar et al. 2010a, 2010b).

3.7 Correlations of Unconfined Compressive Strength (UCS) and Wave Velocity

Figure 12 shows the relationship between the P-wave velocities and UCS. A good correlation factor (0.8972) demonstrates that the non-destructive P-wave measurements can be considered as an indicator for UCS and can be used in the field to assess the mechanical behavior of soil/lime as shown by several authors for other soil mixtures (Consoli et al. 2012; Rios et al. 2016).

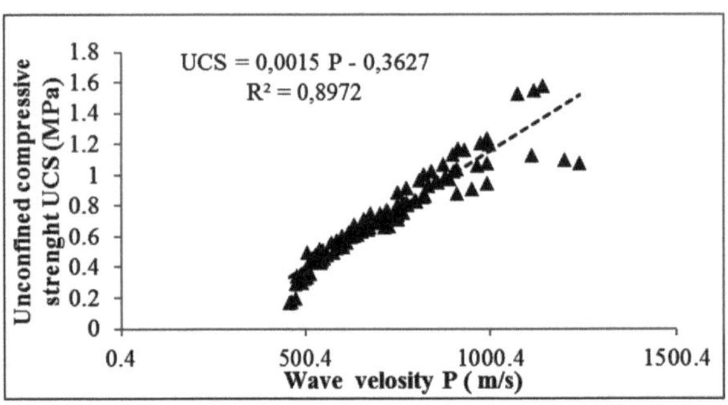

Fig. 12. Correlations of Unconfined compressive strength (UCS) and Wave velocity (P).

The correlation is a linear type:

$$UCS = A(P) + b \tag{1}$$

Previous researches propose a linear correlation between the mechanical strength and wave velocity (Aldaood and Al-Mukhtaret 2014, Viana da Fonseca et al. 2014).

The correlation proposed from the present study in the form:

$$\text{UCS} = 0.0015(\text{P}) - 0.36271$$
$$\text{R}^2 = 0.8972$$

3.8 Impact of Wetting–Drying Cycles

The specimens are prepared with a standard ***ASTM D559 M.***

The changes in volume of the specimens (Fig. 13) were significantly affected by the increase in the number of cycles of wetting-drying in the presence of lime. After 7 days of treatment, volume change has increased slightly, and the shape of the soil samples was retained until the tenth round for the soil with 1% lime (cycle 12 for specimens with 3, 5 and 7%) lost almost 70% of volume for 1% lime and 32.95, 26.38 and 22.68% for 3%, 5% and 7% respectively.

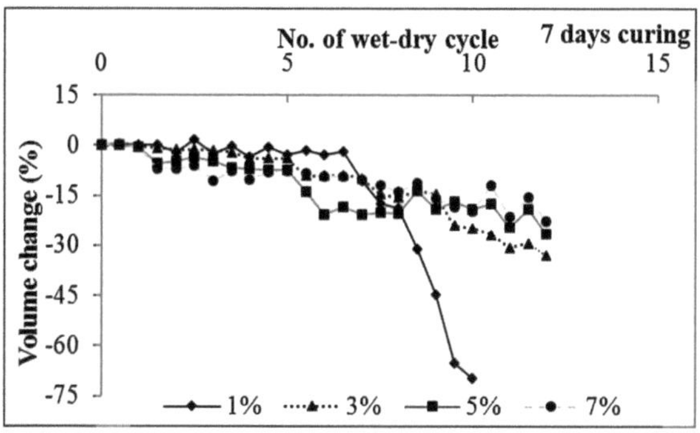

Fig. 13. Volume change with wetting–drying cycles after 7 days curing.

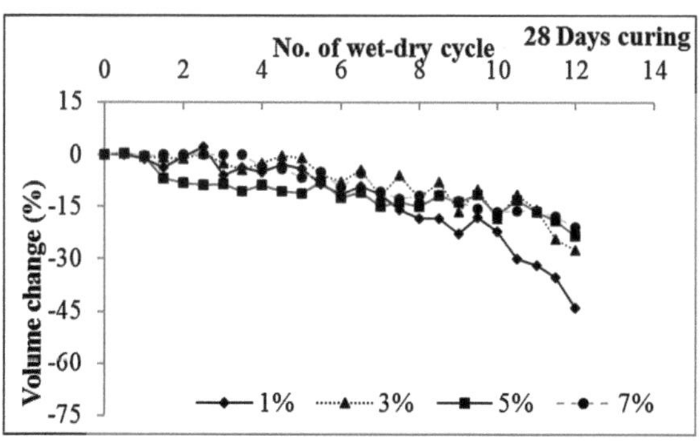

Fig. 14. Volume change with wetting–drying cycles after 28 days curing.

However after 28 days in the treatment sample (Fig. 15) and the shape of soil samples has been preserved up to the last cycle (cycle 12). The specimens lost 43.78, 27.4, 23,405 and 20.99% for 1%, 3%, 5% and 7% respectively. Crack propagation leads to loss in strength and reduction in the volume of soil samples by the disintegration of parts from the cracked samples, later an accelerated deterioration of samples has occurred caused by water penetration into the samples.

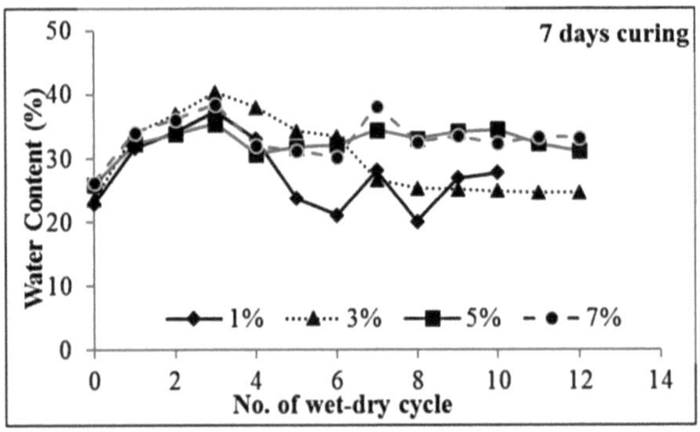

Fig. 15. Water content changes with wetting and drying cycles after 7days curing.

The soil is a porous material and has the capacity of water absorption by capillarity, so to conserve water during the humidification. In this step all empty (micro and macro-pores) were partially saturated, then when the soil samples were subjected to high temperature (70 °C) for 42 h, water inside the pores began escape from samples by evaporation.

The moisture of soils stored for 7 days (Fig. 14) and measured at the end of the drying state was increased during the firsts cycles (4 cycles) that due to the water absorption. From 4th cycle water content remained almost stable for 5% and 7% lime content while it decreased for the initial value in 1% and 3% lime contents.

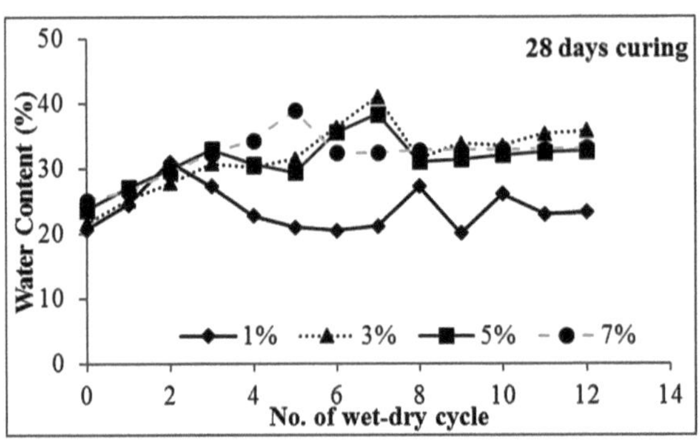

Fig. 16. Water content changes with wetting and drying cycles after 28 days curing.

However, the water content of soils stored at 28 days (Fig. 16) was increased in the first four cycles and even 6 cycles for 7% lime contents. From this cycle the water content remained almost stable for all lime contents except for 1% which decreased to the initial value.

4 Conclusions

After this study, we can draw the following key results: the introduction of lime in the soil has a significant effect on his physical behavior: it was noted a decrease in the plasticity index, a reduction in the Methylene blue value and a flattening of the Proctor curve. The addition of lime brings an immediate improvement in the resistance of the soil by the significant decrease in sensitivity to water. Soil mechanical parameters are clearly influenced by the addition of lime.

- The increase in resistance to compression of treated mixture is well noticed from 28 days. Improving the physical and mechanical behavior can be explained by at short-term (flocculation and agglomeration) and long-term (Pozzolanic reactions).
- The first is a direct, rapid response with immediate effect produced by lime (ground improvement). Lime can then react with clay minerals, causing a flocculation and agglomeration of clays. This flocculation increases the ability to compaction and decreases the plasticity of the soil.
- The second is a slow reaction, which hardens gradually the compacted soil-lime mixture, responsible for long-term (ground stabilization) effects. The alkalinity of the limestone increases the pH of the soil, which releases the aluminates and silicates. These can then react with the calcium provided by lime and water present in the soil which gives two cementation CSH and CAH officers who are responsible to the solidification and hardening the ground.
- The incorporation of lime in the soil increases their behavior with respect to the drying/wetting cycles reducing volume change and keepings the soil shape and the water content.

References

Aldaood, M., Al-Mukhtaret, B.M.: Impact of wetting–drying cycles on the microstructure and mechanical (2014)

Afes, M., Didier, G.: Stabilisation des sols gonflants: cas d'une argile en provenance de Mila (Algérie). Bull. Eng. Geol. Env. **59**, 75–83 (1999)

Al-Mukhtar, M., Lasledj, A., Alcover, J.F.: Behaviour and mineralogy changes in lime-treated expansive soil at 20 °C. Appl. Clay Sci. **50**(2), 191–198 (2010a)

Al-Mukhtar, M., Lasledj, A., Alcover, J.F.: Behaviour and mineralogy changes in lime-treated expansive soil at 50 °C. Appl. Clay Sci. **50**(2), 199–203 (2010b)

Bengraa, H.: Contribution a l'Etude de la stabilisation du gonflement par ajout du sable. Thèse de magister, Université de Mohamed Boudiaf U.S.T.O, Oran (2004)

Chen, F.H.: Foundations on expansive soils. In: Developments in Geotechnical Engineering. Elsevier (1988)

Consoli, N.C., Viana da Fonseca, A., Rios, S., Cruz, R.C., Fonini, A.: Géotechnique **62**(2), 177–183 (2012). http://dx.doi.org/10.1680/geot.8.P.084

Dakshanamurthy, V., Raman, V.: A simple method of identifying an expansive soil. Soils Found. Jpn. Soc. Soil Mech. Found. Eng. **13**(1), 97–104 (1973)

Eades, J.L., Grim, R.E.: Reaction of hydrated lime with pure clay minerals in soil stabilization. Highw. Res. Board Bull. **262**, 51–63 (1960)

Hilt, G.H., Davidson, D.T.: Lime fixation on clayey soils. Highw. Res. Board Bull. **262**, 20–32 (1960)

Ingles, O.G., Metcalf, J.B.: Soil Stabilization. Wiley, New York (1973)

Ingles, O.G., Metcalf, J.B.: Soil Stabilization Principles and Practice. Butterworth, Sydney (1972)

Kolias, S., Kasselouri-Rigopoulou, V., Karahalios, A.: Stabilisation of clayey soils with high calcium fly ash and cement. Cement Concr. Compos. **27**, 301–313 (2005)

Lemaire, K., Deneele, D., Bonnet, S., Legret, M.: Effects of lime and cement treatment on the physicochemical, microstructural and mechanical characteristics of a plastic silt. Eng. Geol. **166**(8), 255–261 (2013)

Little, D.N.: Handbook for Stabilization of Pavement Sub Grade and Base Courses with Lime. Kendall Hunt Publishing Company, Iowa (1995). (by National Lime Association)

Mathew, P.K., Rao, S.N.: Effect of lime on cation exchange capacity of marine clay. J. Geotech. Geoenviron. Eng. ASCE **123**(2), 183–185 (1997)

Necmi, Y., Ekrem, K., Suat, A.: Modification of the geotechnical properties, as influenced by freeze–thaw, of granular soils with waste additives. Cold Reg. Sci. Technol. **48**, 44–54 (2007)

Paige-Green, P.: The durability of stabilized materials. In: Proceedings of the 12th International Conference of International Association for Computer Methods and Advances in Geo-mechanics, pp. 3787–3792. Curran Associates, Inc., Goa, India (2008)

Rios, S., Cristelo, N., Viana da Fonseca, A., Ferreira, C.: Int. J. Geomech. (2016). doi:10.1061/(ASCE)GM.1943-5622.0000783

Viana da Fonseca, A., Amaral, M.F., Panico, F., Rios, S.: Indexation of dynamic and static geomechanical properties of a cemented aggregate for transportation engineering. Transp. Geotech. **1**(1), 31–44 (2014). doi:10.1016/j.trgeo.2014.02.001

Yi, C., Bin, S., Charles, W., Chao-Sheng, T.: Effect of polypropylene fibre and lime admixture on engineering properties of clayey soil. Eng. Geol. **87**, 230–240 (2006)

Yucel, G., Dursun, S., Murat Cetin, M.: Impact of cyclic wetting–drying %Mon swelling behaviour of lime-stabilized soil. Build. Environ. **42**, 681–688 (2007)

Quantitative Research on the Evolution of Mesoparameters at Out of the Shear Band of the Air-Dried Clay

Wei Wang(✉), Pei-ling He, and Bing-hua Zhao

Institute of Civil Engineering and Architecture, Nanjing Institute of Technology, Nanjing, China
ww1177114@163.com

Abstract. Based on the axial symmetry principle, three undisturbed and three remoulded cylindrical specimens of the air-dried clay were made into six half-cylinders through the axial symmetry plane, which were carried out on the unconfined compression tests. In the meantime, to investigate the mesostructures, the CCD camera system was developed with the long work distance telecentric microscope lens. By using the image collection system, a series of mesoimages of six clay samples at out of the shear band were captured under different loading states. According to these images, the evolution of mesoparameters of the air-dried clay was studied along increasing loads. Then, through images processing, the mesoimages were transformed into binary images (white areas were congregate particles while black areas were voids). So, the five mesoparameters such as area porosity, particles circularity, particles fractal dimension, particles orientation and Euler number extracted from the series of binary images were defined and analyzed quantitatively using the statistical method. The results show that: (1) the evolution curve shows that the preferable correlation exists in area porosity, particles circularity, particles fractal dimension and Euler number for undisturbed and remoulded clay while the correlation of other parameters does not exist. (2) Especially, with regard to the undisturbed clay, the correlation between particles circularity and particles fractal dimension can be fitted by the quadratic polynomial curve. Moreover, the correlation for the remoulded clay between particles circularity and particles fractal dimension is linear relationship. (3) This study also indicates that there isn't defined and uniform quantitative regression formula between five mesoparameters and the stress of samples in view of the complexity of clay actual mesostructues. It is gratifying that the evolution of some mesoparameters at out of the shear band of the air-dried clay along increasing loads had been qualitatively drawn from the regression analysis.

1 Introduction

It is well-known that soil mesostructures are a key to control engineering behaviors of foundation and the root of engineering disaster. So, the mesostructures in this area have being paid a good deal of attention by geotechnology researchers with great achievements. According to research methods of soil mesostructures, most people used the

D.N. Singh and A. Galaa (eds.), *Contemporary Issues in Geoenvironmental Engineering*, Sustainable Civil Infrastructures,
DOI 10.1007/978-3-319-61612-4_6

DEM to simulate soil samples with spherical particles. Due to restrictions of experiment apparatus, DEM numerical tests (Ding et al. 2012, Mahmood and Iwashita 2011, Pena et al. 2007, Pietruszczak and Guo 2013) show greater advantages for studying the relationship between mesostructures and macroproperties of soil. Nicot et al. 2015 firstly performed the numerical simulations of granular materials subjected to proportional strain loading paths using the DEM and secondly corroborated the DEM results with two micromechanical models. Zhang et al. 2016 conducted a series of biaxial tests based on DEM and investigated the variations of mesostructure during tests. While, the numerical models used in the above researches were generated randomly and they didn't reflect the actual mesostructure of geomaterials. It is believed that the models with random aggregate structure cannot exactly represent the actual geomaterial inhomogeneities and microstructures. Moreover, the particles adopted in the models were assumed to be rigid, which was different from the actual geomaterials. However, soil mesostructures are a comprehensive product of various natural actions. It is hard to quantify the structure factors because of obvious heterogeneity and undefination. But, people have being had a strong desire to make a great progress about mesostructure research with the development of digital image technology (Ammouche et al. 2001, Hu et al. 1994, Obaidat et al. 1998, Ulusoy and Igathinathane 2016). So, some people explored the macroscopic mechanism from the mesoscopic scale using the digital image processing technology. The digital image analysis can obtain the distribution information of soil mesostructures through SEM, CT or camera (Darma et al. 2013, Ge et al. 2001, Houben et al. 2014, Ketcham 2005, Li et al. 2016). For example, Zhang and Kong 2013 studied the microscopic pore characteristics and analyzed the relationship between the microstructure and physical parameters using the scanning electron microscopy (SEM), mercury intrusion porosimetry (MIP) and nitrogen adsorption (NA). Bo et al. 2016 determined the mineralogical properties of the ultrasoft soils using X-ray diffraction and SEM techniques. However, the soil samples applied on the SEM test were prepared complicated and the experimental environment was demanded strictly. In addition, many scholars have proposed several qualitative and quantitative description methods for geomaterials based on CT. In particular, Wang et al. 2014 investigated the damage cracking characteristic of rock and soil aggregate by X-ray CT under uniaxial compressive loads. Cui et al. 2016 simulated the mesoscopic seepage field of real soil based on the data structure generated by CT scanned slice. While, the price of SEM or CT apparatus was more expensive compare with common camera. For this reason, some researchers developed geomaterials mesostructures through cameras or microscope lens. Yue 2006, 2003 successfully carried out quantitative investigations of the orientations, distributions and shapes of aggregates about asphalt concrete using a conventional camera or a digital camera. Zhang et al. 2010 developed three new porosity models for determining the fractal dimension of rocks and soils.

Overall, the researches mentioned above were mainly concentrated on the mesostructures of geomaterials whether numerical simulations or laboratory experiments. However, the mesoparameters evolution about geomaterials fully analyzed has not been found in public literatures. So, the purpose of this study was to explore the relationship between macrocosmic properties (stress or strain limited) and mesoparameters. At first, experiment method was studied. Subsequently, mesoparameters describing the mesostructures of the air-dried clay were developed. Moreover, the evolution of mesoparameters at out of the shear band was further demonstrated.

2 Experimental Device and Methods

It was generally known that the evolution of clay mesoparameters at out of the shear band was different from that located in the shear band. For the indetermination of the shear location, this study mainly focused on the local regions that located in out of the shear band judged from the knowledge of shear failure about 45 degrees. To capture the mesoimages under every level of pressure, the optical microscope was prepared. The experimental devices were shown in Fig. 1.

Firstly, based on the axial symmetry principle, six specimens of air-dried clay including three undisturbed samples and three remoulded samples were cut into half-cylinders (39.1 mm diameter and 80 mm height) respectively. The half-cylinders unspoiled in the making process were saved as the test specimens. The symmetry planes exposed were the regions of interest for grabbing mesoimages. The specimens were tested on the unconfined compression instrument in sequence. At the same time, the optical microscope called the long work distance telecentric microscope was faced toward the sample plane. Before capturing the mesoimages, the calibration work about the microscope was done using a ruler. The coefficient of the microscope magnification used in this study was $K = 0.0014\,mm/pixel$. Secondly, the mesoimages of local regions at out of the shear band were collected at every level load, until the clay samples were broken. And, the captured mesoimages of a clay sample must be the same region when loads were applied from initial state to terminal state. The pictures of the six broken samples were shown in Fig. 2. Through observation, the failure mode of most samples belonged to shear failure. Before loading, the six initial mesoimages captured were shown in Fig. 3. Obviously, the mesostructures distribution was very different among them. Finally, the image processing technology was utilized to process

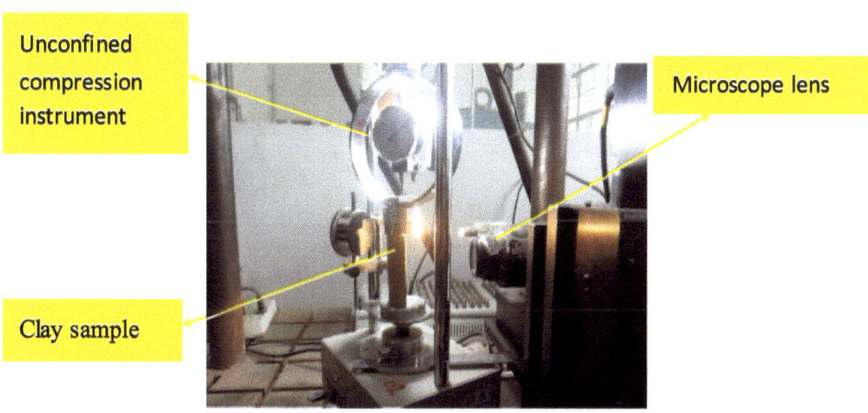

Fig. 1. Experimental equipment

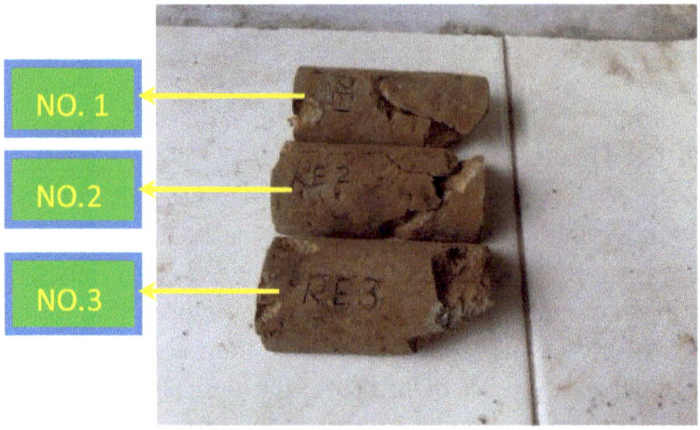

(a) The picture of three broken undisturbed clay samples

(b) The picture of three broken remoulded clay samples

Fig. 2. The pictures of six broken samples

these images with morphologic operation for quantitative analyzing the relationship between mesostructures and macroscopic properties of the air-dried clay.

3 Mesoparameters Description

Modern concept about soil mesostructures can be understood as area porosity, particles circularity, particles fractal dimension, particles orientation and Euler number, which was listed in Fig. 4.

According to the air-dried clay, the mesoimages illustrated the soil mesostructures distribution composed of aggregate particles and voids. Then, the quantitative assessment of mesoparameters describing the soil mesostructures was calculated by

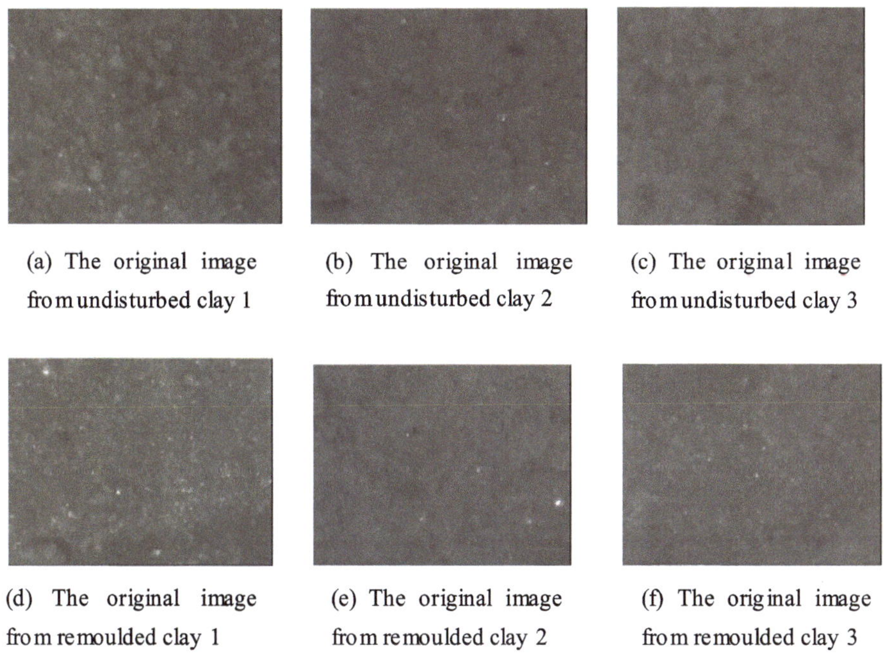

Fig. 3. The six original images

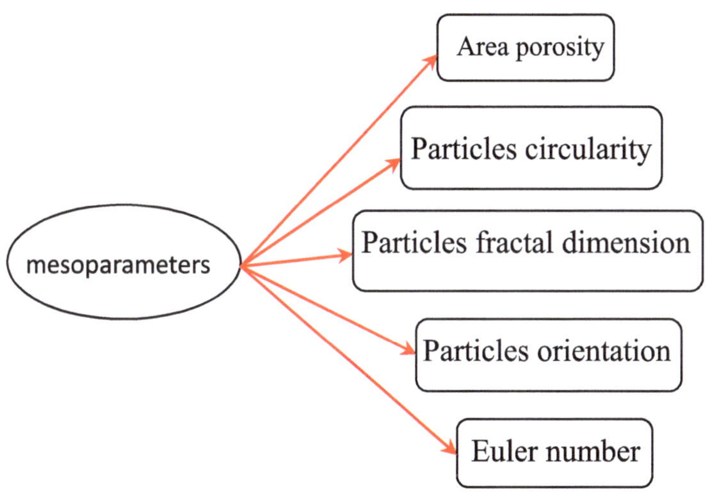

Fig. 4. Five mesoparameters proposed

image processing methods. Through the Otsu's method, the mesoimages were transformed into binary images (white areas were congregate particles while black areas were voids). In fact, the Otsu's method is an image binarization technique. The method not only avoids the interference of artificial factor, but also automatically calculates the maximum threshold of intra-class variance. In the way, the real boundary between congregate particles and voids could be realized. The threshold of every mesoimage was obtained from the formula (1).

$$Q(k) = \frac{(Ave \times W(k) - Aver(k))^2}{W(k) \times (1 - W(k))} \quad (1)$$

Where,

$Q(k)$: The separation index of gray class

Ave: Average of gray-value, $Ave = \sum_{i=0}^{255} (i - 1)Phs(i)$

Aver: The mean of gray level class, $Aver(k) = \sum_{i=0}^{k} (i + 1)Phs(i)$

$W(k)$: The sum of gray class histograms, $W(k) = \sum_{i=1}^{k} Phs(i)$

Phs(i): The probability function of gray level

i: Gray level, $0 \leq i \leq 255$

When the $Q(k)$ value was the biggest, the optimal threshold (T) was found. That was $T = k - 1$. Obviously, the threshold values of mesoimages were entirely different. So, the mesoimages could be processed using the Otsu's method after median filtering and the binary images were obtained successfully. Due to space limitations, the example of binary images obtained by image processing technology was shown in Fig. 5, which demonstrated the mesostructures of the undisturbed clay 1. Based on binary images, the calculation formulas of five mesoparameters were defined shown in Table 1.

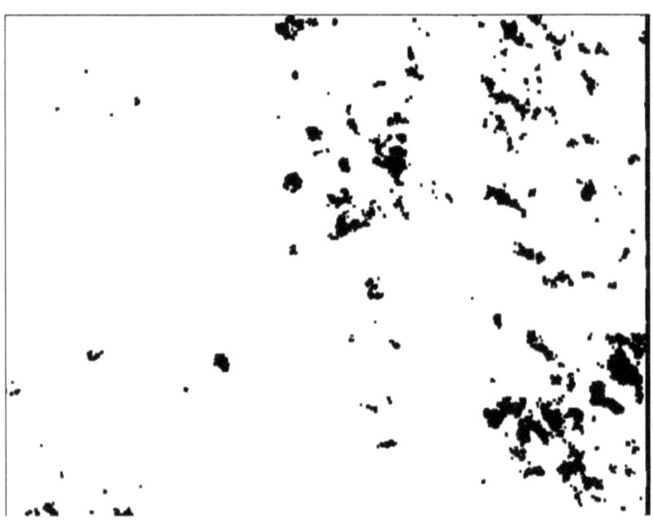

Fig. 5. The binary image of the undisturbed clay labeled 1 processed through the Otsu's method

Table 1. Five mesoparameters formulas

Mesoparameter	Formula
Area porosity (AP)	$n_v = \frac{A_v}{A} \times 100\%$ n_v: area porosity, A_v: voids area of binary image, A: the whole area of binary image
Particles circularity (PC)	$R = \frac{4\pi A}{L^2}$ R: particles circularity, A: particles area of binary image, L: particles perimeter of binary image
Particles fractal dimension (FD)	$D = \frac{2\times \ln(L)}{\ln(A)}$ D: particles fractal dimension, L: particles perimeter of binary image, A: particles area of binary image
Particles orientation (PO)	$H = -\sum_{i=1}^{n} P_i(\alpha) \lg P_i(\alpha)$ H: particles orientation, α: the direction angle of the longest chord of particles in the binary image, $P_i(\alpha)$: probability of the i unit at α range, n: the interval number of direction angle
Euler number (EN)	$E_n = N_C - N_H$ E_n: Euler number, N_C: the total number of objects in the image, N_H: the total number of holes in those objects

Then, the five mesoparameters extracted from the binary-image sequences were analyzed quantitatively using the statistical method.

4 Mesoparameters Evolution in Deformation Process

Aim to find out the evolution of clay mesostructures under the different load increments, we specially made three undisturbed samples and three remoulded samples according to the standard for soil test method. The method of experiment was common compression test from 0 to the maximum pressure, until the samples were broken. In this way, the relevant relationship between mesoparameters and changeable pressure was more clear and comprehensive. Based on the image analysis algorithm and the formulas defined in Table 1, we got the relevant values of mesoparameters under different loading states. The loads were exerted in sequence until the strength of samples reached the limit.

The purpose of the research was to analyze the correlation among five mesoparameters and the evolution of five mesoparameters during the unconfined compression process on the specimens. Then, by analyzing the mesoparameters value obtained from mesoimages, we got some important results as follows.

1. In order to study the relevance among five mesoparameters, the data obtained by the image processing technology were analyzed using the method of linear fitting. The results show that the preferable correlation exists in area porosity, particles circularity, particles fractal dimension and Euler number for undisturbed and remoulded clay while the correlation of other parameters does not exist. The relationship diagrams were plotted in Fig. 6 only taking the undisturbed sample labeled 3 as an example. Obviously, from the squared values of correlation coefficient (R^2) are 0.663, 0.688 and 0.753 respectively shown in Fig. 6(a), (b) and (c), we find that area porosity can be linearly related to particles circularity, particles fractal

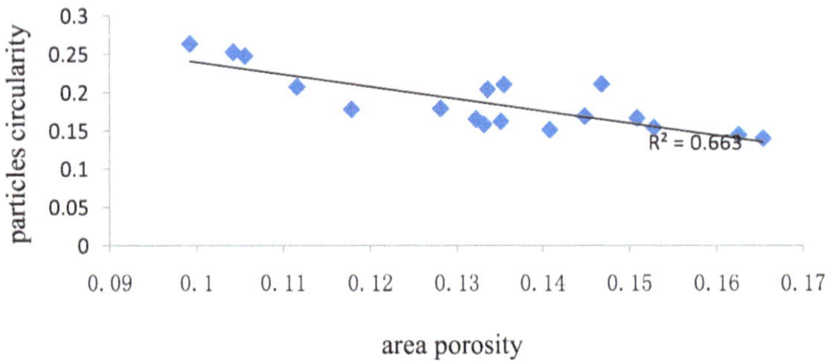

(a) Area porosity-particles circularity relationship

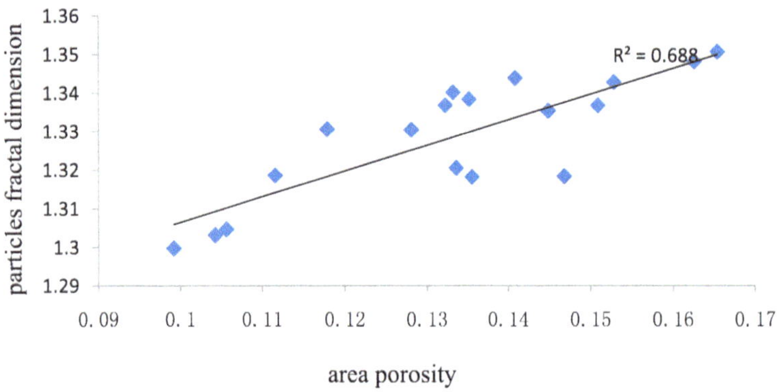

(b) Area porosity-particles fractal dimension relationship

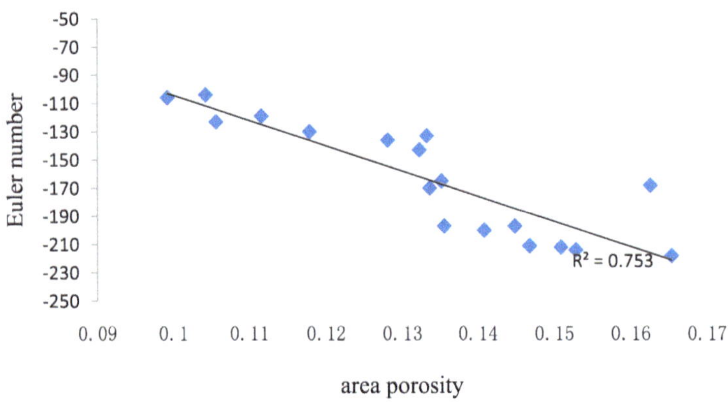

(c) Area porosity-Euler number relationship

Fig. 6. The relationship between parameters diagrams (R: correlation coefficient)

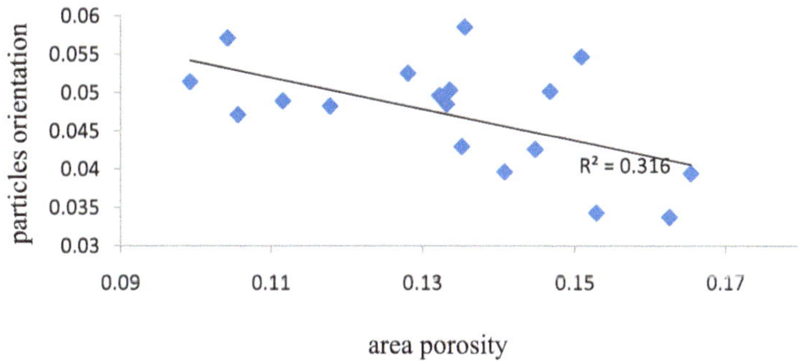

(d) Area porosity-particles orientation relationship

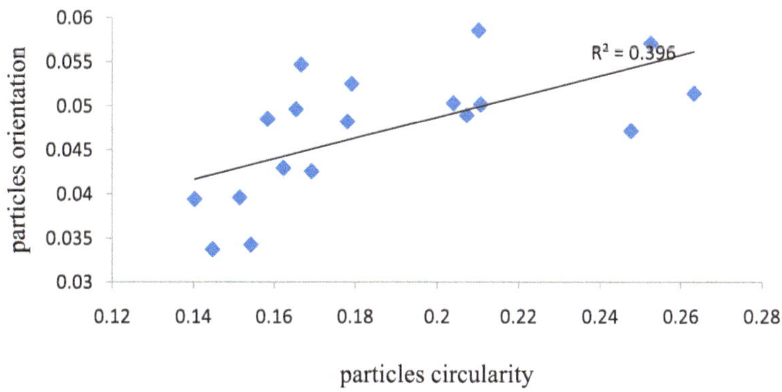

(e) Particles circularity-particles orientation relationship

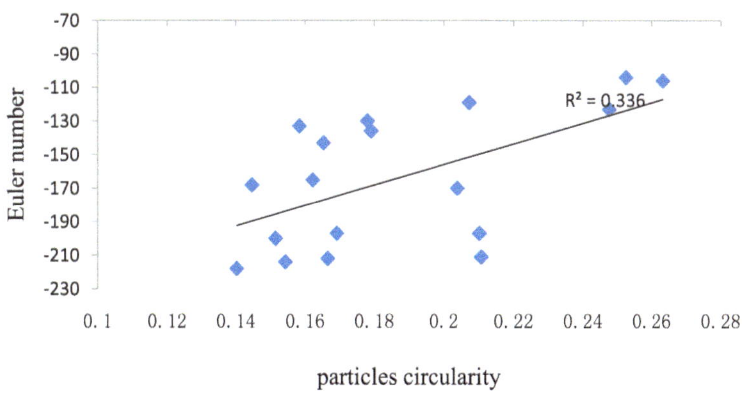

(f) Particles circularity-Euler number relationship

Fig. 6. (continued)

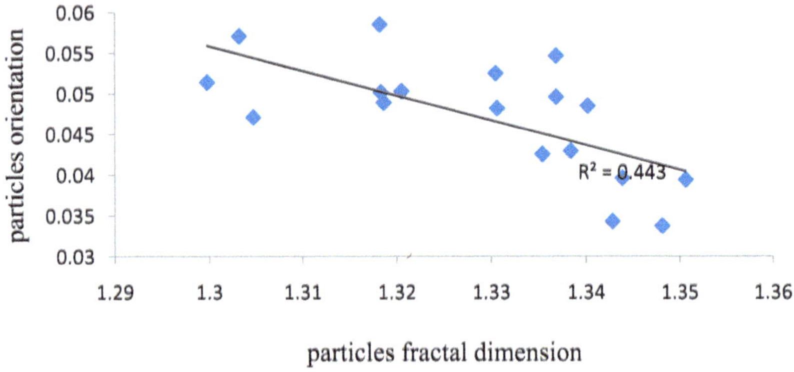

(g) Particles fractal dimension-particles orientation relationship

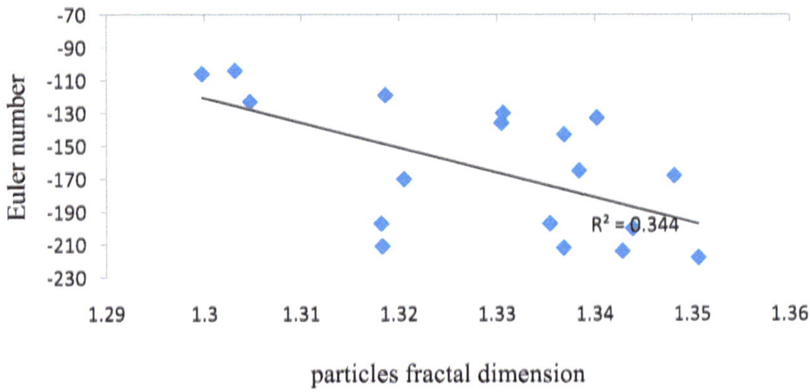

(h) Particles fractal dimension-Euler number relationship

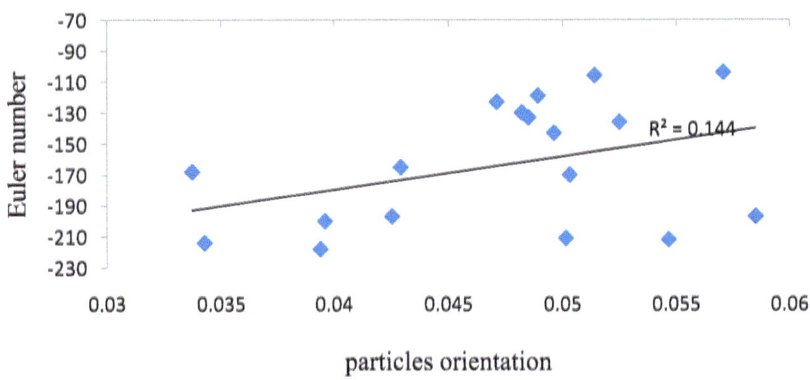

(i) Particles orientation-Euler number relationship

Fig. 6. (continued)

dimension and Euler number basically. Moreover, we also learn that the decreasing of area porosity with exerting pressure will give rise to the increasing of particles circularity and Euler number, while the same decreasing of particles fractal dimension. However, the squared values of correlation coefficient shown in Fig. 6 (d), (e), (f), (g), (h) and (i) are all below 0.5. So, the linear fitting relationship for other parameters is not ideal. In other words, it is not too much significant for studying the relevance between other parameters except the relationship between particles circularity and particles fractal dimension introduced below.

2. Especially, with regard to three undisturbed samples, the correlation between particles circularity and particles fractal dimension can be fitted by the quadratic polynomial curve. Moreover, the correlation for the remoulded ones between particles circularity and particles fractal dimension is linear relationship. The relationship of particles circularity and particles fractal dimension of the undisturbed sample labeled 1 and the remoulded sample labeled 1 was shown in Fig. 7(a) and Fig. 7(b) respectively. The fact tells us that particles circularity is related closely to particles fractal dimension in spite of different fitting formulas shown in Fig. 7(a) and (b). According to the discrete data shown in Fig. 7(a), the fitting curve can be expressed by the parabolic Eq. (2).

$$y = 0.236x^2 - 0.403x + 1.390 \tag{2}$$

Where,

y: Particles fractal dimension of the undisturbed sample labeled 1
x: Particles circularity of the undisturbed sample labeled 1

We find that the Eq. (2) agrees well with the data, through examining the correlation coefficient whose squared value (R^2) is 0.998. Meanwhile, the fitting curve for the data shown in Fig. 7(b) is described by the linear Eq. (3).

$$y = -0.498x + 1.419 \tag{3}$$

Where,

y: Particles fractal dimension of the remoulded sample labeled 1
x: Particles circularity of the remould sample labeled 1

Similarly, the Eq. (3) can achieve the better fitting effect, because the squared value of correlation coefficient is 0.988.

From the two curves, we also learn that particles fractal dimension decreases with the increase of particles circularity. The values of particles fractal dimension are between 1.2–1.32 for undisturbed samples while between 1.31–1.37 for remoulded ones, which tell us that the cohesion of undisturbed clay is bigger than that of remoulded one. Obviously, the bearing capacity of undisturbed clay is also stronger than that of remoulded clay under same loadings and same constraint conditions.

3. As shown in Table 2, we find that the regression equations between mesoparameters and stress of six clay samples are very different and they can't be unitized. Stepwise linear regression reveals that there isn't defined quantitative relationship

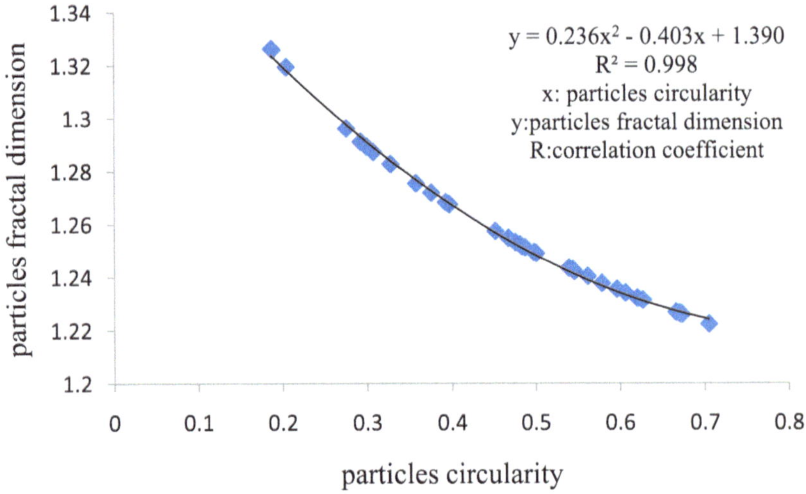

(a) Particles circularity-particles fractal dimension relationship of the undisturbed sample labeled 1

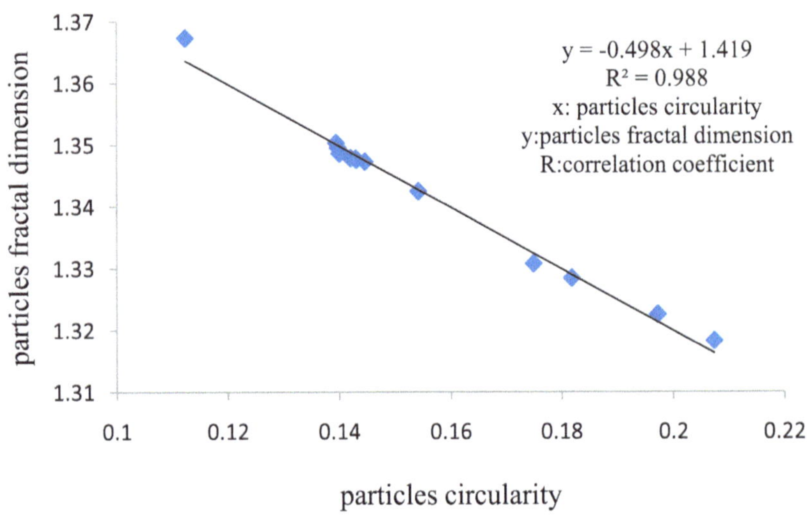

(b) Particles circularity-particles fractal dimension relationship of the remoulded sample labeled 1

Fig. 7. The relationship between particles circularity and particles fractal dimension diagrams

between mesoparameters and stress (or strain). Still, some qualitative evaluation has been done. The major features are as follows:

(1) The values of area porosity about undisturbed clay samples are almost less than 10% from statistical results, which indicates the compression strength of

Table 2. Regression equations of six clay samples

Lable		Regression equation
Undisturbed clay	NO. 1	Stress = –2009.7*AP + 211.04
	NO. 2	Stress = –9585.6*AP-5.39*EN + 275.3
	NO. 3	Stress = –3465.97*AP-2.41*EN + 189.87
Unmoulded clay	NO. 1	/
	NO. 2	/
	NO. 3	Stress = 783.2*PC-89.01

undisturbed clay is higher than that of remoulded clay. And, the value of area porosity about remoulded ones basically exceeds 10%.

(2) With regard to particles circularity, the bigger the initial particles circularity, the more the strain of clay samples. Results processed show that the initial particles circularity of undisturbed clay samples is higher than that of remoulded ones.

(3) For particles orientation, there is no discernible trend along increasing loads. Sometimes, it is going to increase. Sometimes, it is going to decrease. That is, we can't qualitatively analyze because of little relationship between particles orientation and strain of samples.

(4) The absolute value of Euler number about undisturbed clay samples is less than that of remoulded ones, which illustrates the pore connectivity of remoulded clay is good and the clay is breakable.

5 Conclusions

Facts have proved that the image processing technology can be used as the effective measure to get quantitative information on clay mesoparameters. Five mesoparameters describing clay mesostructures were proposed and the quantitative relationship among these parameters was also analyzed.

Generally speaking, the decrease of area porosity with exerting pressure will lead to the decrease of particles fractal dimension but lead to the increase of particles circularity and Euler number. Furthermore, the perfect fitted curve between particles circularity and particles fractal dimension has been pointed out in spite of undisturbed clay or remoulded clay. That is, the tight relationship exists in particles circularity and particles fractal dimension.

On the other hand, the evolutions of five mesoparameters proposed reflect the relationship between mesostructures and macro-mechanical properties of air-dried clay through the analysis of regression. However, the quantitative relation between mesoparameters and stress depends on the different specimen and the coefficients in regression equations are completely different as well. The complexity of clay mesostructures is proved again like other researches on soil mesostructures throughout the open literatures.

According to five mesoparameters, some quantitative relations really exist. This explains that parts of five mesoparameters are not isolated, which inducts us to find a major meso-factor that affects the macro-mechanical properties of clay dramatically. It still needs a lot of tests and data that are our next work.

Acknowledgements. Authors are wishing to acknowledge the financial support from the Science Research Fund of Nanjing Institute of Technology (No. CKJB201310) and the Qingnian Xueshu Gugan Teacher Fund of Nanjing Institute of Technology.

References

Ammouche, A., Riss, J., Breysse, D., Marchand, J.: Image analysis for the automated study of microcracks in concrete. Cement Concr. Compos. **23**(2–3), 267–278 (2001). doi:10.1016/S0958-9465(00)00054-8

Bo, M.W., Arulrajah, A., Sukmak, P., Horpibulsuk, S., Leong, M.: Mineralogy and geotechnical properties of ultrasoft soil from a nearshore mine tailings sedimentation pond. Mar. Georesour. Geotechnol. **34**(8), 782–791 (2016). doi:10.1080/1064119X.2015.1094158

Cui, G.Z., Shen, L.F., Wang, Z.L., Tang, Z.G., Xu, Z.M.: Numerical simulation of mesoscopic seepage field of soil CT scanned slice based on Iattice Boltzmann method. Rock Soil Mech. **37**(5), 1497–1502 (2016). doi:10.16285/j.rsm.2016.05.034

Darma, I.S., Sugiyama, T., Promentilla, M.A.B.: Application of X-Ray CT to study diffusivity in cracked concrete through the observation of tracer transport. J. Adv. Concr. Technol. **11**(10), 266–281 (2013). doi:10.3151/jact.11.266

Ding, X.L., Zhang, H.M., Huang, S.L., Lu, B., Zhang, Q.: Research on mechanical characteristics of unsaturated soil-rock mixture based on numerical experiments of mesostructure. Yanshilixue Yu Gongcheng Xuebao/Chin. J. Rock Mech. Eng. **31**(8), 1553–1566 (2012)

Ge, X., Ren, J., Pu, Y., Ma, W., Zhu, Y.: Real in time CT test of the rock meso-damage propagation law. Sci. China Ser. B: Chem. **44**(3), 328–336 (2001)

Houben, M.E., Desbois, G., Urai, J.L.: A comparative study of representative 2D microstructures in shaly and sandy facies of Opalinus Clay (Mont Terri, Switzerland) inferred from BIB-SEM and MIP methods. Mar. Pet. Geol. **49**, 143–161 (2014). doi:10.1016/j.marpetgeo.2013.10.009

Hu, R.L., Li, X.Q., Guan, G.L., Ye, H.: Fractal characters of loess microstructure in deformation process and its engineering significance. In: Seventh International Congress International Association of Engineering Geology, pp. 3411–3417 (1994)

Landis, E.N., Nagy, E.N.: Three-dimensional work of fracture for mortar in compression. Eng. Fract. Mech. **65**(2), 223–234 (2000). doi:10.1016/S0013-7944(99)00124-1

Ketcham, R.A.: Three-dimensional grain fabric measurements using high resolution X-ray computed tomography. J. Struct. Geol. **27**(7), 1217–1228 (2005). doi:10.1016/j.jsg.2005.02.006

Li, J., Qian, Z.Q.: The application of image acquisition and analysis techniques to the field of drying. Food Eng. Rev. 1–23 (2016). doi:10.1007/s12393-016-9146-2

Mahmood, Z., Iwashita, K.: A simulation study of microstructure evolution inside the shear band in biaxial compression test. Int. J. Numer. Anal. Meth. Geomech. **35**(6), 652–667 (2011). doi:10.1002/nag.917

Nicot, F., Sibille, L., Hicher, P.Y.: Micro-macro analysis of granular material behavior along proportional strain paths. Continuum Mech. Thermodyn **27**, 173–193 (2015)

Obaidat, M.T., Ai-Masaeid, H.R., Gharayben, F., Khedaywi, T.S.: An innovative digital image analysis approach to quantify the percentage of voids in mineral aggregates of bituminous mixtures. Can. J. Civ. Eng. **25**(6), 1041–1049 (1998)

Pena, A.A., Garcia-Rojo, R., Herrmann, H.J.: Influence of particle shape on sheared dense granular media. Granular Matter **9**(3–4), 279–291 (2007). doi:10.1007/s10035-007-0038-2

Pietruszczak, S., Guo, P.: Description of deformation process in inherently anisotropic granular materials. Int. J. Numer. Anal. Meth. Geomech. **37**(5), 478–490 (2013). doi:10.1002/nag.1106

Ulusoy, U., Igathinathane, C.: Particle size distribution modeling of milled coals by dynamic image analysis and mechanical sieving. Fuel Process. Technol. **143**, 100–109 (2016). doi:10.1016/j.fuproc.2015.11.007

Wang, Y., Li, X., Zhang, B., Wu, Y.F.: Meso-damage cracking characteristic analysis for rock and soil aggregate with CT test. Sci. China Technol. Sci. **57**(7), 1361–1371 (2014). doi:10.1007/s11431-014-5578-1

Yue, Z.Q.: Digital representation of mesogeomaterial spatial distribution and associated numerical analysis of geomechanics: methods, applications and developments. Chin. J. Rock Mech. Eng. **5**, 875–888 (2006)

Yue, Z.Q., Chen, S., Tham, L.G.: Finite element modeling of geomaterials using digital image processing. Comput. Geotech. **30**, 375–397 (2003)

Zhang, H.Y., Xu, W.J., Yu, Y.Z.: Numerical analysis of soil-rock mixture's meso-mechanics based on biaxial test. J. Cent. South Univ. **23**(3), 685–700 (2016). doi:10.1007/s11771-016-3114-0

Zhang, J.R., Tao, G.L., Huang, L., Yuan, L.: Porosity models for determining the pore-size distribution of rocks and soils and their applications. Chin. Sci. Bull. **55**(34), 3960–3970 (2010). doi:10.1007/s11434-010-4111-6

Zhang, X.W., Kong, L.W.: Study of pore characteristics of offshore clay by SEM and MIP and NA methods. Yantu Lixue/Rock Soil Mech. **34**(S2), 134–142 (2013)

Effect of Quarry Dust on Compaction Characteristics of Clay

Malaya Chetia[1(✉)], Manash P. Baruah[1], and Asuri Sridharan[2]

[1] Assam Engineering College, Guwahati, Assam, India
mchetia.ce@aec.ac.in, baruahmanash01@gmail.com
[2] Department of Civil Engineering, Indian National Science Academy, Indian Institute of Science Bangalore, Bangalore, India
sridharanasuri@yahoo.com

Abstract. Urbanization and industrialization has lead to scarcity of land with good bearing capacities thus forcing the construction over sites deemed unsuitable for such activities. To improve the geotechnical properties of soil of such sites soil stabilization methods are adopted. Use of graded aggregates as additives to such soil followed by compaction is an old technique of mechanical soil stabilization increases the support quality of soil and also provides material economy. Sand is used for the purpose but is getting scares day by day due to its large field of application. Rock quarry dust is a solid waste product produced in large scale during crushing operations of stones and rocks to obtain aggregates. Similarity in mineral composition and inertness in structure has made it possible to use them as the graded aggregate material and suitable alternative to sand for geotechnical operations. Also bulk utilization of this waste has become necessary due to geo environmental problems related to its disposal. The objective of this paper is to investigate the effect of quarry dust addition of different gradations on the compaction behavior of clay of different compressibility. A series of light compaction tests were carried out on soil mixes prepared with different quarry dust content with a locally available clayey soil and commercially available bentonite clay. Variation in the maximum dry density (MDD) and optimum moisture content (OMC) values has been obtained with increasing content and use of different gradations of quarry dust in the two clay types. Also the compaction characteristics are compared to those of clay-sand mixes by using sand samples having similar gradation to that of quarry dust. An effort has also been made to compare the MDD-OMC relationship of soil mixes of present study to that of naturally occurring soil found in literature.

1 Introduction

Addition of graded aggregate materials to soil followed by compaction has always been one of the usual methods of mechanical stabilization of weak soil deposits. This common practice lends desirable attributes to the soil, such as increased strength, decreased plasticity and also provides material economy and improves support capability of soil. This purpose is well served by naturally available sand for decades. But since the use of sand is not only limited to geotechnical applications but also in construction industry, so efforts have been made to find suitable substitute for it.

D.N. Singh and A. Galaa (eds.), *Contemporary Issues in Geoenvironmental Engineering*, Sustainable Civil Infrastructures,
DOI 10.1007/978-3-319-61612-4_7

Researchers have come out with the idea of utilization of quarry dust, for partial or full replacement of sand in soil stabilization and other applications. Quarry dust is the waste product obtained during crushing operations of stones to obtain aggregates and its disposal causes many geo environmental problems. The annual production of this waste is roughly estimated to be around 200 million tons (Soosan et al. (2001)). Due to comparable mineral composition to that of sand and inertness of coarse grained structure many researches have been carried out on bulk and effective utilization of this waste alone or with some admixtures with a motive to stabilize soil. Compaction of soil is carried out to improve soil densification. Compaction reduces volume of air voids in the soil that leads to an increase in the shear strength and a decrease in the consolidation and the permeability characteristics of soils. The compaction parameters, maximum dry density (MDD) and optimum moisture content (OMC) are important in most of the practical situations needing densification of soil, such as embankments, highway and railway sub grades and foundation soils. Most of the earlier studies were concentrated on the effect of dust content on moisture content and dry density of soils. Soosan et al. (2005) have studied the effect of different percentages of quarry dust on compaction characteristics of three different soils and observed an increase in MDD values and decrease in OMC values with increase in dust content in the soil-dust mixes. The optimum range of quarry dust content in soil was found to be 40–60%. Sridharan et al. (2006) found similar results of increase in MDD and decrease in OMC with increase in quarry dust content in his study on variation of shear strength characteristics of soil on addition of quarry dust. Sarvade and Nayak (2014) in their investigation on behavior of lithomargic clay stabilized with quarry dust and cement obtained similar results. Srikanth and Mishra (2016) investigated the effect of sand size on various sand-bentonite mixes and found MDD increase to be more in the mixes prepared with medium sand to that of fine sand. A significant volume of research work has been done by various researchers on the various aspects of the compaction properties of clay-quarry dust mixes. However comprehensive studies on the effect of gradation on laboratory maximum density and optimum moisture content of clay-quarry dust appeared to be very limited and are not compared with that of clay-sand mixes. It seems important, for practical purposes, to know the effects of grain size of quarry dust on compaction properties of such mixes for a minimum compacting effort. The objective of this paper is to investigate the influence of quarry dust content and gradation on the compaction parameters of clay quarry dust mixtures and to compare them with clay-sand mixes. Present study indicates a significant effect of grain size and content on the OMC and MDD of clay-quarry dust mixtures. Gurtug and Sridharan (2004) in their study, on various types of soils heavily used for constructional works ($w_L \leq 100\%$) and from extensive results from published literatures suggested correlations between compaction parameters for different compaction levels. It was found that good correlation exists between MDD-OMC for all compaction energy levels with better correlation coefficients and even clear relationship between the compaction parameters was obtained irrespective of compaction energy and soil type. But no comparison was made in relation to the MDD-OMC relationships obtained with that proposed for natural soils. The MDD-OMC relationship obtained in this investigation with quarry dust as an additive has been compared with those obtained for sand as an additive.

2 Experimental Investigations

Initially two clay samples, locally available clay (denoted by C) and commercially available bentonite (denoted by B), rock quarry dust (denoted by Q1) and sand (denoted by S1) samples are collected. The samples were subjected to different physical tests by following the guidelines provided by the respective Indian standard (IS) code, the results of which are presented in Table 1. According to IS soil classification system, the quarry dust and sand samples were poorly graded (SP) and clay sample was inorganic clay of medium compressibility (CI), and bentonite was inorganic clay of high compressibility (CH). Shape and mineralogical characterization of the quarry dust and sand samples were carried out from the petrographical images (BS 1994) taken with the help of electron microscope (25x) which indicate that the main mineral present in the samples was quartz. In addition to quartz significant amount of feldspar is present in quarry dust sample. Particles of quarry dust sample exhibits sub angular to angular shape whereas that of sand exhibits sub rounded to rounded shape.

Table 1. Physical properties and classification of the collected samples used in the study

Property	Sample			
	Clay (C)	Bentonite (B)	Quarry dust (Q1)	Sand (S1)
Specific gravity	2.61	2.67	2.71	2.65
Particle size characteristics (%)				
Sand (4.75–0.075 mm)	12.5	0.70	79.2	96.90
• Coarse 7sand (4.75–2.00 mm)	2.80	–	10.33	7.05
• Medium sand (2–0.425 mm)	5.45	–	28.67	60.7
• Fine sand (0.425–0.075 mm)	4.25	0.70	40.2	29.15
Fines (< 0.075 mm)	87.5	99.30	13.6	0.60
• Silt (0.075–0.002 mm)	65.67	34.59	13.6	0.60
• Clay (< 0.002 mm)	21.83	64.71	–	–
Uniformity coefficient (C_u)	–	–	8.83	3.23
Coefficient of curvature (C_c)	–	–	0.91	1.08
Atterberg limits (%)				
Liquid limit (w_L)	42.4	130.5	–	–
Plastic limit (w_P)	20.2	48.9	–	–
Plasticity index (I_P)	22.2	81.6	–	–
IS classification	CI	CH	SP	SP

In order to achieve the objectives of present study efforts have been taken to prepare a quarry dust sample Q_{S1} having grain size similar to sand sample S1. Overall two more samples of sand and five samples of quarry dust were prepared from samples S1 and Q1 respectively. The sand denoted by S2 and quarry dust samples denoted by Q2 and Q_{S2} consist only fine sand fraction (grain size from 425 μ to 75 μ). Sand sample S3 and quarry dust samples Q3 and Q_{S3} consist of medium sand fraction (grain size from

Table 2. Designation of soil mixes prepared

C-Q Mixes	C-S Mixes	C-Q_S Mixes	B-Q_{S1} Mixes	B-S1 Mixes
C-Q1 (90:10)	C-S1 (90:10)	C-Q_{S1} (90:10)	B- Q_{S1} (80:15)	B-S1 (85:15)
C-Q1 (80:20)	C-S1 (80:20)	C-Q_{S1} (80:20)	B- Q_{S1} (70:30)	B-S1 (70:30)
C-Q1 (70:30)	C-S1 (70:30)	C-Q_{S1} (70:30)	B- Q_{S1} (55:45)	B-S1(55:45)
C-Q1 (60:40)	C-S1 (60:40)	C-Q_{S1} (60:40)	Q_S is quarry dust having similar gradation as sand 'S'	
C-Q2 (90:10)	C-S2 (90:10)	C-Q_{S2} (90:10)		
C-Q2 (80:20)	C-S2 (80:20)	C-Q_{S2} (80:20)	Q_{S1} is quarry dust having similar gradation as sand 'S1'	
C-Q2 (70:30)	C-S2 (70:30)	C-Q_{S2} (70:30)		
C-Q2 (60:40)	C-S2 (60:40)	C-Q_{S2} (60:40)		
C-Q3 (90:10)	C-S3 (90:10)	C-Q_{S3} (90:10)		
C-Q3 (80:20)	C-S3 (80:20)	C-Q_{S3} (80:20)		
C-Q3 (70:30)	C-S3 (70:30)	C-Q_{S3} (70:30)		
C-Q3 (60:40)	C-S3 (60:40)	C-Q_{S3} (60:40)		

2 mm to 425 μ), where quarry dust sample Q_{S2} and Q_{S3} are having similar gradation to that of S2 and S3 samples respectively.

The desired soil mixes for the study were prepared by adding quarry dust and sand samples to both clays in different proportions of dry weight of the soil mixes in dry

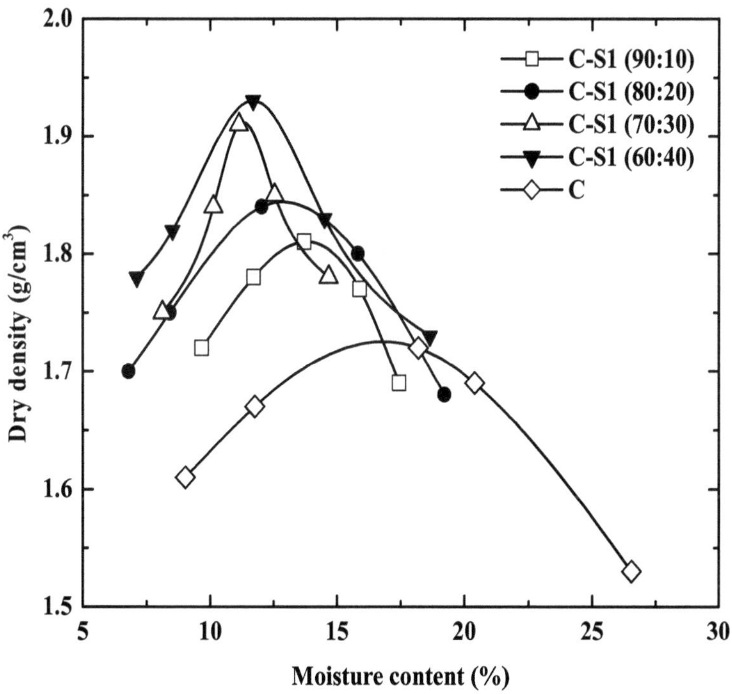

Fig. 1. Compaction curves for C and C-S1 mixes

condition. The soil mixes prepared are presented in Table 2. A soil mix prepared with nine parts of clay 'C' and one part sand 'S1' by dry weight of mix is designated as C-S1 (90:10). In similar manner other soil mixes as presented in Table 2 are designated. 36 soil mixes prepared with low compressibility clay C were kept for curing duration of 24 h and 6 soil mixes prepared with high compressibility clay B were kept for 3 days for curing prior to light compaction test as per IS (1974) along with C and B samples (Fig. 1).

3 Results and Discussion

Figures 2, 3, 4, 5 and 6 show the moisture content versus dry density plots for C, C-Q1, C-Q2, C-Q3, C-Q_{S1}, C-Q_{S2} and C-Q_{S3} mixes. Moisture content and density plots for C-S1, C-S2 and C-S3 are shown in Figs. 7, 8 and 9 respectively. Compaction curves for B-QS1 and B-S1 are shown in Figs. 10 and 11 respectively. Table 3 presents the compaction characteristics obtained from compaction curves of various soil mixes.

3.1 MDD and OMC for Intermediate Compressibility Clay (C) - Quarry Dust Mixes

Figure 12(a) and (b) represent the variation of MDD and OMC respectively with quarry dust content of different gradations for various intermediate compressibility

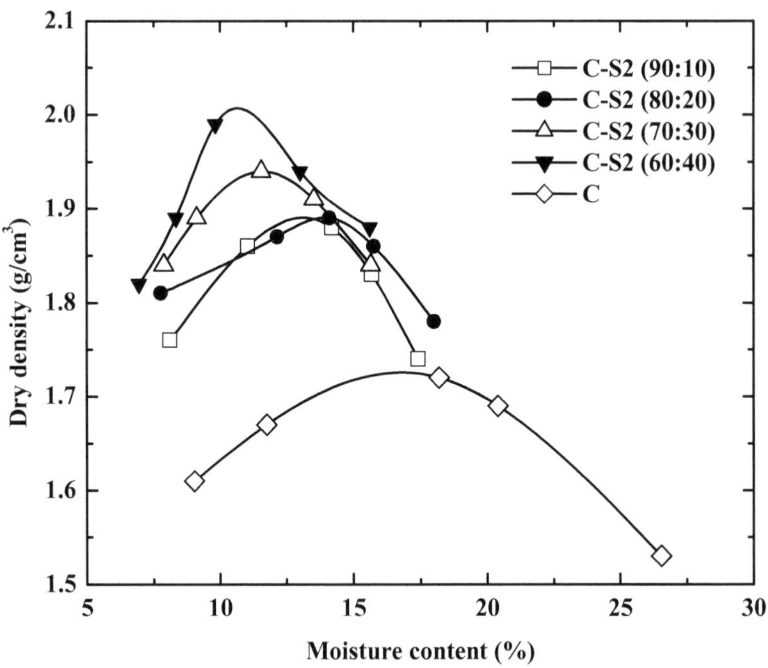

Fig. 2. Compaction curves for C and C-S2 mixes

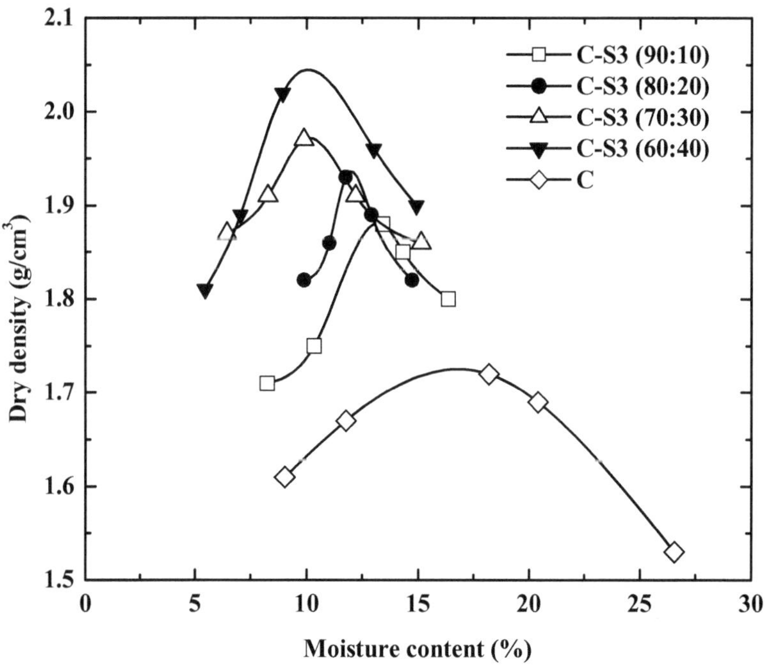

Fig. 3. Compaction curves for C and C-S3 mixes

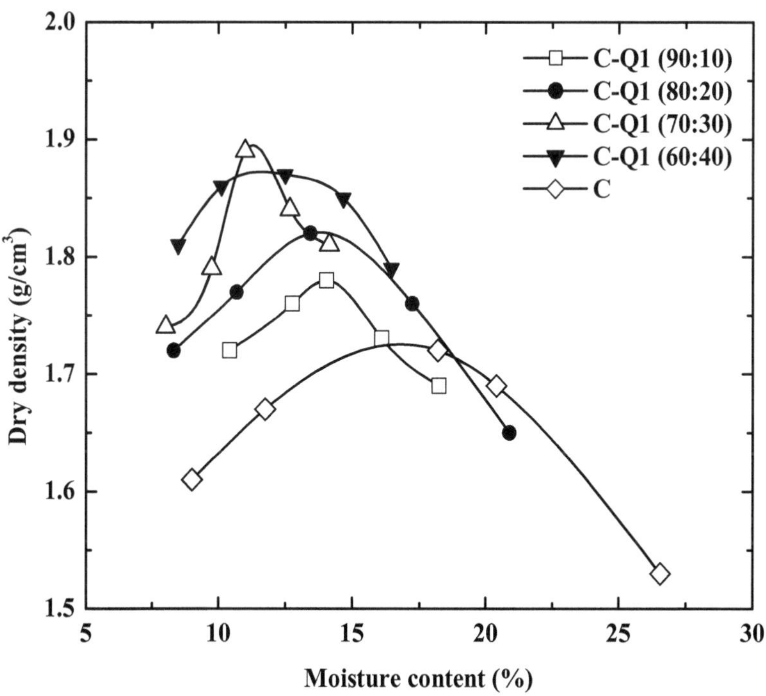

Fig. 4. Compaction curves for C and C-Q1 mixes

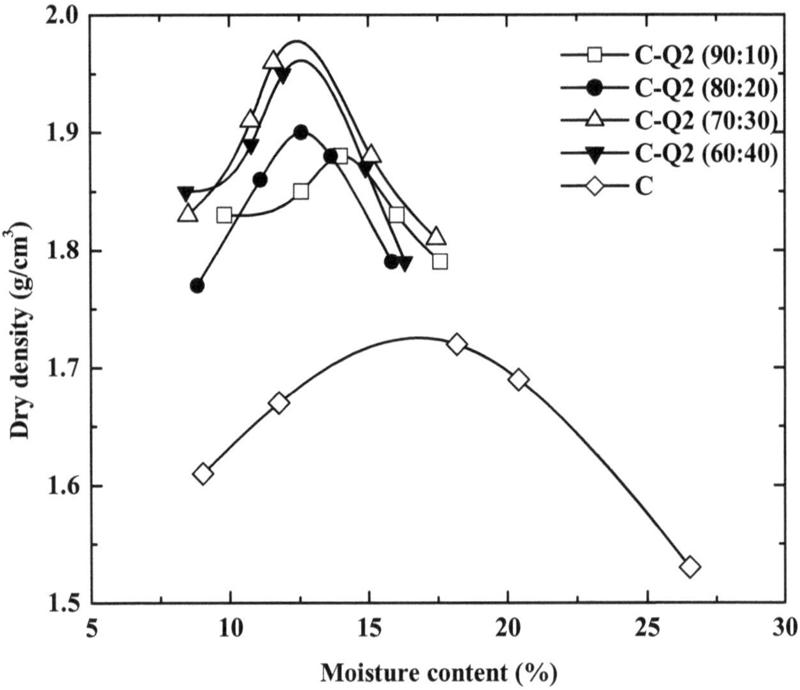

Fig. 5. Compaction curves for C and C-Q2 mixes

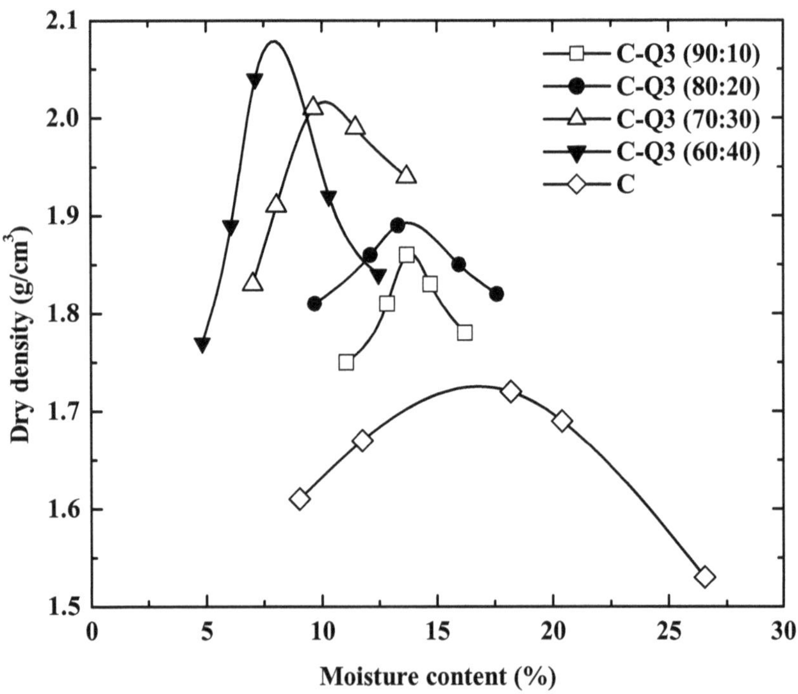

Fig. 6. Compaction curves for C and C-Q3 mixes

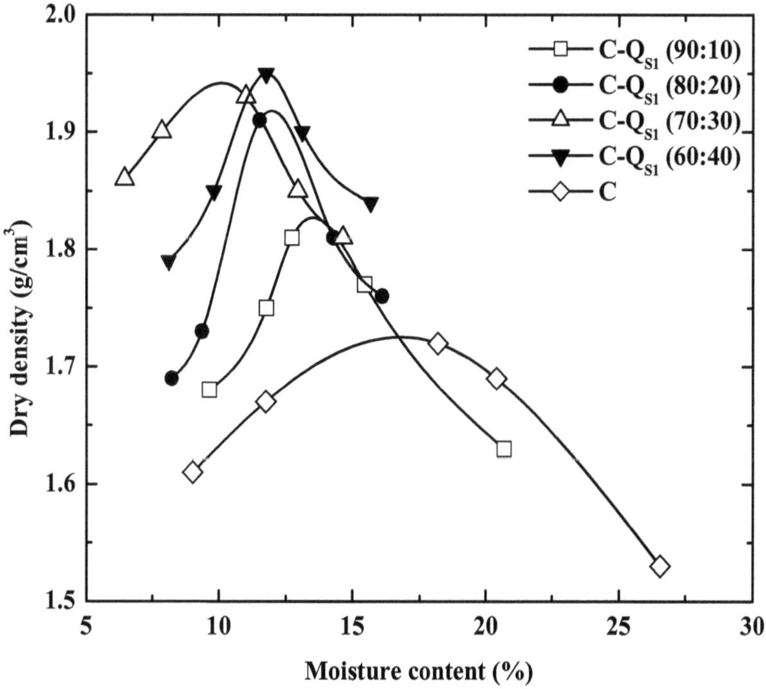

Fig. 7. Compaction curves for C and C-Q_{S1} mixes

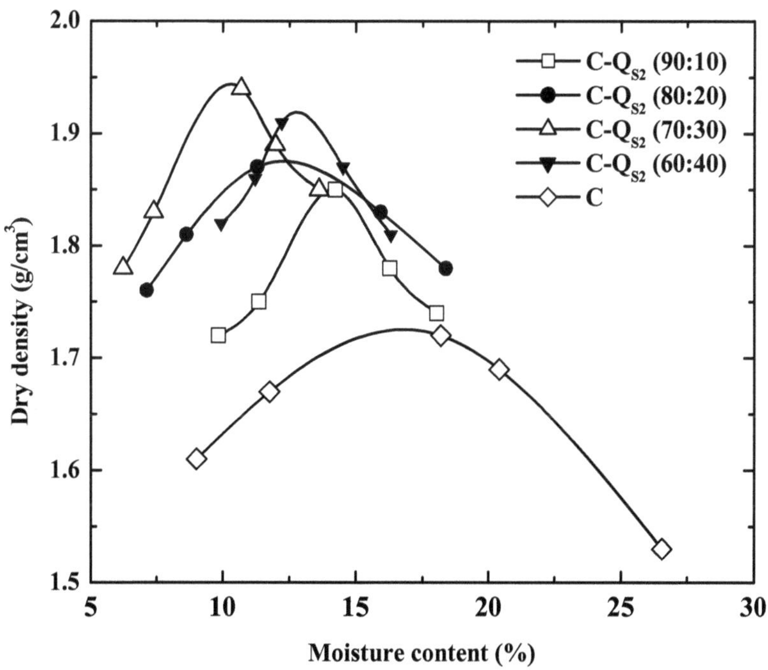

Fig. 8. Compaction curves for C and C-Q_{S2} mixes

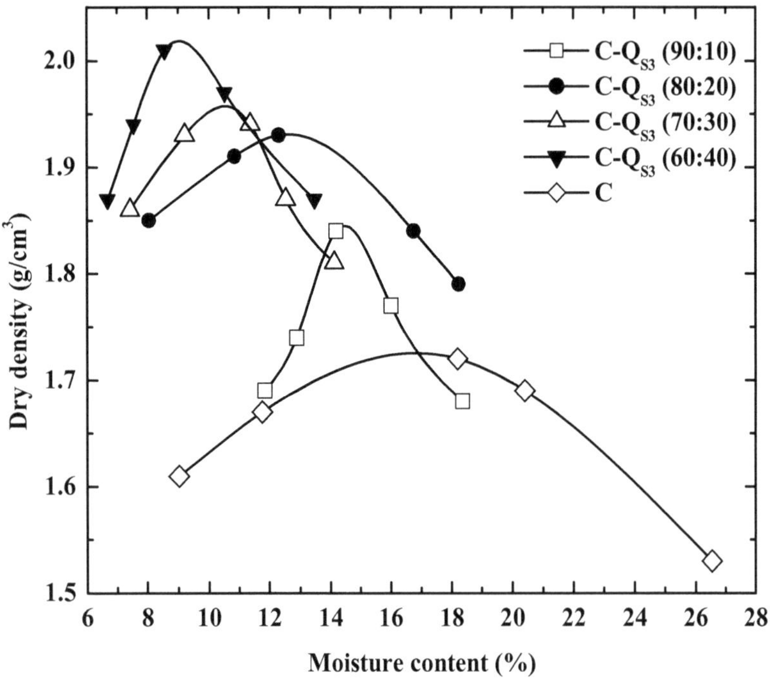

Fig. 9. Compaction curves for C and C-Q_{S3} mixes

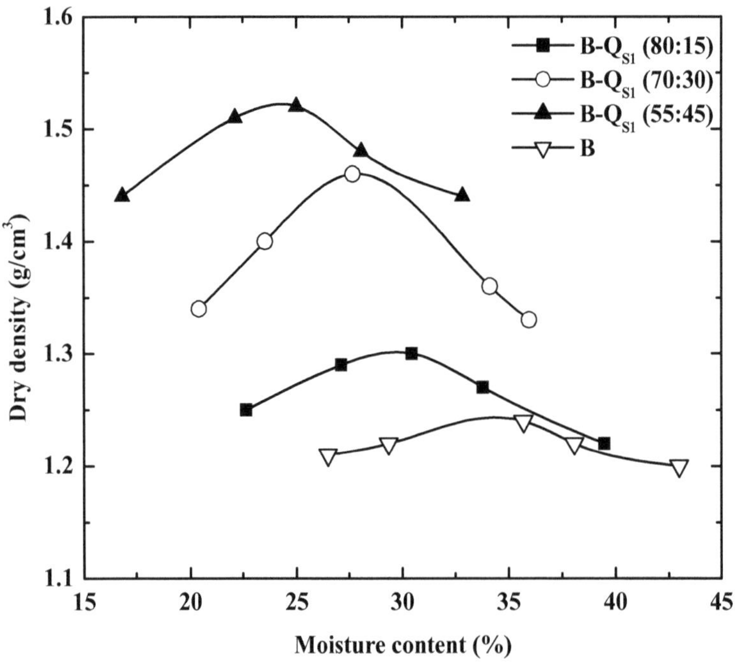

Fig. 10. Compaction curves for B and B-Q_{S1} mixes

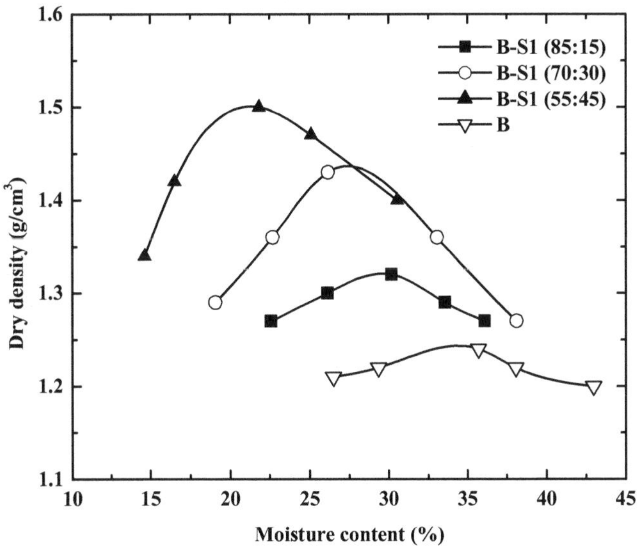

Fig. 11. Compaction curves for B and B-S1 mixes

Table 3. Compaction characteristics of soil mixes

Soil/Soil mix	OMC (%)	MDD (g/cm^3)	Soil/Soil mix	OMC (%)	MDD (g/cm^3)
C	18.20	1.73	B	35.69	1.24
C-S1 (90:10)	13.77	1.81	B-S1 (85:15)	30.17	1.32
C-S1 (80:20)	12.50	1.84	B-S1 (70:30)	26.16	1.43
C-S1 (70:30)	11.14	1.91	B-S1 (55:45)	21.83	1.50
C-S1 (60:40)	11.70	1.93	B-Q_{S1} (85:15)	30.45	1.30
C-S2 (90:10)	14.28	1.86	B-Q_{S1} (70:30)	27.68	1.46
C-S2 (80:20)	14.11	1.89	B-Q_{S1} (55:45)	25.00	1.52
C-S2 (70:30)	11.55	1.94			
C-S2 (60:40)	9.83	1.99			
C-S3 (90:10)	13.42	1.88			
C-S3 (80:20)	11.75	1.93			
C-S3 (70:30)	9.88	1.97			
C-S3 (60:40)	8.93	2.02			
C-Q1 (90:10)	14.05	1.78			
C-Q1 (80:20)	13.44	1.82			
C-Q1 (70:30)	11.01	1.89			
C-Q1 (60:40)	12.50	1.87			

(continued)

Table 3. (*continued*)

Soil/Soil mix	OMC (%)	MDD (g/cm^3)	Soil/Soil mix	OMC (%)	MDD (g/cm^3)
C-Q2 (90:10)	13.99	1.88			
C-Q2 (80:20)	12.57	1.90			
C-Q2 (70:30)	11.56	1.96			
C-Q2 (60:40)	11.92	1.94			
C-Q3 (90:10)	13.7	1.86			
C-Q3 (80:20)	13.31	1.89			
C-Q3 (70:30)	9.65	2.01			
C-Q3 (60:40)	7.14	2.04			
C-Q_{S1} (90:10)	12.74	1.82			
C-Q_{S1} (80:20)	11.54	1.91			
C-Q_{S1} (70:30)	11.03	1.93			
C-Q_{S1} (60:40)	11.78	1.95			
C-Q_{S2} (90:10)	14.24	1.85			
C-Q_{S2} (80:20)	11.30	1.87			
C-Q_{S2} (70:30)	10.71	1.94			
C-Q_{S2} (60:40)	12.21	1.92			
C-Q_{S3} (90:10)	14.19	1.84			
C-Q_{S3} (80:20)	12.31	1.93			
C-Q_{S3} (70:30)	11.38	1.94			
C-Q_{S3} (60:40)	8.57	2.01			

clay-quarry dust mixes. It was observed that with increase in respective quarry dust content in mixes C-Q1, C-Q3 and C-Q_{S3} MDD values are increasing and OMC values are decreasing. However in case of mixes C-Q2, C-Q_{S1} and C-Q_{S3}, the MDD values are increasing till 30% quarry dust content and decreases at 40% content. Also for the above mixes including mix C-Q1 the OMC values are decreasing up to quarry dust content of 30% and an increase is observed at 40% quarry dust content in the mixes. The increase can be attributed to the fact that the gradation of the mix getting well graded with incorporation of granular material which results in higher density. Also the OMC values are decreasing with increase in quarry dust content of the mixes as a result of decreasing in clay fraction in the mixes. But at respective quarry dust content of 40% in the mixes the observed decrease in MDD and increase in OMC may be due to the accumulation of uniform size granular particles decreasing the compaction density and resulting in requirement of more water with increase in amount of fines which may be either due crushing of particles at higher quarry dust content in the mixes.

The effect of gradation of the quarry dust samples can be seen on the compaction properties of the mixes. Mix C-Q1 is showing the lowest MDD values for all proportions which may be due to presence of appreciable amount of fines in Q1 sample.

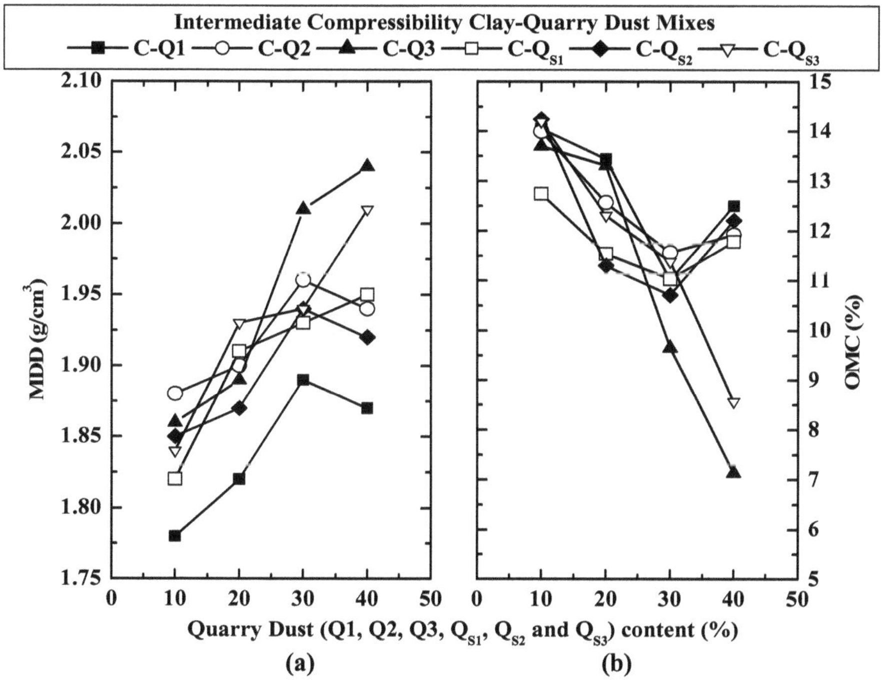

Fig. 12. (a) Variation of MDD with quarry dust content (b) Variation of OMC with quarry dust content for intermediate compressibility clay-quarry dust mixes

C-Q_{S1} mix showing improved properties in comparison to C-Q1 mix for all proportion which is due to the presence of negligible amount of fines present in sample Q_{S1}. The mixes with quarry dust samples in the fine range are showing improved compaction properties for content up to 20% and mixes with quarry dust samples in medium sand range are showing more MDD values and less OMC values for higher quarry dust contents in comparison to other mixes. Though it can be observed that the improvement for mixes C-Q2 and C-Q3 are comparably more than C-Q_{S2} and C-Q_{S3}, which can be due to the presence of larger size particles in more quantity within the particular sand range of the provided quarry dust sample. Above discussions present a clear effect of gradation of quarry dust samples on compaction properties of intermediate compressibility clay-quarry dust mixes.

3.2 MDD and OMC for Intermediate Compressibility Clay (C) - Sand Mixes

Figure 13(a) and (b) represent the respective variation of MDD and OMC of various intermediate compressibility clay-sand mixes with sand content of different gradations. Increase in MDD values with increase in respective sand content have been observed

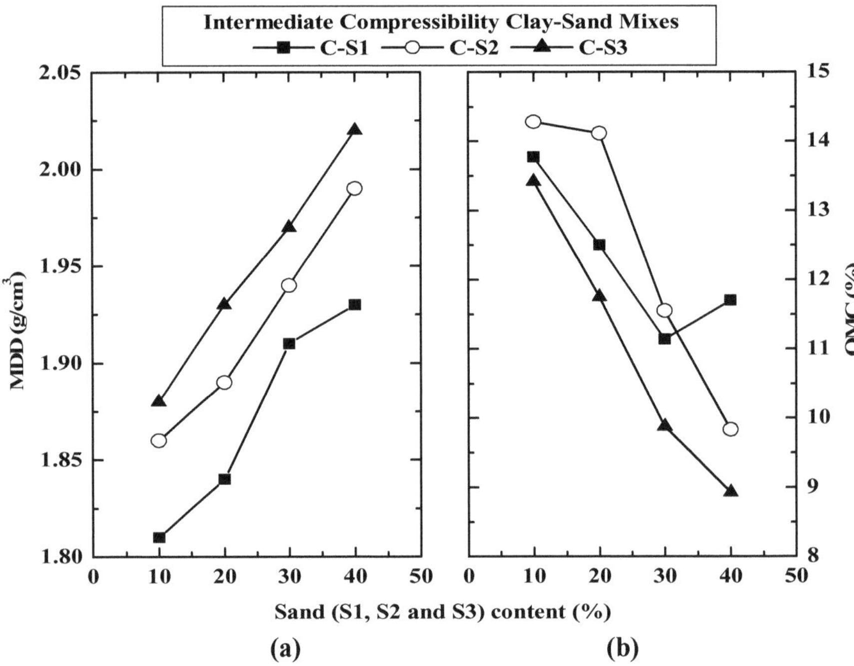

Fig. 13. (a) Variation of MDD with sand content (b) Variation of OMC with sand content for intermediate compressibility clay-sand mixes

for all the mixes, which is due to the resulting well graded nature of mix with inclusion of granular material to clay. The OMC values have been found to decrease for mixes C-S2 and C-S2 with increase in respective sand content but an increase in OMC value within 1% is observed for mix C-S1 for 40% S1 content after a gradual decrease of OMC up to 30%. The decrease in OMC values can be attributed to the decrease in clay content in the mixes.

Figure 13(a) shows that the mixes prepared with sand S1 gives lower MDD values for each proportion in comparison to C-S2 and C-S3 mixes. Intermediate MDD values are observed for mixes prepared with fine sand fraction S2 and higher MDD values are given by mixes prepared with medium sand fraction S3 for each proportion of sand content in the respective mixes. MDD increases with inclusion of larger size particles in case of clay sand mixes. As presented in Fig. 13(b), intermediate OMC values are given by C-S1 mixes, lower values are observed for C-S3 mixes and higher values of OMC are obtained for C-S2 mixes for same proportion of respective sand sample in the mixes with increase in particle size overall surface area decreases with increase in content resulting in lower requirement of water to attain maximum densification due to which mixes prepared with medium sand fraction show lower OMC values in comparison to others. The influence of gradation is clearly indicated upon the compaction properties of clay-sand mixes as discussed above.

3.3 MDD and OMC for High Compressibility Clay (B) - Quarry Dust Mixes

The variation of MDD with increase in quarry dust content for B-Q_{S1} mix is presented in Fig. 14(a). With increase in Q_{S1} content, increase in MDD value have been observed, the reason being the addition of bigger size particles to the fines resulting in a well graded mix. Figure 14(b) represents the decrease in OMC values with increasing Q_{S1} content in B-Q_{S1} mix, which is due to decrease in clay content.

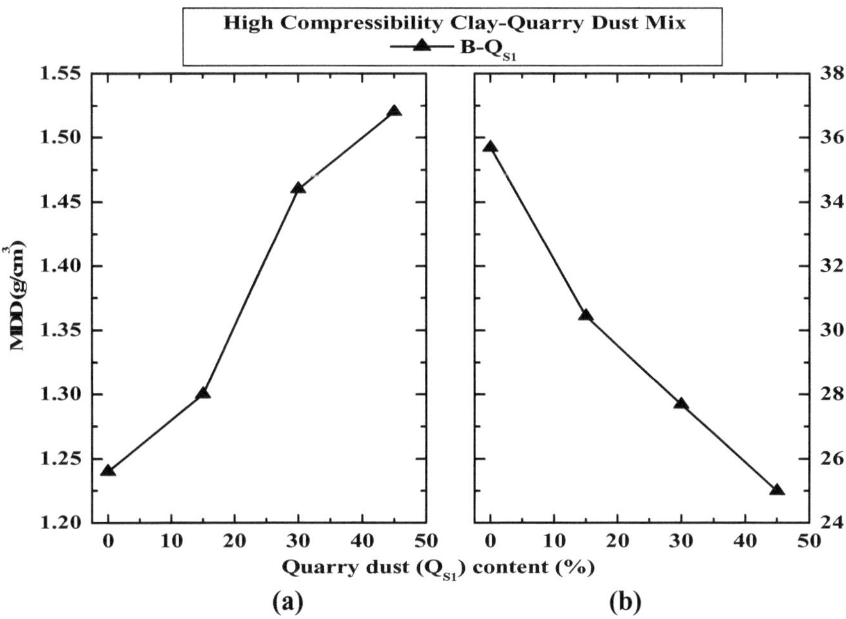

Fig. 14. (a) Variation of MDD with quarry dust content (b) Variation of OMC with quarry dust content for high compressibility clay-quarry dust mix

3.4 MDD and OMC for High Compressibility Clay (B) - Sand Mixes

Variation of MDD and OMC with increasing sand content in B-S1 mix is presented in Fig. 15(a) and (b) respectively. Increase in MDD and decrease in OMC is observed with increase in sand content in the mix. The reason being same as discussed above.

3.5 Comparison of MDD and OMC Variation of Clay Sand and Clay Quarry Dust Mixes

From Fig. 16(a), it can be observed that mix prepared with quarry dust sample Q1 shows lower values of MDD in comparison to C-Q_{S1} and C-S1 mixes for each percentage of the additive sample due to presence of more amount of quarry dust fines. For mixes prepared with similar gradation of quarry dust and sand, C-Q_{S1} mix shows

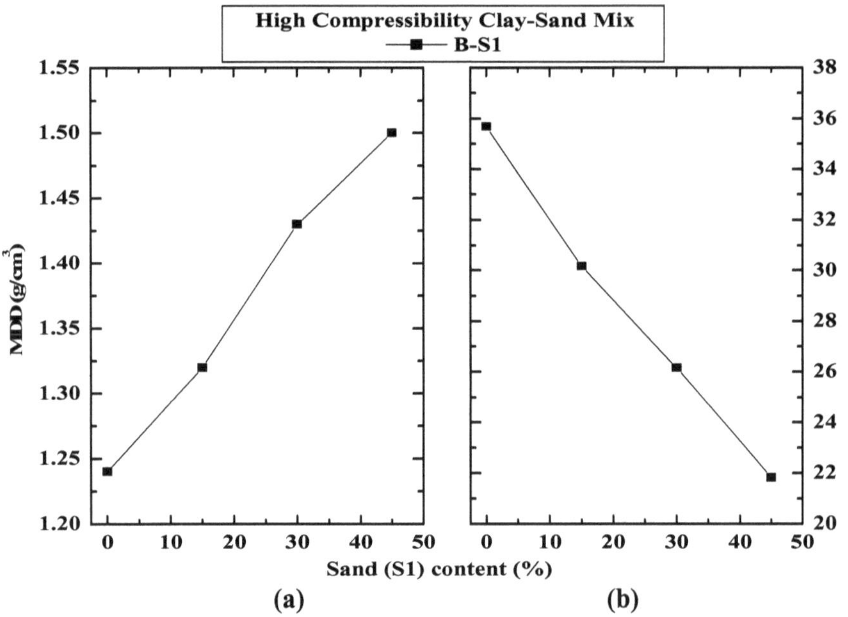

Fig. 15. (a) Variation of MDD with sand content (b) Variation of OMC with sand content for high compressibility clay-sand mix

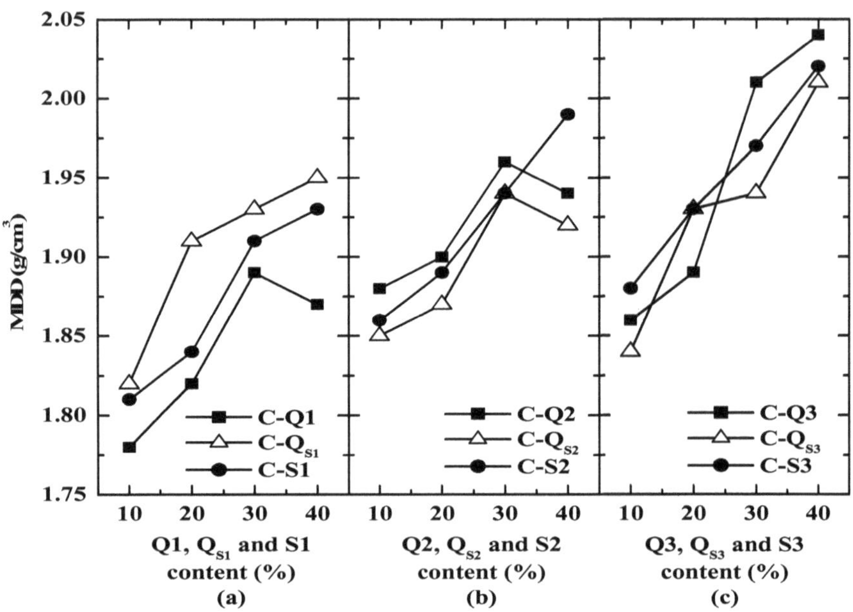

Fig. 16. (a) Comparison of MDD of C-Q1, C-Q_{S1} and C-S1 mixes (b) Comparison of MDD of C-Q2, C-Q_{S2} and C-S2 mixes (c) Comparison of MDD of C-Q3, C-Q_{S3} and C-S3 mixes

higher values of MDD in comparison to C-S1 mix which might be due to greater interlocking of various sizes of quarry dust particles. From Fig. 16(b) and (c), it is observed that for similar gradation of sand and quarry dust in fine sand range and medium sand range the MDD values are on the higher side for mixes prepared with sand in comparison to those of prepared with quarry dust, the reason may be angular shape of quarry dust particles resulting in lower densities.

It was found that minimum OMC values for each proportion are given by C-Q3 mixes as evident from Fig. 17(a). It can be seen from Fig. 17(b) OMC values for C-Q_{S2} mix are less than that of C-S2 mix. However that additive content of 30% in C-Q2 and C-Q_{S2} mix the OMC value increases which may be due crushing of quarry dust particles to fines increasing the water requirement. For mixes prepared with medium sand fractions of additive contents, the OMC values obtained for C-QS3 mix are higher than that of C-S3 mixes as presented in Fig. 17(c). The inference drawn from above comparison is that for intermediate compressibility clay, incorporation of quarry dust having similar gradation to sand results in improved compaction characteristics in comparison to that of sand. For particles in fine sand range of quarry dust, improvement is effective up to 30% content and use of medium sand range quarry dust will result in comparably similar values of compaction parameters.

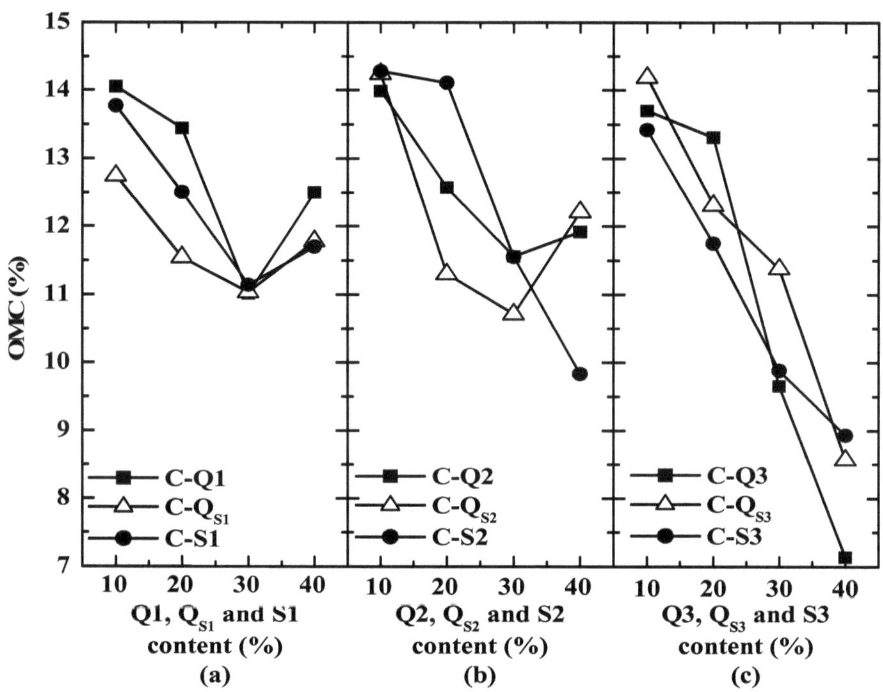

Fig. 17. (a) Comparison of OMC of C-Q1, C-Q_{S1} and C-S1 mixes (b) Comparison of OMC of C-Q2, C-Q_{S2} and C-S2 mixes (c) Comparison of OMC of C-Q3, C-Q_{S3} and C-S3 mixes

The comparison for MDD and OMC with additive content for mixes B-S1 and B-S1 as evident in Fig. 18(a) and (b) reveals that in case of high compressibility clay higher MDD values are obtained with incorporation of quarry dust in the sand range in comparison to that of sand at higher additive content. Also the OMC requirement for clay deceases with increasing quarry dust content though it is less compared to that of increasing sand content.

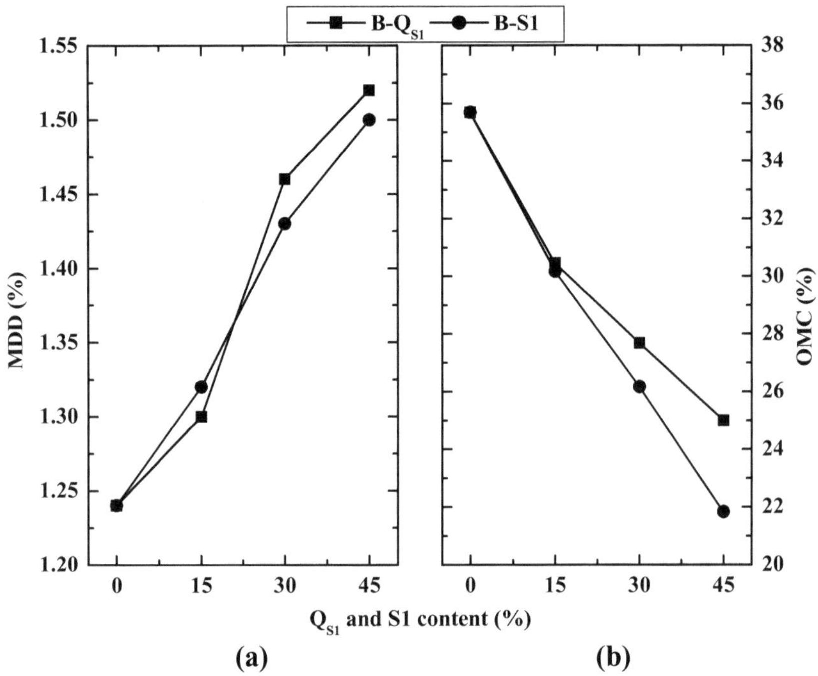

Fig. 18. (a) Comparison of MDD of B-Q_{S1} and B-S1 mixes (b) Comparison of OMC of B-Q_{S1} and B-S1 mixes

3.6 Comparison of MDD-OMC Relationships

Figures 19 and 20 present the linear correlation between OMC and MDD for the mixes of present investigation with regression coefficient R^2. Figure 19(a) represents the linear MDD-OMC relationship for all mixes of intermediate compressibility clay with the quarry dust samples Q1, Q2 and Q3 irrespective of the gradation i.e. mixes C-Q1, C-Q2 and C-Q3 taken together. Similarly Fig. 19(b) represents the same for C-Q_{S1}, C-Q_{S2} and C-Q_{S3} taken together. Figure 19(c) indicates the variation of MDD with OMC for all clay 'C'- quarry dust mixes and the same is shown by Fig. 19(c) for all clay 'C' and sand mixes irrespective of gradation of sand. The correlation between MDD and OMC of B-Q_{S1} and B-Q_{S2} mixes are shown in Fig. 20(a) and (b) respectively.

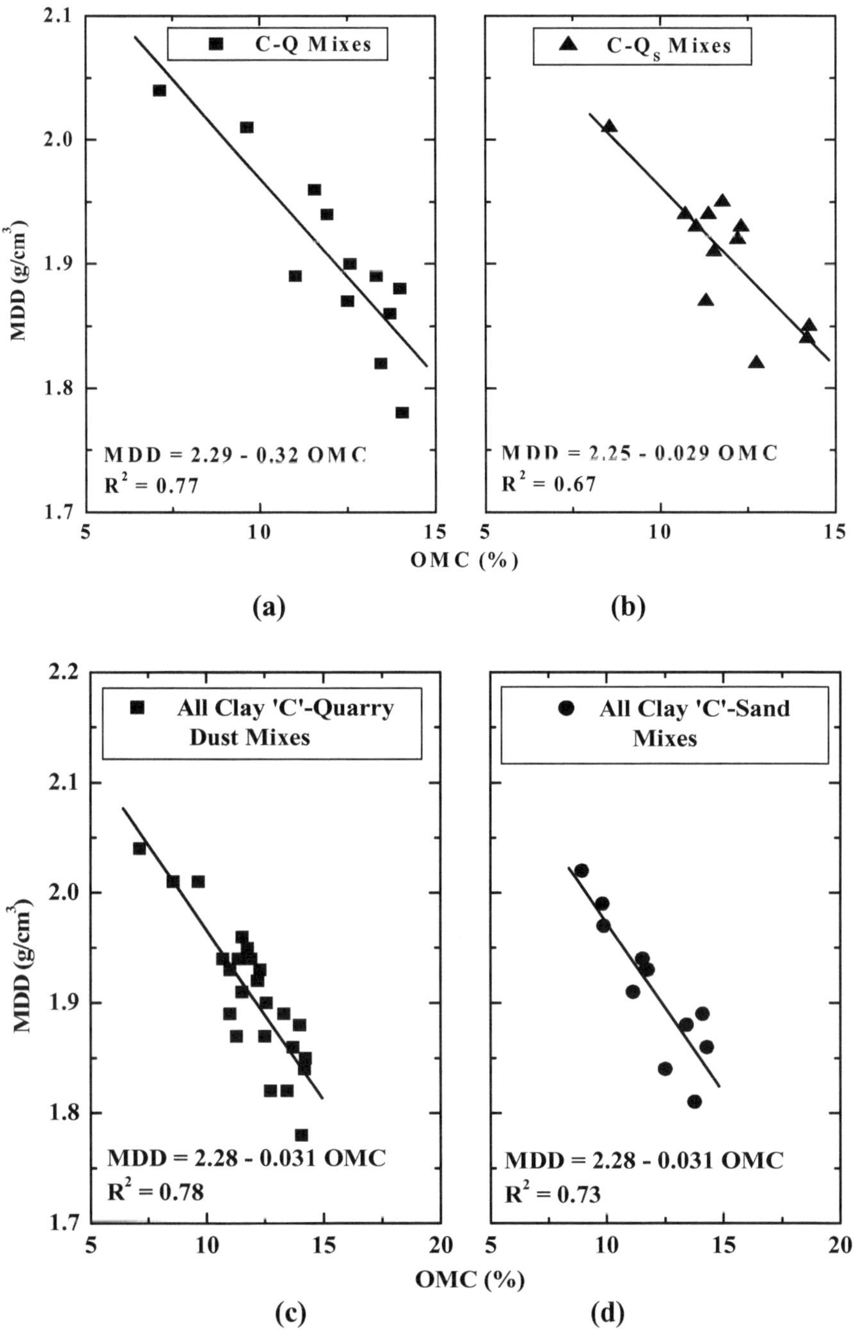

Fig. 19. (a) Correlation of MDD with OMC for all C-Q mixes (b) Correlation of MDD with OMC for all C-Q_S mixes (c) Correlation of MDD with OMC for all intermediate compressibility clay-quarry dust mixes (d) Correlation of MDD with OMC for all intermediate compressibility clay-sand mixes

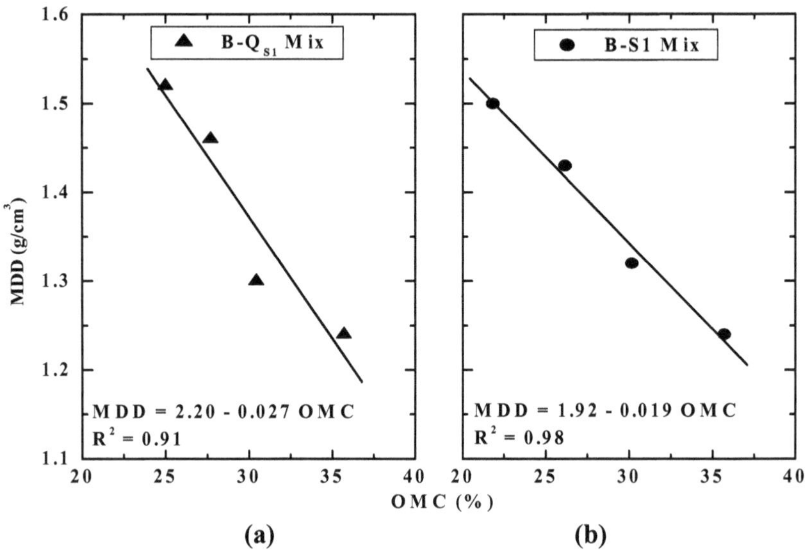

Fig. 20. (a) Correlation of MDD with OMC for B-Q_{S1} mix (b) Correlation of MDD with OMC for B-S1 mix

Influence of clay type on MDD-OMC relationship. The influence of clay type on MDD-OMC relationship of mixes has been presented in Fig. 21(a) and (b) for OMC range of 5 to 30%. The comparison of correlations for all clay 'C'-quarry dust mixes and B-Q_S mix shows that the variation in MDD with OMC are compatibly same. But for mix B-S1 the rate of increase in MDD with decrease in OMC is less compared to that of all clay-sand mixes. Use of quarry dust is as effective in high compressibility soil as in intermediate compressibility soil.

Influence of additives of similar gradation on MDD-OMC relationship. Comparison have been made in between the MDD-OMC relationships obtained for C-Q_S and C-S mixes and in between that of B-Q_{S1} and B-S1 mixes as shown in Fig. 22(a) and (b) respectively. It was found that for intermediate compressibility clay mixes prepared with similar gradation of quarry dust and sand, the correlations obtained between MDD and OMC are almost similar within the experimental range of OMC. For mixes prepared with high compressibility clay with similar gradation of additive samples the increase in MDD with decrease in OMC is more in case of quarry dust addition compared to addition of sand.

Comparison of MDD-OMC relationships of present investigation and existing relationship in literature. Figure 23(a) shows the comparison of variation of MDD-OMC correlation of all clay 'C'-sand and clay 'C'-quarry dust mixes of present investigation to that obtained for naturally occurring soils for Standard Proctor energy level and all compaction energy levels within an OMC range of 5 to 15%. The trend lines for all clay-sand and all clay-quarry dust mixes of present investigation are almost parallel and above the trend line observed for Standard Proctor energy level and below

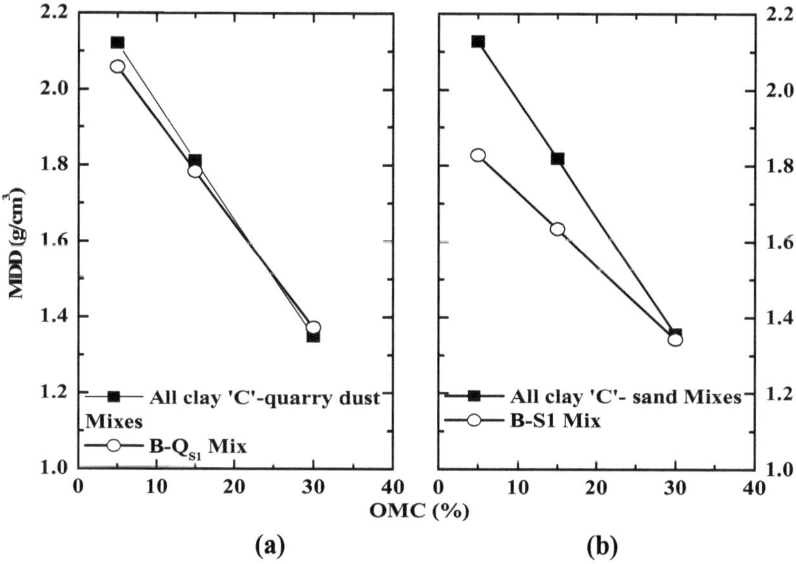

Fig. 21. Influence of clay type on soil mixes prepared with quarry dust

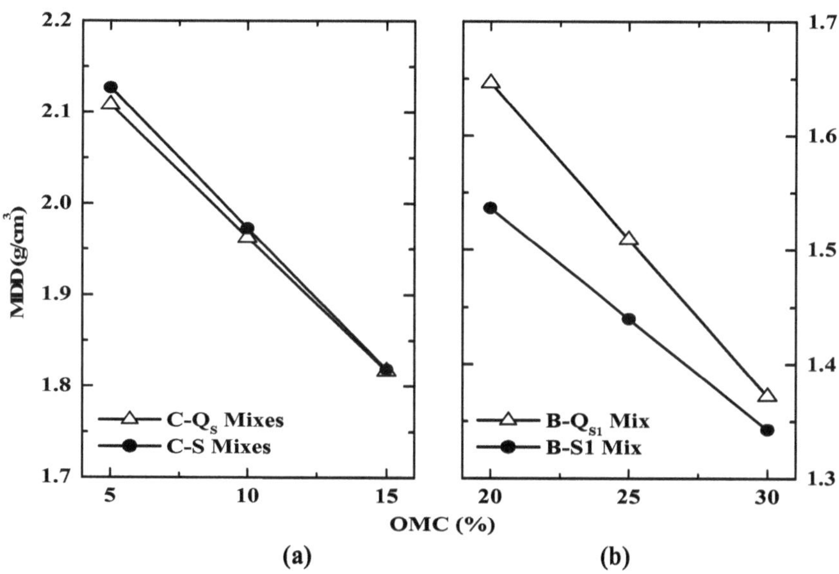

Fig. 22. (a) Comparison of MDD-OMC correlations of all intermediate compressibility clay-quarry dust and clay-sand mixes for similar gradation of additive samples (b) Comparison of MDD-OMC correlations of high compressibility clay-quarry dust and clay-sand mixes for similar gradation of additive samples

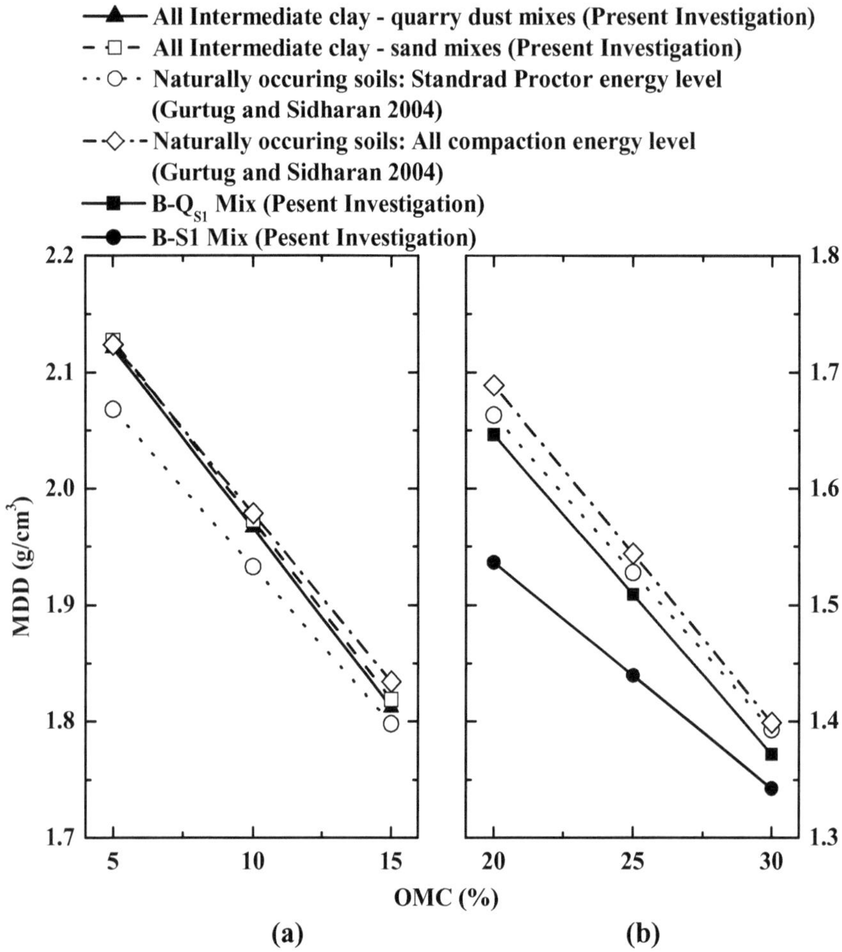

Fig. 23. (a) Comparison of MDD-OMC correlations of all intermediate compressibility clay-quarry dust and clay-sand mixes of present investigation with that of naturally occurring soils proposed by Gurtug and Sridharan (2004) (b) Comparison of MDD-OMC correlations of all high compressibility clay-quarry dust and clay-sand mixes of present investigation with that of naturally occurring soils proposed by Gurtug and Sridharan (2004)

the one observed for all compaction energy levels, however the difference is marginal (within 0.1 g/cm^3). The comparison reveals the clay-quarry dust mixes follows similar trends of MDD-OMC correlation as that of clay-sand mixes and that of naturally occurring soils which also supports the use of quarry dust in replacement of sand in compaction operations of clays of intermediate compressibility.

Figure 23(b) shows the comparison of variation of MDD-OMC correlation of bentonite-sand 'B-SS' and bentonite-quarry dust 'B-QS' mixes to those proposed by Gurtug and Sridharan (2004) within OMC range of 20 to 30%. Results indicate that for Standard Proctor energy level the correlation obtained for B-QS1 mix is similar to that of naturally occurring soils. The trend line observed for B-S1 mix is much below the trendlines for naturally occurring soils for Standard Proctor energy level and all compaction energy levels. In case of high compressibility clays like bentonite the results obtained for mix prepared with quarry dust are more superior in compaction behavior compared to that prepared with sand.

4 Conclusions

This study investigates the influence of quarry dust addition of different gradations on the compaction behavior of clay of different compressibility. Light compaction tests were carried out on soil mixes prepared with different quarry dust content with two different clays of different compressibility. Compaction characteristics of these soil mixes are compared to those of clay-sand mixes by using sand samples having similar gradation to that of quarry dust. The MDD-OMC relationship of soil mixes of present study have been compared to that of naturally occurring soil found in literature. The following conclusions can be drawn from this study:

1. MDD increases and OMC decreases with increase in quarry dust content of any gradation in intermediate compressibility clay-quarry dust mixes. The increase in MDD and decrease in OMC is very effective up to a quarry dust content of 30%. Similar behavior is observed for high compressibility-quarry dust mixes.
2. For any given proportion of intermediate clay-quarry dust mixes, mixes prepared with medium range particles shows higher MDD values and lower OMC values at higher quarry dust contents.
3. For mixes prepared with intermediate compressibility clay with quarry dust samples having similar gradation to that of sand sample, for all proportions, C-Q_{S1} mixes show higher MDD values in comparison to that of C-S1 mixes, comparable MDD values are obtained for C-Q_{S2} mixes and C-S2 mixes up to additive content of 30% and for C-Q_{S3} and C-S3 mixes MDD values obtained are similar leading to increased possibility of replacement of sand with similar gradation of quarry dust.
4. OMC observed for C-Q_{S1} and C-Q_{S2} mixes are relatively more than those obtained for C-S1 and C-S2 mixes respectively at low additive content. For C-Q_{S3} and C-S3 mixes OMC is more or less same upto 30%, after which significant decrease in OMC is observed for C-Q_{S3} mix.
5. For mixes prepared with similar gradation of sand and quarry dust with high compressibility clay for all proportions, MDD values obtained are comparably similar but significant increase in OMC is observed for higher content of quarry dust in the mix in comparison to similar content of sand in the mix.
6. Similar MDD-OMC relationships are obtained for both clay types where quarry dust is used as additive. However the increase in MDD is more with decrease in OMC for intermediate clay-sand mix in comparison to that of high compressibility clay-sand mix.
7. MDD-OMC relationship obtained for mixes prepared with intermediate compressibility clay and quarry dust is similar to that of same prepared with sand irrespective of gradation of additives. For high compressibility clay mixes with

quarry dust, the MDD increase is more with decrease in OMC in comparison to that of prepared with sand.

8. MDD-OMC relationships obtained for all clay-sand mixes, all clay quarry dust mixes and bentonite-quarry dust mixes are found to lie in between those as suggested by researchers for natural soils for standard Proctor energy level and all compaction energy levels. It also signifies irrespective of gradation of quarry dust, it can replace sand to increase the compaction properties in both intermediate and high compressibility clays.
9. This investigation brings out that quarry dust which is a waste material could replace sand, which has become very costly, to improve the engineering properties of fine grained soils including highly plastic bentonite clay. This has been brought out in this investigation for different gradations of the quarry dust as well.

References

BS 812-104. Testing aggregates. Method for Qualitative and Quantitative Petrographic Examination of Aggregates, British Standards Institution (1994)

Gurtug, Y., Sridharan, A.: Compaction behavior and prediction of its characteristics of fine grained soils with particular reference to compaction energy. Soils Found. Sci. Direct. (2004). doi:10.3208/sandf.44.5_27

IS 2720-III. Determination of specific gravity of soils. Indian Standard Methods of Test for Soils, Bureau of Indian Standards, New Delhi (1980)

IS 2720-IV. Grain-size analysis. Indian Standard Methods of Test for Soils, Bureau of Indian Standards, New Delhi (1975)

IS 2720-V. Determination of liquid limit and plastic limit. Indian Standard Methods of Test for Soils, Bureau of Indian Standards, New Delhi (1985)

IS 1498. Indian standard classification and identification of soils for general engineering purposes. Indian Standard Methods of Test for Soils, Bureau of Indian Standards, New Delhi (1970)

IS 2720-VII (1974). Determination of water content-dry density relation using light compaction. Indian Standard Methods of Test for Soils, Bureau of Indian Standards, New Delhi (1974)

Sarvade, P.G., Nayak, S.: Studies on utilization of quarry dust to improve the geotechnical properties of lithomargic clay. Int. J. Adv. Struct. Geotech. Eng. (2014). ISSN 2319-5347

Soosan, T.G., Jose, B.T., Abraham B.M.: Use of quarry dust in embankment and highway construction. In: Proceedings of Indian Geo-technical Conference, Indore, pp. 274–277, December 2001

Soosan, T.G., Sridharan, A., Jose, B.T., Abraham, B.M.: Utilization of quarry dust to improve the geotechnical properties of soils in highway construction. Geotech. Test. J., ASTM (2005). doi:10.1520/GTJ11768

Sridharan, A., Soosan, T.G., Jose, B.T., Abraham, B.M.: Shear strength studies on soil-quarry dust mixtures. Geotech. Geol. Eng. Springer (2006). doi:10.1007/s10706-005-1216-9

Srikanth, V., Mishra, A.K.: A laboratory study on the geotechnical characteristics of sand bentonite mixtures and the role of particle size of sand. Int. J. Geosynthetics Ground Eng. (2016). doi:10.1007/s40891-015-0043-1

Effect of Temperature on Strength and Microstructure of Lime Stabilized Flyash

Aparupa Pani(✉) and S.P. Singh

Department of Civil Engineering, National Institute of Technology, Rourkela, India
aparupapani@gmail.com, spsingh@nitrkl.ac.in

Abstract. Production of extensive amount of flyash from Coal-Based thermal power plants occupies huge areas of highly serviceable land which covers a massive area in measurement and it also creates enormous environmental hazards. Utilization of fly-ash in geotechnical constructions like highway and railway embankments, landfill, road base and sub bases etc. is gaining momentum as it proves to be an effective and effectual mean of bulk utilization of these kinds of materials. For increasing use of flyash as a construction material, it is a requisite to amplify the geotechnical properties of flyash with a suitable stabilizing agent like lime. This paper describes the effect of curing temperature on strength properties and microstructural changes of compacted flyash treated with lime. Specimens are prepared with different percentages of lime compacted to densities corresponding to light and heavy compaction energies at their respective OMC. These specimens are cured at different temperatures such as 10 °C, 25 °C, 45 °C and 90 °C and the unconfined compressive strength are determined after curing periods of 0, 7, 15, 30 and 60 days. The test result showed that an increase of either lime content or curing period results in increase in the strength. The elevated temperature is found to influence both the early and later age strengths. Samples cured at low temperature show rising trends of strength even up to 60 days of curing whereas the strength of high temperature cured specimens stabilize much earlier. A crossover effect of strength is noticed between low and high temperature cured specimens. Also it is found that the incorporation of lime significantly increases the compressive strength of compacted flyash. From the microstructural study it was found that the addition of lime along with curing temperature might generate more amorphous hydration products and accelerate the reaction rate of raw materials.

1 Introduction

Rapid growth of industrialization; scarceness in power and electricity is the leading cause of exculpate huge amount of fly ash by thermal power plants where coal is used as fuel. The class-F fly ash generates major environmental complications such as leaching and dusting of atmospheric air. To mitigate these problems, fly ash can be transformed to a construction worthy material (Subbarao and Ghosh 1997). Stabilization of the low lime fly ash with lime and gypsum gives effective control over hydraulic conductivity and leachate characteristics (Ghosh and Subbarao 2007). The strength enhancement depends on the pozzolanic reaction occuring between lime and flyash (Joshi et al. 1975).

D.N. Singh and A. Galaa (eds.), *Contemporary Issues in Geoenvironmental Engineering*, Sustainable Civil Infrastructures,
DOI 10.1007/978-3-319-61612-4_8

Gray and Lin (1972) studied the gain in strength and the ultimate strength of lime-treated fly ash and showed that it is sensitive to curing temperature. The lime-pozzolana reaction required very long curing period to attain appreciable strength under ambient temperature conditions (Reddy and Gourav 2011). The rate and mechanism of pozzolanic reaction of fly ash vary with curing temperature. Elevated curing temperatures accelerate the rate of main reaction linearly (Narmluk and Nawa 2014). Formation of 'calcium silicate hydrate' as a result of reaction product of lime-rice husk ash was shown by Ali et al. (1992) through XRD analysis. Ghosh and Subbarao (2001) studied the surface morphology of lime stabilized flyash specimen and found that formation of gel like substance of C-S-H and C-A-H found at higher doses of lime. Also, needle like structures of ettringite is visible at long curing period. X-ray diffraction (XRD) studies are generally employed to distinguish the crystalline phases in the stabilized fly ash. Croft (1964) observed the appearance of gel-like varieties of CSH and CAH.

In the present analysis, an effort has been made to study the effect of elevated curing temperature on the lime-stabilized flyash. This paper elucidates the reason of improvement in properties of lime-stabilized fly ash through mechanical and microstructural studies.

2 Materials

Fly ash used for this investigation was obtained from captive power plant of Rourkela steel plant (RSP), India in dry condition. The chemical composition of the flyash is given in Table 1. The major elements of this fly ash are silica, alumina and iron oxide.

Table 1. Chemical composition of flyash

Elements	SiO_2	Al_2O_3	Fe_2O_3	CaO	MgO	SO_4	Unburnt carbon	Others
Composition (%)	59.2	17.9	9.5	3.2	1.3	1.2	7.0	0.7

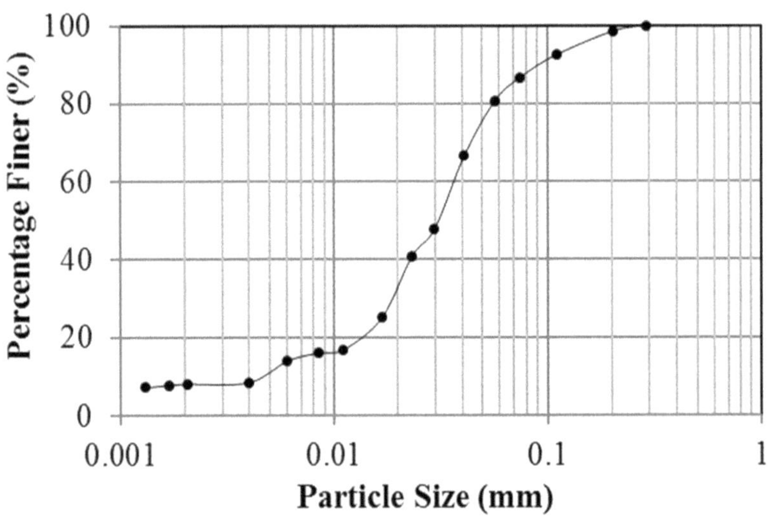

Fig. 1. Grain size distribution curve of fly ash

Calcium present in the fly ash is less than 20%. So, according to ASTM specification C 618-89 (1992), this fly ash belongs to a Class-F category. The specific gravity of flyash was found to be 2.38. The grain size distribution of the fly ash showed that it contained 88% fines (<75 μm). The Cu and Cc values were found to be 8.34 & 2.08 respectively; indicating uniform gradation of the sample. Its gradation curve is given in Fig. 1. Commercial grade hydrated lime of 90.2% purity was used in this study.

3 Methodology

The raw flyash blended with different percentages of lime content such as (0%, 2%, 4%, 8% and 12%) of its dry weight. These samples were condensed with two different compactive energies i.e. 593 kJ/m^3 and 2483 kJ/m^3. The lime stabilized flyash (FA) specimens prepared at compaction energy of 593 kJ/m^3 were designated as L0, L2, L4, L8, L12 and at compaction energy of 2483 kJ/m^3 were designated as H0, H2, H4, H8, H12.

Laboratory compaction tests the moisture content, dry density relationships were found as per IS: 2720 (Part VII) 1980. From the dry density and moisture content relationship, optimum moisture content (OMC) and maximum dry density (MDD) were determined for both compaction energies with varying percentage of lime are reported in Table 2.

Table 2. OMC and MDD of lime treated fly ash

Lime content (%)	Compactive effort: 593 kJ/m^3		Compactive effort: 2483 kJ/m^3	
	Maximum Dry Density, MDD (g/cc)	Optimum Moisture Content, OMC (%)	Maximum Dry Density, MDD (g/cc)	Optimum Moisture Content, OMC (%)
0	1.12	40.5	1.236	33
2	1.085	43	1.206	35.8
4	1.089	42	1.237	35
8	1.097	41.5	1.244	34.8
12	1.108	41.3	1.25	34.5

Unconfined samples were prepared in sealed condition at compaction energy of 2483 kJ/m^3 to their corresponding MDD at OMC. Sealed specimens were made to maintain the actual moisture content during the course of curing. Samples were wax coated for 10 °C, 25 °C, 45 °C temperature and covered with heat resistant wrapper for 90 °C temperature and cured for 0, 7, 15, 30, and 60 days in a temperature controller unit as shown in Fig. 2. These test specimens (50 mm in diameter and 100 mm in height) were cured at a temperature of 10 °C, 25 °C, 45 °C and 90 °C and sheared at an axial strain rate of 1.25 mm/min till failure of the sample after specified curing periods. Three identical specimens were tested for each condition and the average value was specified.

Fig. 2. Curing of unconfined samples in temperature controlled chamber

For CBR study the test specimens were prepared for both soaked and unsoaked condition according to IS: 2720(Part XVI)-1987 in a rigid metallic cylindrical mould having diameter of 150 mm and a height of 175 mm. The samples were prepared for both the compactive energies with separate percentage of lime at their respective MDD and OMC cured for 7 and 30 days (4 days of soaking before testing for soaked samples) before testing. Furthermore, the specimens were tested after specified curing periods in a mechanical loading machine equipped with a movable base that moves at a uniform rate of 1.2 mm/min.

The X-ray diffraction tests were used to determine the hydration product appeared in the lime treated specimen after specified curing periods by using Philips X'PERT high score. The samples were prepared from the fractured unconfined specimens and soaked in an anhydrous ethanol to stop further hydration. Microscopic study were undertaken to examine the morphology and microstructure of hydrated specimens. These were done by a JEOL 6480 LV scanning electron microscope.

4 Results and Discussions

4.1 Moisture Content and Dry Density Relationship

The variation of maximum dry density with their respective optimum moisture content at different compactive energy and lime content are presented in Figs. 3 and 4. Initially with increase in compaction energy and lime content maximum dry density decreases and optimum moisture content increases but after certain point of deflection the maximum dry density increases and optimum moisture content decreases. Formerly, illustrated that in Fig. 1 the fly ash mostly comprises of fine sand to silt size particles with uniform gradation therefore, the particle-particle inter connectivity pores gets filled by lime water and minor flocculation occurs resulting higher initial dry density and lower optimum moisture content.

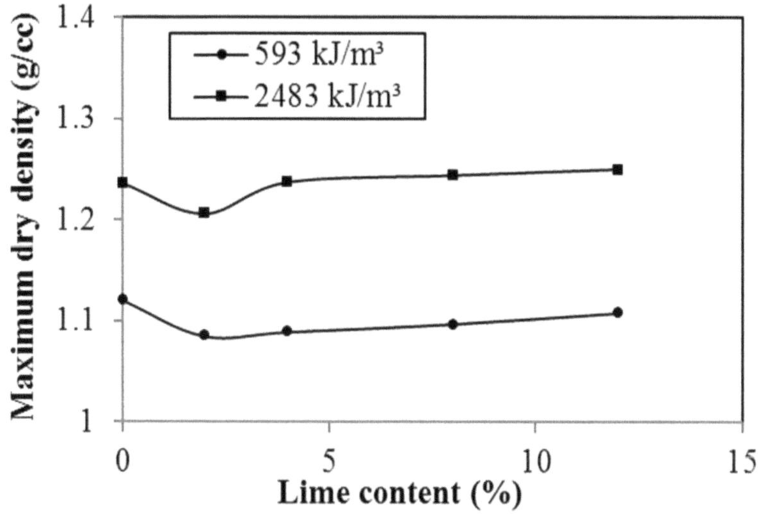

Fig. 3. Variation of MDD with lime content

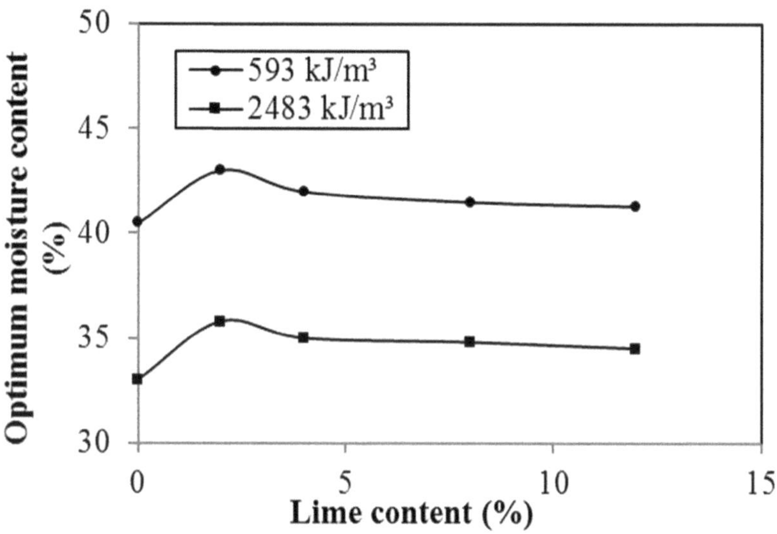

Fig. 4. Variation of OMC with lime content

4.2 Unconfined Compressive Strength

4.2.1 Effects of Compaction Energy

Unconfined compressive strength tests were carried out on treated flyash specimens compacted to their corresponding maximum dry density at optimum moisture content with both the compaction energy. The relation between compactive energy and strength are directly proportional to each other. The increase in unconfined compressive strength of specimen with increase in compactive effort is ascribed to the closer packing of particles, resulting in increased interlocking among particles. A closer packing is also responsible in increasing the cohesion component in the sample resulting in gain in strength. Additionally, the unconfined compressive strength depends on the doses of lime as well as the curing period. From Fig. 5 it is seen that as the curing period increases from 7 days to 60 days at 2% lime there is not much gain in strength observed at both compaction energies. Furthermore, at 2483 kJ/m^3 compaction energy with 8% lime content the achieved strength gives better result than 12% lime at 595 kJ/m^3 with longer curing period.

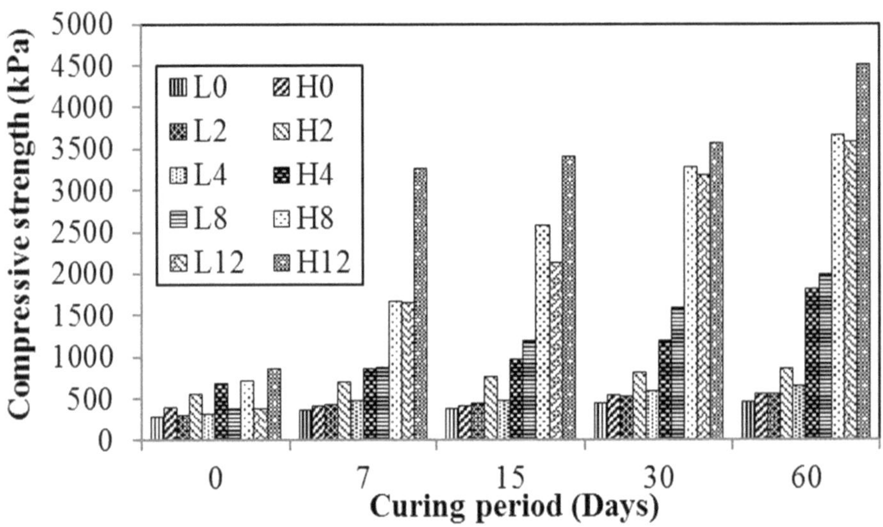

Fig. 5. Comparisons between different compaction energy at different lime content and curing period

4.2.2 Effects of Curing Temperature

The UCS value is found to change from 300 to 520.5 kPa with change in lime content from 0 to 2% indicating that the gain in strength is not so remarkable with smaller amount of lime; the strength enhancement is practically insignificant, even if cured for longer period of time. But a higher dose of lime that is beyond 2% enhances the unconfined strength by many folds. This conveys that about 2% of lime is used for colloidal type of reaction and lime in excess to this amount is utilized for pozzolanic reaction. As the lime percentage and curing period increases this facilitates the pozzolanic reaction that form cementitious gel of CSH and CASH that binds the particles. The process of pozzolanic reaction is improved with longer curing period which results higher strength.

The elevated temperature is very much susceptible to unconfined compressive strength; it is found to influence the sealed unconfined specimens. At higher temperature with additional percentage of lime the UCS shows a remarkable increase in strength. However, flyash added with lower percentage of lime does not show this trend. This indicates that higher temperature favors a better pozzolanic reaction than a lower temperature especially when the lime content is high. Higher curing temperature (90 °C) favours an early strength gain, but the strength at a later age is found to be lower than the samples cured at moderate temperatures like 25 °C and 45 °C. Samples

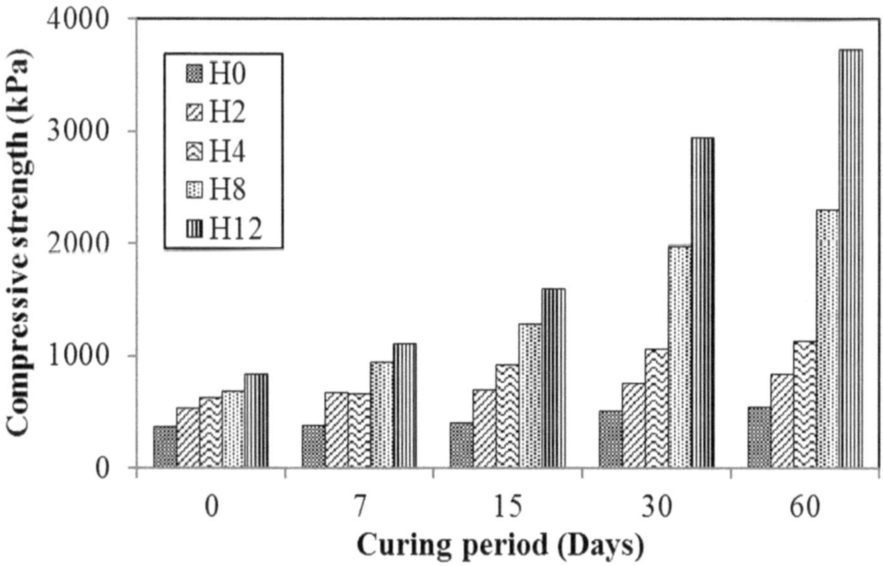

Fig. 6. Unconfined compressive strength of specimens cured at 10 °C

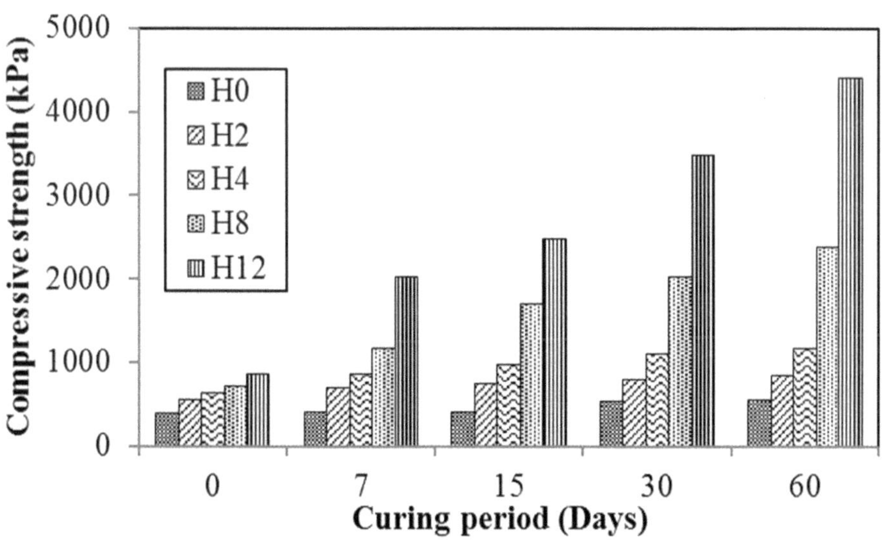

Fig. 7. Unconfined compressive strength of specimens cured at 25 °C

cured at low temperature such as at 10 °C shows rising trends of strength even up to 60 days of curing whereas the strength of high temperature cured specimens stabilize much earlier. A crossover effect of strength is noticed between low and high temperature cured specimens. This indicates that more the lime content more the water needed to complete the pozzolanic reaction. Hence it is recommended that ash samples stabilized with higher amount of lime should either be compacted wet of OMC or sufficient water be added subsequently for proper curing. Shows in Figs. 6, 7, 8 and 9.

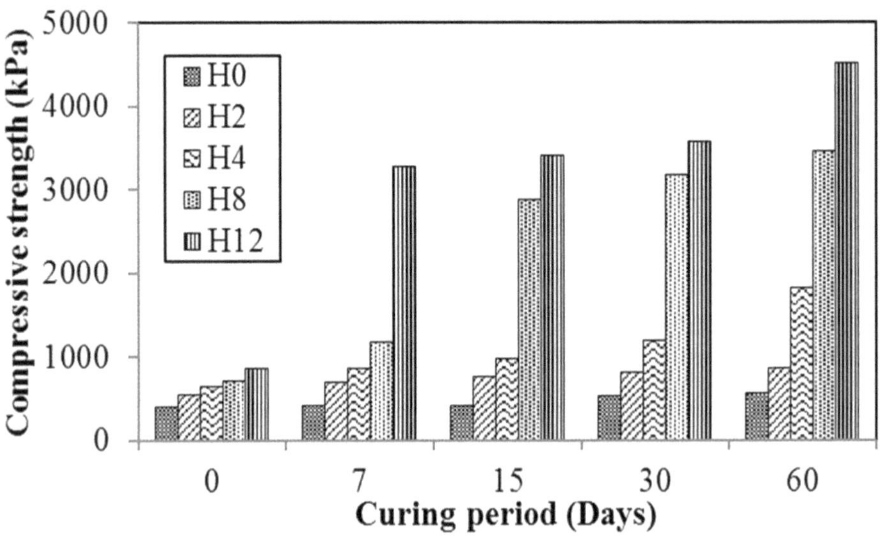

Fig. 8. Unconfined compressive strength of specimens cured at 45 °C

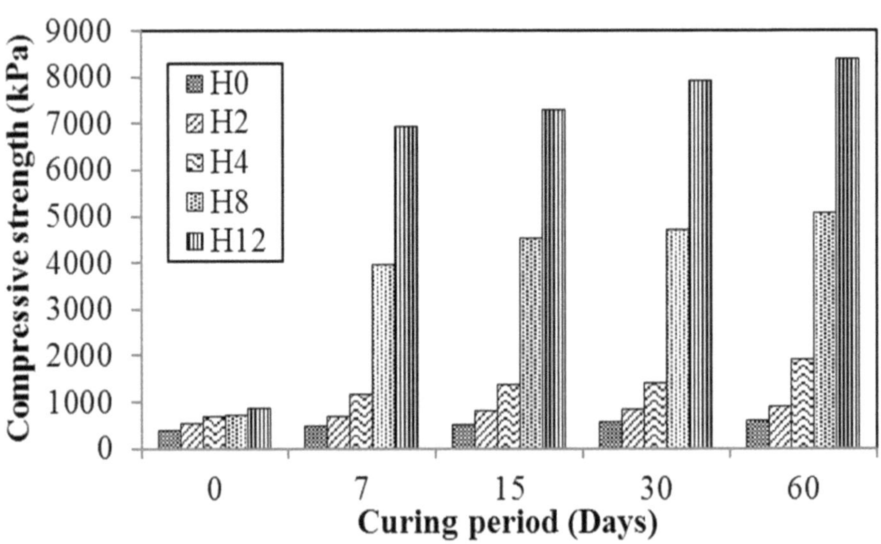

Fig. 9. Unconfined compressive strength of specimens cured at 90 °C

4.3 California Bearing Ratio

4.3.1 Effects of Curing Period and Lime Content

The variation of CBR value due to increase in lime content and curing period with both the compactive energy are presented in Figs. 10, 11, 12 and 13. It is clearly visible that CBR value increases with increase in compaction energy. The unsoaked and soaked

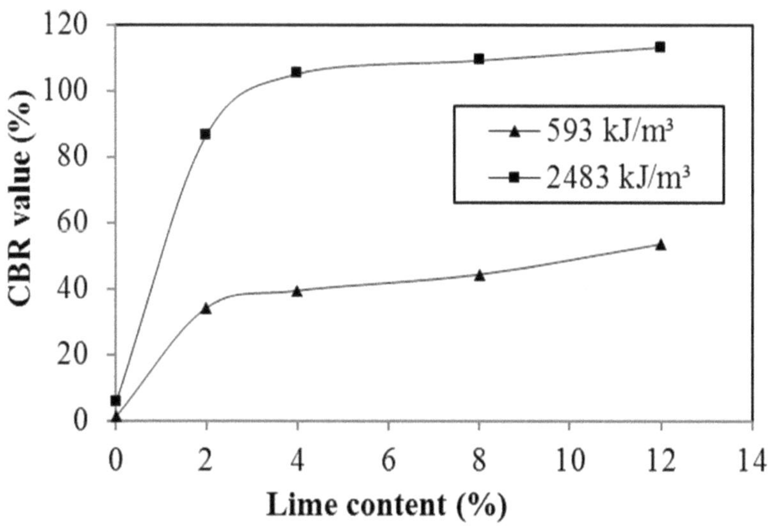

Fig. 10. Variation of soaked CBR with lime content for 7 days cured specimen

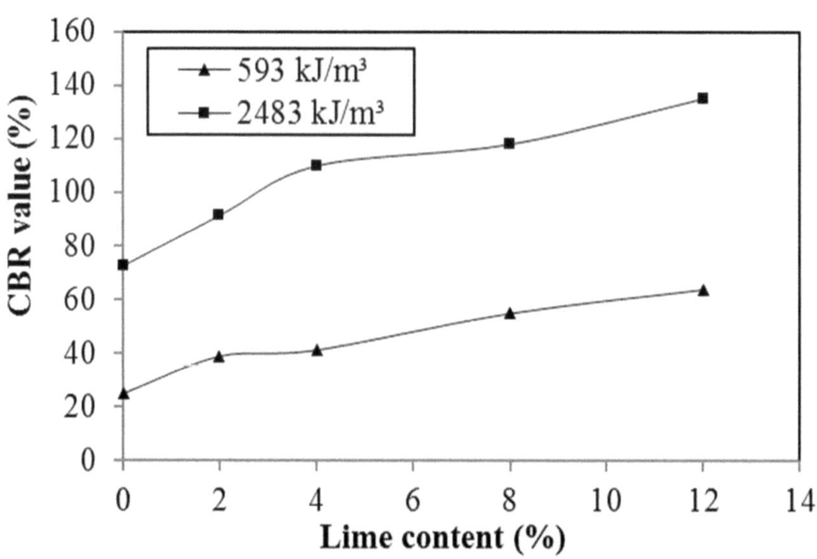

Fig. 11. Variation of unsoaked CBR with lime content for 7 days cured specimens

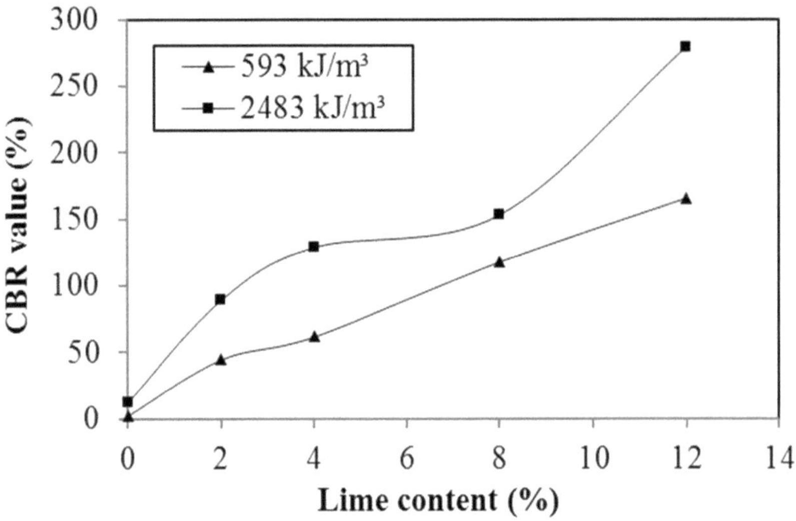

Fig. 12. Variation of soaked CBR with lime content for 30 days cured specimens

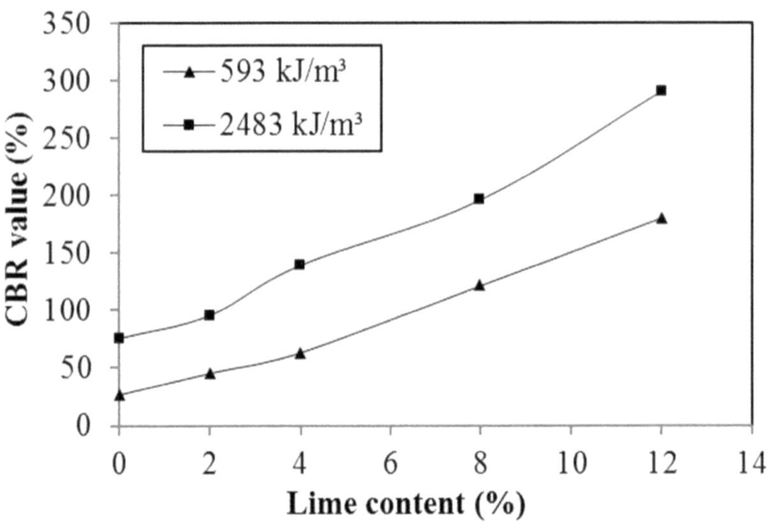

Fig. 13. Variation of unsoaked CBR with lime content for 30 days cured specimens

CBR values are found to increase with increase in lime content beyond 4% which gives marginal strength. Also, the Soaked and unsoaked CBR value gives lesser value when cured for 7 days; as curing period increase up to 30 days these values are slightly increased due to presence of some short of cementing material (free lime). So, higher doses of lime treatment needed for better strength.

The soaked CBR value of flyash is relatively low ranging from 1.3% to 5.8% as compaction energy increases from 595 to 2483 kJ/m3. However, the flyash treated with 12% lime gives comparatively higher CBR value of 44.2% after 7 days in soaked condition. When the sample subjected to a curing period of 26 days and a soaking period of 4 days, CBR value considerably increases due to pozzolanic reaction of lime. This phenomenon occur mainly because fly ash, a fine-grained material, when placed at 95% of Proctor maximum dry density and corresponding water content, exhibits capillary forces, in addition to friction resisting the penetration of the plunger and thus high values of CBR are obtained. On the contrary, when the same fly ash samples are soaked for 24 h maintaining the same placement conditions, they exhibited very low values of CBR. This can be attributed to the destruction of capillary forces under soaked conditions.

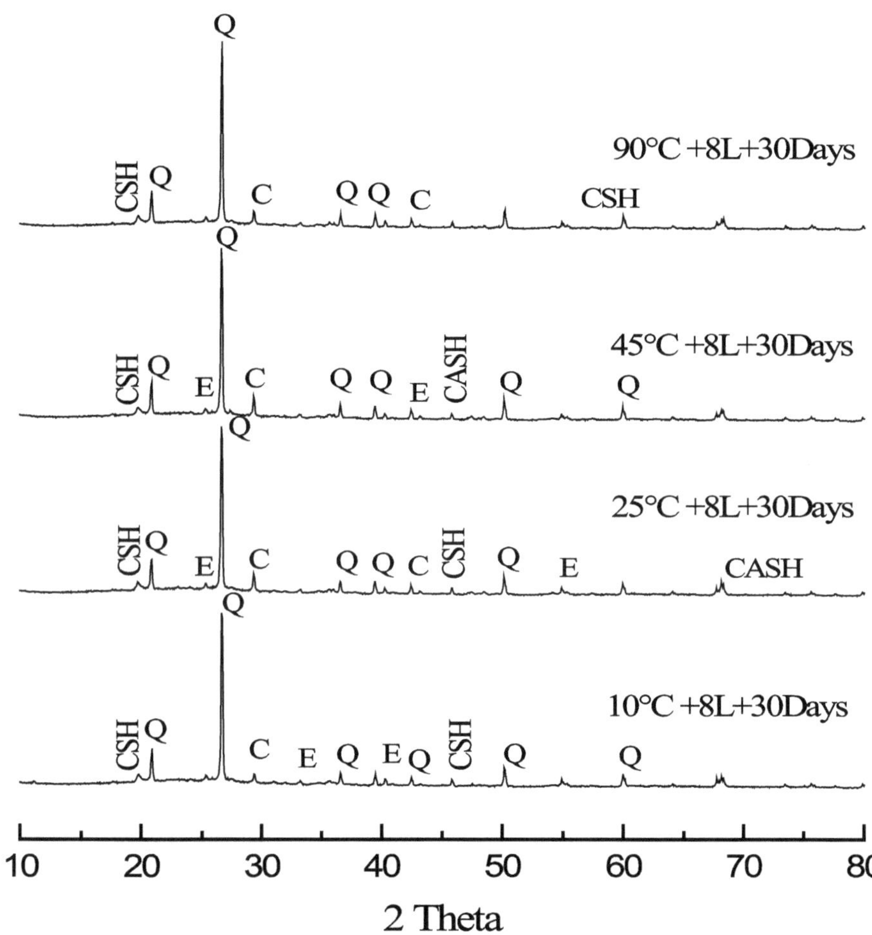

Fig. 14. X-ray diffraction pattern of lime stabilized flyash specimens cured at different temperature (Q – Quartz, E – Ettringite, C – Calcium carbonate, CASH – Calcium aluminium silicate hydrate, CSH – Calcium silicate hydrate)

4.4 Chemical Composition and Micro-Structure

The X-ray diffractographs and scanning electron micrographs of fly ash specimen treated with 8% lime cured at 10 °C, 25 °C, 45 °C and 90 °C for 30 days are presented in Figs. 14 and 15. It is observed that the samples stabilized at lower temperature i.e. 10 °C give more number of ettringite peaks at (2Θ = 33.2°, 40.37° and d-spacing = 2.696Å, 2.23 Å) and calcium aluminum silicate hydrate (2Θ = 47.55° and d-spacing = 1.912Å). Increasing the curing temperature from 25 °C to 45 °C resulted supplementary peaks of low intensity of ettringite. Those peaks of ettringite get disappeared when the treated specimen cured at 90 °C and more CSH/CASH peaks were appeared (d-spacing = 1.541Å, 2Θ = 60.003).

An analysis of the micrographs showed that an interlocking network of the pozzolanic reaction products has been formed in the specimens containing 8% lime cured at 90 °C shown in Fig. 15(d). The needle-like crystals of ettringite exist in the hydration product at lower temperature of 10 °C which get gradually disappeared with increase in temperature from 25 °C to 45 °C and then 90 °C. At higher curing temperature the ettringite get wrapped with a cementitious substance of calcium silicate

(a) 10°C+8%Lime+30Days (b) 25°C+8%Lime+30Days

(c) 45°C+8%Lime+30Days (d) 90°C+8%Lime+30Days

Fig. 15. Microstructure of lime treated flyash specimens cured at different temperatures hydrate or calcium aluminium silicate hydrate gel which gives early gain in strength. These compounds are already found in texture analysis. Extension of the curing temperature enhanced the pozzolanic reaction and formation of hydration products.

5 Conclusions

- Maximum dry density of compacted specimens is found to change from 1.12 to 1.236 g/cc with change in compaction energy from 595 kJ/m^3 to 2483 kJ/m^3, whereas the OMC is found to decrease from 40.5 to 33%. This shows that fly ash sample responds very poorly to the compaction energy.
- Increase in curing period of lime treated fly ash specimen shows improvement in the unconfined compressive strength. But with smaller amount of lime that is 1%–2% the strength improvement is practically insignificant, even if cured for long time. This is similar to the colloidal reaction with lime, which is mainly responsible in modifying the physical properties not the mechanical strength. As the lime percentage increases this facilitates the pozzolanic reaction that form cementitious gel that binds the particles. The process of pozzolanic reaction is improved with curing period which results higher strength.
- The unconfined compressive strength of flyash added with higher percentage of lime show a remarkable increase in strength with increase curing temperature. However flyash added with lower percentage of lime does not show this trend. This indicates that a higher temperature favours early strength gain and better pozzolanic reaction than a lower temperature especially when the lime content is high.
- Both the unsoaked and soaked CBR values are found to increase with lime content up to 4% beyond which the increment is marginal. This trend is observed for specimens cured for 7 days. However specimens cured for 30 days showed a continuous increase in CBR value with lime content. This indicates that the reaction of lime with flyash is slow and a higher curing period is needed to complete the pozzolanic reaction.
- From the diffraction study the new amorphous humps of calcium aluminium silicate hydrate, calcium silicate hydrate and ettringite with low intensity are detected. Abundance of needle-like structures of ettringite is found in the specimen at lower temperature. As curing temperature proceeds up to 90 °C the needle shaped crystals are seen wrapped with gel like substances of calcium silicate hydrate (CSH) and calcium aluminium silicate hydrate (CASH) gel which takes a major role in strength enhancement.

References

Ali, F.H., Adnan, A., Choy, C.K.: Use of rice husk ash to enhance lime treatment of soil. Can. Geotech. J. **29**(8), 843–852 (1992). Ottawa

Croft, J.B.: The pozzolanic reactivities of some New South Wales fly ashes and their application to soil stabilization. Aust. Road Res. Board **2**(2), 1144–1168 (1964). Vermont South, Victoria

Ghosh, A., Subbarao, C.: Hydraulic conductivity and leachate characteristics of stabilized fly ash. J. Environ. Eng. ASCE (1998). doi: 10.1061/(ASCE)0733-9372(1998)124:9(812)
Ghosh, A., Subbarao, C.: Microstructural development in fly ash modified with lime and gypsum. J. Mater. Civ. Eng. ASCE (2001). doi:10.1061/(ASCE)0899-1561(2001)13:1(65)
Ghosh, A., Subbarao, C.: Strength characteristics of class F fly ash modified with lime and gypsum. J. Geotech. Geoenvironmental Eng. ASCE (2007). doi:10.1061/(ASCE)1090-0241(2007)133:7(757)
Ghosh, A., Subbarao, C.: Deformation modulus of fly ash modified with lime and gypsum. Geotech. Geol. Eng. Springer (2012). doi:10.1007/s10706-011-9468-z
Gray, D.H., Lin, Y.K.: Engineering properties of compacted fly ash. J. Soil Mech. Found. Div. **98**(4), 361–380 (1972). ASCE
Joshi, R.C., McMaster, H.M., Duncan, D.M.: New and conventional engineering uses of fly ash. J. Transp. Eng. **101**(4), 791–806 (1975). ASCE
Kumar, B., Ghosh, A.: Compaction characteristics of pond ash. J. Mater. Civ. Eng. ASCE (2007). doi:10.1061/(ASCE)0899-1561(2007)19:4(349)
Lav, A.H., Lav, M.A.: Microstructural development of stabilized fly ash as pavement base material. J. Mater. Civ. Eng., 157–163. ASCE (2000). doi:10.1061/(ASCE)0899-1561(2000)12:2(157)
Maitra, S., Ahmad, F., Das, A.K., Das, S., Dutta, B.K.: Effect of curing conditions and ionic additives on properties of fly ash-lime compacts. Bull. Mater. Sci. Springer (2010). doi:10.1007/s12034-010-0026-8
Narmluk, M., Nawa, T.: Effect of curing temperature on pozzolanic reaction of fly ash in blended cement paste. Int. J. Chem. Eng. Appl. IJCEA (2014). doi:10.7763/IJCEA-2014-V5-346
Pal, S.K., Ghosh, A.: Hydraulic conductivity of fly ash-montmorillonite clay mixtures. Indian Geotech. J. Springer (2013). doi:10.1007/s40098-012-0033-3
Venkatarama Reddy, B.V., Gaurav, K.: Strength of lime–fly ash compacts using different curing techniques and gypsum additive. Mater. Struct. Springer (2011). doi:10.1617/s11527-011-9738-5
Subbarao, C., Ghosh, A.: Fly ash management by stabilization. J. Solid Waste Technol. Manage. **3**, 126–130 (1997). doi:10.1109/2015.738862824

Flow of Microbial Suspension Through Porous Media

B.S. Shashank(✉), Shetty Rakshith, Jeevan Joseph, Arif Mohammad, and D.N. Singh

Department of Civil Engineering, Indian Institute of Technology Bombay, Mumbai 400076, India
bs_shashank@yahoo.co.in, shettyrakshith@gmail.com, jeev.jsph@gmail.com, arif.juce93@gmail.com, dns@civil.iitb.ac.in

Abstract. Biological methods of soil remediation and improvement are being widely considered and accepted as environmentally sustainable and economically feasible solution as compared to conventional mechanical/physical or chemical methods. Application of these methods in *in-situ* soil conditions involves the injection of bio-grouts or microbial suspensions in to the soil matrix. Flow of such suspensions in a porous soil matrix depends on various parameters such as, the applied grouting pressure, microbial cell density, viscosity of the injected fluid, and in-situ density of the soil mass. In the present study, the microbial suspension with different cell densities were used to conduct hydraulic conductivity experiments to ascertain the effect of above mentioned parameters. Based on the intrinsic permeability of the soil matrix and hydraulic conductivity for flow of the microbial suspensions through the soil mass, the flow nature of microbial suspension in a porous matrix has been discussed. The study further investigates the rheological behaviour of these microbial suspensions. The study adds to the understanding of flow of microbial suspension through geomaterials having practical significance in bio-grout employed for soil remediation/stabilisation.

Keywords: Soil remediation · Bio-grouting · Rheology · Microbial suspension · Porous media

1 Introduction

The ground conditions which pose a threat to safety and integrity of any construction activity on it or to the environment mandates remediation. Such remediation measures aim at either improving the engineering properties of the soil mass, or mitigating the spread of contaminants or decontamination of the soil mass. Ground remediation for improving the engineering properties include several physical or chemical methods (Ou et al. 1996; Horpibulsuk et al. 2004; Zayyat et al. 2005; Jayanthi and Singh 2016). The increased use of chemical methods for these applications has led to the pollution of soil and the ground water and hence necessitates the use of environmentally sustainable and economically feasible methods. Among these broad areas of geo-environmental issues, one such issue is the stabilisation of loose coarse-grained deposits. These deposits

D.N. Singh and A. Galaa (eds.), *Contemporary Issues in Geoenvironmental Engineering*, Sustainable Civil Infrastructures,
DOI 10.1007/978-3-319-61612-4_9

impose challenges to practising engineers to develop methodologies to address various issues such as, salt water infiltration in coastal areas, creation of water retaining structures, slope stability of loose fill materials, creation of helipads in desert areas etc. Such a need has resulted in extensive research being carried out in the area of stabilization of coarse-grained geo-materials based on various concepts arising from several perspectives. Biology-inspired microbial stabilization is one such measure wherein microbes can be used to bind together loose materials by virtue of their extracellular secretions called Extracellular Polysaccharides (EPS) (DeJong et al. 2013), or due to precipitation of binding materials such as calcite in presence of suitable catalysts (Cheng and Cord-Ruwisch 2012; Montoya et al. 2013). The technique is gaining significance due to its environmental and economic advantage over other stabilization techniques such as mechanical and chemical methods.

The stabilization by means of microbial measures majorly involves microbial suspension being injected in the soil mass by means of grouting or gravity percolation. In this context, the applicability of biological methods in different types of geomaterials such as coarse and fine-grained soils depend primarily on the physical compatibility of biological grouts/fluids through the porous media (i.e., flow through porous media). The efficient execution of which requires a thorough scientific knowledge of flow of suspension through porous matrix, the effects of viscosity of suspension on the process, hydraulic conductivity of the media and its variability due to viscosity changes. To this effect, in the present study, rheological properties of microbial suspensions at different cell densities and hydraulic conductivity of coarse-grained soil matrix under saturated conditions have been investigated. In subsequent sections, test methodologies adopted and materials used, discussion on the result obtained and major outcome from the study have been elaborated.

2 Materials and Methods

The study was conducted on Indian Standard Sand, Grade-II and the microorganism used in the study was *Sporosarcina pasteurii*. The details of these and the test methodology adopted are presented in the following sections.

2.1 Microbial Suspension

The bacterial strain, *S. pasteurii*, DSM-33, procured from Deutsche Sammlung von Mikroorganismen und Zellkulturen (DSMZ), Germany. The stock solution of bacteria was prepared by, inoculating DSM-33 in a nutrient broth, Brain Heart Infusion (BHI), along with 20 g/L of urea, at 30 °C. The stock solution so prepared was stored at 4 °C prior to use. Further, 0.1% v/v of the stock solution was inoculated in BHI broth, and incubated for 24 h in an orbital shaker. Subsequently, these cultures were harvested by centrifuging at 6000 rpm for 8 min. The supernatant was discarded and the cell palette was rinsed with tap water in order to remove the traces of nutrient medium in the cell palette. The cell palette so obtained was re-suspended in tap water and the concentration of the cells in the suspension was varied as desired. The cell density was measured through optical density at 600 nm (OD_{600}) (i.e., the absorbance in the medium at 600 nm wavelength). OD is the representative of cell concentration and increases with increase in cell concentration. Different cell concentrations considered in this study are OD_{600} of 0.5, 1.0, 2.0 and 3.0. The harvested microbial cultures were used for testing immediately.

3 Rheological Testing of Microbial Suspension

The rheological properties of the microbial suspensions were determined using co-axial geometry of Anton Paar™ modular compact rheometer. The cell suspensions were dynamically sheared from a shear rate of 0.1 s^{-1} to 500 s^{-1} at 25 °C. The flow curves (plot of shear stress v/s shear rate) so obtained were used in calculating the viscosity of the cell suspensions based on Bingham model which is given by Eq. 1.

$$\tau = \tau_y + \mu\dot{\gamma} \tag{1}$$

where, τ is the shear stress, τ_y is the yield stress, μ is the viscosity, and $\dot{\gamma}$ is the strain rate.

4 Flow Studies of Microbial Suspension

The coarse-grained material used in this study was Grade-II Indian Standard Sand, confirming to IS650:1991. The index properties of the sand are presented in Table 1 and the particle size distribution characteristics are presented in Fig. 1. Prior to the experiment, sand was washed with DI water to remove dust or other impurities, followed by drying in a hot air oven at 105 °C for 24 h. The sample was then cooled to room temperature prior to use. The oven dried sand was packed in a column of length 480 mm and internal diameter 40 mm by sand raining method. The dry density of the packed sand column was noted to be 1.7 g/cc. The flow properties of the tap water and the microbial suspensions in the sand column were studied under a pressure of 12 kPa so as to maintain laminar flow, i.e., $R_e < 10$ in case of soils (Czachor 2011).

Table 1. Index properties of Indian standard sand Grade-II

Parameter	Value
Specific Gravity, G	2.7
Coefficient of uniformity, C_u	2.02
Coefficient of curvature, C_c	1.02

The flow properties of each test fluid in the sand column was studied by adopting the common procedure as discussed below using the setup as shown in Fig. 2. The sand-column was initially saturated with the test fluid and subsequently the test fluid was injected into the column under a pressure of 12 kPa. The fluid discharge through the column was measured for a time period of 30 s and the exercise was repeated multiple times. The hydraulic conductivity of the sand column was calculated based on the Darcy's law.

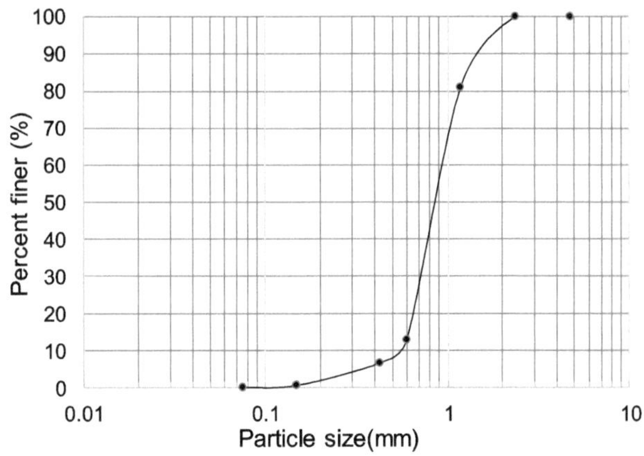

Fig. 1. Particle size distribution of Indian standard sand Grade-II

5 Results and Discussions

The flow curves, obtained from rheological testing of tap water and bacterial strain *Sporosarcina pasteurii* suspension of different cell concentrations, are presented in Fig. 3. The variation in viscosity of these bacterial suspensions is shown in Fig. 3. From Fig. 3 it should be noted that these suspensions of *S. pasteurii* are shear thickening in nature. It could be seen that at higher strain rates (>100) the slope of the flow curves are steeper showing increased resistance to flow. The shear thickening nature of the suspensions, might due to the rupture of cells at higher shear rates and release of cellular matrix in to the solution phase (Al-Asheh et al. 2002). This signifies the existence of optimal flow-rate of these suspensions. However, it has to be borne in mind that the rheological behaviour of microbial suspensions and value of such optimal flow-rates may vary significantly based on the bacterial strain along with various other factors. Furthermore, it has to be noted that in case of *S. pasteurii* the viscosity of the suspension increases with an increase in the OD, as shown in Fig. 4. This may result in delayed or no access to the pores wherein large depths of soil mass are encountered and when a higher OD suspension is used.

The hydraulic conductivity (K) for water, and microbial suspensions of various concentrations in the sand column is presented in Fig. 5. It has been observed that K decreases with an increase the cell density. The trends are consistent with the observations from rheological parameters. As viscosity is greater at higher cell concentrations, naturally there has been a decrease in the respective hydraulic conductivities. This phenomena could be attributed to the clogging of pores in the soil matrix. In addition, the 'intrinsic permeability' (k) of the sand matrix has been determined by employing Eq. 2 and its variation is depicted in Fig. 6. The intrinsic permeability being governed by the pore characteristics of the matrix exhibited a decreasing trend. A fair justification for this phenomenon might be the clogging of pores due to the presence of

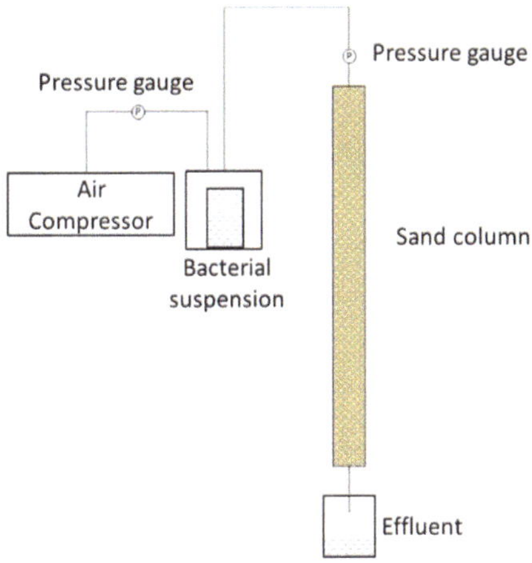

Fig. 2. Schematic representation of test setup

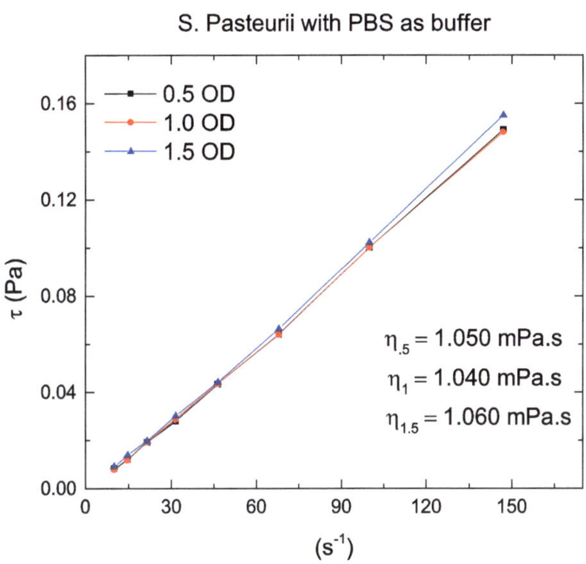

Fig. 3. Flow curves for different concentrations of *Sporosarcina pasteurii* suspensions

micron sized bacteria or their colonies, acting as a particulate matter. This also explains the reason for the trend that is being observed in fluid conductivity values.

$$k = K \times \frac{\mu}{\rho g} \tag{2}$$

Further, to understand the variation of intrinsic permeability, k, with OD, the fluid conductivity, K, has been plotted against ($\rho g/\mu$) and the same is depicted in Fig. 6.

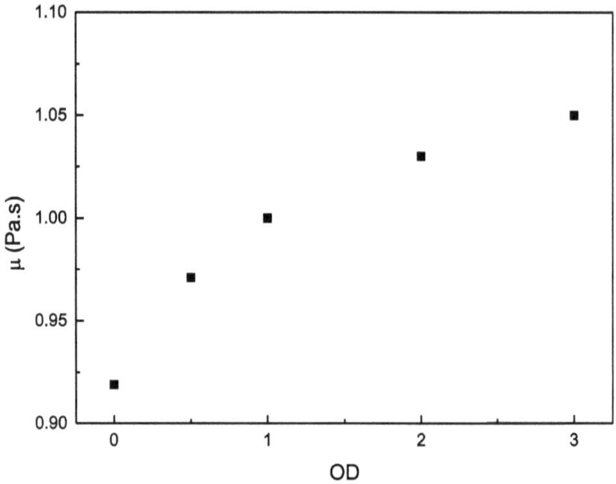

Fig. 4. Variation in viscosity of *Sporosarcina pasteurii* cultures at different cell concentrations

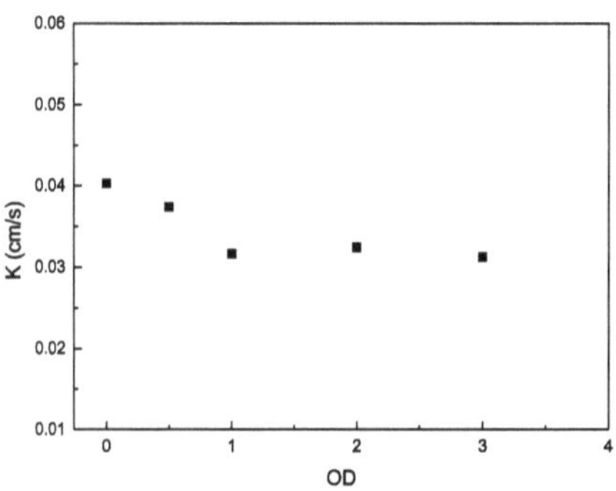

Fig. 5. Variation of hydraulic conductivity with OD.

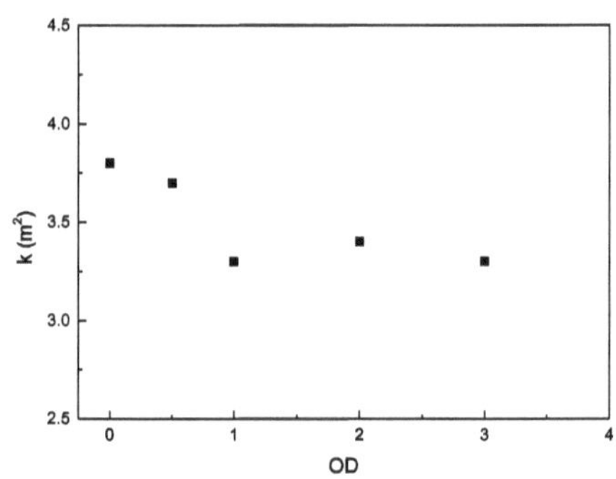

Fig. 6. Variation of intrinsic permeability of sand column for different OD.

The intrinsic permeability being a constant for a porous matrix, K should vary linearly with ($\rho g/\mu$) of the fluid percolating through the porous media. However, the variation observed in Fig. 7 is very minimal for it to be considered as non-linear variation and to be able to draw conclusions based on the variations observed. On the other hand, the variations observed in k and K of the matrix during flow of bacterial suspensions of different cell concentrations (refer Figs. 5 and 6) indicates that the microbes clogging the pores of the sand column leading to a decrease in these parameters. This warrants further investigations into the phenomenon occurring during flow of microbial suspensions through porous media.

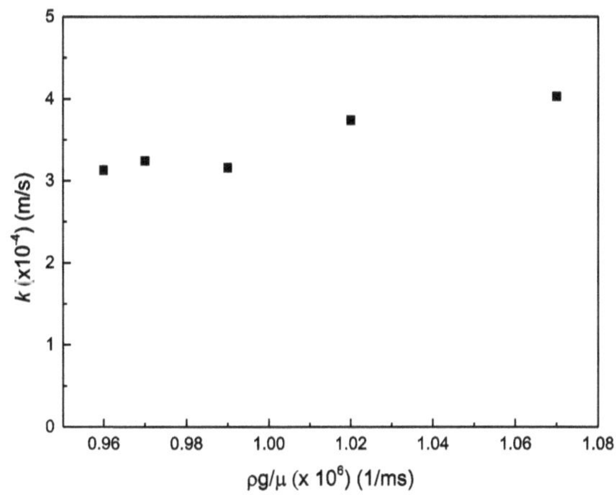

Fig. 7. Variation of hydraulic conductivity with different fluid properties.

6 Conclusions

Rheological properties of the microbial suspensions and their flow behaviour in porous media have been studied at varying cell concentrations. The shear thickening behaviour of the bacterial suspensions indicates the existence of critical shear rates beyond which the flow is likely to be restricted inside the soil matrix. Further, the hydraulic conductivity of bacterial suspensions with different cell concentrations is observed to decrease and the observed reduction in hydraulic conductivity is primarily due to the clogging of pores by the bacterium in the suspension, which is also evident from the variation in intrinsic permeability of the sand column. However, the non-linear behaviour of the intrinsic permeability needs to be validated for higher OD ranges and their correlations should be developed incorporating the variation in the pore characteristics of the matrix. Further, the impact of these observations on the microbial process/vactivity inside the soil matrix needs to be investigated.

References

Al-Asheh, S., et al.: Rheological characteristics of microbial suspensions of Pseudomonas aeruginosa and Bacillus cereus. Int. J. Biol. Macromol. **30**(2), 67–74 (2002). doi:10.1016/S0141-8130(02)00006-5

Cheng, L., Cord-Ruwisch, R.: In situ soil cementation with ureolytic bacteria by surface percolation. Ecol. Eng. **42**, 64–72 (2012). doi:10.1016/j.ecoleng.2012.01.013

Czachor, H.: Laminar and turbulent flow in soils. In: Gliński, J., Horabik, J., Lipiec, J. (eds.) Encyclopedia of Agrophysics. Springer, Netherlands (2011). doi:10.1007/978-90-481-3585-1_80

DeJong, J.T., et al.: Biogeochemical processes and geotechnical applications: progress, opportunities and challenges. Géotechnique **63**(4), 287–301 (2013). doi:10.1680/geot.SIP13.P.017

Horpibulsuk, S., et al.: Analysis of strength development in deep mixing: a field study. In: Proceedings of the Institution of Civil Engineers - Ground Improvement, vol. 8, no. 2, pp. 59–68 (2004). doi:10.1680/grim.2004.8.2.59

Jayanthi, N.V.P., Singh, D.N.: Utilization of sustainable materials for soil stabilization: state-of-the-art. Adv. Civ. Eng. Mater. ASTM International (2016, in Press). doi:10.1520/ACEM20150013

Montoya, B., et al.: Dynamic response of liquefiable sand improved by microbial-induced calcite precipitation. Géotechnique **63**(4), 302–312 (2013). doi:10.1680/geot.SIP13.P.019

Ou, C.-Y., et al.: Analysis of deep excavation with column type of ground improvement in soft clay. J. Geotech. Eng. **122**(9), 709–716 (1996). doi:10.1061/(ASCE)0733-9410(1996)122:9(709)

Zayyat, M.M., et al.: Foundation soil improvement using vibro compaction combined with geo grids. In: Innovations in Grouting and Soil Improvement, Reston, VA, pp. 1–10. American Society of Civil Engineers (2005). doi:10.1061/40783(162)25

Investigations on the Effects of Microorganisms on Coarse Grained Soil

S. Sowmya(✉), R. Asha Latha, P.S. Meenu, and D.N. Singh

Department of Civil Engineering, Indian Institute of Technology Bombay, Powai, Mumbai 400076, India
itssowmya@gmail.com, asha491991@gmail.com, meeps91@gmail.com, dns@civil.iitb.ac.in

Abstract. The association of soil and microorganisms is known to be existent since the formation of earth. However, conventional geotechnical engineering practices has only considered mechanical loading on the soil without taking into account of the biological aspects that might be associated. Keeping this in view, the present research was designed to study the impact of biological loading on the geo-material. Bacterium *Proteus vulgaris* was inoculated and incubated with two coarse grained soil samples namely, standard sand and aeolian sand and the effects of bacteria on soil samples were observed. Also, efforts were made to develop a protocol that could be used to establish bio-geo interface in coarse-grained geo-materials. However, long term and in-depth study is essential in this aspect to have a clear and broader image of the interaction, which could have prospective application in geo-technical engineering.

Keywords: Bio-geo interface · Coarse grained soils · Aeolian sand · SEM · *Proteus vulgaris*

1 Introduction

Characterization of soils and geomaterials form a prerequisite in the field of geotechnical engineering. Conventional geomechanics addressed various aspects of physical, chemical, mechanical and mineralogical properties in material characterization without considering the biological entity entangled within the geosystem. However, from time immemorial, living system, particularly microorganisms have been associated with geoenvironment [1–3]. These tiny living creatures release various metabolites, enzymes and extracellular secretions that might alter the chemical or mineralogical characteristics of their surroundings [3, 4]. They play vital roles in areas of element biotransformations and biogeochemical cycling, metal and mineral transformations, decomposition, bioweathering and formation of soil and sediment [4]. Also, recent advances in inter-disciplinary research involving microbiology in civil engineering have reported the influence of microorganisms on engineering properties of the geosystem [5–7]. Biological activity such as formation of biofilm was seen to alter the hydraulic conductivity of the porous media [8, 9]. Microbial involvement is also evidenced in carbonate precipitation and mineralization [10, 11]. In addition, bacterial metabolites namely expolysaccharides, enzymes and gases may either bind/degrade the geomaterials or alter

D.N. Singh and A. Galaa (eds.), *Contemporary Issues in Geoenvironmental Engineering*, Sustainable Civil Infrastructures,
DOI 10.1007/978-3-319-61612-4_10

pore fluid, thereby, changing the engineering properties of the geosystem such as shear strength, stiffness, compressibility [12, 13]. On a holistic view, all these biomechanisms render soil as a dynamic multiphase system consisting of a living entity along with other three phases of matter (i.e., soil solids, water and air) [7]. Further, biomediated mechanisms seem to be an environmental friendly and sustainable way for the modification of soil characteristics alternative to the traditional soil improvement techniques [14, 15]. In view of these, deciphering the fundamental and complex interactions within the bio-geo interface will play a significant role for better understanding of the intricacies involved and providing new directions in geotechnical practice. With this background, the present study attempts to develop a protocol to establish the bio-geo interface and study the effects of interaction between coarse grained soils and bacteria.

2 Materials and Methods

The bacterial species used in the study, *Proteus vulgaris* (MTCC 742) was purchased from the Microbial Type Culture Collection (MTCC), Chandigarh, India. Aeolian sand collected from the Thar Desert, Jaisalmer, Rajasthan and commercially available standard sand were used in the study. Bacterial growth media (nutrient agar, nutrient broth) were obtained from Himedia and other chemicals were obtained from standard manufacturers.

3 Experimental Methodology

3.1 Characterization of Soil

Specific Gravity
The specific gravity, G, of the coarse grained sands were determined using an ULTRA-PYCONOMETER (Quanta-chrome, USA), which employs inert helium gas as the displacing fluid according to the guidelines provided by ASTM D 5550-14 [16].

pH, Electrical Conductivity, Total Dissolved Solids and Salinity
Water quality analyzer (Oakton PC 2700) was used to determine pH, Electrical Conductivity (EC), Total Dissolved Solids (TDS), and salinity of the samples. Sand samples were stirred in distilled water for 1 h, maintaining liquid to solid (L/S) ratio of 20. The mixture was then filtered, and the filtrate was used to determine the parameters [17].

Chemical Composition
The chemical composition of the samples in the form of oxides was determined using X-ray Fluorescence. Details of the standard method for sample preparation are as follows: 4 g of the finely powdered soil sample was mixed with 1 g of methyl cellulose powder and isopropyl alcohol and kept under infrared lamp for drying. The powder was then poured into an aluminium dish with inner diameter of 33 mm and height of 12 mm containing methyl cellulose up to two third of its volume. The pellets were prepared using hydraulic press by applying a load of 15 T for 60 s. Subsequently,

the sample was mounted on the monochromatic sample holder for determining its chemical composition [18].

Mineralogical Characterization
The mineralogical composition of the sample was determined using an X-ray diffraction (XRD) spectrometer (X'Pert PRO, PANalytical, The Netherlands) available in the Department of Metallurgy and Material Sciences (MEMS), IIT Bombay, India, which utilizes a graphite monochromator and Cu-Kα radiation. The samples were scanned for 2θ ranging between 5° and 120°. Qualitative search-match phase analysis of the peak patterns were carried out using the ICDD database [18].

Morphological Characterization
In order to study the surface of the coarse grained particles after the establishment of the bio-geo interface, Scanning Electron Microscopy (SEM) available at the Sophisticated Analytical and Instrumentation Facility, IIT Bombay, India and Cryo FEG-SEM, model no. JSM-7600F were used. The grains of sand samples were placed on stub, sputter coated with platinum and analysed to obtain micrographs [17, 18].

3.2 Bacterial Culture Used in the Study

The bacterium, *Proteus vulgaris* obtained in the lyophilized form was inoculated to nutrient broth (NB) and incubated at an optimum temperature. After 24 h, the broth culture was plated on nutrient agar (NA) plates and incubated for 24 h to obtain individual colonies. Subsequently, the pure culture obtained was stored in 15% glycerol broth at −80 °C following the standard protocol [3].

3.3 Sterilization of the Soil

The sand sample used in the experiment already contained different strains of bacteria and fungus. Hence, it was necessary to eliminate the indigenous microorganisms present to study the effect of a particular strain. Based on the preliminary sterilization studies, autoclaving the soil was considered as an effective sterilization procedure. The sand samples were heated in a conical flask to 121 °C at 15 psi for 15–20 min in an autoclave, which were then cooled, preserved in sterile plastics and used for analyses. Table 1 gives the details of the experimental samples used in the study.

3.4 Preparation of Bacteria Amended Sand Samples

The bacterium, *Proteus vulgaris* was inoculated in NB media and incubated in shaker incubator at 37 °C for 24 h. The culture broth was then centrifuged for 15 min at 5000 rpm and the pellets obtained were washed two to three times in 0.85% NaCl. Subsequently, the pellets were added separately to NB media and minimal salt media (MSM) and optical density at 600 nm (OD_{600}) was adjusted to 1 using spectrophotometer. 15 ml of this culture broth was then added to 30 g of autoclaved soils taken in

Table 1. Designation of 12 experimental samples used in the study. SS: standard sand; AS: aeolian sand

Sl. no.	Samples	Designation
1	SS non autoclaved	SSN
2	SS autoclaved	SS
3	SS + nutrient broth	SN
4	SS + nutrient broth + bacteria	SNP
5	SS + minimal salt media	SM
6	SS + minimal salt media + bacteria	SMP
7	AS nonautoclved	ASN
8	AS autoclaved	AS
9	AS + nutrient broth	AN
10	AS + nutrient borth + bacteria	ANP
11	AS + minimal salt media	AM
12	AS + minimal salt media + bacteria	AMP

a sterile conical flask and incubated in shaker at 37 °C for interaction between bacteria and sand samples.

3.5 Analysis of the Samples

Sub-samples were aseptically taken at regular intervals from experimental flasks, serially diluted, inoculated onto agar plates and incubated at optimum temperatures to ensure viability and purity of bacterial culture. After 30 days of experimentation, bacterial inoculated sand samples were used for various soil characterizations as discussed above.

4 Results and Discussions

Table 2 shows the results of characterization of the 12 experimental sand samples. The specific gravity of the coarse grained soil samples studied were around 2.6 for standard sand and 2.7 for aeolian sand in the present study. The autoclaved sand (SS and AS) samples did not show significant variation in specific gravity compared to their respective non autoclaved samples (SSN and ASN). However, changes are seen in autoclaved samples containing media with/without bacteria (SM, SMP, AN, ANP, AM, AMP) in comparison to their respective autoclaved controls (SS, AS). This variation may be due to the presence of media components or bacteria. It is observed that the pH of the samples is in the range of 6.5 to 7.8 and did not vary due to autoclaving or bacterial treatment. Autoclaved standard sand (SS) showed the electrical conductivity (EC) of 97.6 µs/cm in comparison to the non-autoclaved sample (SSN), which showed EC of 55.6 µs/cm. Similarly, autoclaved aeolian sand (AS) showed increased EC of 111 µs/cm whereas non-autoclaved aeolian sand (ASN) showed EC of 88.4 µs/cm. In addition, it is observed that incubation of these sand samples with media (namely SN, SM, AN and AM as in

Table 2. Chemical characteristics of the sand samples used in the study

Sl. no.	Samples	Specific gravity	pH	EC (μs/cm)	TDS (ppm)	Salinity (ppm)
1	SSN	2.62	7.79	55.58	34.68	38.12
2	SS	2.66	7.34	97.59	63.21	64.25
3	SN	2.66	6.55	639.3	513.9	505.6
4	SNP	2.66	6.78	555.5	448.3	440.4
5	SM	2.82	6.8	851.1	685.7	683.7
6	SMP	2.48	6.82	830.7	670.4	668.6
7	ASN	2.73	7.72	88.40	55.59	56.30
8	AS	2.77	7.99	111.0	71.90	72.15
9	AN	2.80	7.23	445.3	359.5	350.7
10	ANP	2.83	6.55	541.1	436.2	427.7
11	AM	2.82	7.36	886.1	714.7	714.3
12	AMP	2.78	7.22	667.6	538.5	532.4

Table 1) and bacterial culture (SNP, SMP, ANP, AMP as in Table 1) has significantly increased their electrical conductivity, which is due to the presence of media components. It is also seen that samples inoculated with MSM has higher EC (SM, SMP, AM and AMP as in Table 1) than NB treated samples (SN, SNP, AN and ANP as in Table 1). This may be due the presence of higher concentration of salts in MSM. Similarly, concentrations of total dissolved solids and salinity have increased on autoclaving, which are further increased in samples treated with media namely NB and MSM.

Chemical composition of the standard sand and aeolian sand samples are presented in Table 3. It is seen that autoclaving has not increased the concentration of the tested elements. However, sodium, potassium and phosphorus concentrations have seen to be increased in cases of samples inoculated with minimal salt media with or without bacteria, which may be due to the salts and nutrients present in the culture media. XRD data

Table 3. Chemical compositions of the sand samples used in the study

Sample	Oxide % by weight												
	SiO_2	TiO_2	Al_2O_3	Fe_2O_3	MnO	MgO	CaO	Na_2O	K_2O	P_2O_5	Cr_2O_3	ZnO	NiO
SSN	91.4	0.02	6.17	0.96	0.08	0.11	0.04	1.21	0.04	0.00	0.00	0.00	0.00
SS	93.2	0.02	4.76	0.89	0.07	0.03	0.03	1.00	0.03	0.00	0.02	0.00	0.00
SN	81.5	0.14	9.30	2.84	0.13	0.14	0.41	4.62	0.36	0.50	0.02	0.01	0.00
SNP	82.2	0.14	9.67	3.01	0.13	0.24	0.38	3.55	0.28	0.34	0.34	0.02	0.01
SM	80.2	0.09	6.70	1.94	0.09	0.10	0.24	1.70	3.38	5.56	0.02	0.01	0.00
SMP	76.8	0.08	8.82	2.31	0.13	0.10	0.08	2.39	3.68	5.56	0.02	0.01	0.00
ASN	64.5	0.18	14.50	6.56	0.10	0.81	6.45	3.85	2.76	0.25	0.01	0.00	0.00
AS	63.9	0.11	13.94	7.01	0.08	0.84	6.96	3.93	2.88	0.28	0.00	0.00	0.00
AN	58.8	0.12	16.73	7.68	0.12	0.55	6.37	6.03	3.09	0.48	0.03	0.01	0.00
ANP	54.3	0.03	17.63	9.05	0.15	0.53	7.4	6.80	3.58	0.45	0.02	0.01	0.00
AM	44.9	0.02	15.80	8.83	0.15	0.48	5.52	3.79	9.37	11.08	0.00	0.01	0.01
AMP	53.7	0.09	15.65	7.03	0.12	0.52	5.54	7.28	2.68	7.41	0.00	0.01	0.00

presents quartz as the major mineral in all the samples, which shows that autoclaving or bacterial action for 30 days did not result in mineralogical changes in the tested samples.

Figures 1, 2 and 3 shows the SEM micrographs (E-SEM and cryo SEM) of the soil samples, bacterium *Proteus vulgaris* and the soil samples after bacterial interaction respectively. *Proteus vulgaris* is a gram negative rod shaped bacterium usually 0.5 to 1.0 by 1.0 to 3.0 μ in size, occurring singly, in pairs or in long chains. Attachment of bacteria to the sand sample was clearly observed in case of aeolian sand when cryo

Fig. 1. E-SEM micrographs of (A) standard sand (B) aeolian sand

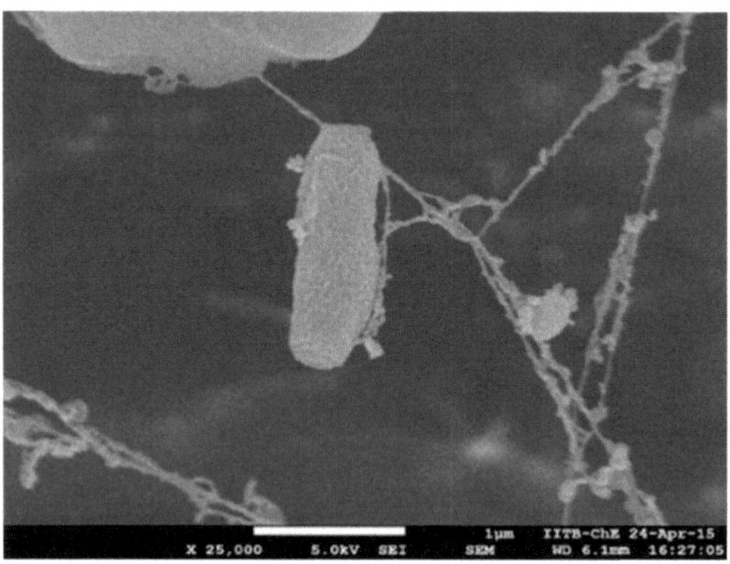

Fig. 2. Cryo-SEM micrographs of the bacterium Proteus vulgaris used in the study

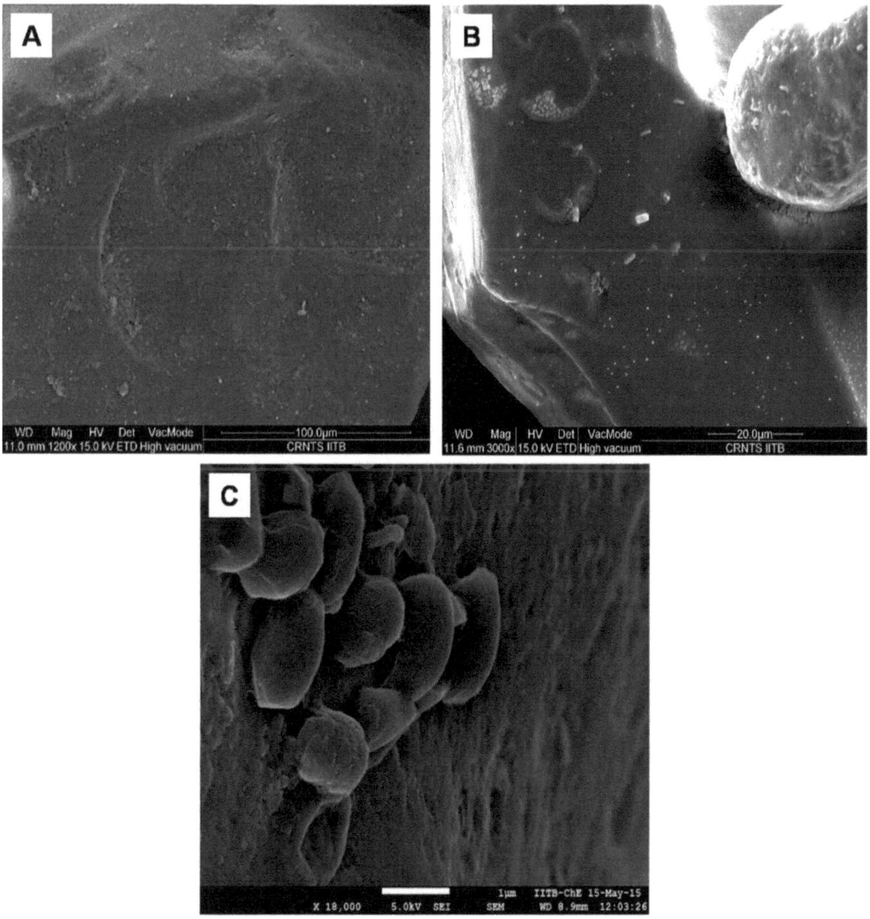

Fig. 3. SEM micrographs of aeolian sand (A) E-SEM micrographs before treatment with Proteus vulgaris (B) E-SEM micrographs after treatment with Proteus vulgaris (C) Cryo-SEM micrographs of aeolian sand after treatment with Proteus vulgaris

SEM was conducted. However, in case of standard sand, clear cryo SEM images were not obtained. It is known that bacteria produce exopolysaccharides and other metabolites, which help in its attachment and interaction with soil and geomaterials. Also, few microorganisms including bacteria and fungi known as lithotrophs are known to survive on rocks or sand due to their specific characteristics. The data obtained shows that there are no significant changes in the chemical mineralogical or structural parameters of the sand due to microbial action. This may be due to the short term interaction. However, long term exposure and extensive studies might lead to changes in the system and helps in understanding microbial mechanism and their interaction with geomaterials.

5 Conclusions

Research on impact of biological activity on geomaterials has gained importance in the recent years leading to a relatively new branch of geotechnical engineering, called biogeotechnology. Biological processes particularly, microbial and plant activities are being employed for development of environmental friendly, sustainable and cost effective geotechnical solutions. However, various factors are collectively involved in the overall success of biomediated processes such as concentration and speciation of microbes, pathogenicity associated with the organism and environmental conditions. Hence, working with these dynamic and diverse living entities require thorough insight into the various mechanisms prevailing within the bio-geo system. In this direction, a protocol was developed to establish bio-geo interface in coarse grained soil, which can be used to understand the behaviour of geomaterials under bacterial influence. Microbial activities being species specific, sterilization by autoclaving the soil samples could lead to better understanding of the relation between the particular microbe and the soil. The comparative study of the soil properties carried out prior, post autoclaving and bacterial doping showed no substantial changes in mineralogy of materials. However, changes in chemical characteristics of materials and attachment of bacteria were observed. Due to the slower rate of microbial activity in the present experimental conditions, long term and in-depth studies are demanded that could lead to observable results.

Acknowledgement. The authors would like to acknowledge Dr. Kiran Kondabagil and Dr. Prashant Phale, Department of Biosciences and Bioengineering, IIT Bombay, India for providing the lab facilities and their valuable suggestions.

References

1. Katsuji, W., Susumu, A., Koichi, H.: Evaluation of extracellular protease activities of soil bacteria. Soil Biol. Biochem. **26**, 479–482 (1994). doi:10.1016/0038-0717(94)90180-5
2. Kowalchuk, G.A., Buma, D.S., de Boer, W., Klinkhamer, P.G., van Veen, J.A.: Effects of above-ground plant species composition and diversity on the diversity of soil-borne microorganisms. A. Van. Leeuw. J. Microb. **81**, 509–520 (2002). doi:10.1023/A:1020565523615
3. Shreedhar, S., Devasya, R.P., Naregundi, K., Young, C.C., Bhagwath, A.A.: Phosphate solubilizing uranium tolerant bacteria associated with monazite sand of a natural background radiation site in South-West coast of India. Ann. Microbiol. **64**, 1683–1689 (2014). doi:10.1007/s13213-014-0812-4
4. Gadd, G.M.: Metals, minerals and microbes: geomicrobiology and bioremediation. Microbiology **156**, 609–643 (2010). doi:10.1099/mic.0.037143-0
5. Mitchell, J.K., Santamarina, J.C.: Biological considerations in geotechnical engineering. J. Geotech. Geoenviron. **131**, 1222–1233 (2005). doi:10.1061/(ASCE)1090-0241(2005)131:10(1222)
6. Morgan, P., Atlas, R.M.: Hydrocarbon degradation in soils and methods for soil biotreatment. Crit. Rev. Biotechnol. **8**, 305–333 (1989). doi:10.3109/07388558909148196

7. Shashank, B.S., Sharma, S., Sowmya, S., Latha, R.A., Meenu, P.S., Singh, D.N.: State-of-the-art on geotechnical engineering perspective on bio-mediated processes. Environ. Earth Sci. **75**, 1–16 (2016). doi:10.1007/s12665-015-5071-6
8. Kanmani, S., Gandhimathi, R., Muthukkumaran, K.: Bioclogging in porous media: influence in reduction of hydraulic conductivity and organic contaminants during synthetic leachate permeation. J. Environ. Health Sci. Eng. **12**, 1–11 (2014). doi:10.1186/s40201-014-0126-2
9. Dennis, M.L., Turner, J.P.: Hydraulic conductivity of compacted soil treated with biofilm. J. Geotech. Geoenviron. **124**, 120–127 (1998). doi:10.1061/(ASCE)1090-0241(1998)124:2(120)
10. Dupraz, C., Reid, R.P., Braissant, O., Decho, A.W., Norman, R.S., Visscher, P.T.: Processes of carbonate precipitation in modern microbial mats. Earth Sci. Rev. **96**, 141–162 (2009). doi:10.1016/j.earscirev.2008.10.005
11. De Muynck, W., De Belie, N., Verstraete, W.: Microbial carbonate precipitation in construction materials: a review. Ecol. Eng. **36**, 118–136 (2010). doi:10.1016/j.ecoleng.2009.02.006
12. DeJong, J.T., Fritzges, M.B., Nüsslein, K.: Microbially induced cementation to control sand response to undrained shear. J. Geotech. Geoenviron. **132**, 1381–1392 (2006). doi:10.1061/(ASCE)1090-0241(2006)132:11(1382)
13. DeJong, J.T., Mortensen, B.M., Martinez, B.C., Nelson, D.C.: Bio-mediated soil improvement. Ecol. Eng. **36**, 197–210 (2010). doi:10.1016/j.ecoleng.2008.12.029
14. Jonkers, H.M., Thijssen, A., Muyzer, G., Copuroglu, O., Schlangen, E.: Application of bacteria as self-healing agent for the development of sustainable concrete. Ecol. Eng. **36**, 230–235 (2010). doi:10.1016/j.ecoleng.2008.12.036
15. Lambert, J.W.M., Novakowski, K., Blauw, M., Latil, M.N., Knight, L., Bayona, L.: Pamper bacteria, they will help us: application of biochemical mechanisms in geo-environmental engineering. In: GeoFlorida 2010: Advances in Analysis, Modeling & Design, pp. 618–627 (2010). doi:10.1061/41095(365)59
16. ASTM D5550-14. Standard Test Method for Specific Gravity of Soil Solids by Gas Pycnometer, ASTM International, West Conshohocken, PA (2014)
17. Sharma, S., Meenu, P.S., Latha, R.A., Shashank, B.S., Singh, D.N.: Characterization of sediments from the sewage disposal lagoons for sustainable development. Adv. Civ. Eng. Mater. **5**, 1–23 (2016). doi:10.1520/ACEM20150010
18. Koshy, N., Jha, B., Kadali, S., Singh, D.N.: Synthesis and characterization of Ca and Na zeolites (non-pozzolanic materials) obtained from fly Ash–$Ca(OH)_2$ interaction. Mater. Perfor. Character. **4**, 87–102 (2015). doi:10.1520/MPC20140053

Stabilization of Dispersive Soil Using Biopolymer

Kajal Swain(✉), Mahasakti Mahamaya, Shamshad Alam, and Sarat Kumar Das

Department of Civil Engineering, NIT, Rourkela 769008, India
kajal.swain23@gmail.com, mahamaya.2008@gmail.com,
safzal88@gmail.com, saratdas@rediffmail.com

Abstract. Dispersive soils are considered as unstable, as these are easily erodible by wind and water. Structures, such as embankments, channels and foundations are vulnerable to severe erosion when this type of soil is used. Dispersive soil exhibits no strength resistance against environmental factors like air and water forces and fails rapidly during soaking as per crumb test and double hydrometer test. In the present study two biopolymers; xanthan gum in 1.0%, 2.0%, 3.0% and guar gum in of 0.5%, 1.0% and 2.0% are used for stabilization of the dispersive soil. The effects of biopolymer stabilizing agents are studied through crumb test and cylinder dispersion test for erosion control and other geotechnical properties like compaction and unconfined compression test (UCS). The optimum moisture content (OMC) of the bio treated samples increased and maximum dry density (MDD) decreased with increase in percentage of biopolymer. The bio treated samples cured in sun light for a period of 3 days shows higher unconfined compressive strength (UCS) value compared to as compacted sample and cured under ambient condition. Microstructural studies like scanning electron microscope (SEM), x-ray diffraction (XRD) and energy dispersive x-ray (EDX) tests were also conducted. The SEM analysis showed that the particles of biopolymer modified dispersive soil are bonded together by gum layer. This may be the cause of decreases in the dispersivity of the soil. Guar gum was found to be more effective than xanthan gum.

1 Introduction

Soils in which the clay particles get separated instinctively from each other and formed into suspension in quiet water are called dispersive soils (Yong and Sethi 1977; Mitchell 1993). Dispersivity is a physico-chemical process, which is mostly affected by the type of soil minerals and chemical properties of the soil pore fluid (Yong and Warkentin 1966; Sherad et al. 1976). The Dispersive soil (DS) causes dust problem in its dry state and leads to failure of dam and embankment due to piping in saturated condition (Sherad et al. 1976). The dispersiveness of soil is mainly due to the presence of exchangeable sodium ion. The formation of dispersivity may cause the formation of piping phenomena in earth dams (Fell et al. 1992), deterioration and demolition of roads (Nevels 1993), and the erosion of the compacted soils of landfill clay liners (Tin 1984; Ouhadi and Goodarzi 2003). These soils are found as both residual and transported soil and

D.N. Singh and A. Galaa (eds.), *Contemporary Issues in Geoenvironmental Engineering*, Sustainable Civil Infrastructures,
DOI 10.1007/978-3-319-61612-4_11

found to exist in various types of climates in various locations in Australia, Brazil, Iran, New Zealand, the United States, and many other countries (Sherad et al. 1976; Ludwig 1979; Goodarzi 2003). Thus, small seepage velocity of seepage can erode the dispersive soil, which cause harmful to the stability of earth dam and earth retaining wall. These soils are also identified with negative differential free swell values. The repulsive force is dominant over attractive force among the soil particles under saturated condition, which causes segregation of soil to get, disperse. The alum (aluminium sulphate) was found to be effective to control the dispersivity of bentonite, which is a dispersive soil (Ouhadi and Goodarzi 2006). Based on pinhole tests, double hydrometer and crumb test (Bhuvaneshwari and Soundara 2007) observed that dispersive soil could be stabilized with lime and fly ash mixture. Lime and cement improves the strength and reduce the dispersivity of the soil (Umesh et al. 2009, Vakili et al. 2013) and chemical agents including sodium chloride, sodium carbonate, sodium sulfate and sodium polyphosphate are also found to be effective for reducing dispersivity of natural soil (Abbasi and Nazifi 2013). The magnesium chloride ($MgCl_2$) solution is effective in reducing dispersive and expansiveness of clay soils (Turkoz et al. 2014). The industrial by-product, granulated blast furnace slag (GBFS) and basic oxygen furnace slag (BOFS) also can be used to reduce the dispersivity of soil (Goodarzi and Salimi 2015).

However, the chemicals used for such stabilization may not be suitable for the ecosystem. The erosion control of canal banks, due to dispersive soils using chemical admixtures may not be in coherence to sustainable development. The microorganisms present in the soil can be helpful to achieve some modification in soil (Ivanov and Chu 2008). From various researches it has been found that microorganisms produce exopolysaccharides that helps in soil aggregation, bioclogging, bio cementation and can be helpful in stabilization, mitigation of liquefaction potential, strengthening tailing dams, binding etc. (Ivanov and Chu 2008). Chen et al. (2013) used xanthn gum and guar gum for stabilization of mine tailing. Hence, in the present study, two biopolymers named xanthan gum and guar gum are used to stabilize the dispersive soil.

2 Materials

In the present study, dispersive soil is stabilized using two biopolymers named xanthan gum and guar gum to reduce its erosion. For this purpose, dispersive soil was collected from campus of NIT, Rourkela, India, which is a residual soil and the stabilizing agent xanthan gum and guar gum used which are commercially available in market.

3 Methods

Different geotechnical engineering properties like grain size, liquid limit, plastic limit, compaction characteristic and unconfined compressive strength tests were conducted as per relevant ASTM standards (ASTM D422, ASTM D4318, ASTM D1557 and ASTM D2166). The depressiveness of the soil was studied by double hydrometer test (ASTM D4221-99), crumb test (ASTM D6572-13e2) and cylinder dispersion test (Atkinson et al. 1990). Cylinder dispersion test is an advance process of crumb test, which is consider to observe the characteristics of soil at zero effective stress by immersing a saturated sample in water (Atkinson et al. 1990). Long-term strength and durability property of soil sample was investigated in a specific way as discussed by Al-Kiki et al. (2011). In this process, two sets of stabilized samples were prepared named controlled sample and soaked sample. Particle morphology and mineralogy were studied for dispersive soil to explain some macro properties. The individual morphology and particle chemistry were studied in a Jeol-840-A model scanning electron microscope (SEM) fitted with electron dispersive X-ray (EDX) analyser. The mineralogy of the materials was studied using X-ray diffraction (XRD) analysis on a Rich-Siefert X-ray diffractometer using copper target and Ni filter.

4 Results and Discussion

4.1 Grain Size Distribution, Plasticity and Specific Gravity

Figure 1 Shows the grain size distribution curve of dispersive soil and it can be seen that dispersive soil is fine grained with 95% particles are fine particles (<75 μm). The specific gravity (G) values of the materials are found to be 2.72.

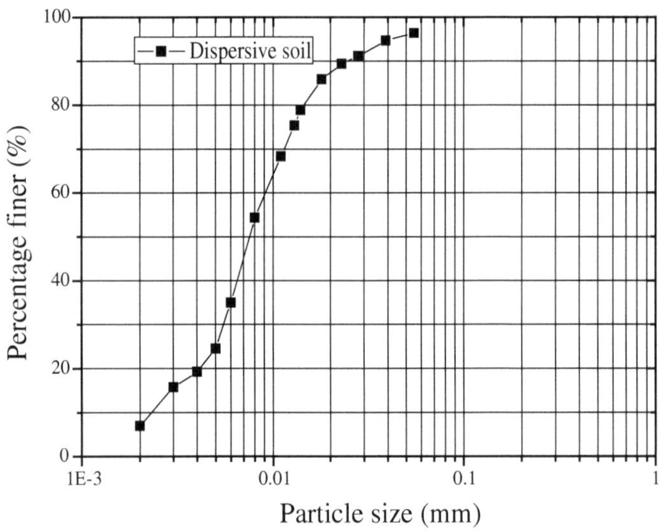

Fig. 1. Grain size distribution of dispersive soil.

The liquid limit and shrinkage limit of dispersive soil were found to be 29% and 20.45%, respectively with plasticity index as 7. Other properties of dispersive soil are shown in Table 1. As per ASTM D2487 classification, the present dispersive soil can be classified as CL-ML (organic clay with high plasticity) or (inorganic silt with low plasticity) category.

Table 1. Properties of dispersive soil

Properties	Value
Specific gravity (G)	2.72
Coefficient of curvature (C_c)	1.46
Coefficient of uniformity (C_u)	0.58
Optimum moisture content (OMC)	15.19%
Maximum dry density (MDD)	17.18 kN/m^3
Liquid limit (LL)	29%
Plastic limit (PL)	22%
Shrinkage limit	20.45%
Plasticity index (PI)	7
Soil classification	CL-ML
(Dispersion ratio) Double hydrometer	89.57%
Differential free swelling	–

4.2 Compaction Characteristics

The heavy compaction (Modified proctor) was conducted as per ASTM D1557 and the maximum dry density (MDD) of dispersive soil was found to be 17.18 kN/m^3, at optimum moisture content (OMC) of 15.19%.

4.3 Dispersion Test

Dispersion ratio of dispersive soil as per double hydrometer test was found to be 89.57%, which is extremely dispersive as per Volk (1937). The crumb test was carried out on above soil and it was observed the soil get disseminated in the water and shows obscure water within a short period. The depressiveness of the soil sample was further investigated by cylinder dispersion test and in every case, the soil sample was found to be highly dispersive.

As the soil was found to be highly dispersive, it needs to be stabilized to decrees its dispersivity and increase the strength and stability. For this purpose, the soil samples were subjected to biopolymer treatment by using two types of biopolymer, i.e. xanthan gum and guar gum in very small quantity. The amount of xanthan gum and guar used to stabilize the soil was 1.0%, 2.0%, 3.0% and 0.5%, 1.0%, 2.0% respectively. The biopolymer in powder form thoroughly mixed with soil and then added required amount of water to mixture. After mixture was prepared it was subjected to various test procedure.

Compaction test were also conducted on dispersive soil and dispersive soil, which was treated with 1.0%, 2.0% and 3.0% of xanthan gum and 0.5%, 1.0% and 2.0% of guar gum. Figure 2 shows MDD and OMC of dispersive soil and dispersive soil treated with different percentage of xanthan gum and guar gum. It is observed from Fig. 2 that the MDD of the bio treated dispersive soils decreases and OMC is increases gradually with increase in concentration of biopolymer as found (Das et al. 2015) but both OMC and MDD are increased with concentration of biopolymer as studied Mahamed et al. (2016).

Dispersive soil treated with 1.0% of xanthan gum shows higher MDD (17.04 kN/m^3) at OMC of 17.18% and lower value of MDD achieved with 3% of xanthan gum is 15.78 kN/m^3 at OMC of 18.38%. In case of guar gum treated dispersive soil, the higher value of MDD was achieved at 0.5% of guar gum is 16.95 kN/m^3 at a OMC of 17.28% and lower value of MDD achieved at 2.0% of guar gum is 16.17 kN/m^3 at a OMC of 17.88%. Crumb test was conducted on dispersive soil and bio treated dispersive soil samples are shown in Fig. 3.

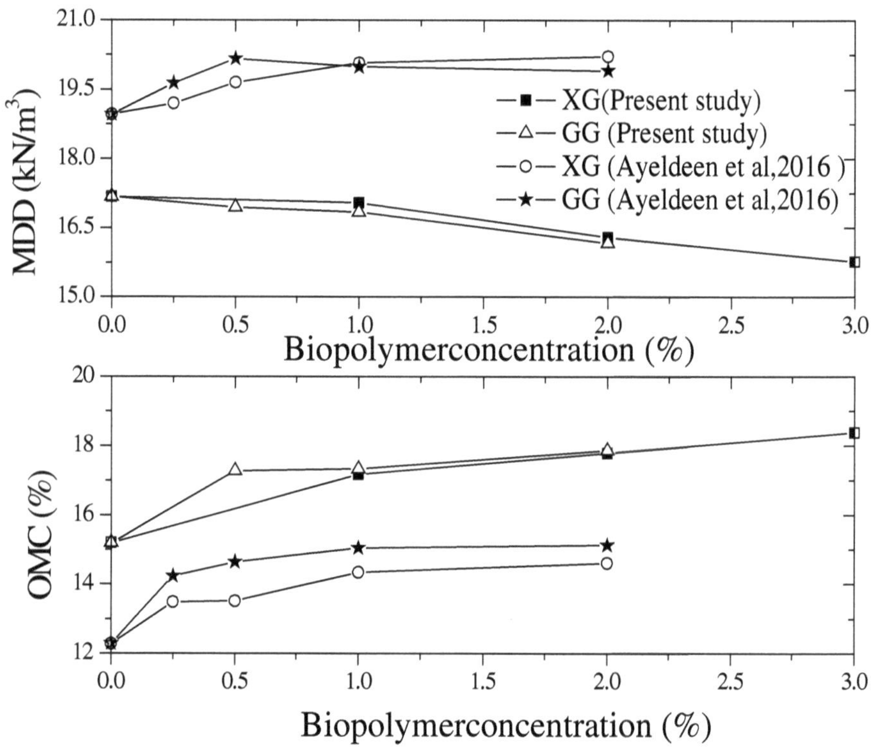

Fig. 2. Compaction characteristics of dispersive soil and dispersive soil treated with different percentage of xanthan gum and guar gum.

Figure 4 shows cylinder dispersion test on dispersive soil and dispersive soil treated with 1.0% of xanthan gum and guar gum. It can be clearly observed that the soil, which is not treated with biopolymer, is dispersed and it formed dark muddy water inside the biker whereas the biopolymer treated soil sample shows clear water around the cylinder. This shows the effectiveness of the xanthan gum and guar gum for stabilization of dispersive soil.

Fig. 3. Crumb test on dispersive soil and dispersive soil treated with different percentage of xanthan gum and guar gum.

Fig. 4. Cylinder dispersion test on dispersive soil and dispersive soil treated with 1.0% of xanthan gum and guar gum.

4.4 Unconfined Compressive Strength Test

To determine the strength of dispersive soil and bio treated dispersive soil unconfined compressive strength (UCS) test were conducted on freshly prepared samples (50 mm diameter and 100 mm height) and the samples after 3 and 7 days of its preparation with curing in normal temperature. The average of three tests results are considered in the

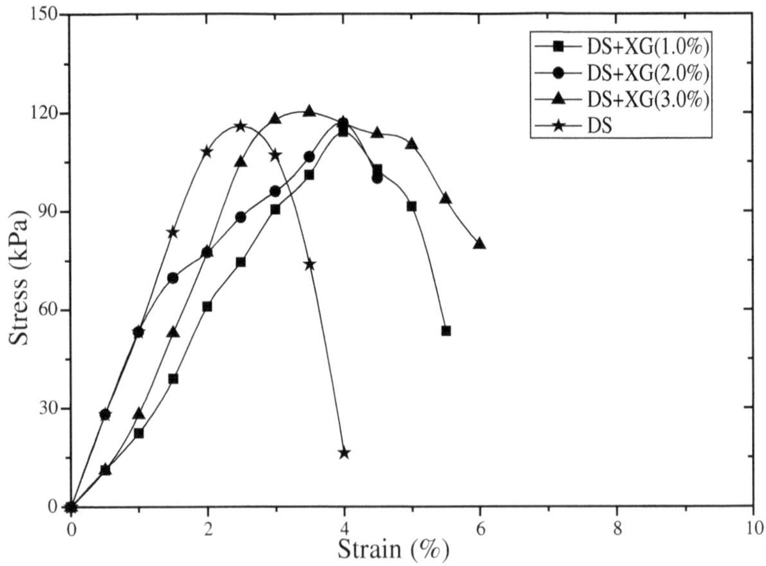

Fig. 5. Unconfined compressive strength curves of dispersive soil sample and dispersive soil sample treated with different percentage of xanthan gum of 3 days curing.

present study. Figure 5 shows stress-strain curves of dispersive soil sample and dispersive soil sample treated with different percentage of xanthan gum after 3 days of curing and it indicated that the sample which is treated with 3% of xanthan gum have higher strength.

Similarly, Fig. 6 shows stress- strain curves of dispersive soil sample and dispersive soil sample treated with different percentage of guar gum of 3 days of curing and maximum strength is found at 2.0% of guar gum. The samples were treated with different percentage of xanthan gum and guar gum and subjected different types of curing condition. Bio treated samples were cured at normal temperature for a period of 3 days, 7 days than it these were subjected to UCS test and another set of bio treated samples were subjected to sun light, allowed to dried for a period of 3 days after that UCS test were conducted which is shown in Figs. 7 and 8. A comparison study on UCS values of different samples is shown in Fig. 9, which specifies the UCS value of dispersive soil and bio treated dispersive soil steadily increases with curing period of 3 days and decreases 7 days than freshly prepared samples and found considerably much higher strength at sundried samples at curing period of 3 days.

4.5 Long-Term Strength with Soaking Test

The durability of bio treated samples was investigated using soaking test as discussed by Al-Kiki et al. (2011). For this process, two sets of stabilized samples were prepared named controlled sample and soaked sample. Cylindrical samples of dispersive soil and bio treated dispersive soil of size (50 mm diameter x 100 mm height) were prepared

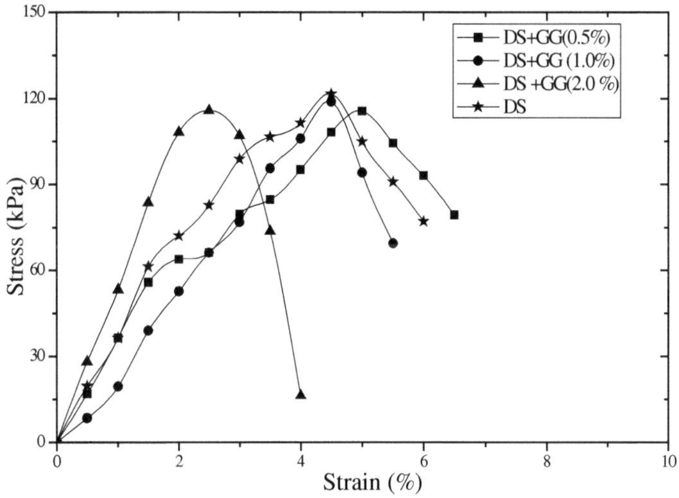

Fig. 6. Unconfined compressive strength curves of dispersive soil sample and dispersive soil sample treated with different percentage of guar gum of 3 days curing.

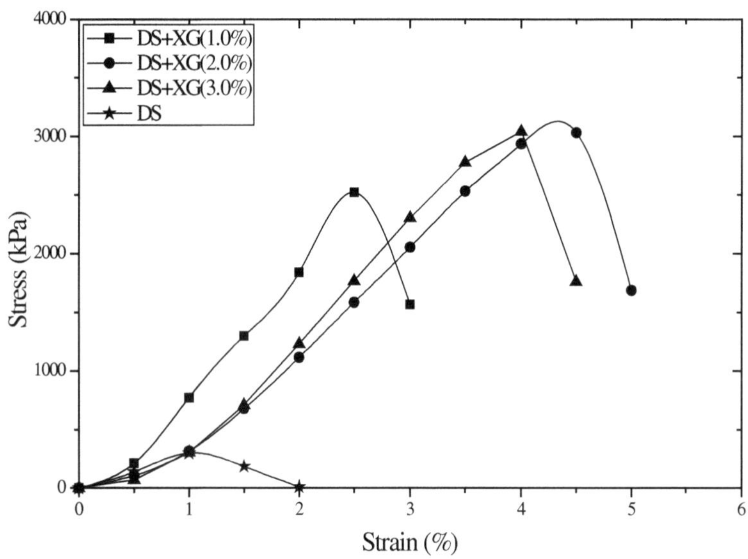

Fig. 7. Unconfined compressive strength curves of dispersive soil sample and dispersive soil sample treated with different percentage of xanthan gum of 3 days curing at sun light.

and cured for 2 days at 49 °C. At the end of the curing period one set of sample were submerged in water considered as soaked sample over a period of 3 days, at the end of the soaking period samples were subjected to UCS test. The second set of samples were subjected to different curing conditions at normal temperature (25 °C ± 3°) and

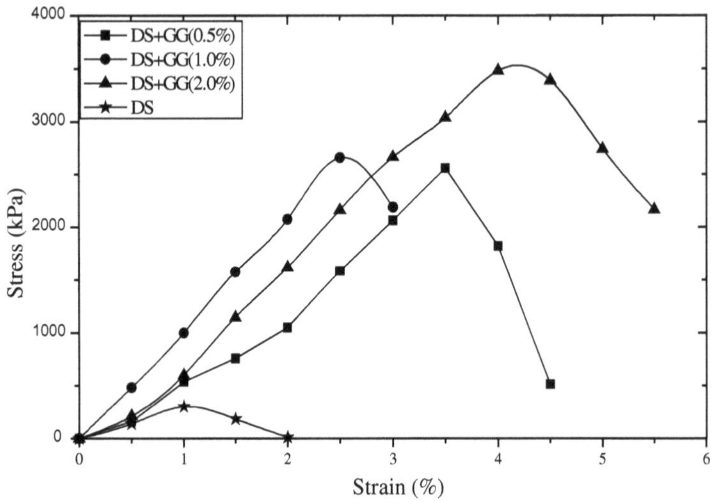

Fig. 8. Unconfined compressive strength curves of dispersive soil sample and dispersive soil sample treated with different percentage of guar gum of 3 days curing at sun light.

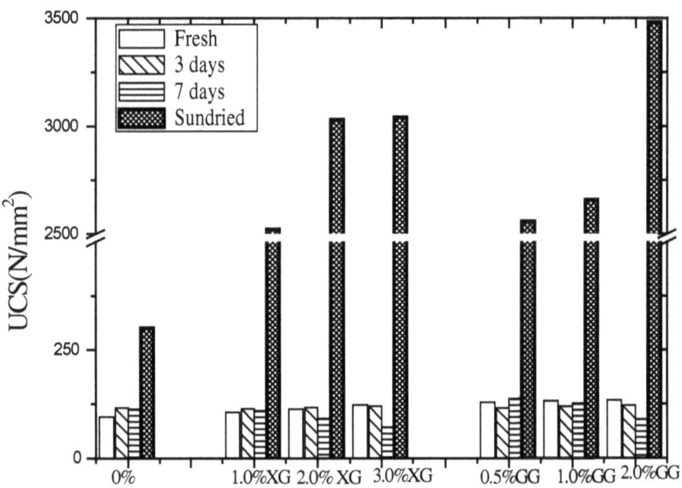

Fig. 9. Unconfined compressive strength curves of value of different soil.

considered as controlled samples with curing period same as soaking period of first set of sample and controlled samples were subjected to immerse in water for a period of 2 days before testing. The resistance to loss in strength was determined as a ratio of UCS of soaked sample to the UCS of controlled sample.

Figures 10, 11, 12 and 13 shows stress-strain curve of bio treated dispersive soil with different percentage of xanthan gum and guar gum for soaked samples and controlled samples. The soaked samples found to have better strength than the

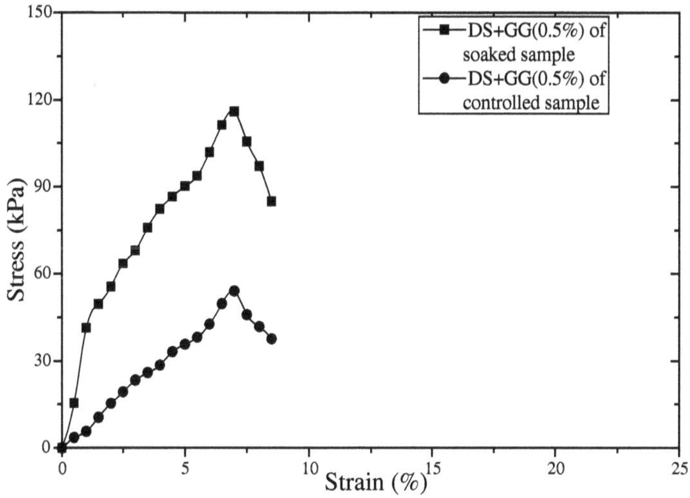

Fig. 10. Unconfined compressive strength curves of dispersive soil sample treated with 0.5% of guar gum of soaked and controlled method.

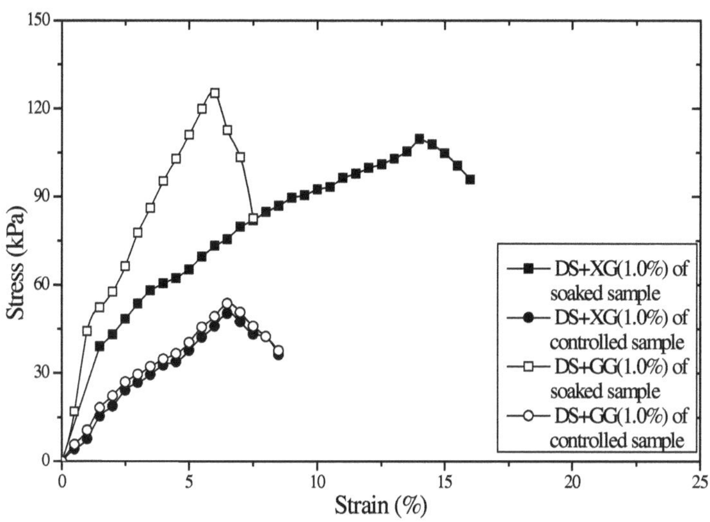

Fig. 11. Unconfined compressive strength curves of dispersive soil sample treated with 1.0% of xanthan gum and guar gum of soaked and controlled method.

controlled samples. A comparison study on resistance to loss in strength (RLS) for dispersive soil sample and bio treated dispersive soil samples with xanthan gum (1.0%, 2.0% and 3.0%) and guar gum (0.5%, 1.0% and 2.0%) are shown in Table 2. It can be seen that dispersive soils have no durability as it loses its shear strength and are dissolved when it comes in contact with water. Whereas, biopolymer modified samples

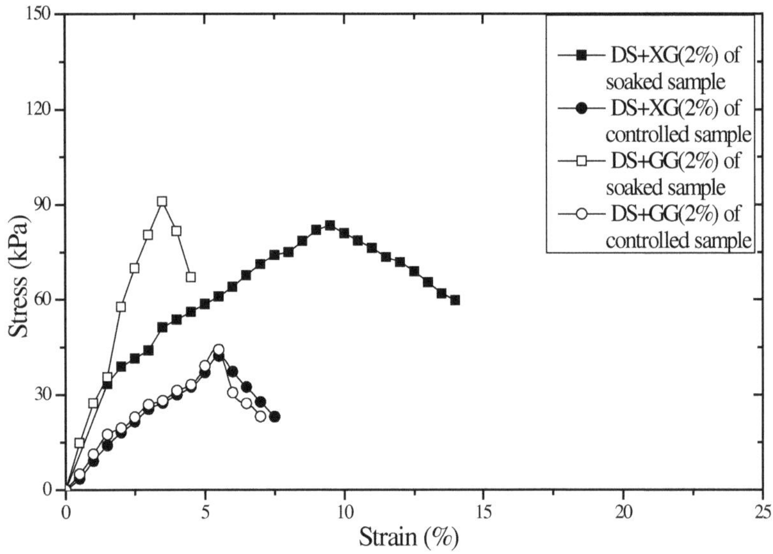

Fig. 12. Unconfined compressive strength curves of dispersive soil sample treated with 2.0% of xanthan gum and guar gum of soaked and controlled method.

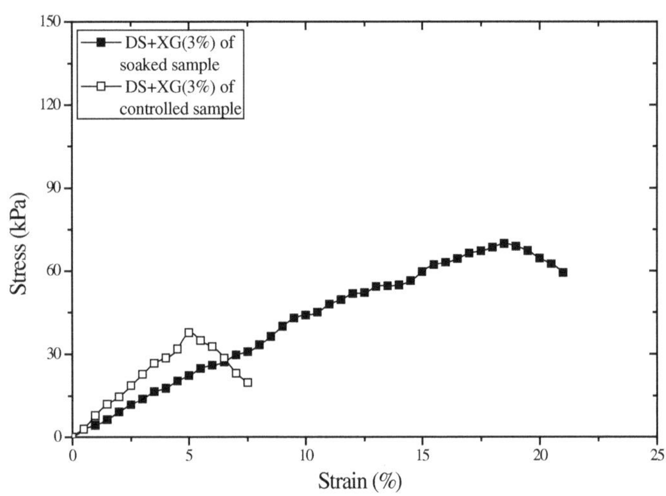

Fig. 13. Unconfined compressive strength curves of dispersive soil sample treated with 3% of xanthan gum of soaked and controlled method.

Table 2. Comparison of resistance to loss in strength for dispersive soil and bio treated dispersive soil

Sample name	UCS of Controlled Sample (UCSC) (kPa)	UCS of Soaked Sample (UCSS) (kPa)	Increase/Decrease in Strength from Controlled Sample (CS) to Soaked Sample (SS)	Resistance to Loss in Strength (RLS = UCSS/UCSC)
DS + XG (1%)	50.37	109.73	Increase	+2.18
DS + XG (2%)	42.20	83.40	Increase	+1.98
DS + XG (3%)	37.71	69.90	Increase	+1.85
DS + GG (0.5%)	54.06	116.02	Increase	+2.15
DS + GG (1%)	53.68	125.27	Increase	+2.33
DS + GG (2%)	44.21	90.98	Increase	+2.06

(a)

(b)

(c)

Fig. 14. Scanning electron micrograph of (a) dispersive soil (b) dispersive soil treated with 1% xanthan gum(c) dispersive soil treated with 1% guar gum.

showed an increase in stress and RLS was found to be maximum at 1.0% guar gum treated samples and minimum at 3.0% of xanthan gum treated samples.

4.6 Morphology of the Material

Scanning electron micrograph (Fig. 14a) shows that dispersive soil consists of is mostly angular and flaky particles of varying size. Figure 14(b) and (c) shows the dispersive soil treated with 1.0% xanthan gum and 1.0% guar gum, respectively. It can be observed that the bio treated materials got denser state by application of gum due to bonding between soil particles with the biopolymer and leads to denser structure, which may cause to increase in surface strength, stability and erosion resistance of dispersive soil in ecofriendly manner.

Figure 15(a) shows EDX pattern of dispersive soil and Fig. 15(b) and (c) shows the EDX patterns of dispersive soils treated with 1.0% of xanthan and 1.0% of guar gum, respectively. Table 3 shows elemental composition of dispersive soil from EDX test.

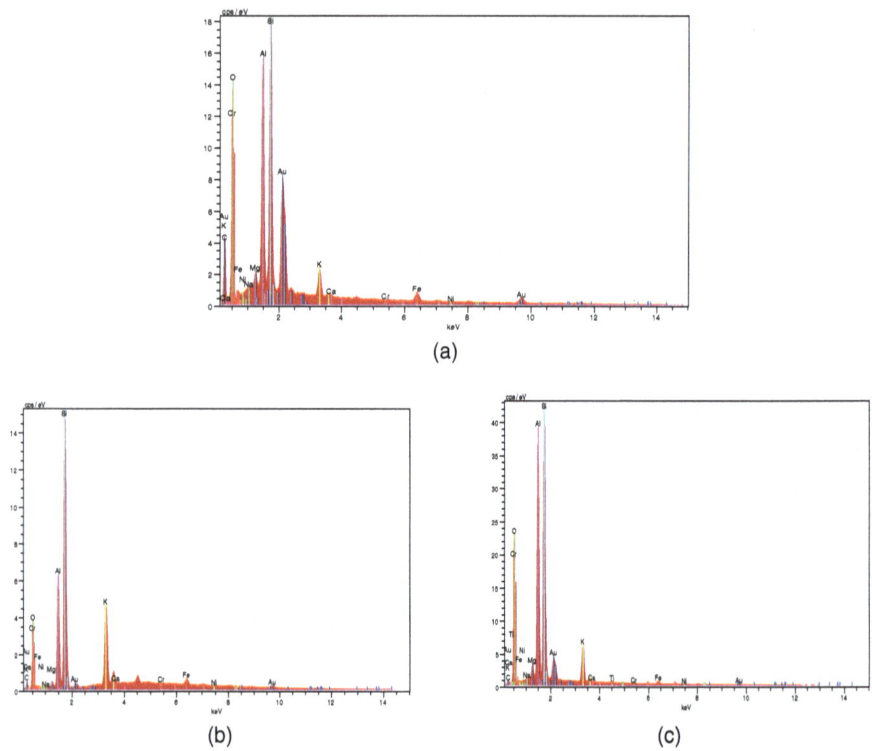

Fig. 15. EDX analysis of (a) dispersive soil (b) dispersive soil treated with 1% of xanthan gum (c) dispersive soil treated with 1% of guar gum.

Table 3. Elemental composition of dispersive soil from EDX test.

Element	Atomic number	Normalized percentage in the geomaterial
O	8	37.78%
C	6	21.31%
Si	14	17.66%
Al	13	12.98%
Fe	26	4.53%
K	19	3.50%
Mg	12	0.92%
Ni	28	0.70%
Ca	20	0.35%
Na	11	0.20%
Cr	24	0.07%
Au	79	0.00%

4.7 Mineralogy of the Material

Based on XRD test, presented in Fig. 16 it was observed that dispersive soil contains quartz and aluminum silicate and was found in crystalline phase and no difference was observed between in XRD pattern of dispersive soil and bio treated dispersive soils.

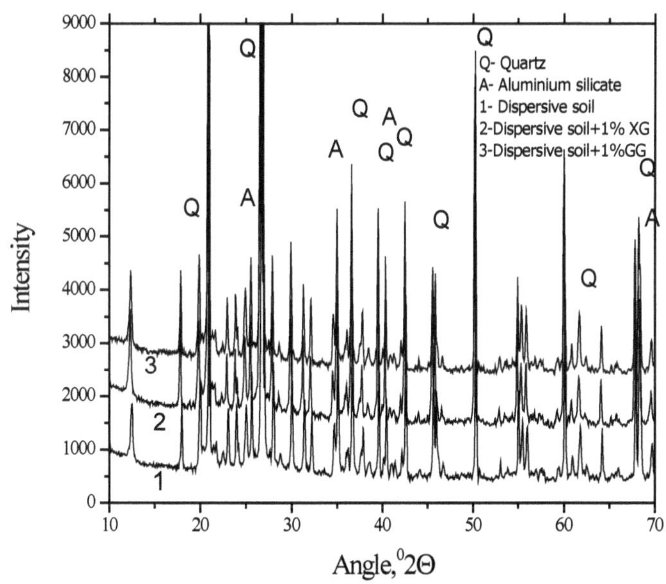

Fig. 16. XRD analysis of dispersive soil and dispersive soil treated with 1% of xanthan gum and guar gum.

5 Conclusions

In the present study, two biopolymers xanthan gum and guar gum are used to stabilize dispersive soil. The following conclusions can be drawn.

1. As per crumb test and double hydrometer test the present soil was found to be highly dispersive and failed rapidly during soaking.
2. The xanthan and guar gum were found to be effective to stabilize the dispersive soil.
3. The OMC of the bio treated samples increased and MDD decreased with increase in percentage of biopolymer.
4. The UCS of guar gum samples shows higher value than xanthan gum samples. The bio treated samples cured in sun light for a period of 3 days shows higher UCS value compared to as compacted sample and cured under ambient condition.
5. The resistance to loss in strength (RLS) value of xanthan gum treated samples gradually decreased with increase in percentage of biopolymer whereas, there is no certain pattern of RLS value in case of guar gum treated samples.
6. The micro structural changes studied by SEM analyses clearly show that the minerals of biopolymer modified dispersive soil, are bonded together by gum layer, which decreases the dispersivity of the soil.
7. The mineralogical study on bio treated samples shows no change in XRD pattern.

References

Abbasi, N., Nazifi, M.H.: Assessment and Modification of Sherard chemical method for evaluation of dispersion potential of soils. Geotech. Geol. Eng. **31**, 337–346 (2013)

Al-Kiki, I.M., et al.: Long term strength and durability of clayey soil stabilized with lime. Eng. Tech. J. **29**(4), 725–735 (2011)

ASTM D 422- 63. Standard Test Method for Particle-Size Analysis of Soils. ASTM International, West Conshohocken, PA (2007)

ASTM D1557-12e1. Standard Test Methods for Laboratory Compaction Characteristics of Soil Using Modified Effort (56,000 ft-lbf/ft^3 (2,700 kN-m/m^3)), ASTM International, West Conshohocken, PA (2012)

ASTM D2166/D2166 M-16. Standard Test Method for Unconfined Compressive Strength of Cohesive Soil, ASTM International, West Conshohocken, PA (2016)

ASTM D2487-11. Standard Practice for Classification of Soils for Engineering Purposes (Unified Soil Classification System), ASTM International, West Conshohocken, PA (2011)

ASTM D4221-99. Standard Test Method for Dispersive Characteristics of Clay Soil by Double Hydrometer, ASTM International, West Conshohocken, PA (1999)

ASTM D4318-10e1. Standard Test Methods for Liquid Limit, Plastic Limit, and Plasticity Index of Soils, ASTM International, West Conshohocken, PA (2010)

ASTM D6572-13e2. Standard Test Methods for Determining Dispersive Characteristics of Clayey Soils by the Crumb Test, ASTM International, West Conshohocken, PA (2013)

ASTM D854-14. Standard Test Methods for Specific Gravity of Soil Solids by Water Pycnometer, ASTM International, West Conshohocken, PA (2014)

Atkinson, J.H., et al.: Examination of erosion resistance of clays in embankment dams. Q. J. Eng. Geol. **23**, 103–108 (1990). London

Ayeldeen, M.K., et al.: Evaluating the physical characteristics of biopolymer/soil mixtures. Arab. J. Geosci. **9**, 371 (2016). doi:10.1007/s12517-016-2366-1

Bhuvaneshwari, S., et al.: Stabilization and microstructural modification of dispersive clayey soils. In: First International Conference on Soil and Rock Engineering, Colombo, Sri Lanka, 5–11 August 2007

Chen, R., et al.: Biopolymer stabilization of mine tailings. J. Geotech. Geoenviron. Eng. **139**(10), 1802–1807 (2013)

Das, S.K., et al.: Stabilization of pond ash using biopolymer. Procedia Earth Planet. Sci. **11**, 254–259 (2015)

Fell, R., et al.: Geotechnical Engineering of Embankment Dams, p. 675. A.A. Balkema Publishers, Brookfield (1992)

Goodarzi, A.R.: Pore fluid effect on the dispersivitybehaviour and its interaction with alum as an improving material with an attention to microstructure aspects. MSc thesis, Civil Engineering Department, University of Bu-Ali Sina, Iran, p. 178 (2003)

Goodarzi, A.R., Salimi, M.: Stabilization treatment of a dispersive clayey soil using granulated blast furnace slag and basic oxygen furnace slag. Appl. Clay Sci. **108**, 61–69 (2015)

Ivanov, V., Chu, J.: Applications of microorganisms to geotechnical engineering for biocloguar guming and biocementation of soil in situ. Rev. Environ. Sci. Biotechnol. **7**(2), 139–153 (2008)

Ludwig, H.: A study of some aspects of disperse clay particle interaction. Master thesis, McGill University, p. 201 (1979)

Mitchell, J.K.: Fundamentals of Soil Behavior, p. 437. Wiley, New York (1993)

Nevels, J.J.B.: Dispersive clay embankment erosion: a case history. Transp. Res. Rec. **1406**, 50–57 (1993)

Ouhadi, V.R., Goodarzi, A.R.: Assessment of the stability of a dispersive soil treated by alum. Eng. Geol. **85**, 91–101 (2006)

Ouhadi, V.R., Goodarzi, A.R.: Pore fluid characteristics effect on the dispersivity behavior of soils from macro and micro structure aspects. In: Proceedings of the 2nd International Symposium on Contaminated Sediments, Quebec City, Canada, pp. 200–206 (2003)

Sherad, J.L., et al.: Identification and nature of dispersive soils. J. Geotech. Eng. ASCE **102**, 298–312 (1976)

Tin, J.E.: Waste management schemes of potash mine in Saskatechwan. MSc thesis, University of Saskatchewan, Canada (1984)

Turkoz, M., et al.: The effect of magnesium chloride solution on the engineering properties of clay soil with expansive and dispersive characteristics. Appl. Clay Sci. **101**, 1–9 (2014)

Umesh, T.S., et al.: Control of dispersivity of soil using lime and cement. Mater. Sci. Appl. **3**, 8–15 (2009)

Vakili, A.H., et.al.: Stabilization of dispersive soils by pozzolan. Forensic Eng. 726–735 (2013)

Volk, G.M.: Method of determination of degree of dispersion of clay fraction of soils. In: Proceedings of Soil Science Society of America, vol. 2, p. 561 (1937)

Yong, R.N., Sethi, A.J.: Turbidity and zeta potential measurements of clay dispersibility. ASTM STP **623**, 419–431 (1977)

Yong, R.N., Warkentin, B.P.: Introduction to Soil Behaviour, p. 451. MacMillan, New York (1966)

Study on the Effects of the Rheological Properties for Flow Behaviour of Assam Crude Oil with Various Chemicals

Shilpi Sarmah(✉), Subrata Borgohain Gogoi, Joyshree Barman, and Debashree Dutta

Department of Petroleum Technology, Dibrugarh University, Dibrugarh, India

Abstract. The crude oil exhibit Non-Newtonian flow behaviour which results in serious problems related to its transportation from oil fields to refineries. The rheological behaviours of different Assam crude oil namely Jorajan, Bhogpara, Lakwa and Kothaloni with or without surfactants were studied in the present work. The crude oil mixtures for the study were prepared by blending with kerosene. Anionic surfactant BL (Black Liquor) and Chrome free lignosulphonate and non-ionic surfactant TX100 (Triton X100) were used. Rheological experiments were carried out by Grace M3600 viscometer. The experimental data's was interpreted with the model of Ostwald-de Waele power-law and Bingham Plastic with the entire range of data at 40 °C and 80 °C. Hence the required shear stress or yield stress was obtained for the crude oil mixtures.

Effect of several parameters like shear rate, temperature, density and rpm and addition of kerosene on viscous behaviours had been studied. Crude oil of Kothaloni oil field having API gravity of 36^0 was mixed with CMC values of different surfactants which are 0.6% Black Liquor, 0.04% TX100, 4% of 1 Molar Chrome free lignosulphonate and 5% Kerosene. On investigation it was observed that there was a significant decrease in dynamic viscosity and increase in shear stress with increase in shear rate after adding non ionic surfactant TX-100, anionic surfactants Black liquor and liqnosulphonate as compared with the pure crude oil. Further, there was significant decrease in dynamic viscosity when 5% kerosene was added with the surfactants. The experimental results showed that viscosity of an anionic surfactant Black liquor when mixed with crude oil decreased by almost 19.10% on addition of 0.6%wt/v of the surfactant at shear rate 1021.3 s^{-1} and at a temperature of 40 °C. The purpose was to observe the viscosity reduction at a maximum speed of 600 rpm and at room temperature. Shear stress of the above mixture was also reduced by almost 15.38%. Viscosity of the mixture was further reduced by almost 70.78% on addition of 5% kerosene to the above mixture. Again, by adding different anionic surfactant chrome free lignosulphonate (4% wt/v) to crude oil a viscosity reduction of 16.85% and shear stress reduction of 14.28% was observed. Viscosity was further reduced to 68.53% and shear stress to 68.13% when kerosene was added to the above mixture. Another non-ionic surfactant TX-100 resulted in a viscosity and shear stress reduction of 12.34% and 12.08% respectively. Further addition of kerosene reduced its viscosity to 65.17% and shear stress 64.83%. It was observed that black liquor generally exhibited better reduction of viscosity than lignosulphonate. Also by mixing two crudes of different API (one lighter, Jorajan 5

D.N. Singh and A. Galaa (eds.), *Contemporary Issues in Geoenvironmental Engineering*, Sustainable Civil Infrastructures,
DOI 10.1007/978-3-319-61612-4_12

and the other heavier, Kothaloni) a reduction of 66.29% viscosity and 65.93% shear stress was observed. This was solely done in order to show that when two pipes carrying crude of different API from different well heads, the crudes gets mixed and flows through one pipe towards the same oil collecting stations, thus facilitating the flow of heavy crude by reducing its specific gravity and enhancing the property of heavier crude oil.

Keywords: Assam crude · Non-Newtonian · Thixotropy · Yield stress · Shear stress · Shear rate · Specific gravity · API and viscosity

1 Introduction

Rheology plays a very important role in the petroleum industry. The focus of this paper is on crude oil rheology related to production and transportation. Crude oil production and its transportation to the processing units are of utmost importance in the petroleum industry. Moreover, due to complex nature of crude oil, there are varieties of difficulties in production, refining and transportation. The main characteristics of viscous oils are discussed to highlight the oil properties that affect their flowability in the processes of recovery and pipeline transportation [1–4]. Hence different methods are used in order to reduce viscosity of crude oil and thereby enhancing flow ability of crude oil through pipeline during the transportation process for mixed crude oil of Assam. These methods although prevalent everywhere, our study was limited to the crude oil samples of Upper Assam Basin.

The crude oil of Assam is mix crude which is highly paraffinic and has high wax content. These result in poor flow behaviour at low temperature. Due to high wax content, it has high pour point and hence the crude oils behave as Non-Newtonian fluid at temperatures below pour point. The wax can crystallize as the crude is cooled to form a gel or a partial gel. Under static conditions a rigid gel is formed, but if the crude is cooled while in motion, the apparent viscosity will increase but the materials remain fluid. Therefore, the rheological properties are functions of temperature, shear rate and shear stress using the application of which experiments were conducted to understand the behaviour of crude oil of Upper Assam Basin.

Rheology is the study of the change in form and flow of matter in term of elasticity, viscosity and plasticity. A clear understanding of the rheological properties of waxy crude is vital in the design and construction of petroleum pipelines and the subsurface and surface production equipment in the oil industry. The rheological behaviour of waxy crude oil is thus considered to have crucial importance in the design of pipeline, flow handling equipment and the processing purposes.

Reduction of viscosity can be achieved by heating, blending heavy oils with lighter oils as well as oil-in-water emulsification. Heating often requires considerable amount of energy and there are some logistic problems in using diluents. Though heating of crude oil, diluents in the form of lighter hydrocarbons and kerosene also resulted in viscosity reduction, there were some disadvantages. Usually, heating has been used for reducing the viscosity, but it has the disadvantage of significant equipment cost and fuel consumption as well as difficulty in restarting after shutdown [4]. It is also considered inconvenient in transporting crude oil in colder region, remote region or underwater transfer [6]. For field production to be transported by pipeline using a diluents, two pipelines would be required that is one for the oil and other for the diluents. It would thus be desirable to develop an alternative method for transportation. A possible alternative is to use surfactants to emulsify the heavy crude oil in water and to transport the low viscosity emulsion in a pipeline [3].

Considering the above factors, the rheological properties of the crude oil was studied by adding three surfactants Triton X-100, Black liquor and Chrome free lignosulphonate which resulted in significant decrease in viscosity of crude oil by reducing the interfacial tension.

2 Experimental Setup

2.1 Materials

Samples specifications are in Table 1 and surfactant specification are in Table 2.

Table 1. Sample specification

Serial no.	Samples	Specific gravity at 35 °C	API at 35 °C
1	Bhogpara	0.894	27
2	Nahorkatiya	0.896	26
3	Lakwa	0.863	32
4	Jorajan	0.955	17
5	Jorajan5	0.826	40
6	Kothaloni	0.844	36
7	Kerosene	0.820	41

Table 2. Surfactant specification

Sl. no.	Surfactants	Type of surfactants	Molecular formula	Viscosity (cp) at (40 °C)	Molecular weight, gm/mol	Source
1	Black Liquor (Sodium – lignosulphonate)	Anionic	$C_{10}H_{14}N_2Na_2O_82H_2O$	2.3	372	Jagiroad Paper Mill.
2	TX-100	Nonionic	$C_{34}H_{62}O_{11}$	–	646.87	HiMedia Laboratories Pvt. Ltd.
3	Chrome free lignosulphonate	Anionic	$C_{20}H_{24}Na_2O_{10}S_2$	2.9	534	Oil India Limited.

2.2 Methods

2.2.1 Determination of Kinematic Viscosity

For the determination of kinematic viscosity the samples were introduced into the viscometer in an inverted fashion which was then immersed into the constant temperature bath, of model no. CT-500 series II made in USA. Each sample was brought to the etched line of the viscometer. The viscometer with the sample was inserted in the constant temperature bath. The sample was allowed to reach equilibrium at the desired temperature and the efflux time was measured by allowing the sample to flow freely between the upper and lower levels of the bulb. The kinematic viscosity was then calculated by multiplying the efflux time by the viscometer constant as per the recommended viscosity ranges for the Cannon-Fenske routine viscometers.

Every Cannon-Fenske tube has its own calibrated constant, which is multiplied by the recorded time in second to obtain kinematic viscosity in cSt, that is:

$$\text{Kinematic viscosity, } V = k * t \tag{1}$$

Where,
k = viscometer constant in cSt/sec.
t = time in secs

2.2.2 Determination of Specific Gravity

For the determination of Specific gravity, the samples were introduced into the measuring cylinders which were then immersed into the constant temperature bath, of model no. CT-500 series II made in USA. Specific gravity was then measured using Specific gravity hydrometer at 20 °C of ranges 800 to 900 and 1000 to 1100 when the required temperature was reached.

2.2.3 Determination of Shear Stress and Dynamic Viscosity

A welcome message is first displayed on the screen of the Grace Instrument M3600-1 viscometer followed by the main menu. By choosing the Test Setup screen, the parameters for the time, temperature and rotational speed are entered. After the test steps have been created, the test is started and the results are obtained on the screen of the instrument which displays the dynamic viscosity in cp and shear stress in dyn/cm^2 respectively against the parameters entered. Figure 1 shows the test setup screen of the instrument.

2.2.4 Calculation of Shear Rate

The shear rate can also be calculated at a distance R from the axis of bob rotation inside the annular gap

Shear rate [31]

$$\gamma = \frac{2\Omega R_1^2 R_2^2}{R^2 X_0^2 (R_1 + R_2)} \tag{2}$$

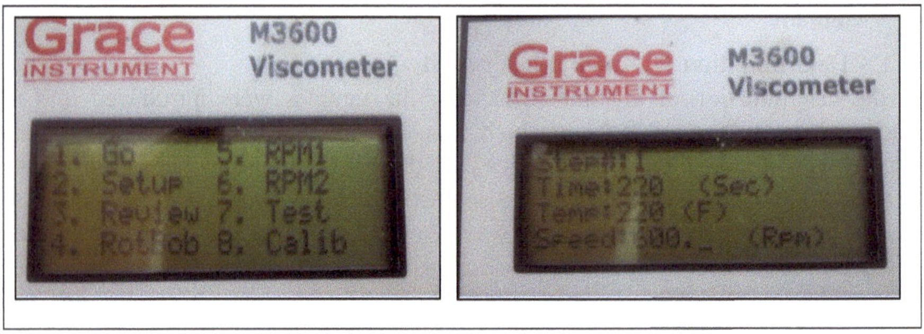

Fig. 1. Test set up screen diagram of the grace M3600 viscometer

Where,

$$\Omega = \frac{RPM}{9.55} = \text{angular bob velocity}$$

$$R_1 = \text{ radius of bob}$$

$$R_2 = \text{radius of cup}$$

$$X_0 = \text{width of the annular gap}$$

The rotor bob assembly that we used in calculating shear rate of the crude oil was as per the measurement range given in Grace Instrument M3600 Viscometer operation manual [32].

Where,

$$R_1 = 1.7245\,\text{cm},$$

$$R_2 = 1.8415\,\text{cm and}$$

$$X_0 = 0.1170\,\text{cm}. \quad (.)$$

3 Results and Discussion

3.1 Viscosity Behaviour of Different Crude Oil Samples Collected from Upper Assam Basin

3.1.1 Effects of Density, Temperature and API on Viscosity

The relation of density, temperature, shear rate and rpm with viscosity were plotted in Figs. 2 to 5 respectively to investigate their effects on the flow behaviour of crude oil at different temperatures (°C). It was observed that all the crude oil samples followed the Non-Newtonian behaviour.

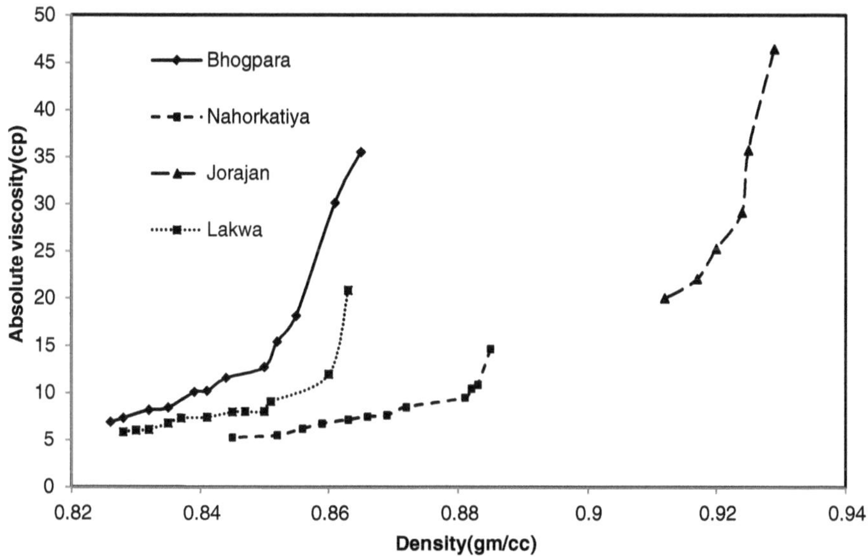

Fig. 2. Variation of viscosity with density for different crude oil sample.

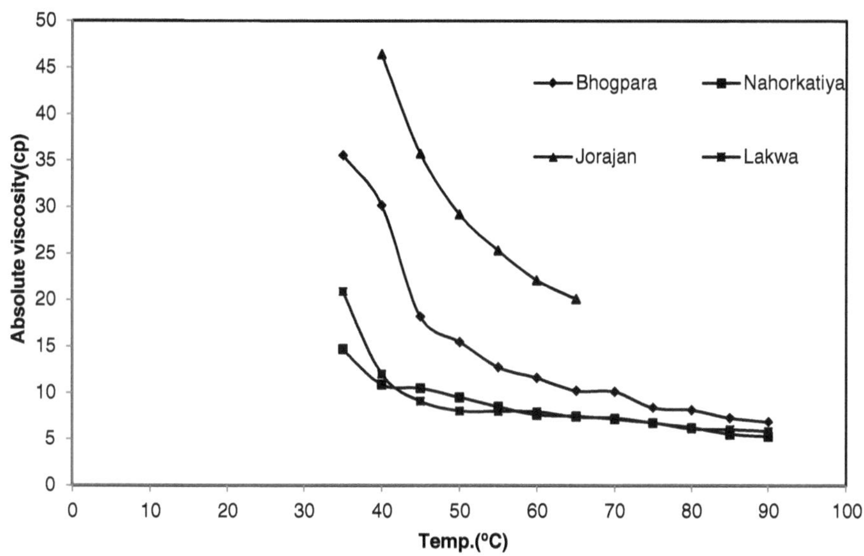

Fig. 3. Variation of viscosity with temperature for different crude oil samples.

There is a direct relationship between the viscosity and density of the fluid. The graph in Fig. 2 shows a useful correlation between density and absolute viscosity of different crude oil samples at different temperatures which showed decrease in viscosity with subsequent decrease in density as the temperature was increased [5].

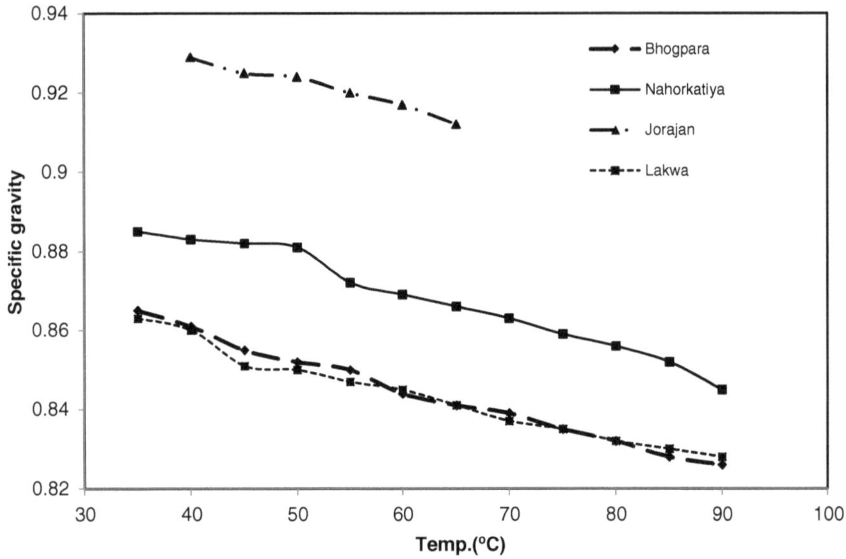

Fig. 4. Variation of specific gravity with temperature for different crude oil samples.

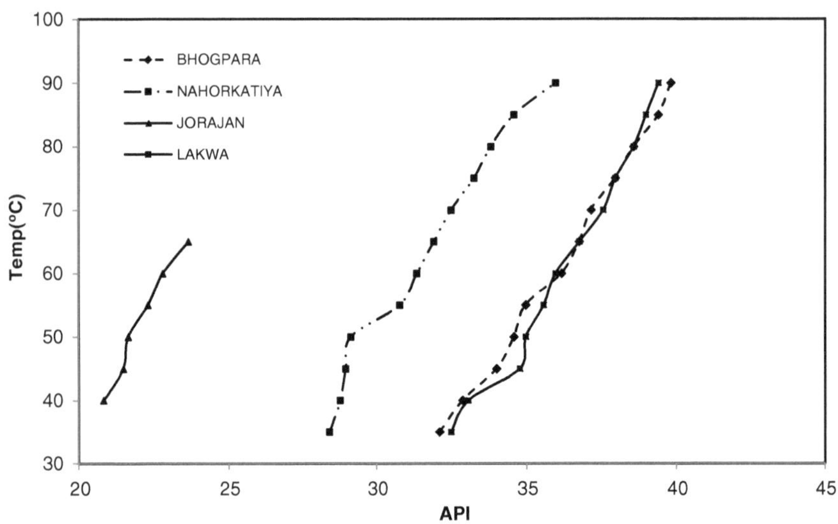

Fig. 5. Variation of temperature with API for different crude oil sample.

The relation between Absolute viscosity and Temperature were found to show an inverse relation with one another. Figure 3 shows that with increase in temperature there is a decrease in viscosity for different crude oil sample. It is further observed that at a temperature of around 45 °C, there is a significant decline in viscosity. Heating the oil at a temperature of 45 °C thus can facilitate in the surface transportation of crude oil thus facilitating smooth flow of crude oils through pipelines. However, the viscosity almost remains constant beyond 60 °C. Maintaining the temperature of the jackets above 60 °C would prove to be uneconomic.

It is well known that as we increase the temperature, the specific gravity of crude oil starts decreasing. The Fig. 4 shows the characteristic trend of four crude oil samples collected from four different wells having different specific gravity. These experiments were done to examine the change of specific gravity at higher temperature. Although, this method of heating process is prevalent for transportation of crude oil yet it offers hindrance as discussed earlier.

The direct variation of API and Temperature is depicted in Fig. 5. With increasing temperature there is an increase in API for different crude oil samples.

3.1.2 Effect of Shear Rate and Shear Stress on Viscosity

On investigating the relation between shear rate and shear stress of crude samples like Jorajan and Lakwa it was found that Jorajan being time dependent was thixotropic in nature whereas Lakwa was found to be pseudoplastic. It was observed from Fig. 6 that Jorajan crude oil sample showed thixotropic behavior but as the temperature increases it starts behaving as a Newtonian fluid. Again from Fig. 7 it was seen that the crude oil sample from Lakwa exhibited Pseudoplastic behaviour.

3.1.3 Effect of Shear Rate and Temperature on Viscosity

The behaviour of absolute viscosity for crude oil samples from Jorajan and Lakwa were investigated under the effects of shear rate at different temperatures as presented in Fig. 8. In general, over a range of shear rate from 170.2 1/s to 1021.3 1/s, the viscosity

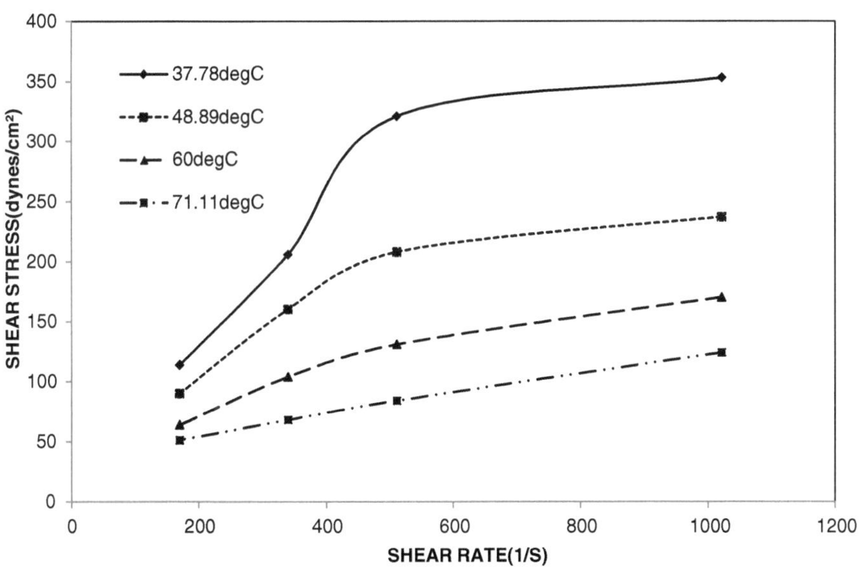

Fig. 6. Non newtonian behavior of a time dependent 17° API upper assam basin crude oil (Jorajan)

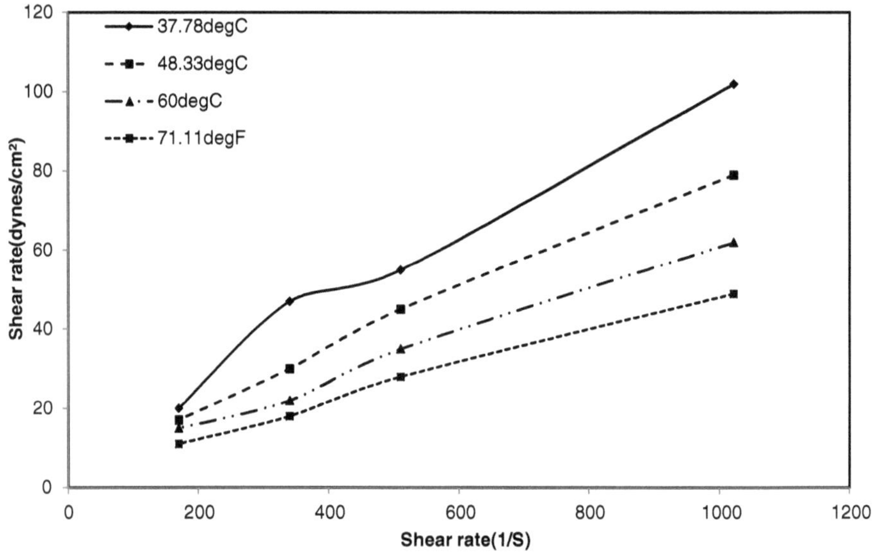

Fig. 7. Non newtonian behaviour of a time independent 32° API upper assam basin crude oil (Lakwa)

was significantly decreased as the shear rate was increased for both the samples. We generally carried out our study at four rpm's i.e. 100, 200, 300 and 600 and thereby manually calculated the shear rates at the four rpm's respectively. Sample from Lakwa was found to be depended on shear rate and followed shear thinning behaviour (pseudo-plastic behaviour) whereas the sample from Jorajan was also found to be depended on shear rate but was thixotropic in nature as we observed a decrease of viscosity with the time of shear [11]. The reason for this behaviour is that, the moment the spindle of the viscometer starts moving, the bonds of the aggregated molecules in the samples starts to get destroyed and deformed and the molecules gets oriented more parallel to the spindle surface. Thus the resistance to spindle rotation (shear rate) is less and therefore viscosity. As the shear rate is increased, the aggregates may be more deformed, breaking down into individual flocks thus further reducing the viscosity.

Since temperature is one of the most important factors which affects the rheological behaviour of crude oil, a small variation will result in significant change in viscosity. Therefore the effect of temperature was investigated as shown in the Fig. 8 below. It was found that with increase in temperature there was a considerable decrease in viscosity.

It is noted that crude oil viscosity decreases considerably with temperature and supports the exponential Arrhenius type relationship between the viscosity and the temperature. It is also noted that Newtonian behaviour occurs at relatively high values of shear rates but with decrease in temperature this becomes limited. The shear thinning behaviour or the pseudoplasticity appears at lower shear rate and becomes more pronounced as the temperature increases [3].

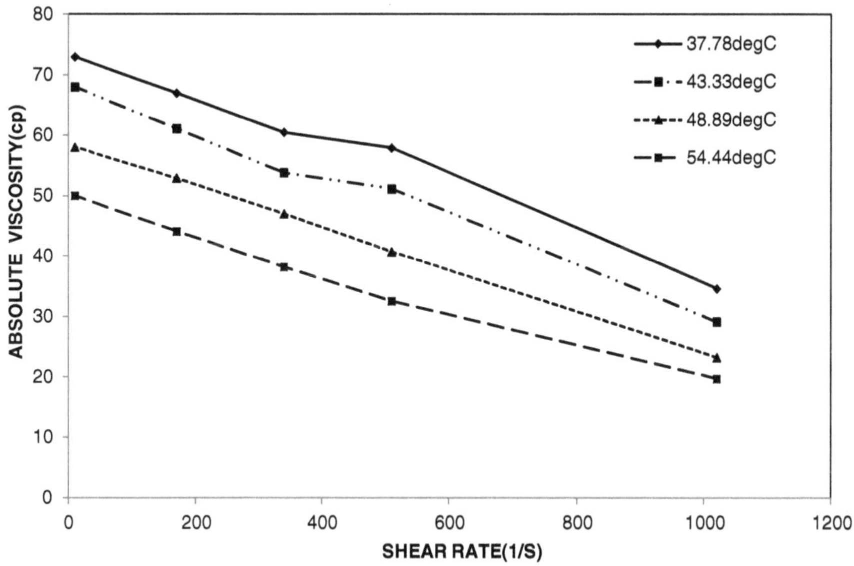

(a)

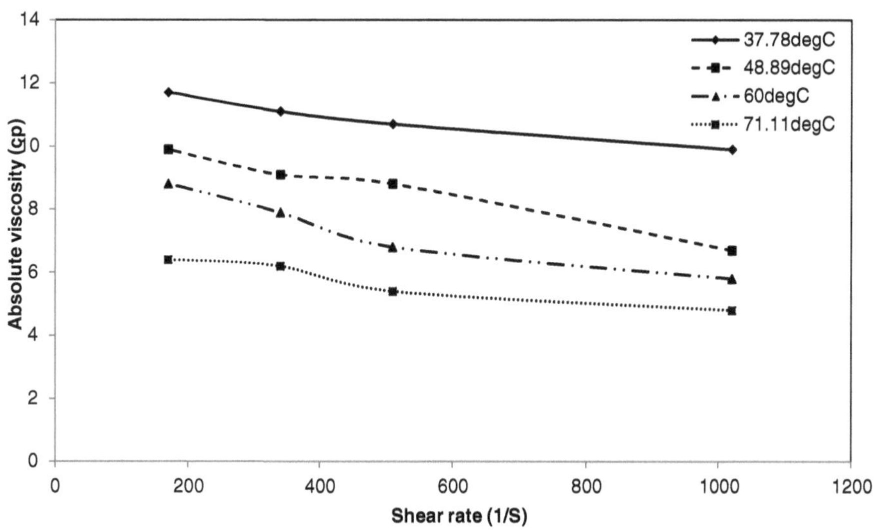

(b)

Fig. 8. Effects of shear rate on viscosity of: (a) Jorajan crude (thixotropic), (b) Lakwa (pseudoplastic)

3.1.4 Effect of Water Cut on Viscosity

The volume fraction of water (dispersed phase) was found to have a significant effect on the emulsion viscosity behaviour of crude oil as shown in Fig. 9 which depicts the

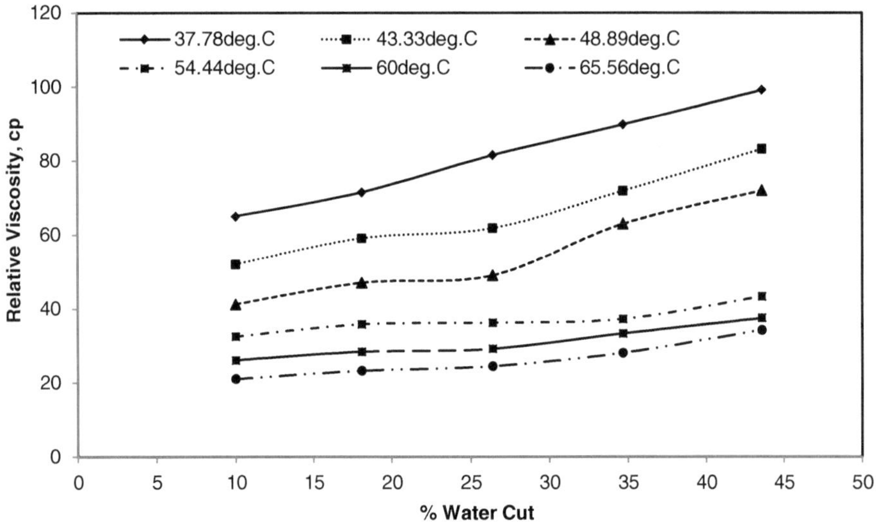

Fig. 9. Correlations for relative viscosity and water cut of Jorajan-5 at different temperature and shear rate of 1021.3 s^{-1}

increase in viscosity with an increase in percentage of water. Moreover it can be noted that water cut with 43.58% characterize higher elastic behaviour and it needs a higher shear rate followed by higher water cut as compared to lower water cut. The reason behind increase of viscosity when the high volume fraction of the dispersed phase (water) was used is due to increase in the number of hydrogen bond that leads to increase in the hydrodynamic forces and hence the viscosity. Thus it needs a higher shear rate to start the deformation [7–9] (Fig. 10).

The volume fraction of water was found to have a significant effect on the relative viscosity of the Crude-Water sample as shown in the Fig. 11 above. In the Fig. 11 data of (18.1–81.9, 26.39–73.61, 34.68–65.32, 43.58–56.42) vol. percent crude-water mixture was taken and the viscosity variation at different rotational rpm (600 and 300 rpm) and at 100 °F temperature. It can be noted that with increase in the volume percent of water resulted in significant increase in relative viscosity (from 65 to 99.2, 69.1 to 102.4) cp which is about 34.4%. From Fig. 11 it was observed that higher rpm lower the viscosity and vice-versa.

3.2 Behaviour of Crude Oil Samples Collected from Upper Assam Basin When Surfactants, Kerosene Was Added

Further study was carried out by mixing the crude oil samples with different surfactants and kerosene to see the impact on rheological behaviour of the crude. Therefore a total of eleven samples were prepared which constituted of (a) pure crude oil sample (Kothaloni), (b) crude oil (Kothaloni) mixed with 4%wt/v surfactant lignosulphonate, (c) Kothaloni crude mixed with 0.04%wt/v surfactant tritonX-100, (d) kothaloni crude

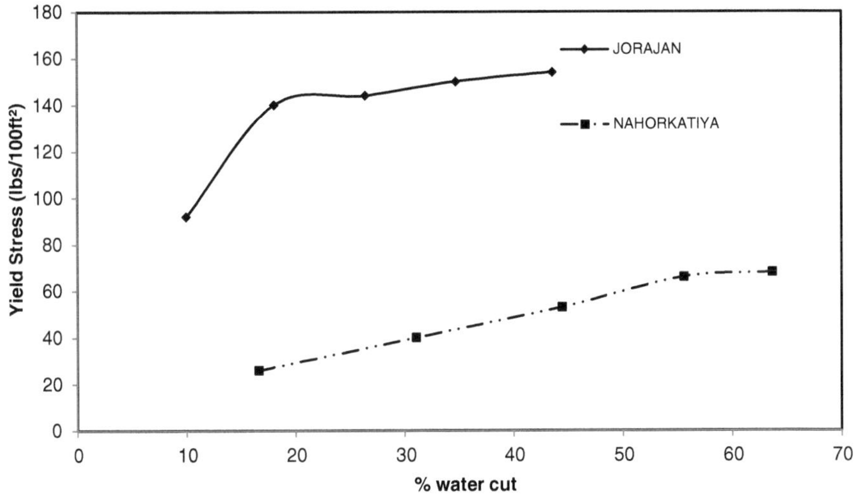

Fig. 10. Variation of yield stress with water cut at same temperature for jorajan and nahorkatiya crude oil sample having an API of 17 and 26 respectively.

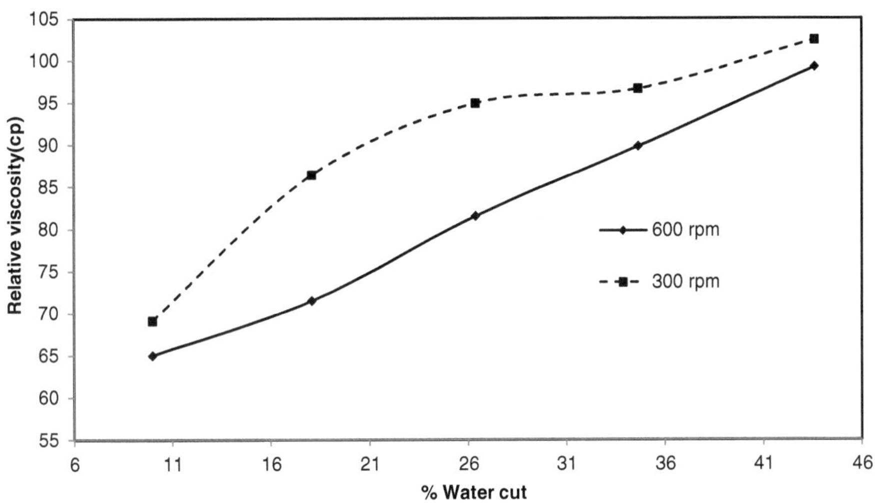

Fig. 11. Effects of the water volume fraction on viscosity of Jorajan-5 crude oil at 600 and 300 rpm and 37.78 °C.

mixed with 0.6%wt/v surfactant black liquor, (e) Kothaloni with 0.04%wt/v TX-100 and 5% kerosene, (f) Kothaloni with 0.6%wt/v BL and 5%wt/v kerosene, (g) Kothaloni with 4%wt/v lignosulphonate and 5% kerosene, (h) blend of heavier crude (Kothaloni) with lighter crude (Jorajan 5), (i) crude oil mixture of Kothaloni and Jorajan5 with surfactant 0.04%wt/v TX-100, (j) crude oil mixture of Kothaloni and Jorajan5 with surfactant 4%wt/v lignosulphonate, (k) crude oil mixture of Kothaloni and jorajan5 with surfactant 0.04%wt/v TX-100 with further blending of kerosene.

3.2.1 Effect of Kothaloni Crude with Different Surfactants at 40 °C

The above experiment conducted in the Grace M-3600 viscometer shows the performance characteristics of the eleven different samples prepared for the rheological study. It has been observed from the experimental results in Fig. 12 that there is a decrease in viscosity with increasing shear rate for all the samples. Further Table 3 shows the percentage reduction of viscosity of the different crude oil mixtures with respect to pure Kothaloni crude. Also, a significant declining trend was observed when Kothaloni crude was mixed with an anionic surfactant BL. A further reduction in viscosity was observed when 5% kerosene was added to Crude oil. The above experimental results conducted at 40 °C shows that addition of a small percentage of kerosene or surfactant significantly reduces viscosity thus enhancing flow ability of crude oil. Hence it can be a major solution in transportation of heavy crude oil through pipelines.

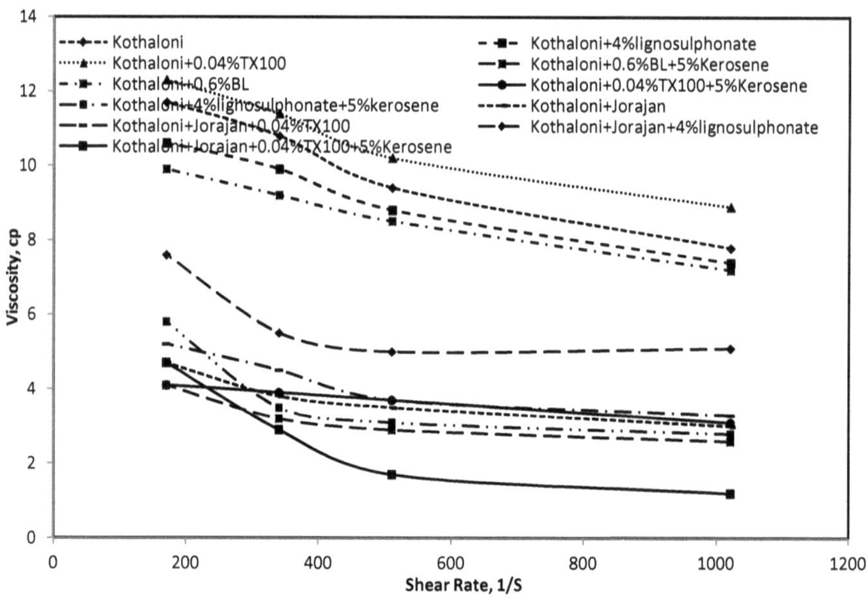

Fig. 12. Effect of viscosity vs. shear rate of different crude oil mixtures at 40 °C

The above characteristics trend Fig. 13 shows that shear stress increases with shear rate. However a reduction in this behavioural trend can be observed when mixing an anionic surfactant BL (which is a waste product of Jagiroad paper mill) and chrome free lignosulphonate. Further reduction in shear stress is observed when Kothaloni crude is blended with 5% kerosene or blended with lighter crude (Jorajan5). The reduction in shear stress implies a reduction in drag force. Thus smooth flow through the pipeline bends can be ensured. Also from Table 4, it can be observed that the

Table 3. Percentage of Viscosity Reduction at constant shear rate of the different crude oil mixtures at 40 °C taking pure kothaloni as reference

Shear Rate, 1/s	% Reduction in viscosity									
	K + 0.04% TX100	K + 4% LS	K + 0.6% BL+5% kero	K + 0.6% BL	K + 0.04% TX100 + 5% kero	K + 4% LS +5% kero	K + J	K + J + 0.04% TX 100	K + J +4% LS	K + J + 0.04% TX100 + 5% kero
1021.3	12.36	16.85	70.78	19.10	65.17	68.53	66.29	62.91	42.69	86.51
510.65	7.84	13.73	71.56	16.67	63.73	69.61	65.68	63.72	50.98	83.33
340.43	5.26	13.16	71.93	19.29	65.79	69.30	66.67	60.53	51.75	74.56
170.22	4.88	13.82	66.67	19.51	66.67	52.85	61.79	57.73	38.21	61.79

K stands for Kothaloni, J stands for Jorajaan, LS stands for Lignosulphonate, BL stands for Black Liquor and Kero stands for kerosene.

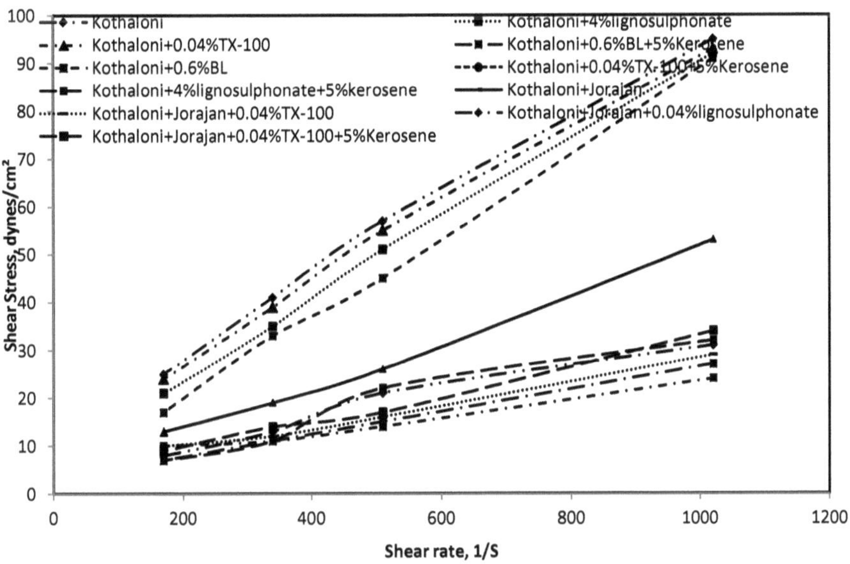

Fig. 13. Effect of shear stress vs. shear rate of different crude oil mixtures at 40 °C

Table 4. Percentage of Shear Stress Reduction at constant shear rate of crude oil at 40 °C taking pure kothaloni as reference

Shear Rate, 1/s	% Reduction in shear stress									
	K + 0.04% TX100	K + 4% LS	K + 0.6% BL+ 5% kero	K + 0.6% BL	K + 0.04% TX100 + 5% kero	K + 4% LS + 5% kero	K + J	K + J +0.04% TX 100	K + J +4% LS	K + J+ 0.04% TX100 + 5% kero
1021.3	12.08	14.28	70.32	15.38	64.83	68.13	65.93	62.63	41.75	73.62
510.65	14.28	19.64	73.21	23.21	60.71	71.43	62.5	69.64	53.57	75
340.43	5.12	12.82	71.79	15.38	71.79	69.23	66.67	64.10	51.28	71.79
170.22	4.76	14.28	66.67	19.05	66.67	52.38	61.90	57.14	38.09	52.38

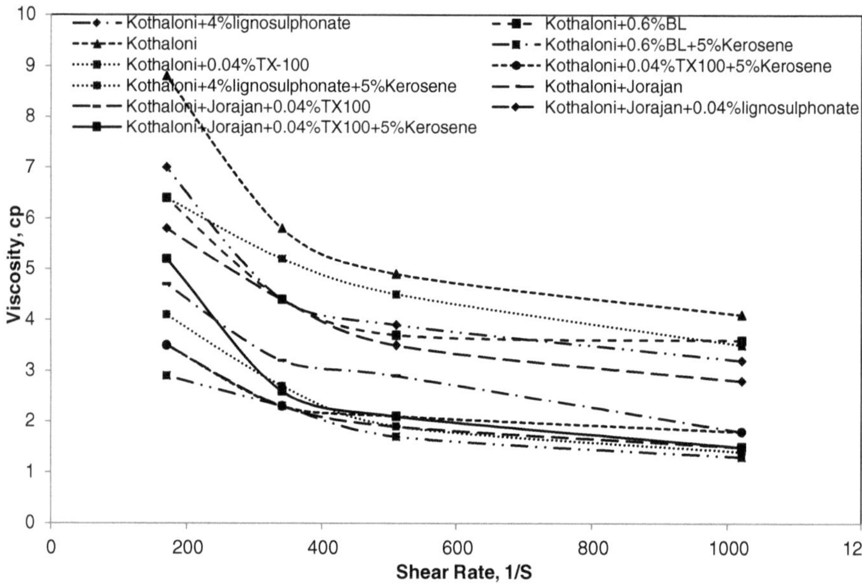

Fig. 14. Effect of viscosity vs. shear rate of Jorajan5 crude with different surfactants at 80 °C.

percentage reduction of shear stress with respect to the pure crude increases on mixing BL. There resulted in much more decrease when kerosene and lighter crude was added (Fig. 14).

3.2.2 Effect of Jorajan-5 Crude with Different Surfactants at 80 °C

The above characteristics trends were found when the Viscosity Vs shear rate was studied on further heating upto 80 °C. It was found that there was a further reduction of viscosity as well as reduction in shear stress. The viscosity Vs shear rate were found to decrease for the crude oil samples mixed with anionic surfactant BL and chrome free lignosulphonate at their CMC values to a much higher extent on heating upto 80 °C.

Figure 15 shows the shear stress Vs shear rate performance of crude oil at a temperature of 80 °C. The decrease in shear stress of crude oil with addition of surfactant at the reservoir condition can signify fewer obstacles in the flow of hydrocarbons from reservoir to well head (Tables 5 and 6).

3.2.3 Modelling Analysis

In this study modelling analysis was carried out to model the experimental data. Two models namely Ostwald de Waele power law and generalized Bingham plastic were used as given in Eqs. 3 and 4 and described by fitting the experimentally measured shear stress-shear rate data to the model. Jorajan crude with different surfactants and blending with kerosene and lighter crude oil Kothaloni can be characterized by the calculation of model parameter, (behavior index, n and consistency factor, k). The results of the modeling analysis were presented in Table 7.

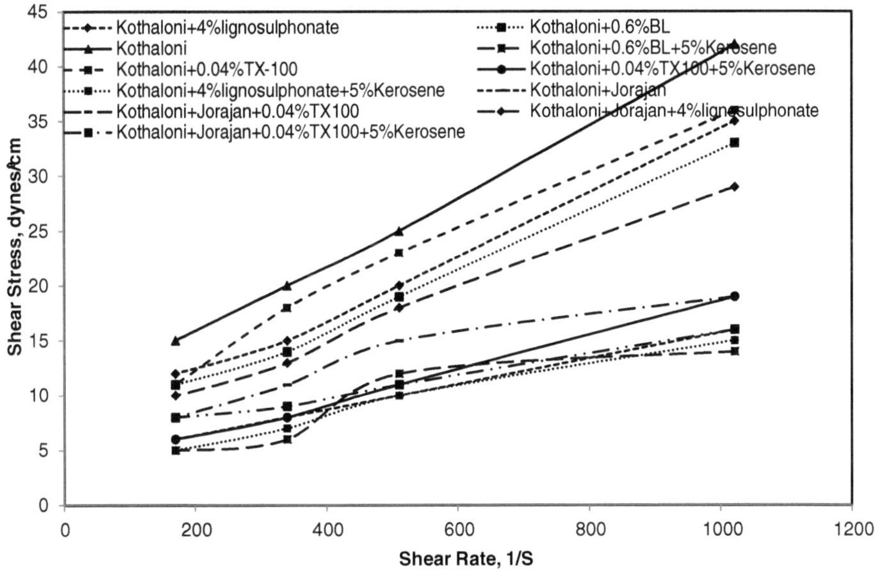

Fig. 15. Effect of shear stress vs. Shear rate of Jorajan5 crude with different surfactants at 80 °C.

Table 5. Percentage of Viscosity Reduction at constant shear rate of the different crude oil mixtures at 80 °C taking pure kothaloni as reference

Shear Rate, 1/s	% Reduction in viscosity									
	K + 0.04% TX100	K + 4% LS	K + 0.6% BL + 5% kero	K + 0.6% BL	K + 0.04% TX100 + 5% kero	K + 4% LS + 5% kero	K + J	K + J + 0.04% TX 100	K + J +4% LS	K + J + 0.04% TX100 + 5% kero
1021.3	47.44	60.25	83.33	61.54	76.92	82.05	80.77	76.92	64.10	80.77
510.65	47.80	60.64	81.91	62.76	77.66	79.78	79.79	69.15	62.77	77.66
340.43	46.29	62.04	78.70	61.11	78.70	75.00	78.70	70.37	63.89	75.93
170.22	24.78	45.29	75.21	44.44	70.08	64.76	70.08	59.83	50.43	55.56

Table 6. Percentage of Shear Stress Reduction at constant shear rate of crude oil at 80 °C by using taking pure kothaloni as reference

Shear Rate, 1/s	% Reduction in viscosity									
	K + 0.04% TX100	K + 4% LS	K + 0.6% BL +5% kero	K + 0.6% BL	K + 0.04% TX100 + 5% kero	K + 4% LS + 5% kero	K + J	K + J + 0.04% TX 100	K + J +4% LS	K + J + 0.04% TX100 + 5% kero
1021.3	47.50	60.00	82.50	61.25	76.25	81.25	80.00	76.25	63.75	80.00
510.65	47.92	60.42	75.00	62.50	77.08	79.17	79.17	68.75	62.50	77.08
340.43	45.95	62.16	83.78	64.86	78.38	81.08	78.38	70.27	64.86	75.67
170.22	25	45.00	75.00	50.00	70.00	75.00	70.00	60	50.00	60.00

Table 7. Calculated shear stress for starting non-linear region in order to predict non-Newtonian flow (Bingham type or pseudoplastic type) at 40 °C

Samples	Bingham		Power law	
	R^2	Y = mX + C	R^2	$Y = a\,X^b$
Kothaloni	0.996	0.081x + 12.92	0.999	$0.531x^{0.748}$
Kothaloni + 4% lignosulphonate	0.999	0.083x + 7.107	0.997	$0.289x^{0.829}$
Kothaloni + 0.04% TX-100	0.997	0.080x + 11.60	0.998	$0.047x^{0.760}$
Kothaloni + 0.6% BL + 5% kerosene	1	0.023x + 3	0.995	$0.140x^{0.754}$
Kothaloni + 0.6% BL	0.998	0.086x + 2.357	0.998	$0.144x^{0.928}$
Kothaloni + 0.04%TX-100 + 5% kerosene	0.978	0.017x + 5.964	0.891	$0.705x^{0.493}$
Kothaloni + 4% lignosulphonate + 5% kerosene	0.942	0.029x + 2.785	0.957	$0.071x^{0.890}$
Kothaloni + Jorajan +0.04% TX-100	0.961	0.025x + 4.75	0.946	$0.247x^{0.688}$
Kothaloni + Jorajan +4% lignosulphonate	0.991	.0.029x + 3.5	0.976	$0.202x^{0.976}$
Kothaloni + Jorajan + 0.04% TX100 + 5% kerosene	0.993	0.047x + 3.321	0.974	$0.213x^{0.783}$

$$\tau = K(\gamma)^n \tag{3}$$

Where,
τ = Shear stress, dynes/cm^2
γ = Shear rate, 1/s
n = Flow behavior index (n < 1)
K = Consistency factor

$$\tau - \tau_y = K(\gamma)^n \tag{4}$$

Where,
τ_y = yield stress, dynes/cm^2
τ = Shear stress, dynes/cm^2
γ = Shear rate, 1/s
n = Flow behavior index (n < 1)
K = Consistency factor

Tables 7 and 8 show measurement of yield stress of the crude oil mixtures using two above models at 40 °C and 80 °C. Validations of the two models were done on the basis of the highest regression correlation coefficient, R^2 whose value is 1.

3.2.4 Calculation of Shear Stress for Starting Non-Linear Region to Predict Non-Newtonian Flow

From the plotting of graph (not shown) yield stress was determined for different samples for starting non-linear region to predict non Newtonian flow. It was done by extrapolating the plot of the measured shear stress as a function of shear rate.

Table 8. Calculated shear stress for starting non-linear region in order to predict non-Newtonian flow (Bingham type or pseudoplastic type) at 80 °C

Samples	Bingham		Power law	
	R^2	Y = Mx + C	R^2	$Y = a X^b$
Kothaloni	0.994	0.027x + 6.357	0.954	$0.0496x^{0.602}$
Kothaloni + 4% lignosulphonate	0.995	0.026x + 5.75	0.961	$0.418x^{0.62}$
Kothaloni + 0.04% TX-100	0.999	0.031x + 9.314	0.976	$0.744x^{0.573}$
Kothaloni + 0.6% BL + 5% kerosene	0.821	0.010x + 3.678	0.869	$0.188x^{0.629}$
Kothaloni + 0.6% BL	0.988	0.028x + 7.428	0.999	$0.377x^{0.658}$
Kothaloni + 4% lignosulphonate + 5% kerosene	0.986	0.011x + 3.25	0.989	$0.194x^{0.626}$
Kothaloni + Jorajan	1	0.011x + 4	0.983	$0.344x^{0.547}$
Kothaloni + Jorajan +0.04% TX-100	0.935	0.012x + 6.821	0.981	$0.629x^{0.496}$
Kothaloni + Jorajan +4% lignosulphonate	0.996	0.022x + 5.928	0.976	$0.421x^{0.604}$
Kothaloni + Jorajan + 0.04% TX100 + 5% kerosene	.994	.009x + 6.071	0.937	$.996x^{.392}$

After extrapolating the experimental data some of the flow curve made intercept (Bingham type behaviour) and some passed through origin and convex to shear stress axis (Pseuodoplastic type of behavior $n < 1$). The parameters standard deviation co-efficient (R^2) and standard straight line equation ($Y = mX + C$) obtained from graphs was shown in Tables 7 and 8. From Tables 7 and 8 it was observed that the required shear stress or yield stress decreased with introduction of different chemicals. A significant decrease was observed with addition of black liquor and kerosene.

4 Conclusion

1. Crude oil samples followed Non-Newtonian behaviour. Also, an increase in temperature up to 60 °C had positive effect on the flow ability of crude oil as it decreased the viscosity and yield stress.
2. The crude oil samples Jorajan of the Upper Assam Basin was observed to be time dependent fluid which was thixotropic in nature. Whereas, the crude oil sample Lakwa was found to be pseudoplastic.
3. The crude oil viscosity decreased considerably with temperature and supports the exponential Arrhenius type relationship between the viscosity and the temperature. It was also noted that Newtonian behaviour occurs at relatively high values of shear rates but with decrease in temperature this becomes limited. The shear thinning behaviour or the pseudoplasticity appears at lower shear rate and becomes more pronounced as the temperature increased.
4. The volume fraction of water (dispersed phase) was found to have a significant effect on the emulsion viscosity behaviour of crude oil. It can be noted that with

increase in the volume percent of water resulted in significant increase in relative viscosity (from 65 to 99.2, 69.1 to 102.4) cp which is about 34.4%.

5. The viscosity reduction induced by Kothaloni crude oil using two anionic and a non-ionic surfactant namely Black Liquor, Chrome free Lignosulphonate and Triton X-100 was experimentally investigated at 40 °C. Based on the experimental findings, the following can be concluded:
   i. The experimental results showed that viscosity of crude oil mixture decreased by almost 19.10% on addition of 0.6%wt/v anionic surfactant black liquor at shear rate 1021.3 s^{-1} and at a temperature of 40 °C. Shear stress of mixture was also reduced by almost 15.38%. Viscosity of the mixture was further reduced by almost 70.78% on addition of 5% kerosene to the above mixture.
   ii. By adding a different anionic surfactant chrome free lignosulphonate (4% wt/v) to crude oil a viscosity reduction of 16.85% and shear stress reduction of 14.28% was observed. Viscosity was further reduced to 68.53% and shear stress to 68.13% when kerosene was added to the above mixture. It was observed that black liquor generally exhibited better reduction of viscosity than lignosulphonate.
   iii. Also by mixing two crudes of different API (one lighter, Jorajan 5 and the other heavier, Kothaloni) a reduction of 66.29% viscosity and 65.93% shear stress was observed.
   iv. Another non-ionic surfactant TX-100 resulted in a viscosity and shear stress reduction of 12.34% and 12.08% respectively. Further addition of kerosene reduced its viscosity to 65.17% and shear stress 64.83%.
6. The viscosity reduction induced by Kothaloni crude oil using two anionic and a non-ionic surfactant namely Black Liquor, Chrome free Lignosulphonate and Triton X-100 was further investigated at 80 °C. Based on the experimental findings, the following can be concluded:
   i. Further on heating the same crude oil samples to 80 °C the experimental results showed that viscosity of crude oil mixture decreased by almost 61.54% on addition of 0.6%wt/v anionic surfactant black liquor at shear rate 1021.3 s^{-1} and at a temperature of 80 °C. Shear stress of mixture was also reduced by almost 61.25%. Viscosity of the mixture was further reduced by almost 83.33% on addition of 5% kerosene to the above mixture.
   ii. By adding a different anionic surfactant chrome free lignosulphonate (4% wt/v) to crude oil a viscosity reduction of 60.25% and shear stress reduction of 60% was observed. Viscosity was further reduced to 82.05% and shear stress to 81.25% when kerosene was added to the above mixture. It was observed that black liquor generally exhibited better reduction of viscosity than lignosulphonate.
   iii. Also by mixing two crudes of different API (one lighter, Jorajan 5 and the other heavier, Kothaloni) a reduction of 80.77% viscosity and 80% shear stress was observed.
   iv. Another non-ionic surfactant TX-100 resulted in a viscosity and shear stress reduction of 47.44% and 47.50% respectively. Further addition of kerosene reduced its viscosity to 76.92% and shear stress 76.25%.

From the above study an overall crude oil characteristics of some oil fields of Upper Assam Basin could be understood which can play a vital role in the design and construction of petroleum pipelines and subsurface and surface production equipments in the oil industry. Although different methods to facilitate transportation of crude oils (waxy crude) were discussed, based on the economic feasibility, addition of an anionic surfactant Black liquor which is a waste product of Jagiroad Paper mill proved to be a promising alternative for improving transportation of crude oil through pipelines.

References

1. Santos, R.G., Loh, W., Bannwart, A.C., Trevisan, O.V.: An overview of heavy oil properties and its recovery and transportation methods. Braz. J. Chem. Eng. (2014). doi:10.1590/0104-6632.20140313s00001853
2. Crandall, G.R., Wise, T.H.: Availability of diluents may inhibit heavy oil exports. Can. Pet. **25**, 37–40 (1984)
3. Al-Roomi, Y., George, R., Elgibaly, A., Elkamel, A.: Use of a novel surfactant for improving the transportability/transportation of heavy/viscous crude oils. J. Petrol. Sci. Eng. (2004). doi:10.1016/j.petrol.2003.12.014
4. Kumar, R., Mohapatra, S., Mandal, A., Naiya, T.K.: Studies on the effect of surfactants on rheology of synthetic crude. J. Pet. Sci. Res. (JPSR) (2014). doi:10.14355/jpsr.2014.0302.06
5. Ronningsen, H.P.: Rheology of petroleum fluids. Ann. Trans. Nord. Rheol. Soc. **20**, 11–18 (2012)
6. Ghannam, M.T., Hasan, S.W., Abu-Jdayil, B., Esmail, N.: Rheological properties of Heavy and Light crude oil mixtures for improving Flowability. J. Petrol. Sci. Eng. (2012). doi:10.1016/j.petrol.2011.12.024. Elsevier
7. Abd, R.M., Nour, A.H., Sulaiman, A.Z.: Experimental investigation on dynamic viscosity and rheology of water-crude oil two phases flow behavior at different water volume fractions. Am. J. Eng. Res. (AJER) **3**(3), 113–120 (2014)
8. Sunil, L.K.: Crude oil emulsions. In: Larry, W.L., John, R.F. (eds.) Petroleum Engineering Handbook: General Engineering, vol. 1, (U.S. Texas 2006)
9. Ilia Anisa, A.N., Abdurhman, H.N., Azhary, H.N.: Catastrophic and transitional phase inversion of w/o emulsion for heavy and light crude oil. J. Appl. Sci. **10**(23), 2076–3083 (2010)
10. Al-Wahaibi, T., et al. Experimental Investigation of the Effects of Various Parameters on Viscosity Reduction of Heavy Crude by Oil-Water Emulsion. Springer (2015). doi:10.1007/s12182-014-0009-2
11. Boger, D.V., Halmos, A.L.: Department of Chemical Engineering, Monash University, Clayton, Victoria 3168, Australia, Non-Newtonian Flow I- Characterization of Fluid Behaviour. Module C2.1
12. Langevin, D., et al.: Crude oil emulsion properties and their application to heavy oil transportation. Oil Gas Sci. Technol. (2004). Pipeline Transportation of Heavy Oils
13. Hart, A.: A review of technologies for transporting heavy crude oil and bitumen via pipelines. J. Pet. Explor. Prod. Technol. (2014). doi:10.1007/s13202-013-0086-6. Springer
14. Abdul-Hadi, A.A., Khadom, A.A.: Studying the effect of some surfactants on drag reduction of crude oil flow. Chin. J. Eng. (2013). doi:10.1155/2013/321908. Hindawi
15. Yaghi, B.M., Al-Bemani, A.: Heavy crude oil viscosity reduction for pipeline transportation. J. Ener. Sources (2010). doi:10.1080/00908310252774417

16. Alomair, O.A., Almusallam, A.S.: Energy & fuels. Ener. Fuels **27**, 7267–7276 (2013). doi:10.1021/ef4015636. ACS Publications
17. Suriya, T., Ariffin, T., Yahya, E., Husin, H.: The rheology of light crude oil and water –in – oil – emulsion. Procedia Eng. **148**(2016), 1149–1155 (2016). Elesvier
18. Wang, W., Wang, P., Li, K., Duan, J., Wu, K., Gong, J.: Prediction of the apparent viscosity of non – Newtonian water- in – crude oil emulsions. Sci. Dir. **40**(1), 130–133 (2013)
19. Hassanean, M.H., Awad, M.E., Marwan, H., Bhran, A.A., Kaoud, M.: Studying the rheological properties and the influence of drag reduction on a waxy crude oil in pipeline flow. Egypt. J. Pet. (2015). Elsevier
20. Sun, G., Zhang, J., Maa, C., Wang, X.: Start-up flow behavior of pipelines transporting waxy crude oil emulsion. J. Pet. Sci. Eng. (2016). doi:10.1016/j.petrol.2016.10.00. Elesvier
21. Wena, J., et al.: Effective viscosity prediction of crude oil-water mixtures with high water fraction. J. Pet. Sci. Eng. (2016). doi:10.1016/j.petrol.2016.09.052. Elesvier
22. Ghannam, M.T., Hasan, S.W., Abu-Jdayil, B., Esmail, N.: Rheological properties of heavy & light crude oil mixtures for improving flowability. J. Pet. Sci. Eng. (2011). doi:10.1016/j.petrol.2011.12.02. Elesvier
23. Ghannam, M.T., Hasan, S.W., Abu-Jdayil, B., Esmail, N.: Rheological properties of heavy & light crude oil mixtures for improving flowability. J. Pet. Sci. Eng. (2011). doi:10.1016/j.petrol.2011.12.024. Elesvier
24. Martínez-Palou, R., de Lourdes, M., Mosqueira, B.Z.-R., Mar-Juárez, E., Bernal-Huicochea, C., de la Cruz, J., Clavel-López, J.A.: Transportation of heavy and extra-heavy crude oil by pipeline: a review. J. Pet. Sci. Eng. (2010). doi:10.1016/j.petrol.2010.11.020. Elesvier
25. Ahmadpour, A., Sadeghy, K., Maddah-Sadatieh, S.-R.: The effect of a variable plastic viscosity on the restart problem of pipelines filled with gelled waxy crude oils. J. Nonnewton. Fluid Mech. doi:10.1016/j.jnnfm.2014.01.005. Elsevier
26. Gogoi, S.B.: Effluent as surfactant for enhanced oil recovery. Innovative Ener. Policies (2014). doi:10.4172/2090-5009.1000109
27. Gogoi, S.B.: Adsorption – Desorption of surfactant for Enhanced Oil Recovery. Transport in Porous Media. Springer Science (2011). doi:10.1007/s11242-011-9805-y
28. Das, B.M., Gogoi, S.B.: Adsorption study of a natural product in to the oil producing porous media of upper Assam Basin. Int. J. Res. Eng. Appl. Sci. (2015). ISSN 2249-3905
29. Das, B.M., Gogoi, S.B.: Relating IFT with oil recovery with special reference to bhogpara porous media of upper assam basin. J. Pet. Eng. Technol. **5**(1), 1–8 (2015). SJIF = 3.932 (2015). ISSN: 2231-1785
30. Hazarika, K., Gogoi, S.B.: Comparative study of an Enhanced Oil Recovery process with various chemicals for Nahorkatiya oil field. Int. J. Appl. Sci. Biotechnol. **2**(4), 432–436 (2014). IF = 1.422; *SIF = 3.419, ISSN 2091-2609
31. Triantafillopoulos, N.: Measurement of Fluid Rheology and Interpretation of Rheograms, 2nd Edn. Kaltec Scientific, Inc.
32. Grace Instrument M3600 Viscometer operation manual, US Patent: 6,571,609, Grace Instrument Company, 1660 Townhurst Drive, Building G, Houston, Texas 77043 U.S.A

Spatial Imaging of the Water Resource and Natural Environment Coastal Gharb Basin: Downstream Part of the Hydraulic Basin of the Sebou River (Morocco)

Nezha El Mahmouhi, Mohamed El Wartiti, Rajae Azelmad, Ridouane Sadki(✉), and Sanae Berred

Laboratory of Applied Geology of the Faculty of Sciences, University of Mohamed V, Rabat, Morocco

Abstract. This research will be further developed with the integration of high resolution remote sensing data, aerial photographs, ground surveys and meteorological data to better define the functional models for these hazardous processes in the Sebou River basin with special emphasis on desertification and land degradation.

The physico-chemical characteristics of groundwater of the aquifer system contained in the plio-quaternary geological of the gharb (Morocco) indicate that a hydrochemical diversity exists depending on the nature of the aquifer, which itself related to the lithology in this semi-arid zone of Morocco.

The salinity of water increases, following a W-E section going from the coastal zone towards the centre, South East and North East of the basin. In the centre and East, the shallow aquifer exhibits cation and anion values going beyond the morocco norm of drinking water. This salinity is due to the abundance of under ground salty domes. The deep aquifer protected by a layer of clay more or less thick presents water of good quality except for the coastal zone where the deep formations appear. Hence, the irrational use of fertilizers influences the geochemical zoning system.

These deductions are based on preliminary studies to determine the principal water-rock interactions which control the composition of groundwater of the Plio-quaternary aquifer of the Gharb and to identify the natural process and the human impact which influence the hydrochemical zoning of this semi-arid zone of north-west of Morocco.

Keywords: SIG · Gharb · Geochemistry · Groundwaters · Morocco

1 Introduction

The Gharb plain located in the Atlantic coastal zone in the junction of two structural domains of north-western Morocco: Meso-Cenozoic Rif domain belonging to the Alpine cycle, and the Mesetien domain corresponding to the septentrional Hercynids. The Gharb plain extends on a surface of 4000 km^2; two important geomorphological units can be distinguished in the Gharb basin (Bryssing 1967):

D.N. Singh and A. Galaa (eds.), *Contemporary Issues in Geoenvironmental Engineering*, Sustainable Civil Infrastructures,
DOI 10.1007/978-3-319-61612-4_13

- The plain which has is triangular and crossed by the Oued Sebou and their affluents (Ouergha, Beht); this plain represents 90% of the basin.
- The coast formed by a system of sub-parallel dune cords in the coast separates the plain from the Atlantic Ocean. These cords are interrupted in the south by the Oued Sebou.

In the region studied, groundwater constitutes frequently the main resource exploited. Two aquifer units have been distinguished:

The shallow aquifer spreads over all the plain (about 3500 km^2), except at the borders of the basin in which deep formation appear. This aquifer is contained in superficial silty clay with little permeable lithologies.

The superficial part of the reservoir is formed by very thick soils "tirs" with a permeability comprised between 1.10–7 and 5.10–7 or by "Dehs" with a permeability comprised between 1.10^{-7} and 4.10^{-6} m/s (DRPE 1986).

Previous hydrogeological studies of the Gharb region showed that in the course of the time the hydrogeological catactsristics of the area didn't changed (Combe 1967; Kacimi 1992).

The deep aquifer has a multi-layer system. This aquifer circulates mainly of the eastern side of the plain in the coarse levels of gravels, and sands whose permeability varies between 1.10–4 and 6.10–3 m/s (DRPE 1994), whereas those of the west show a finer granulometry essentially formed of sands, silt and sandy clays whose permeability is estimated between 1.10–5 and 5.10–5 m/s (DRPE 1989). The multi-layer character is due to the presence of lenses of permeable coarse sandy material with pebbles and sandstone, sometimes separated by benches of clay. The thickness of these benches varies from some millimetres to about ten meters. The flow of the deep aquifer is mainly towards the centre of the plain due to its pan shape, and then the flow imprints the direction of the Atlantic (Fig. 1).

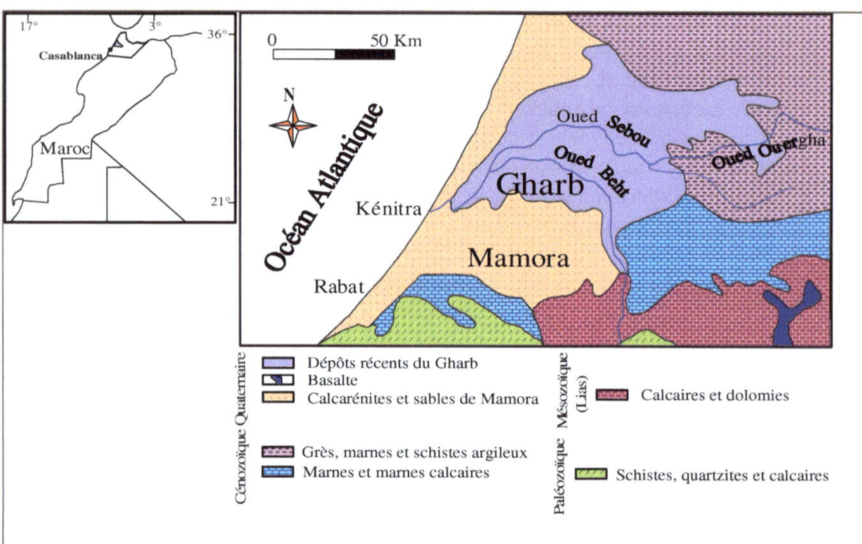

Fig. 1. Carte de la situation et la géologie de la plaine de Gharb et ses bordures

Except for the borders of the basin where deep formations appear, the deep aquifer is covered in its quasi-totality by clay. The alimentation of the aquifer is therefore mainly made via the basin borders.

2 Socio-Economic Aspects and Theirs Influence on the Water's Quality

According to the 1994 census, the population of the Gharb is 1.243 787 habitants with 691.704 habitants in the rural areas. Between 1982 and 1994 the rhythm of the average growth was evaluated to 3%, this rate is higher than the national average fixed at 2%. This increase in the population, mainly rural, deprived of cleansing means frequently leads to an intense pollution of groundwater.

Benefiting of a privileged geographical situation, very fertile soils and important hydraulic resources, the Gharb plain is considered amongst the main plains of the Kingdom. The economic development of the region is based especially on agriculture. The main types of cultivation are cereals (mainly rice), and sugar cane (ORMVAG 1998). In addition to this agricultural activity, the Gharb plain is also an industrial zone. The industry in the zone is represented by 5 sugar factories, a distillery of alcohol, a paper-mill, and an oil refinery. The wastes are often poured in the Oued Sebou, or in uncontrolled discharges. The polluting substances contained in the Oued Sebou are diverse. Generally, they are composed of mineral or organic matter, toxic substances, heavy metals, hydrocarbons (Saddiki 2002), polluting agricultural (fertilizing) and pesticides.

3 Hydrogeochemistry

3.1 Sampling and Analytical Methods

The choice of sampling points is based on the follow-up of the general direction of hydraulic (DGH) so as to undertake a comparative study. Moreover, 22 wells that have been sampled in September 2001 were distributed in a uniformly in the plain.

Temperature, pH, redox potential (Eh), conductivity, and alkalinity were measured in site. Nitrogen species were measured by spectrophotometer. The major components were determined by ion chromatography and ICP-OES. Mercury, antimony, arsenic and trace elements were determined by ICP-MS.

4 Discussion

In order to classify water and so as to establish the relationship between the rock and groundwater in the aquifer complex of the Gharb, the chemical composition of waters in different sampled wells are shown first on the Piper diagram. On these diagrams are distinguished wells that exploit the deep aquifer from those that exploit the shallow aquifer.

The water of different aquifers from the Gharb varies between bicarbonate type and the chloride one for anions, and between Ca type and Na for cations. Most of the sampled waters of the deep aquifer are characterized by a majority of bicarbonate for anions and calcium for cations. As such, they can be classified as calcium bicarbonate water; some samples are richer in sodium, another one is mainly alkaline, either are richer in chlorides On the other hand, samples of the shallow aquifer can be classified of type alkaline-chloride with a variable alkaline-earth ion contribution between 20 and 30%.

The variation of the concentration of major elements in relation to the total salinity (TDS: total dissolved solid) in the Gharb aquifer shows that the water of the deep aquifer is less salty (TDS: 0.24–0.9 g/l). Bicarbonate for anion and calcium for cation give the major contribution to salinity. On the contrary, the water of the shallow aquifer is the more salty one (TDS: 0.7–1.8 g/l), chloride and sodium the salt which predominates. Sulphate and magnesium are always subordinated to other ions. to notice also that dissolved calcium and bicarbonate increase, from 25 to 172 mg/l and from 43 to 55 mg/l, respectively, parallel with increases TDS from 0.8 g/l to nearly 2 g/l. The concentrations of chloride and sodium present a clear correlation, and Br is also correlated with Cl in the deep water, while the Cl/Br ratio in the shallow water is raised than the sea water.

4.1 Deep Aquifer

The chemistry of calcium bicarbonate of the deep aquifer water is derived from the interaction with calcarenites, the carbonated sandstone, and the plio-quaternary carbonates. In the coastal zone the calcium concentration varies between 71 mg/l and 150 mg/l, the presence of this ion has as origins sandy cords. The salt water of the deep aquifer has been sampled in the western to north-western sectors of the basin. High levels of nitrate up to 180 mg/l. were found (Fig. 2).

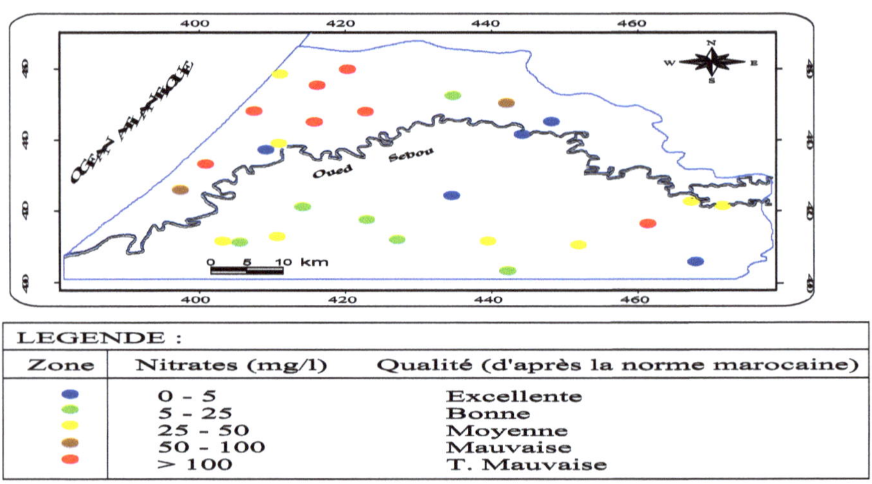

Fig. 2. Carte de répartition des nitrates dans les eaux souterraine de Gharb

The southern part of the basin, is less salty: to alkalino-chloride when the interaction makes with silico-clasts formations and to calcium bicarbonate where the interaction with sandstone and sands to carbonated cement in the centre of the basin where the deep aquifer is protected by a layer of clay, water presents the same chemistry.

4.2 The Shallow Aquifer

Six water samples of the shallow aquifer were studied the principal differences between the chemistry of deep water and shallow aquifer: They present an alkaline-chloride character and high TDS (up to 1.85 g/l) is due to the great concentration in Na and Cl. This can be explained by the dissolution of salt domes dispersed in the basin (Burger et al. 1960) during the aquifer recharge or by a weak circulation of the aquifer in the centre of the basin. The high chloride rates are recorded in wells of different depths; this can be explained, by the communication and/or exchange between the deep and shallow aquifer or, by the existence of ground salty domes. The water enrichment in Na can provide also from cation exchange between clay and water according to chemical reaction $Ca2^{+} + Na^{+} - X$ ½ Ca-X +Na^{+}. Salinity increases from the south west to the centre and the East of the basin.

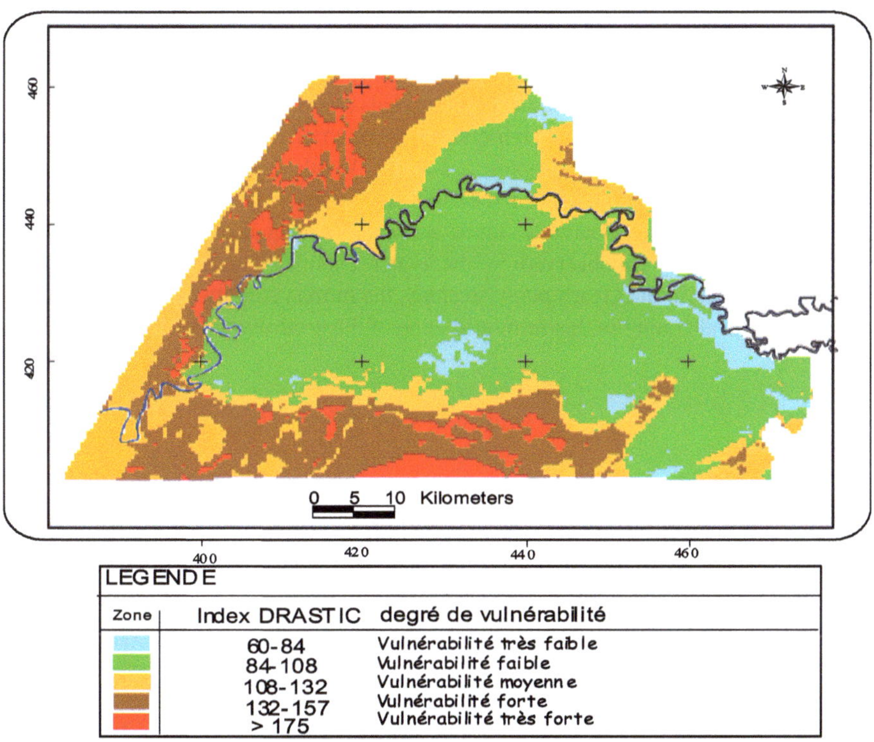

Fig. 3. Map of vulnerability of the Gharb aquifer.

Despite the fact that alkalino-chloride is the dominant salt, the shallow aquifer of the Gharb also records concentrations of bicarbonate and calcium greater than those observed in the deep aquifer. The strong contents are identified mainly in the centre and the east of the basin. In the centre of the basin where soils of clay types "dess" and "tirs"; the presence of calcium in water can have multiple origins. Soil minerals could contribute to the enrichment of water in calcium. The Oued Sebou which penetrates in the basin from the eastern part contributes to the input of terrigenous minerals (Snoussi 1988), the tributaries of Sebou, such as Oued Beht, Smento and others less important coming from the south and north of the basin. The presence of bicarbonate and calcium can also derived from the erosion or leaching the mesozoic and cenozoic formations who surround the basin. We noted that as the TDS increases by 0.8 till about 2 g/l, bicarbonate and calcium remains nearly constant: it can be the result of mineralogical control by the saturation in calcite.

5 Conclusion

We have deduced that there are two families of water that are representative of two types of aquifers: shallow aquifer and deep aquifer.

The contents relatively high in Ca, Na, Cl, HCO_3 in groundwater of the Gharb aquifer reflect the effect of the water-rock interactions, but the high concentration in NO_3 in the water of the coastal zone is due to an increase in the use of fertilizers in this Moroccan zone.

The interactions of water with salty domes contributes to an increase of salinity (TDS = 2 g/l) in the shallow aquifer water. This water is characterized by a high concentration in Cl and Na. The high content in sulphate is also detected in the wells which catch the deep aquifer (samples 7, 8 and 17) with content >200 mg/l. Calcium and bicarbonate are also present in the shallow aquifer. Their presence reflects the interaction process with calcarenite, sands and carbonate sandstone.

The spatial variation observed in the water chemistry reflects the lithological diversity occurring in the Gharb plain. Besides, the human activity has an influence on the hydrochemical zoning in this semi-arid zone of the north-western Morocco (Fig. 3).

References

Bryssing, G.: Plain of Gharb. In: Agronomic Research, vol. 24, pp. 266–330. INRA (1967)

Burger, J., House, B., Lévy, R.: Contributions of work of exploration of the company of the sherifs of oils to the knowledge of the Rifian aquifer. In: Book Memoiry P. Fallot, Mem, H. Ser. Soc. Geol. Fr., vol. 1, pp. 423–430 (1960)

Choubert, G.: Test of classification of quaternary continental of Morocco. C.R. Acad. Sci. **243**, 504–506 (1956). Paris

DRPE: Electric prospection of the zone of Gharb. African Company of Geophysics (A.C.G.) new report (1989)

Kacimi, I.: Some aspects on the hydrogeology and the hydrochimy of the water aquifer of Gharb – Morocco. The particular problem of iron and manganese. Control in earth sciences, university from Quebec in Montreal, 154 p. (1992)

Michard, A.: Elements de Moroccan Geology. Notes and Mem. Serv. Geol., Morocco, Rabat, vol. 252, 408 p. (1976)

ORMVAG: Report of the ORMVA (Office regional of agricultural development of Gharb) (1998)

Snoussi, M.: Nature, estimate and comparison of matter flows from the basins slopes of Adour (France), Sebou, Oum-Er-Rbia and Sous (Morocco). Impact of the climate on the fluviatile contributions with the ocean. Memories of the Institute of Geology of the Basin of Aquitaine, University of Bordeaux 1, vol. 22, 459 p., 67 fig. (1988)

Thauvin, J.P.: Hydrogeologic Monograph of Mamora. Notes and Mem. Ser. Geol., Morocco, vol. 195, 120 p. (1966)

Evaluating and Characterizing Oil field Produced Water of Upper Assam Basin

Subrata Borgohain Gogoi(✉), Amarjit Rajbongshi, and Shilpi Sarmah

Department of Petroleum Technology, Dibrugarh University, Dibrugarh, India
subrata@dibru.ac.in

Abstract. Large volumes of water are produced (PW) during oil and gas production from the various oil fields of Upper Assam Basin, India. PW is a very complex mixture and contains different toxic compounds that adversely affect the surrounding environment, the reservoir environment and also the economics of oil and gas production. PW separated from oil and gas is either injected into the reservoir for pressure maintenance or discharged to the environment. Therefore evaluation of PW and its characterization is very important and essential for environment and reservoir management. This study deals with the complete evaluation and characterization of PW generated from different depths and horizons of oil reservoirs. The evaluation included the analysis of pH, specific gravity, salinity, total dissolved solids (TDS), cations, anions, organics and minerals. The values of specific gravity, salinity, TDS, and some ions such as sodium and chloride of oilfield PW were increased with increasing the reservoir depths. Useful relationships were observed between salinity, TDS, cations, anions, organics and minerals. These results can be used in the design, operation, and optimization of wastewater treatment systems of petroleum industries for future studies.

Keywords: Evaluation · Characterization · Environment · Produced water

1 Introduction

Currently, it has been well known that the oil industry has to handle more produced water (PW) than oil which makes the oil industry looks more like a "water industry" [1]. In the United States, oil wells produce about 7 to 10 barrels of water for each barrel of oil [2]. It was estimated that, about 210 million bbl of water was produced each day in 1999 worldwide and is an integral and unavoidable part of oil and gas recovery processes [1, 3, 4]. This PW includes the formation water and water resulting from drilling, production and workover activities [5, 6]. Out of various options for management of PW, mostly adopted options by the hydrocarbon exploration industries are underground injection for increasing oil recovery in the deep reservoir zone and the injection to non pay zones in the porous rocks for mere disposal purpose [7, 8]. PW is sometimes discharged to the surrounding environment or may be used for water flooding or reservoir pressure maintenance [2, 5]. The quality of PW vary considerably depending on the geographic location of the reservoir oilfield, the geochemistry of the producing formation, the type of hydrocarbon product being produced and the characteristics and type of the producing well [2, 5, 9, 10].

D.N. Singh and A. Galaa (eds.), *Contemporary Issues in Geoenvironmental Engineering*, Sustainable Civil Infrastructures,
DOI 10.1007/978-3-319-61612-4_14

The North-East India oil field PWs are the by-products generated during various operations in the hydrocarbon exploration industries are major sources of pollution in the area. These pollutants find their way and traverse to various environmental receptors causing potential danger of polluting the same. The water bodies, both surface and underground are more susceptible to such pollution in oil fields. Management of oil field PW is a key issue for the oil and gas producers because of its large volume of generation and high handling cost. The aging oilfields of Upper Assam basin belonging to Oil India Limited, Duliajan and Oil and Natural Gas Limited, Nazira, produces crude oil with very high water cuts of more than 90%. Production of PW from these aging oilfields is also in increasing trend. This PW is highly saline, very rich in dissolved minerals and has a high temperature than the normal atmospheric water temperature. The salinities of the PWs from these oil fields range from 1000–3000 mg/l [7] i.e., 1000–3000 ppm. The oil field in the study area has disposed in an average 3000 m^3 of PW on a daily basis from its various oil fields which are an environmental concern since the injected PW has a potential to migrate and contaminate the ground water of the area.

2 Experimental

2.1 Materials

Oil samples from Moran, Naharkatiya and Dikom Oil fields were collected from well heads along with formation water from a depth of more than 3000 m. n-Hexane (C_6H_{14}), NaCl, KCl, Petroleum ether, Conc. HCl, $BaCl_2$ were purchased from Merck Chemical India, Mumbai.

2.2 Methods

Characterization of PW by determination of the physical properties:

The PWs were characterized by determining experimentally the physical properties like pH, salinity by Mettler Toledo pH meter, conductivity and resistivity by Ecotester, quantifying the crude oil presence in PW, total dissolved solids and presence of sulphate by analytical methods and inorganics by Induced coupled plasma optical emission spectrometry (ICP-OES). The determinations of organic components in PWs were done by Thin-layer chromatography and flame ionization detection (TLC-FID) and GC-MS (Gas Chromatography Mass Spectroscopy).

Characterization of PW by Geochemical properties of formation water by core sample analyses:

To understand the relationship of the components in PWs with the subsurface geological formation of the oil producing reservoirs a study on the clay minerals and other minerals present in the core samples obtained from M#1 (depth = 3928.2 m) and M#2 (depth = 3905.3 m) were studied. Clay minerals were examined by X-ray diffraction (XRD) and minerals by microscopically examining thin slides of the core samples (Table 3).

3 Results and Discussion

Samples of crude oil and formation water were collected from Moran, Naharkatiya and Dikom Oil Fields of Upper Assam Basin. It was observed that the wells M#1 (depth = 3928.2 m) and M#2 (depth = 3905.3 m) belonging to Moran oil field have been producing crude oil with 98% water cut. Therefore the PW from these wells were separated easily in separating funnels by ethyl ether and all further experiments were done in these PWs. pH was found to be 6.89 and 6.76 for M#1 and M#2 respectively and the salinity was found to be 2760 ppm (2760 mg/l) and 3020 ppm (3020 mg/l) for M#1 and M#2 respectively. The conductivity was 13.3 mS/cm and 12.9 mS/cm for M#1 and M#2 respectively, while the resistivity was 75.2 Ohm-cm and 77.5 Ohm-cm for M#1 and M#2 respectively. Since PW has ions so the conductivity is well above pure water whose conductivity is 0.55 mS/cm. The crude oil fraction per ml of PW was 0.01676 gm and 0.01712 gm for M#1 and M#2 respectively. While the total dissolved solids (TDS) per ml of PW were 0.00286 gm and 0.00477 gm for M#1 and M#2 respectively. The inorganics per litre of PW were found to be Ca = 7.817 mg & 6.792 mg; Fe = 0.016 mg & 0.19 mg; Mg = 2.285 mg & 3.104 mg; Mn = 0 & 0; Mo = 0 & 0.013 mg; Ni = 0.010 mg & 0.017 mg; Pb = 0 & 0; Sr = 0.402 mg & 0.398 and Zn = 0.006 mg & 0.007 mg for M#1 and M#2 respectively. The total inorganics was 10.536 mg/l and 10.521 mg/l for M#1 and M#2 respectively.

(a) The results of the organics in PW measured by TLC-FID are shown in Table 2

Determinations of organics in PWs are shown in Figure 1 which shows the majority of n-alkane distributions for whole-oil gas chromatograms. A broad spectrum of n-alkanes ranging from n-C_{10} through n-C_{35} was observed. Modest odd-carbon preference in the n-C_{15} through n-C_{19} range was also observed. The presence of isoprenoidspristane ($C_{19}H_{40}$) and phytane ($C_{20}H_{42}$) was also detected. The heavy hydrocarbons from C_{18} to C_{20} can also be seen.

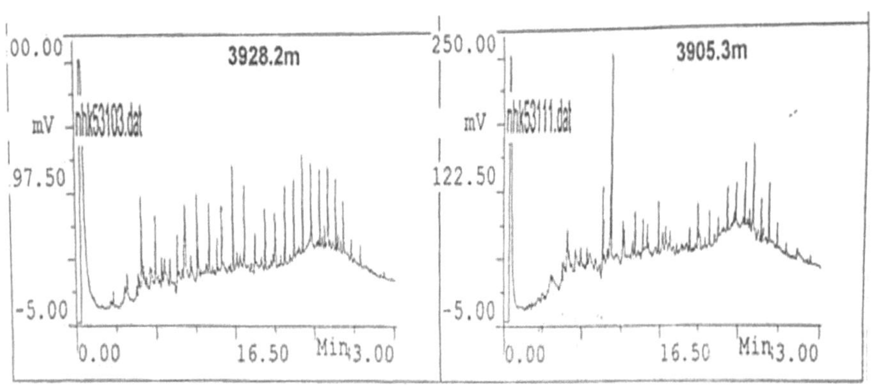

Fig. 1. Shows the GC fingerprints of PW from M#1 & M#2.

(b) To understand the relationship of the components in PWs with the subsurface geology the results of the XRD and thin slides examination are in Table 2 and the pictures of the slides is in Fig. 2.

Photomicrographs showing M#1& M#2 dissolution of quartz and feldspar by calcite cement

Fig. 2 Showing the pictures of the thin slides

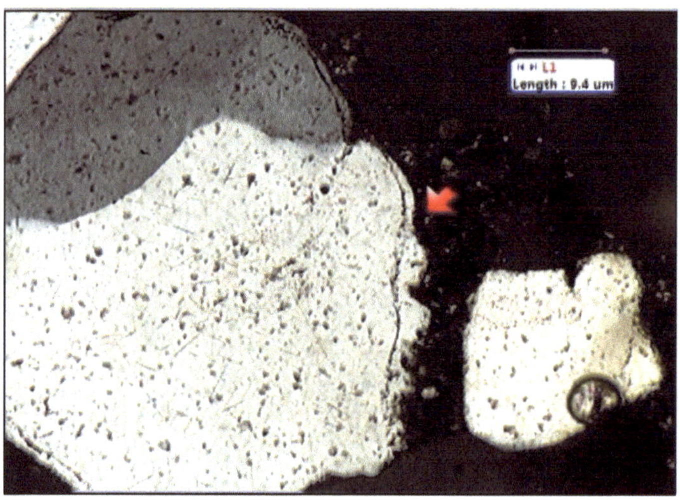

Fracturing of mica in M#1 and in M#2 quartz overgrowth

Tendency of calcite cementto form crystalsin M#1 and in M#2 haematite cement in Barail Sandstones

Fig. 2 (continued)

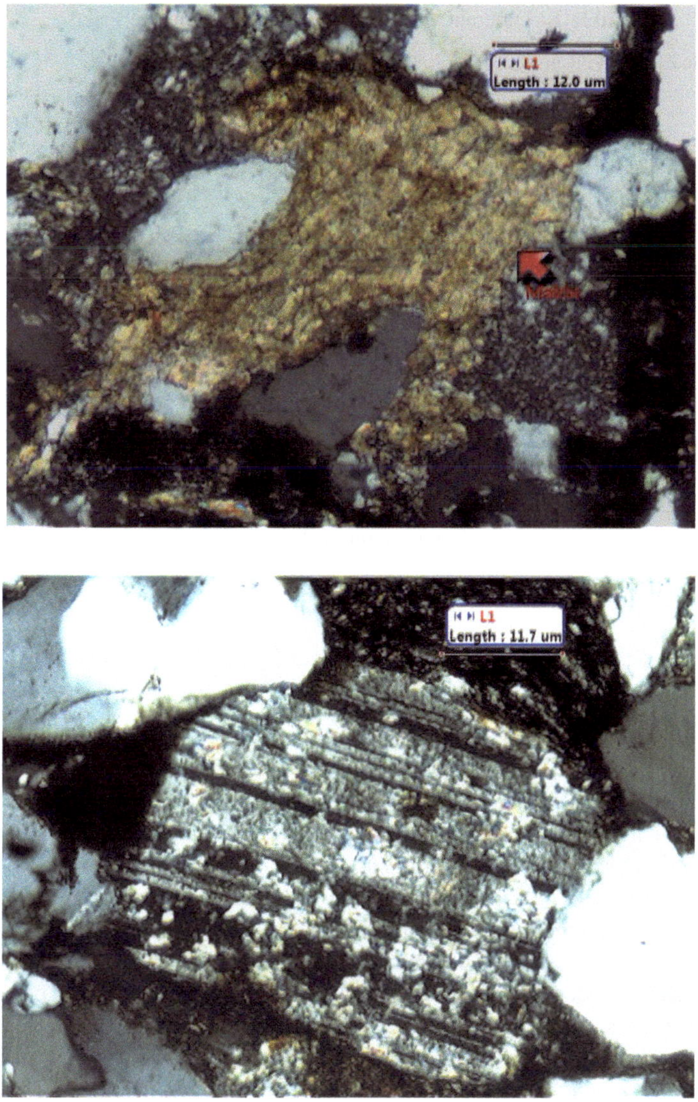

M#1 isolated patches of clay minerals and in M#2 the grain in the middle of the slide is a feldspar grain which has started getting sericitised. Sericite is a very fine grained clay which is formed by diagenetic transformation of feldspar *Magnification was 100x in all the above thin slides*

Fig. 2 (continued)

Table 1. Different water samples collected from Oil Fields

Sample No./ Oil field	Well no./ Year of producing	Geologic formation/depth (m)	Res. pressure (kg/cm^2)	Res. Temp. (°C)	Water cut i.e. water in crude oil (%)
IV/Dikom	D#1/1993	Eocene/3800	240	100	40
III/Moran	M#1/1965	Barail/3100	200	88	98
III/Moran	M#2/1963	Barail/3250	190	85	98
I/Moran	M#3/1969	Barail/3200	240	87	26
II/Nahorkatiya	N#1/2006	Eocene/4400	450	110	2
CW	Chemically treated water collected from the manifold of Moran OCS[a]				
F	Pond water from Moran OCS[a]				
V(W)	Chemically treated water collected from the manifold of Dikom OCS[a]				

[a] Oil Collecting Station

Table 2. Composition of PW using TLC-FID

Depth (m)	%Saturates	%Aromatics	%Res + Asphalts	Interpretation
3928.2	21.9	52.7	25.4	Heavy oil
3905.3	28.2	58.4	13.4	Heavy oil

(c) Pollutants in terms of toxicity

pH effects the chemical and biological properties of water. PWs pH is governed by the carbon dioxide/ bicarbonate/ carbonate equilibria and ranges between 4.5 and 8.5 although mostly basic. pH of the PWs obtained was almost neutral. Electrical Conductivity, which is a measure of the ionic properties of the water body, showed values much higher than pure water. Total hardness, predominantly caused by divalent cations such as Calcium, Magnesium, and others is the sum of total cations present in the water, which is perhaps due to minerals coming from the drilling activities and from underground reservoir. The concentration of sodium in water in terms of salinity is about 3000 mg/l, which is detrimental for the nearby agricultural fields. The aromatics detected in TC-FID (Table 1) behave as a bioaccumulate in marine organisms and may threaten the environment. A broad spectrum of n-alkanes ranging from n-C_{10} through n-C_{35} was observed in GCMS (Fig. 1) where the modest odd-carbon preference in the n-C_{15} through n-C_{19} range is present. The presence of isoprenoidspristane ($C_{19}H_{40}$) and phytane ($C_{20}H_{42}$) was also detected in GCMS. The heavy hydrocarbons from C_{18} to C_{20} can also be seen. These molecules are used as biomarkers and their analysis is applied in studies of biodegradation.

Table 3. Experimental results of XRD and Thin slides

Sl no	Sample (Well no.)	Mineral	Chemical formula
1	M#1 & M#2	Clay types	
		Kaolinite	$Al_2Si_2O_5(OH)_4$
		Illite	$(K,H_3O)(Al,Mg,Fe)_2(Si,Al)_4O_{10}[(OH)_2,(H_2O)]$,
		Chlorite present as	$NaClO_2$ or $Mg(ClO_2)_2$
2	M#1 & M#2	Quartz	SiO_2
		Feldspar	$KAlSi_3O_8 - NaAlSi_3O_8 - CaAl_2Si_2O_8$
3	M#1	Mica	$X_2Y_{4-6}Z_8O_{20}(OH,F)_4$ in which X is K, Na, or Ca or less commonly Ba, Rb, or Cs Y is Al, Mg, or Fe or less commonly Mn, Cr, Ti, Li Z is chiefly Si or Al, but also may include Fe^{3+} or Ti
4	M#2	Calcite cement crystals	Calcite and Aragonite. Aragonite is a carbonate mineral, one of the two common, naturally occurring, crystal forms of calcium carbonate, $CaCO_3$ (the other form being the mineral calcite). It is formed by biological and physical processes, including precipitation from marine and freshwater environments
5	M#2	Haematite cement	Hematite is the mineral form of iron (III) oxide (Fe_2O_3), one of several iron oxides
6	M#1	Isolated patches of clay minerals	As in Sl.no. 1
7	M#2	Feldspar grains	As in Sl.no. 2, Feldspar grains which has started getting sericitised. Sericite is a very fine grained clay which is formed by diagenetic transformation of feldspar

4 Conclusion

PWs obtained was almost neutral and conductivity higher than water. Hardness was due to divalent cations which may be due to drilling activities. The salinity and minerals are detrimental for agricultural fields. The aromatics may behave as a bioaccumulate in marine organisms and threaten environment. A broad spectrum of n-alkanes ranging from n-C_{10} through n-C_{35} together with isoprenoidspristane ($C_{19}H_{40}$) and phytane ($C_{20}H_{42}$) and heavy hydrocarbons (C_{18} to C_{20}) were observed. The heavy hydrocarbons can be used as biomarkers and their analysis will be applied in studies of biodegradation.

Acknowledgment. The authors are also greatly indebted to the anonymous reviewers whose comments will significantly improve this manuscript. This work was funded by the Department of Biotechnology, Government of India in the form of Joint R&D activities under the twinning programme for scientists working in Dibrugarh University in the form of a Twinning project no. DBT-ENV/2013/222 (Twin 2013).

Nomenclature

PW	Formation Water
TDS	Total Dissolved Solid
RSM	Response Surface Methodology
XRD	X - Ray Diffraction
ICP-OES	Induced Coupled Plasma Optical Emission Spectrophotometer
TLC-FID	Thin Layer Chromatography and Flame Ionization Detection
GC	Gas Chromatography
GC-MS	Gas Chromatography Mass Spectroscopy

References

1. Du, Y., Guan, L., Liang, H.: Advances of produced water management. Can. Int. Pet. Conf. Petroleum Society of Canada (2005). doi:10.2118/2005-060
2. Guerra, K., Dahm, K., Dundorf, S.: Oil and gas produced water management and beneficial use in the Western United States U.S. Department of the Interior Bureau of Reclamation. Science and Technology Program Report No. 157, Technical Service Center, September 2011
3. Khatib, Z., Verbeek, P.: Water to value – produced water management for sustainable field development of mature and green fields. Society of Petroleum Engineers. doi:10.2118/73853-MS
4. Gogoi, S.B., Tamuli, A.S.: Recent advances in treating oil field produced water: a review. J. Pet. Eng. Technol. **4**(2), 366–370 (2014)
5. Veil, J.A., Pruder, M.G., Elcock, D., Redweik, R.J.: A white paper describing produced water from production of crude oil, natural gas, and coal bed methane prepared by Argonne National Laboratory for U.S. Department of Energy. National Energy Technology Laboratory, January 2004
6. Al-Haleem, A.A., Abdulah, H.H., Saeed, E.A.: Components and treatments of oilfield produced water. Al-Khwarizmi Eng. J. **6**(1), 24–30 (2010)
7. Patel, H., Borgohain, D., Jena, S.: Hydro geochemistry of Upper Assam Oil field. J. Pet. Technol. **2**, 11–14 (2003)
8. Stone, B.: Management of Produced Water, pp. 37–41. Oxford Publications, Washington (2003)
9. Joel, O.F., Amajuoyi, C.A., Nwokoye, C.U.: Characterization of formation water constituents and the effect of fresh water dilution from land rig location of the Niger Delta. Nigeria. J. Appl. Sci. Environ. Manag. (2010). doi:10.4314/jasem.v14i2.57832
10. Jacobs, R.P.W.M., Grant, E., Kwant, J., Marqueine, J.M., Mentzer, E.: The composition of produced water from shell operated oil and gas production in the North Sea. Technol./Environ. Issues Solutions. Springer (1992). doi:10.1007/978-1-4615-2902-6_2

Remediation of Petroleum Hydrocarbons in a Tropical Sand Tank Model

Olaolu G. Fadugba[1(✉)], Samuel A. Ola[2], and Oluwapelumi O. Ojuri[1]

[1] Department of Civil and Environmental Engineering, The Federal University of Technology, Akure, Ondo State, Nigeria
ogfadugba@futa.edu.ng, olaolufadugba@yahoo.com
[2] Department of Civil Engineering, Afe Babalola University, Ado-Ekiti, Ekiti State, Nigeria

Abstract. This paper evaluates In-Situ Chemical Oxidation (ISCO) remediation technique for the remediation of hydrocarbon contaminated soil and groundwater in a laboratory three Dimensional (3D) sand tank model. The sand tank model was used to study natural attenuation and chemical oxidation of petroleum hydrocarbon in groundwater within a Tropical aquifer (porous media) material. Detailed geotechnical, hydraulic and geochemical characterization of the chosen laboratory experimental porous media material were performed. 60 ml of unleaded gasoline was introduced as a contaminant of concern into the 3-D sand tank via the injection wells. Thereafter, samples were taken periodically for 24 days for Benzene, Toluene, Ethyl benzene, and Xylene (BTEX) analysis was done for Natural Attenuation case study. Sand replacement was done for case study of Insitu chemical oxidation (ISCO I and ISCO II), the same volume of gasoline was injected as pollutant, Potassium permanganate ($KMnO_4$) was chosen as oxidizing agent with varied percentages (1.5% and 5%). Samples were taken periodically for 24 days for BTEX test. The results of BTEX for Natural Attenuation (NA), ISCO I and ISCO II, i.e., 15 g/l and 50 g/l of $KMnO_4$ were analyzed and compared. Modeling simulation carried out on Natural attenuation of the same Contaminant of concern (COC) revealed a clean-up period of about 45 years. The clean-up period for ISCO I will take about 69 months while the clean-up period for ISCO II will take about 35 months.

Keywords: BTEX · Natural attenuation · Potassium permanganate · Gasoline · Tropical soil

1 Introduction

Groundwater quality is of great public interest. Groundwater contamination can occur in various ways, from the ground surface (by infiltration of polluted surface water, spills, waste dumps, tailings facilities, airborne particles, agricultural activities, etc.), from above the water table (leaks in pipelines or underground storage tanks, leachates from landfills, etc.), or from below the water table (mines, drainage wells, canals, groundwater withdrawal, etc.) (US Office of Water 2012). Different remediation options exist for contaminated groundwater: containment of the contaminants to prevent migration (e.g. using geotechnical measures), removal of the pollution (e.g. by excavation and

D.N. Singh and A. Galaa (eds.), *Contemporary Issues in Geoenvironmental Engineering*, Sustainable Civil Infrastructures,
DOI 10.1007/978-3-319-61612-4_15

subsequent soil washing), groundwater treatment at the point of use, removal of the groundwater for treatment and replacement, or remediation of the groundwater in situ.

Information on a wider range of remediation technologies can be found in the literature (Khan et al. 2004, USEPA 1995). The most common physical and chemical remediation technologies are pump and treat, air sparging, in-situ flushing, insitu chemical oxidation (ISCO) and permeable reactive barriers. The most common biotechnologies include monitored natural attenuation, bioremediation, and phyto-remediation. Containment methods such as slurry walls and grout curtains are also used to control contaminant plumes within groundwater (USEPA 1995). Containment methods such as these are often used as interim measures prior to the final selection and implementation of a remedial method.

Cleanup of contaminated subsurface environments is complicated both by the physical nature of the formation and the behavior of contaminants introduced to the formation. Soils and many aquifers are porous media comprising of solid material and pore spaces which are occupied by fluids and/or gases. In addition to a complex physical structure, subsurface formations are hidden from view and highly inaccessible with accessibility limited to localized points (boreholes) or complete disruption (ex-cavation), remediation of the subsurface environment represents a formidable challenge.

In-situ chemical oxidation (ISCO) is one of several innovative technologies that showed promise in destroying or degrading an extensive variety of hazardous wastes in ground water, sediment, and soil. The oxidants used are readily available, and treatment time is usually measured in months rather than years, making the process economically feasible, also, in-situ chemical oxidation (ISCO) has been widely used for source zone remediation of Dense Non-aqueous Phase Liquids (DNAPLs) in subsurface environ-ments. Permanganate (MnO_{4-}) is available both $NaMnO_4$ and $KMnO_4$, the sodium form is distributed as a liquid available at 40% concentration by weight. Potassium permanganate is most commonly available in a crystalline form, and has a solubility of approximately 6% by weight in water at 20 °C. While permanganate is most com-monly delivered as an aqueous solution, it may also be mechanically mixed into soil or introduced via pneumatic or hydraulic fracturing.

The aim of this research is to study contaminant transport in a tropical aquifer material and to remediate the hydrocarbon contaminated soil and groundwater in a three–dimensional sand tank model using Insitu chemical oxidation (ISCO).

2 Materials and Method

2.1 Laboratory Tests

Sieve Analysis Test: Sieve Analysis test was carried out to know the characteristics of the porous media and flow characteristics of the material, the test was conducted in accordance with the British standards. BSI (1990) 1377 (Code of Practice).

Bulk Density Test: Bulk density is defined as the mass of particles of the material divided by the total volume they occupy. The total volume includes particle volume, inter-particle void volume and internal pore volume. BSI (1990) 1377-2.

Other tests carried out were to indicate the characterization of the geotechnical, hydraulic and geochemical properties of the porous media and determine the properties such as pH, Oxidation/Redox potential and Hydraulic Gradient.

Organic Parameter Test: Benzene, Toluene, Ethylbenzene, and xylene (BTEX) test was done with Gas Chromatography Apparatus (HP5890) with the aid of flame ionization detector (GC-FID) analysis was carried out.

2.2 Equipment Used for the Research

3-D Sand Tank Model

(Ojuri and Ola 2010), described the Configuration of the 3-D sand tank model, Fig. 1 shows the isometric and sectional view of the 3-D laboratory sand tank model.

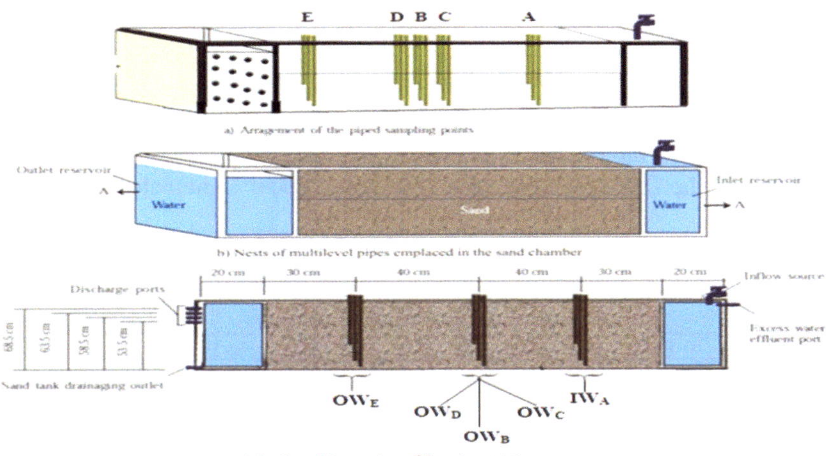

Fig. 1. Isometric and sectional view of 3D Laboratory Sand Tank Model (Ojuri and Ola 2010). KEY: IW_A – Injection Well A, OW_B – Observation Well B, OW_C – Observation Well C, OW_D – Observation Well D, OW_E – Observation Well E

2.3 Laboratory Procedure

Sand Tank Experiment (STE) for the Natural Attenuation of the Contaminants of Concern (COC)

60 ml of unleaded gasoline was injected using a 20 ml syringe through the multilevel injection well (Well A.) in Fig. 1. The flow rate was set at the calibrated 58 ml/min

value and hydraulic gradient was 0.0321. Contaminated Groundwater samples were collected at 8 h (1st Sampling period) after the injection of unleaded gasoline in the sand tank, contaminated water samples were taken periodically at 8th day, 15th day and 24th day from all the multilevel observation wells (OW_B, OW_C, OW_D and OW_E) into a 40 ml glass bottles with a Teflon siphon. The contaminated water samples were kept below 4 °C in the refrigerator and later transfered to the laboratory for organic analyses such as Benzene, Toluene, Ethylbeneze, and Xylene (BTEX) and concentration profile of the samples collected were measured at 1,8,15 and 24 days.

Sand Tank Experiment (STE) for the Remediation of Contaminant of Concern (COC) with 15 g/l and 50 g/l of $KMnO_4$

60 ml of unleaded gasoline was injected via a 20 ml syringe through the middle (30 ml) and bottom (30 ml) (IWA_{middle} and IWA_{bottom}) multilevel injection well (Well A.), 60 min after the injection of the unleaded gasoline, 15 g/l of $KMnO_4$ was injected through the Top Level well (IWA_{Top}) of the same injection wells (Well A) (Fig. 1), the flow rate was set at the calibrated 58 ml/min value (Port 1) and hydraulic gradient was 0.0321. All other procedures are as stated in (STE for the Natural Attenuation of the Contaminant of Concern). The same procedural steps were taken using 50 g/l of $KMnO_4$.

3 Results and Discussions

3.1 Results of the Porous Media and Flow Characteristics

Characterization of the coastal **river sand**

PSD curve for the Owena sand adopted from (Ojuri 2009) is presented in Fig. 2.

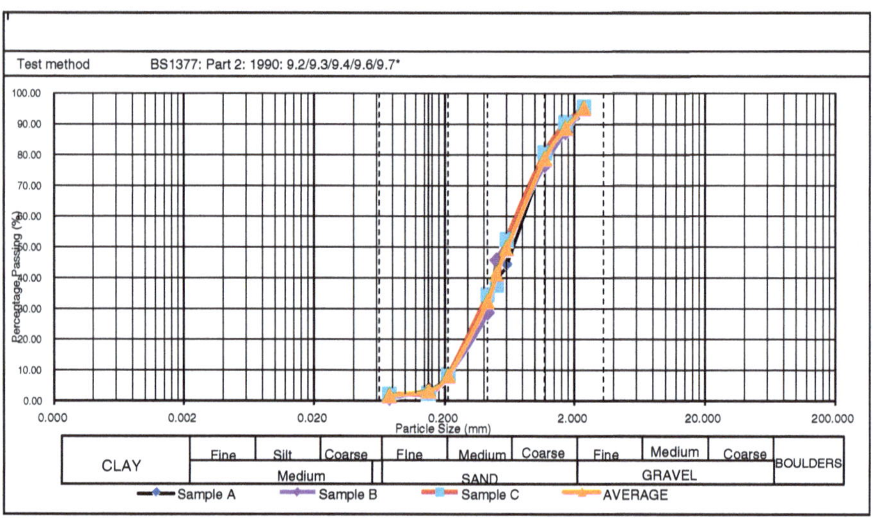

Fig. 2. Particle size distribution curve for samples A, B and C.

$$\text{Coefficient of Uniformity. } C_u = \frac{D_{60}}{D_{10}} = \frac{0.83}{0.32} = 2.60$$

$$\text{Coefficient of gradation. } C_z = \frac{D_{30}^2}{D_{60} \cdot D_{10}} = \frac{0.53^2}{0.32 \times 0.83} = \frac{0.2809}{0.2656} = 1.06.$$

The River sand is poorly graded.

Characterization of Geotechnical, Hydraulic and Geochemical Properties
Table 1 shows the results of laboratory analyses of the geotechnical, hydraulic and geochemical properties of Owena sand Table 2 indicates the flow characteristics of the sand tank model.

Table 1. Results of owena sand geotechnical, hydraulic and geochemical properties (Ojuri 2009)

S/N	Property	Results
1	Bulk density (Mg/m3)	1.86
2	Porosity	33%
3	Hydraulic conductivity (m/s)	5.75×10^{-4}
4	pH in H_2O	7.54
5	pH in NaCl	6.61
6	Water holding capacity (%)	30.2

Table 2. Sand tank flow characteristics (Ojuri 2009)

S/N	Head (cm)		Head differences (dh)(cm)	Gradient (dh/dl)	Outflow rate (ml/min)	Discharge ports
	Outlet	Inlet				
1	68.5	73	4.5	0.0321	58	1ST PORT
2	63.5	73	9.5	0.0679	160	2ND PORT
3	58.5	73	14.5	0.1036	280	3RD PORT
4	53.5	73	19.5	0.1393	405	4TH PORT

L = 140 cm (Length of middle chamber).

BTEX Analysis for tested water samples from the Monitoring wells
Table 3 and Fig. 3 presented the results of the analysis of the BTEX of tested water samples for the periodic sessions of the remediation.

Prediction of Clean-up Period for BTEX Concentrations for Field-Scale Model Using Reduction Ratio of BTEX Concentration
Table 4 shows the injected BTEX concentrations for field-scale model and simulation, while Table 5 displays the spatial distribution of hydrocarbon concentrations for

Table 3. BTEX concentration from the monitoring wells

Sampling session	Natural attenuation (mg/l)	ISCO (mg/l)	
		15 g/l of $KMnO_4$	50 g/l of $KMnO_4$
1st (Day 1)	2.23	2.17	2.11
2nd (Day 8)	2.07E-01	2.11E-01	2.01E-01
3rd (Day 15)	2.06E-04	2.68E-04	2.13E-04
4th (Day 24)	1.05E-05	1.29E-05	1.00E-05

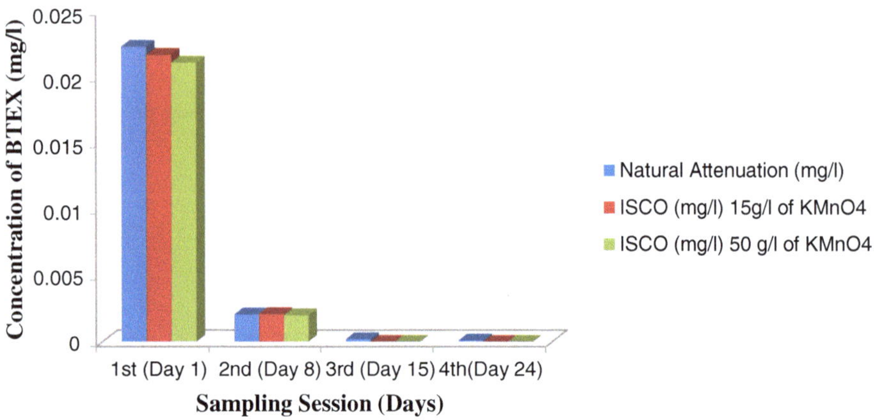

Fig. 3. BTEX analysis for water during remediation

passive natural attenuation and Table 6 displays Correlation of laboratory data with field – scale data for natural attenuation (NA), 15 g/l and 50 g/l of $KMnO_4$.

From Observation Well A (Middle)

The ratio of BTEX concentration for Natural Attenuation at day 24 to BTEX concentration at pure phase hydrocarbon (petrol) is:

$$\text{Reduction Ratio [RR]} = \frac{1.04 \times 10^{-4}}{21.86} = 4.76 \times 10^{-6}$$

- Reduction Ratio for remediating with 15 g/l of $KMnO_4$
  BTEX Concentration for pure phase hydrocarbon is 21.86 mg/l
  BTEX concentration at day 24 after remediation with 15 g/l of $KMnO_4$ is $1.29 \times 10^{-5}\,\text{mg/l}$
  Thus Reduction Ratio [RR] $= \frac{1.29 \times 10^{-5}}{21.86} = 5.90 \times 10^{-7}$
- Reduction Ratio for remediating with 50 g/l of $KMnO_4$
  BTEX concentration at day 24 after remediation with 50 g/l of $KMnO_4$ is $1.00 \times 10^{-5}\,\text{mg/l}$
  Thus Reduction Ratio $= \frac{1.00 \times 10^{-5}}{21.86} = 4.575 \times 10^{-7}$

Table 4. Injected BTEX concentrations for field-scale model and simulation results

Simulation time (years)						
0	Injected concentration	87.2 g/l	436 g/l	872 g/l	4360 g/l	8720 g/l
0.5	Max. observed concentration	105 mg/l	200 mg/l	550 mg/l	1020 mg/l	1800 mg/l
12	Max. observed concentration	18 mg/l	45 mg/l	100 mg/l	430 mg/l	730 mg/l
20	Max. observed concentration			35 mg/l		
40	Max. observed concentration			10 mg/l		
60	Max. observed concentration			0.5 mg/l		

Source: (Ojuri 2009).

Table 5. Spatial Distribution of Hydrocarbon concentrations for the 4th Sampling session (Day 24)

Groundwater sample ID	Benzene (ppm)	Toluene (ppm)	Ethyl benzene (ppm)	Xylene (ppm)	BTEX (ppm)
4AM	1.49E-04	1.05E-06	6.32E-06	4.25E-05	1.04E-04
4BM	1.28E-06	1.20E-05	6.53E-06	4.34E-06	1.10E-04
4EM	4.33E-06	8.84E-07	6.54E-07	4.09E-06	9.07E-04
Control	6.91E-09	2.33E-09	2.93E-08	2.59E-07	2.97E-07

Source: (Ojuri 2009).

3.2 For Field-Scale

The Total injected BTEX Concentration at the initial time is 872 g/l for field – scale model, it would take 540 months for BTEX concentration to deplete to 0.5 mg/l (Ojuri 2009).

Thus, Reduction Ratio $= \frac{0.5}{872000} = 5.734 \times 10^{-7}$

Hence, Prediction of time at which BTEX concentration of 872 g/l at the initial time (t = 0), would deplete to 0.5 mg/l remediating with 15 g/l of $KMnO_4$.

RR of BTEX for NA laboratory ≡ RR of BTEX for NA field
RR of BTEX for 15 g/l $KMnO_4$ laboratory ≡ RR of BTEX for 15 g/l $KMnO_4$ field

Table 6. Correlation of Laboratory Data with Field – Scale Data for Natural Attenuation, 15 g/l and 50 g/l of $KMnO_4$

LABORATORY DATA			FIELD DATA		
Natural Attenuation	15 g/l KMnO4	50 g/l KMnO4	Natural Attenuation	15 g/l KMnO4	50 g/l KMnO4
21.86(mg/l)	21.86(mg/l)	21.86 (mg/l)	872g/l	872g/l	872g/l
24 Days ↓	24 Days ↓	24 Days ↓	540 Months ↓	x_1 Months ↓	x_2 Months ↓
1.04×10^{-4} mg/l (A middle) Ojuri 2009	1.29×10^{-5}mg/l	1.00×10^{-5} mg/l	0.5 mg/l (Ojuri 2009)	0.5 mg/l	0.5 mg/l

$\therefore$ RR of BTEX for 15g/1 $KMnO_4$ field

$$= \frac{\text{RR of BTEX for 15g/1 KMnO}_4\text{ laboratory}}{\text{RR of BTEX for NA laboratory}} \times \text{RR of BTEX for NA field}$$

Reduction Ratio of BTEX Concentration using 15 g/l $KMnO_4$ at day 24 in laboratory = 5.90×10^{-7}
Reduction Ratio of BTEX Concentration for Natural Attenuation at day 24 in laboratory = 4.76×10^{-6}
Reduction Ratio of BTEX Concentration on field-scale = 5.734×10^{-7}

$$\frac{5.90 \times 10^{-7}}{4.76 \times 10^{-6}} \times 5.734 \times 10^{-7} = 7.107 \times 10^{-8}$$

Then correlating laboratory to field scale

$$\text{If } 5.734 \times 10^{-7} \equiv 540\,\text{months}$$
$$7.107 \times 10^{-8} = x_1$$

Where x_1 = Time on field remediating with 15 g/l of $KMnO_4$ for Total BTEX Concentration 872 g/l to deplete to 0.5 mg/l on field-scale

$$x_1 = \frac{7.107 \times 10^{-8} \times 540\,\text{months}}{5.734 \times 10^{-7}}$$
$$x_1 = 66.9\,months$$
$$x_1 \cong 67\,months$$

Hence, x_1 at Observation well A middle is 67 months
Using 50 g/l of $KMnO_4$, time at which total injected BTEX concentration of 872 g/l at initial time to deplete to 0.5 mg/l

Using the Expression

$$\therefore \text{RR of BTEX for } 5\text{g/l KMnO}_4 \text{ field} = \frac{\text{RR of BTEX for } 15\text{g/l KMnO}_4 \text{ laboratory}}{\text{RR of BTEX for NA laboratory}} \times \text{RR of BTEX for NA field}$$

Reduction Ratio of BTEX Concentration using 50 g/l KMnO4 at day 24 in laboratory = 4.575×10^{-7}
Reduction Ratio of BTEX Concentration for Natural Attenuation at day 24 in laboratory = 4.76×10^{-6}

Reduction Ratio of BTEX Concentration on field-scale = 5.734×10^{-7}

$$\frac{4.575 \times 10^{-7}}{4.76 \times 10^{-6}} \times 5.734 \times 10^{-7} = 5.511 \times 10^{-8}$$

Then correlating laboratory to field-scale
If $5.734 \times 10^{-7} \equiv 540$ months

$5.511 \times 10^{-8} = x_2$

Where x_2 = Time on field remediating with 50 g/l of $KMnO_4$ for Total BTEX Concentration of 872 g/l to deplete to 0.5 mg/l on field-scale

$$x_2 = \frac{5.511 \times 10^{-8} \times 540 \text{ months}}{5.734 \times 10^{-7}}$$

$$x_2 = 51.8\, months$$

$$x_2 \cong 52\, months$$

Hence, x_2 at Observation well A middle is 52 months.

Following the same method for the calculations for clean-up time of observation well B middle and observation well E at 24th day will give:

- x_1 at Observation well B middle is 98 months
- x_1 at Observation well E middle is 41 months

Average of x_1 (i.e. Time on field remediating with 15 g/l of $KMnO_4$ for Total BTEX Concentration 872 g/l to deplete to 0.5 mg/l on field-scale) for $OW_{A\ (Middle)}$, $OW_{B\ (Middle)}$ and $OW_{E\ (Middle)}$

x_1 at Observation well A middle is 67 months
x_1 at Observation well B middle is 98 months
x_1 at Observation well E middle is 41 months

$$\text{Average of } x_1 = \frac{OW_{A(Middle)} + OW_{B(Middle)} + OW_{E(Middle)}}{3}$$

$$\text{Average of } x_1 = \frac{67 + 98 + 41}{3} = 68.6\, months$$

$x_1 \cong 69\,months$, i.e., 6 years

Hence, Time on field remediating with 15 g/l of $KMnO_4$ for Total BTEX Concentration 872 g/l to deplete to 0.5 mg/l on field –scale would be 6 years

Average of x_2 (i.e. Time on field remediating with 50 g/l of $KMnO_4$ for Total BTEX Concentration 872 g/l to deplete to 0.5 mg/l on field-scale) for OWA $_{\text{(Middle)}}$, OWB $_{\text{(Middle)}}$ and OWE $_{\text{(Middle)}}$.

x_2 at Observation well A middle is 52 months.
x_2 at Observation well B middle is 49 months
x_2 at Observation well E middle is 4 months

$$\text{Average of } x_1 = \frac{\text{OW}_{\text{A(Middle)}} + \text{OW}_{\text{B(Middle)}} + \text{OW}_{\text{E(Middle)}}}{3}$$

$$\text{Average of } x_2 = \frac{52+49+4}{3} = 35\,months$$

$$x_2 = 35\,months, \text{ i.e., } 3\,\text{years}$$

Hence, Time on field remediating with 50 g/l of $KMnO_4$ for Total BTEX Concentration 872 g/l to deplete to 0.5 mg/l on field –scale would be 3 years

4 Conclusion

From the study, it is obvious that chemical oxidation can be used predominantly to address contaminant in the source area, saturated zone and capillary fringe, while treatment of the source area (saturated zone) requires another remedial technology (e.g. free product recovery) need to be conducted prior to chemical oxidation for the remediation to be safe and cost effective.

From both experimental data and modeling simulation carried out on Natural attenuation of the same Contaminant of Concern gave a clean-up of about 45 years (Ojuri 2009), from the study, empirically, the clean-up for 15 g/l of $KMnO_4$ will take about 6 years while the clean-up period for 50 g/l of $KMnO_4$ will take about 3 years when apply on pilot scale, oxidant should be applied at sufficient number of points/wells such that there would be adequate overlap of effective zones where an oxidant is in contact with contaminants.

Acknowldgement. This research was carried out through the funds given by TETFUND NATIONAL RESEARCH FUND (NRF) NIGERIA. The authors therefore acknowledge the TETFUND National Research Fund of The Prof. S.A. Ola Research Group, Federal University of Technology, Akure referenced TETF/ES/NRF/013/VOL. I for the Research Project titled "Site Remediation in Nigeria: Proven and Innovative Technologies, Recovery of Free Hydrocarbon from Soil/Groundwater".

References

Khan, F.I., Husain, T., Hejazi, R.: An overview and analysis of site remediation technologies. J. Environ. Manag. **71**, 95–122 (2004)

Ojuri, O.O.: Geo-Environmental Assessment and Remediation of Contaminated Sites. A Case Study of Baruwa Community, Lagos, Nigeria, Ph.D. thesis (2009)

Ojuri, O.O., Ola, S.A.: Estimation of contaminant transport parameters for tropical sand in a sand tank model. Int. J. Environ. Sci. Tech. **7**(2), 385–394 (2010). ISSN: 1735-1472

Ola, S.A., Ojuri, O.O.: Remediation of Hydrocarbon Contaminated Sites. Paper presented at the 2nd ISSMGE International Lecture Tour in Environmental Geotechnics (NSE-NGA/ISSMGE), Nigeria, 14–16 January 2008

Suthersan, S.: Remediation Engineering: Design Concepts. Environmental Science and Engineering Series. Lewis, Boca Raton, Florida (1996). Tóth, S.: Soilcrete – Dichtwand als vorsorgliche (1989)

US Office of Water. Containment and Disposal of Large Amounts of Contaminated Water: A Support Guide for Water Utilities, EPA (4608T) EPA 817-B-12-002. September 2012 US Environmental Protection Agency, Washington, D.C. (2012)

Sharma, H.D., Reddy, K.R.: Geoenvironmental Engineering: Site Remediation, Waste Containment, and Emerging Waste Management Technologies. John Wiley & Sons, Hoboken (2004)

USEPA, Assessment of Barrier Containment Technologies, International Containment Technology Workshop, Baltimore, Maryland, 29–31 August 1995

U.S. Environmental Protection Agency. Subsurface Contamination Reference Guide, EPA/540/2-90/011, Office of Emergency and Remedial Response, Washington, D.C. (1990)

Evaluation of Land Subsidence Based on Distributed Monitoring and SEM Analysis

Hongtao Jiang[1], Kai Gu[2,3(✉)], Jianhua Yin[2], Bin Shi[2], and Jiayu Ma[2]

[1] School of Geographic and Oceanographic Sciences, Nanjing University, Nanjing 210023, China
[2] School of Earth Sciences and Engineering, Nanjing University, Nanjing 210023, China
gukai@nju.edu.cn
[3] Suzhou High-Tech Institute, Nanjing University, Suzhou 215123, China

Abstract. A better understanding on the relationship between macro compression behavior and microstructure of soil is essential in order to evaluate the trend of land subsidence. In this paper, distributed fiber optic sensing (DFOS) technology was applied in a borehole located in Guangming Village, Wuxi, China, to obtain the deformation of soil layers after the prohibition of deep groundwater withdrawal in this area. DFOS provided detailed information on the movement of soil layer within any depth range. The microstructure of undisturbed soil samples from the borehole was analyzed using SEM images. With the application of Particle/Pore and Crack Analysis System (PCAS), the micro porosity of samples were characterized by two statistical parameters called area probability distribution index and fractal dimension. Area probability distribution index described the distributions of micro-porosities and fractal dimension described the shape of pores in nature regardless of complexity. Good correlations between the deformation of soil layers and the two parameters about the microstructure characteristics were illustrated.

Keywords: Land subsidence · DFOS · SEM · Compressibility

1 Introduction

Due to the excessive withdrawal of fluid such as groundwater and hydrocarbon resources, land subsidence has widely occurred in China (Chai et al. 2004; Shi et al. 2008, 2007; Wang et al. 2009; Yi et al. 2011). Among others, Suzhou-Wuxi-Changzhou area located in the lower reaches of the Yangtze River with an area over 12,000 km^2, known as Su-Xi-Chang area and one of the most developed region in China, has been suffering from the severe land subsidence due to the extraction of groundwater for domestic and industrial uses since 1960s. In 2002, the maximum cumulative subsidence of Su-Xi-Chang area reached 2.00 m (Hu 2011). In order to slow down and control the regional land subsidence, the local government of Jiangsu prohibited the deep groundwater withdrawal since 2005, since when subsidence in this area has been significantly controlled from the macro viewpoint (Hu 2011). Although the annual subsidence rate has been controlled, the subsidence continues mainly

D.N. Singh and A. Galaa (eds.), *Contemporary Issues in Geoenvironmental Engineering*, Sustainable Civil Infrastructures,
DOI 10.1007/978-3-319-61612-4_16

attributing to the relatively small withdrawal of groundwater for special purposes and the slow deformation of soil layers of low permeability.

The mechanism of land subsidence can be generally summarized as the withdrawal of fluid induces the increase in effective stress on solid skeleton and eventually results in the compression of soil. The whole process of subsidence can be divided into four stages according to the rate of subsidence: (1) initial stage with subsidence rate <10 mm/a; (2) developing stage with subsidence rate >20 mm/a; (3) continuation stage and (4) stable stage when subsidence almost stops. Although there are many methods to obtain the deformation history of soil layers, the evaluation for soil deformation trend is still a huge challenge, due to the lack of accurate description on soil deformation distribution.

Currently, the monitoring techniques for land subsidence can be classified into two main categories (Galloway and Burbey, 2011): (1) Ground-based geodetic surveys and techniques: Precise differential leveling, Global positioning system (GPS), Tripod LiDAR, Extensometry; (2) Remotely sensed geodetic surveys and techniques: InSAR, LiDAR. In most cases, the displacement obtained by these methods (except the layerwise mark method) is the whole movement of the land surface, without telling the specific displacement of each aquitard and aquifer in the system. Recently, Wu et al. (2015) reported the application of distributed fiber optic sensing (DFOS) technologies in land subsidence monitoring by illustrating the sensing cables, the monitoring scheme and the deformation characteristics of soil layers. The result indicated that DFOS is a very advanced technique for monitoring the soil subsidence due to the distributed monitoring pattern and high resolution. With this monitoring pattern, the strain of the optical fiber, which can be transited to the deformation of soil, at any depth of the borehole can be obtained over time. This helps us to better understand the deformation history, the amount of deformation, and the rate of deformation of each soil layer.

This paper reports the preliminary investigation on land subsidence using the combination of DFOS technology and SEM analysis on soil microstructures. A borehole located in Guangming Village, Wuxi was taken an example and the deformation of soil along the whole borehole was obtained. The relationship between the macro compression behavior and the microstructure parameters was discussed.

2 Land Subsidence Monitoring with Distributed Fiber Optic Sensing Technology

A borehole, namely GM03, located in Guangming Village, Wuxi, was drilled for the integrated investigation of land subsidence. During the quaternary, this area underwent alternating periods of marine transgression and regression which induced the coexistence of marine and continental sediments. The complex aquifer system consists of intercalated sands, silts, silty clays and clays. The borehole was 147 m in depth and 129 mm in diameter. Figure 1 illustrated the monitoring scheme in GM03 using DFOS technology. The borehole was drilled until the bedrock emerged. Then the sensing cable was installed in the borehole with the help of a heavy guider, which can drag the cable down to the bottom and also keep the cable tight during the installation. The borehole was then backfilled with a mixture of gravel, sand and clay until the borehole

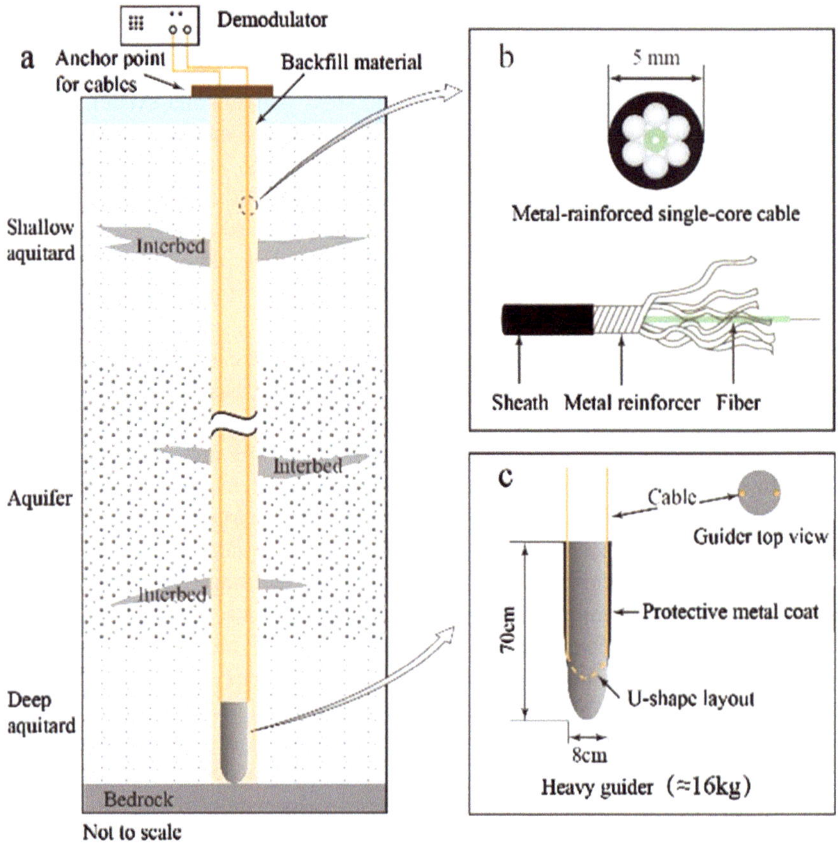

Fig. 1. Monitoring scheme for land subsidence using DFOS technology. (a) Sensing cable installation; (b) Metal-reinforced single core cable; (c) Heavy guider for cable installation

was completely filled. Note that, the cable should always be in a state of tension during this process, which may take up to weeks. As long as the backfill materials in the borehole compacted and reached a stable status, the initial strain of the embedded cables was acquired.

Figure 2 illustrates the strain distribution of monitoring cable in the borehole. Due to the weak coupling between the cable and the soil, strain data in the phreatic layer was not adopted here. As can be seen, there was no clear variation of strain from Ad1 to Af1. The small fluctuation of data can be demonstrated by the noise. Sediment layers deeper than 78.1 m showed slightly compaction. In this depth range, the cable obtained negative strain data.

By integrating the strain data, the deformation of soil within any depth range can be calculated (i.e. d1, d2, and d3 in Fig. 2). For the simplification, however, Fig. 3 only shows the amounts of deformation of each main sediment layer since the monitoring. A clear development of the soil displacement with time can be observed. During spring

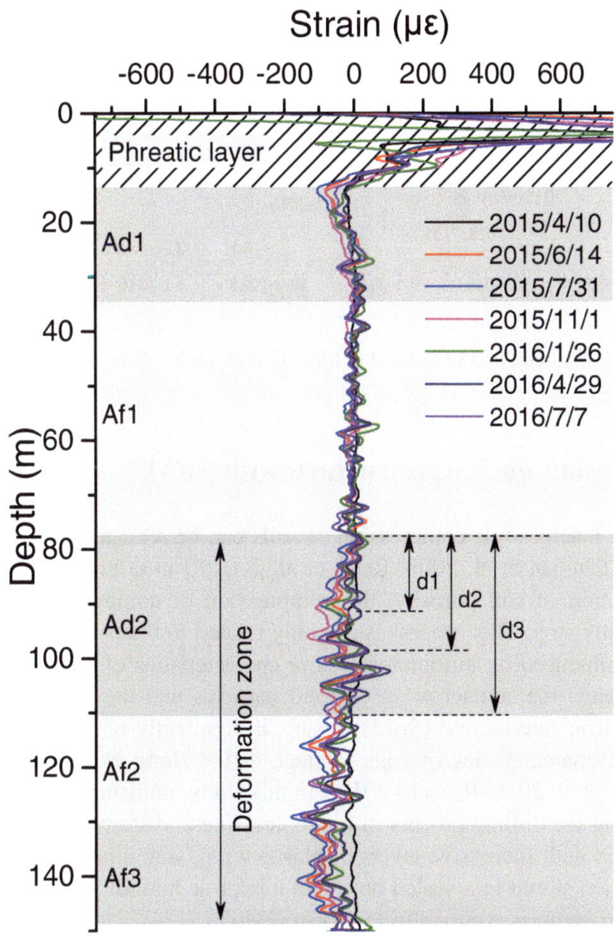

Fig. 2. Strain distribution of optical fibers in GM03

and summer (from April to September), the compression of soil significantly developed in the deeper three layers (Ad2, Af2 and Ad3), which contributed most to the subsidence. Then soil layers gradually rebounded until winter due to the decrease of effective stress on soil skeleton when groundwater level raised. As can be seen, the deformation of some soil layers were reversible while others were not. Ad2 compressed in the period from June 2015 to January 2016, then it rebounded to its initial thickness around April, 2016 and compressed again. The deformations of Af2 and Ad3 also varied with time, but by contrast the compression tends to accumulate in the coming years. Generally, due to the high spatial resolution (up to 5 cm in this study), DFOS provides detailed movement information of soil layers along the borehole.

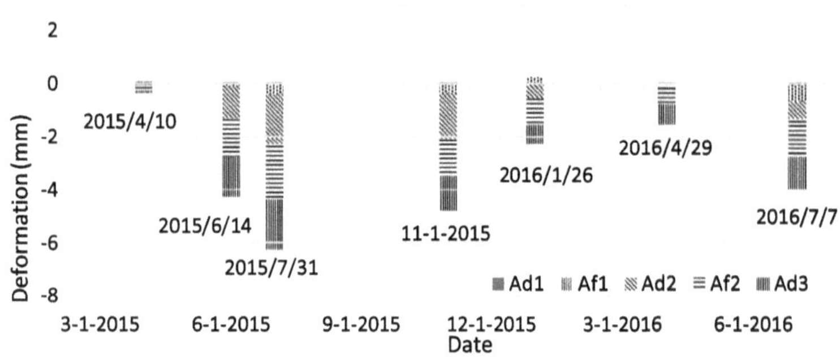

Fig. 3. Total deformation and individual deformation of each soil layer compared to the initial status in GM03

3 Investigation on Microstructure with SEM

Generally, the macroscopic compression of soils can be recognized as the decrease in soil porosity (Chuhan et al. 2003; Revil et al. 2002; Tan et al. 2010). With the irreversible reduction of soil porosity, the compression of compressible soil may slow down and finally stop. This process is not only related to the hydraulic conditions, but also greatly influenced by the microstructure characteristics of soils such as the morphology of pores, the contact ways of solid particles and the pore size distribution. Scanning electron microscope (SEM) images is frequently used to investigate these microstructure characteristics of soils (Delage 2010; Hong et al. 2006; Koliji et al. 2010; Monroy et al. 2010; Romero 2013). In this study, undisturbed soil samples were collected during the drilling process of boreholes. The undisturbed sample blocks were carefully sealed with successive layers of plastic wrap, aluminium foil and wax. These blocks were then stored in a sealed box with a relative humidity of 99 ± 1% for later tests. In all, 26 samples were collected form GM03.

The samples for SEM investigation ($\sim 20 \times 5 \times 5$ mm in size) were cut form the sample blocks with care. Then they were quickly frozen at about −60°C followed by vacuum drying for 48 h. The SEM images were captured with Hitachi SU3500. To quantitatively analyze the pores in these images, we used Particle/Pore and Crack Analysis System (PCAS) to conduct the quantification and characterization of micro porosity. The details of the system can be referred to Liu et al. (2011). Figure 4 shows an example of the SEM image processing using PCAS. A magnification of 1000 of the images was adopted here. At this magnification, the side length of each pixel is 0.1 μm and can ensure most of the pores recognized by the analysis system. During the segmentation of an image, a threshold (T) should be carefully selected to transfer the original image (Fig. 4a) to the binary image (Fig. 4b). As can be seen from Fig. 3, a SEM image is always complex with pores of diverse sizes. At a certain magnification, some very small pores (e.g. intra-aggregate pores) cannot be correctly determined and will also affect the statistics results of the pores in the image, therefore they should be removed. The PCAS allows a selection of a minimum analyzed pore area (so called "minimum area"), under which the pores in the binary SEM image will be removed,

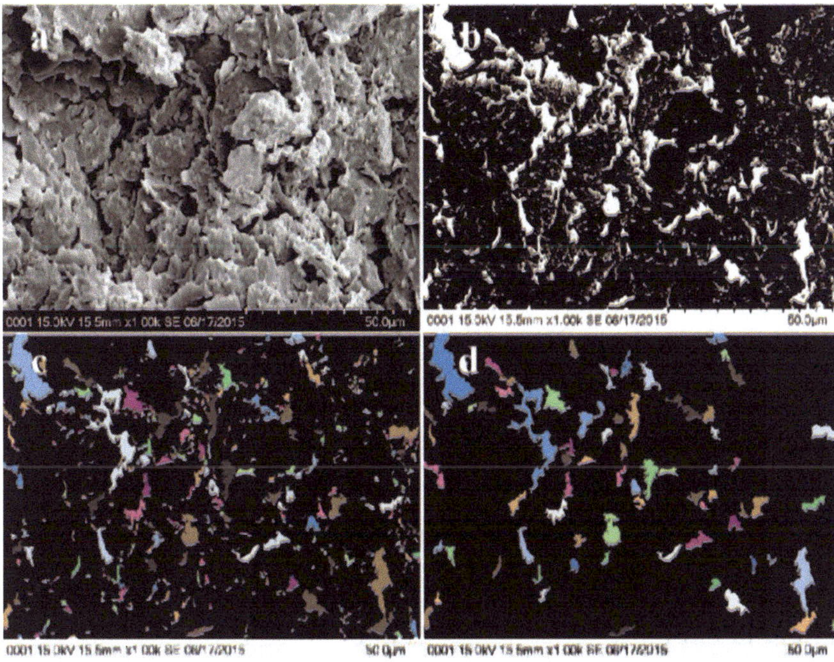

Fig. 4. Original SEM image and PCAS proceeded images. (a) Original SEM image at a magnification of × 1000; (b) Binary image at T = 36, black area for soil particles and white area for pores; (c) Pore characterization with a minimum area of 50 pixels; (d) Pore characterization with a minimum area of 400 pixels.

hence a proportion of small pores which are not significantly affected by the mechanical compression can be removed. Figure 4c and d show an example by changing minimum area from 50 pixels to 400 pixels. Finally, every pore larger than the minimum area can be recognized by the system with the information on area (*S*), perimeter (*C*), length, width, long-axis direction, etc.

With the abovementioned information on every pore, two statistical parameters which were area probability distribution index (*b*) and fractal dimension (D_f) were used to describe the geometric characteristics of the pores in this study. The probability distribution function, *f*(*S*), describes the density of pores within a specific area range as illustrated in Fig. 5. As can be seen, the pore area can be divided into several area range (e.g. 0–a1, a1–a2, a2–a3, etc.). By counting the numbers of pores whose area value fell in each area range, the density of pores within a specific area range can be determined. The distribution of the densities has a power function relationship with the area as expressed by:

$$f(s) = a \cdot s^{-b} \tag{1}$$

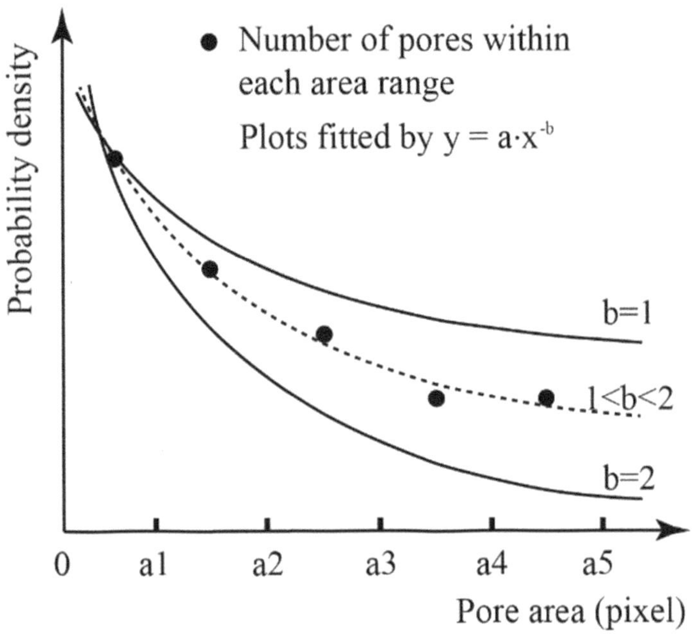

Fig. 5. The concept of area probability distribution index, *b*

Where *b* is area probability distribution index and *a* is determined by *b* according to the following equation:

$$a = \frac{1 - b}{(2^{n(1-b)} - 1) \cdot s_0^{1-b}} \tag{2}$$

D_f was developed with the purpose of describing a shape of pore in nature regardless of complexity (Dathe et al. 2001; Dathe and Thullner 2005; Moore and Donaldson 1995; Xu and Sun 2005). Here D_f of each sample was determined by plotting the perimeter of pore (*C*) and the area of pore (*S*) on a log–log coordinates. The $\log(C) - \log(S)$ data will show a simple linear form, and the slope of the approximated line is $D_f/2$.

Figure 6 presents the two parameters of each sample. The average values within each main layer were also calculated. As can be seen from Fig. 6a, the area probability distribution index (*b*) of samples in Ad2, Af2 and Ad3 are generally larger than that in shallower layers, although the result is somehow scattered due to the complexity of the soil system. Smaller *b* indicates higher probability density of pores within large area range as illustrated in Fig. 5. Usually large pore are easier to be compressed while small pores in soil are not significantly affected by mechanical compression (Delage and Lefebvre 1984; Griffiths and Joshi 1991, 1990, 1989), therefore larger *b* values from SEM images revealed that the soils are more compressible. The larger deformation of Ad2, Af2 and Ad3 obtained by DFOS (Figs. 2 and 3) agrees well with this deduction.

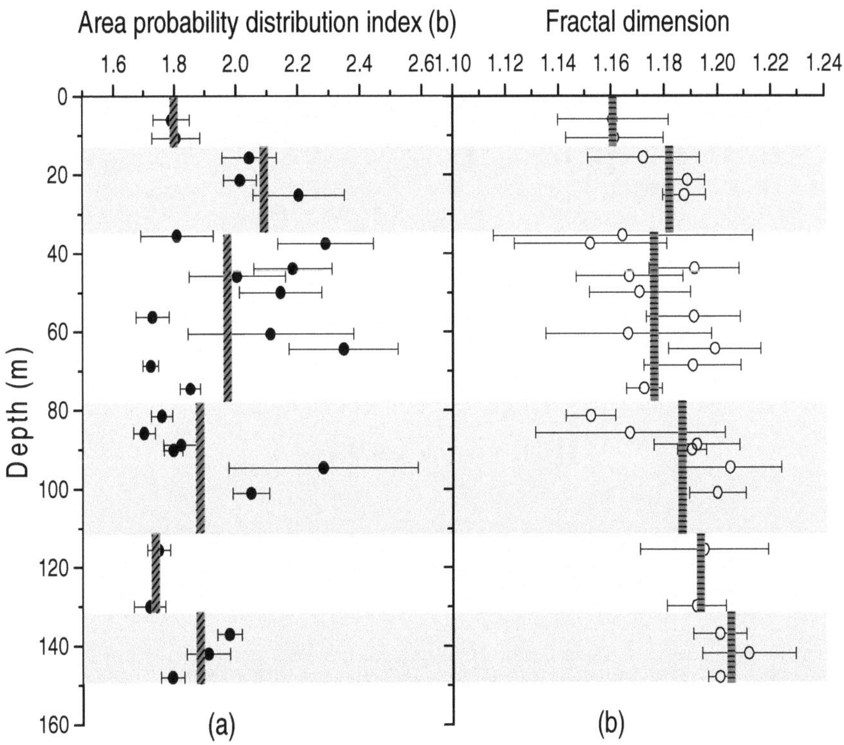

Fig. 6. Variation of area probability distribution index and fractal dimension with depth, average values were calculated within each main soil layer

Figure 6b shows the distribution of D_f with depth. The variation of D_f is generally in the range of 1.14 and 1.21. The average D_f in the lower three layers are slightly larger than the shallower layers, due to the higher density of large pores as explained in Fig. 6a. Vallejo (1996) investigated the fabric changes in a consolidating clay and found that a rougher shape has a larger fractal dimension. Indeed, as can be seen from Fig. 4c and d, larger pores are usually rougher than small pores. By changing the minimum area in binary image using PCAS, the effect of pore size on D_f was quantitatively evaluated and shown in Fig. 7. When the minimum area was selected as 50 pixel, D_f was at a low level (the average is 1.18). The average D_f increased to 1.25 and 1.30 when the minimum area was set as 200 pixels and 400 pixels, respectively.

The microstructure analysis based on SEM images provides a better understanding on the deformation of soil layers as obtained by the DFOS technique. The larger deformation of the deep three layers can be attributed to the more compressible soil structures, which have smaller area probability distribution index and larger fractal dimension. By plotting the maximum compression of each soil layer obtained on 2015/4/29 with the average area probability distribution index with each soil layer, we have a very good linear relationship. Therefore the area probability distribution index could be used to evaluate the potential of compressibility of soil layer. Additionally,

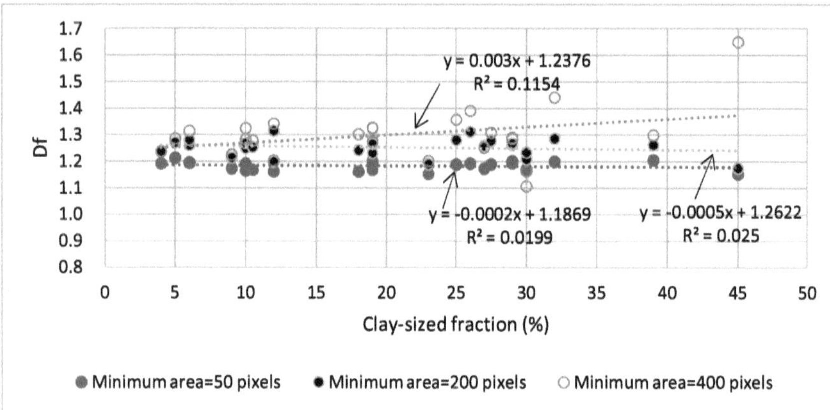

Fig. 7. Effect of pore size on D_f

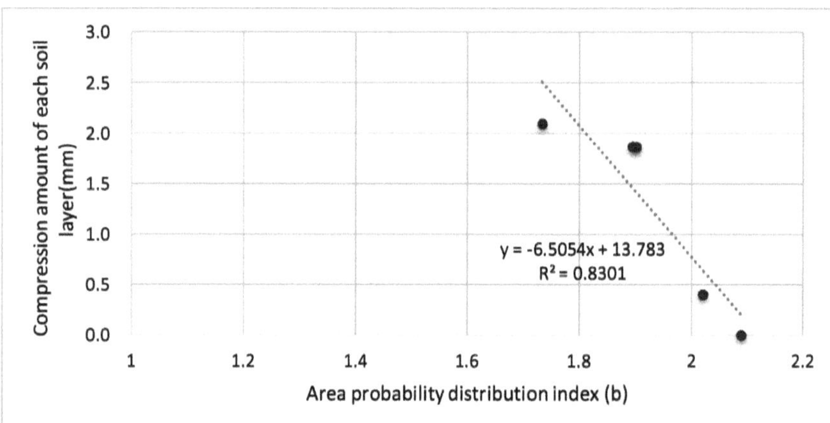

Fig. 8. Relationship between compression amount and area probability distribution index

fractal dimension not only describes the complexity of the pores, but also indirectly reflects the proportion of large pores in the soil (Fig. 8).

4 Conclusions

This paper presents the results of an investigation on land subsidence in Su-Xi-Chang area, China, with distributed fiber optic sensing (DFOS) technology and SEM analysis. The following conclusions can be drawn:

1. Due to the characteristic of "distributed", DFOS can provide detailed information on the movement of soil layer, which allows a better understanding on the movement of soil layers therefore the mechanism and the trend of land subsidence.

2. The macro mechanical behavior of soil is controlled by the microstructure. SEM analysis on the undisturbed soil samples form the borehole indicated the smaller the area probability distribution index is, the more compressible the soil layer is. The large pores tends to be more complex in shape than small pores.
3. Although there are many methods to observe the deformation history of land subsidence, the combination of DFOS technology and microstructure analysis provide an access to the evaluation of the trend of soil layer movement, by illustrating a better picture on the whole process of land subsidence, and it requires further investigations.

Acknowledgments. Financial support from National Natural Science Foundation of China (No. 41372265, 41230636, 41502274), Natural Science Foundation of Jiangsu Province (No. BK20150389), China Postdoctoral Science Foundation (No. 2015M580414), Open foundation of State Key Laboratory of Geohazard Prevention and Geoenvironmental Protection, Chengdu University of Technology (SKLGP2016K010) was greatly appreciated.

References

Chai, J.-C., Shen, S.-L., Zhu, H.-H., Zhang, X.-L.: Land subsidence due to groundwater drawdown in Shanghai. Géotechnique **54**, 143–147 (2004). doi:10.1680/geot.2004.54.2.143

Chuhan, F.A., Kjeldstad, A., Bjørlykke, K., Høeg, K.: Experimental compression of loose sands: relevance to porosity reduction during burial in sedimentary basins. Can. Geotech. J. **40**, 995–1011 (2003). doi:10.1139/t03-050

Dathe, A., Eins, S., Niemeyer, J., Gerold, G.: The surface fractal dimension of the soil-pore interface as measured by image analysis. Geoderma **103**, 203–229 (2001). doi:10.1016/S0016-7061(01)00077-5

Dathe, A., Thullner, M.: The relationship between fractal properties of solid matrix and pore space in porous media. Geoderma **129**, 279–290 (2005). doi:10.1016/j.geoderma.2005.01.003

Delage, P.: A microstructure approach to the sensitivity and compressibility of some Eastern Canada sensitive clays. Géotechnique **60**, 353–368 (2010). doi:10.1680/geot.2010.60.5.353

Delage, P., Lefebvre, G.: Study of the structure of a sensitive Champlain clay and of its evolution during consolidation. Can. Geotech. J. **21**, 21–35 (1984). doi:10.1139/t84-003

Galloway, D.L., Burbey, T.J.: Review: regional land subsidence accompanying groundwater extraction. Hydrogeol. J. **19**, 1459–1486 (2011). doi:10.1007/s10040-011-0775-5

Griffiths, F.J., Joshi, R.C.: Change in pore size distribution owing to secondary consolidation of clays. Can. Geotech. J. **28**, 20–24 (1991). doi:10.1139/t91-003

Griffiths, F.J., Joshi, R.C.: Clay fabric response to consolidation. Appl. Clay Sci. **5**, 37–66 (1990). doi:10.1016/0169-1317(90)90005-A

Griffiths, F.J., Joshi, R.C.: Change in pore size distribution due to consolidation of clays. Géotechnique **40**, 303–309 (1989). doi:10.1680/geot.1990.40.2.303

Hong, Z., Tateishi, Y., Han, J.: Experimental study of macro- and microbehavior of natural Diatomite. J. Geotech. Geoenviron. Eng. **132**, 603–610 (2006). doi:10.1061/(ASCE)1090-0241(2006)132:5(603)

Hu, J.P.: A study on the land subsidence effect after prohibiting extraction of groundwater in Suzhou- Wuxi-Changzhou area. Ph.D. thesis, Nanjing University (2011)

Koliji, A., Vulliet, L., Laloui, L.: Structural characterization of unsaturated aggregated soil. Can. Geotech. J. **47**, 297–311 (2010). doi:10.1139/T09-089

Liu, C., Shi, B., Zhou, J., Tang, C.: Quantification and characterization of microporosity by image processing, geometric measurement and statistical methods: application on SEM images of clay materials. Appl. Clay Sci. **54**, 97–106 (2011). doi:10.1016/j.clay.2011.07.022

Monroy, R., Zdravkovic, L., Ridley, A.: Evolution of microstructure in compacted London Clay during wetting and loading. Géotechnique **60**, 105–119 (2010). doi:10.1680/geot.8.P.125

Moore, C.A., Donaldson, C.F.: Quantifying soil microstructure using fractals. Géotechnique **45**, 105–116 (1995). doi:10.1680/geot.1995.45.1.105

Revil, A., Grauls, D., Brevart, O.: Mechanical compaction of sand/clay mixtures. J. Geophys. Res. **107**, 2293 (2002). doi:10.1029/2001JB000318

Romero, E.: A microstructural insight into compacted clayey soils and their hydraulic properties. Eng. Geol. **165**, 3–19 (2013). doi:10.1016/j.enggeo.2013.05.024

Shi, X., Wu, J., Ye, S., Zhang, Y., Xue, Y., Wei, Z., Li, Q., Yu, J.: Regional land subsidence simulation in Su-Xi-Chang area and Shanghai City, China. Eng. Geol. **100**, 27–42 (2008). doi:10.1016/j.enggeo.2008.02.011

Shi, X.Q., Xue, Y.Q., Ye, S.J., Wu, J.C., Zhang, Y., Yu, J.: Characterization of land subsidence induced by groundwater withdrawals in Su-Xi-Chang area, China. Environ. Geol. **52**, 27–40 (2007). doi:10.1007/s00254-006-0446-3

Tan, Z., Kong, L., Guo, A., Wan, Z.: Research on effect of compaction on pore size distribution of laterite soil. Rock Soil Mech. **31**, 1427–1430 (2010). (in Chinese)

Vallejo, L.E.: Fractal analysis of the fabric changes in a consolidating clay. Eng. Geol. **43**, 281–290 (1996). doi:10.1016/S0013-7952(96)00038-5

Wang, G.Y., You, G., Shi, B., Yu, J., Tuck, M.: Long-term land subsidence and strata compression in Changzhou, China. Eng. Geol. **104**, 109–118 (2009). doi:10.1016/j.enggeo.2008.09.001

Wu, J., Jiang, H., Su, J., Shi, B., Jiang, Y., Gu, K.: Application of distributed fiber optic sensing technique in land subsidence monitoring. J. Civ. Struct. Heal. Monit. (2015). doi:10.1007/s13349-015-0133-8

Xu, Y.F., Sun, D.A.: Correlation of surface fractal dimension with frictional angle at critical state of sands. Géotechnique **55**, 691–695 (2005). doi:10.1680/geot.2005.55.9.691

Yi, L., Zhang, F., Xu, H., Chen, S., Wang, W., Yu, Q.: Land subsidence in Tianjin, China. Environ. Earth Sci. **62**, 1151–1161 (2011). doi:10.1007/s12665-010-0604-5

An Assessment of Crude Oil Leakage Beneath a Tank in a Refinery in the Niger Delta Sub-region Nigeria, Through Geotechnical Techniques

So-ngo Clifford Teme[1(✉)], Oruene Olem-Ogoin[2], and Prosper Ogbunie[2]

[1] Geotechnical Division, Department of Geology and Institute of Geosciences and Space Technology, Rivers State University of Science and Technology, Port Harcourt, Nigeria
cliffordteme@yahoo.com
[2] Teks Geotechnical Consultancy (Nigeria) Limited, Port Harcourt, Nigeria
teksgeotech@yahoo.com

Abstract. A substantial amount of crude oil was alleged to have leaked beneath a Tank at a refinery situated within the Niger Delta Sub-region of Nigeria. Our geotechnical consultancy firm was engaged to find a method of estimating the amount of crude oil lost through leakage beneath this Tank. A geotechnical technique that involved a very careful manual drilling of Standard Penetration Test (SPT) holes within the confinement of this 64-meter diameter tank to depths of 20 meters was carried out. The manual drilling was undertaken to avoid any incidences of fire outbreak that may arise from the use of petrol engines commonly used in percussion drillings in the presence of the flammable crude oil remnants within this crude-oil drained tank, during project execution. This technique was used to retrieve crude-oil contaminated soil samples within the x4No.borings carried out within the crude oil tank. Applying the **"4-Phase Concept of Air/Crude Oil/Water/Soils"** after Otunyo (1993), Teme and Otunyo (1999), an estimated ***Total Crude Oil Equivalent*** of ***about 545.44 US Barrels of crude oil,*** was found to have leaked beneath the crude oil Tank. Soil profiles encountered during the boring with their engineering properties included (i) Brownish Lateritic Clayey Sands ***(SC)***; with friction angle (ϕ_u) of = 8°; unconfined Cohesive strength (C_u) of 14.4 kPa and average bearing pressures $(q_{u\ (ave.)})$ of 566.54 kPa; (ii) Darkish well-graded Sands (SW) with friction angle (ϕ_u) of = 34°; unconfined Cohesive strength (C_u) of 0.00 kPa and average bearing pressures $(q_{u\ (ave.)})$ of 129.83 kPa; (iii) Brownish Lateritic Silty Sandy Clays (SM-SC) with friction angles (ϕ_u) of = 8–14°; unconfined Cohesive strength (C_u) of 24.8-26.60 kPa and average bearing pressures $(q_{u\ (ave.)})$ of *151.472 kPa* and (iv) Reddish-brown lateritic Silty Clays (ML) with friction angles (ϕ_u) of = 4–8°; unconfined Cohesive strength (C_u) of 28.6–35.20 kPa and average bearing pressures $(q_{u\ (ave.)})$ of *179.012 kPa*. Additionally, a 20-Ton CPT Machine was employed to evaluate the soil bearing capacities surrounding the almost 49-year old Crude Oil Tank. The values of the maximum soil bearing pressures at 6-equidistant positions (60°) outside the tank varied from 2300–3500 kg/cm^2 [225.56 – 343.25] kPa, while the maximum sleeve-frictions varied

D.N. Singh and A. Galaa (eds.), *Contemporary Issues in Geoenvironmental Engineering*, Sustainable Civil Infrastructures,
DOI 10.1007/978-3-319-61612-4_17

from 100.00–420.00 kg/cm^2 [96.177–41.19] kPa, at depths varying from 1.00 m to 16.80 m. This paper describes the detailed methodology of planning, execution and computations of the various parameters involved in this novel study.

Keywords: Total crude oil equivalent content · Geotechnical techniques · Sub-soil geotechnical engineering properties

1 Introduction

The Niger Delta sub-region is the southernmost portion of Nigeria that produces over 90% of the crude oil in the country. It follows logically that refineries to process he crude oil are situated within this zone.

The objectives of the study are to carry out detailed soil tests (geotechnical investigation) in order to investigate the incidence of **crude oil leakages beneath the Tank # 50TK01A** with a view to determining if crude oil had leaked and if so the amount of crude oil that has leaked into the sub-soil beneath the tank. It was also to obtain engineering properties of sub-soils beneath and around the tank's external peri**meter** in order to effect an efficient rehabilitation of the tank structures.

The Oil Refinery is located within the marginal lands of the Niger Delta sub-region between latitudes **N4° 45'44.1"and N4° 45'48.7"** and Longitudes **E007° 06' 26.1"–E007° 06'30.2"** (Fig. 1a).

1.1 Local Geology

The site falls within the Niger delta region, which is made up of thick clastic sedimentary sequence with ages ranging from Eocene to Recent. It consists in ascending order, of the Akata Formation, Agbada Formation and Benin Formation (Short and Stauble 1967). The site sits astride the Niger flood plains, which overlies the Benin Formation that is often called the Coastal Plain Sands (Qp) (Tattam 1943); this formation consists of soft clayey sands, very poorly sorted sands, pebbly sands and gravels. The geological map of the area is as shown in Fig. 1b.

2 Materials and Methods

In order to gain access into the project site, a temporary road was constructed through the bond wall surrounding the crude oil tank (Fig. 2a) and entry into the empty tank was gained through a side entry point (Fig. 2b)

2.1 Field Investigations

A geotechnical investigation technique that involved a very careful manual drilling of Standard Penetration Test (SPT) borings within the confinement of the 64-meter

(a)

(b)

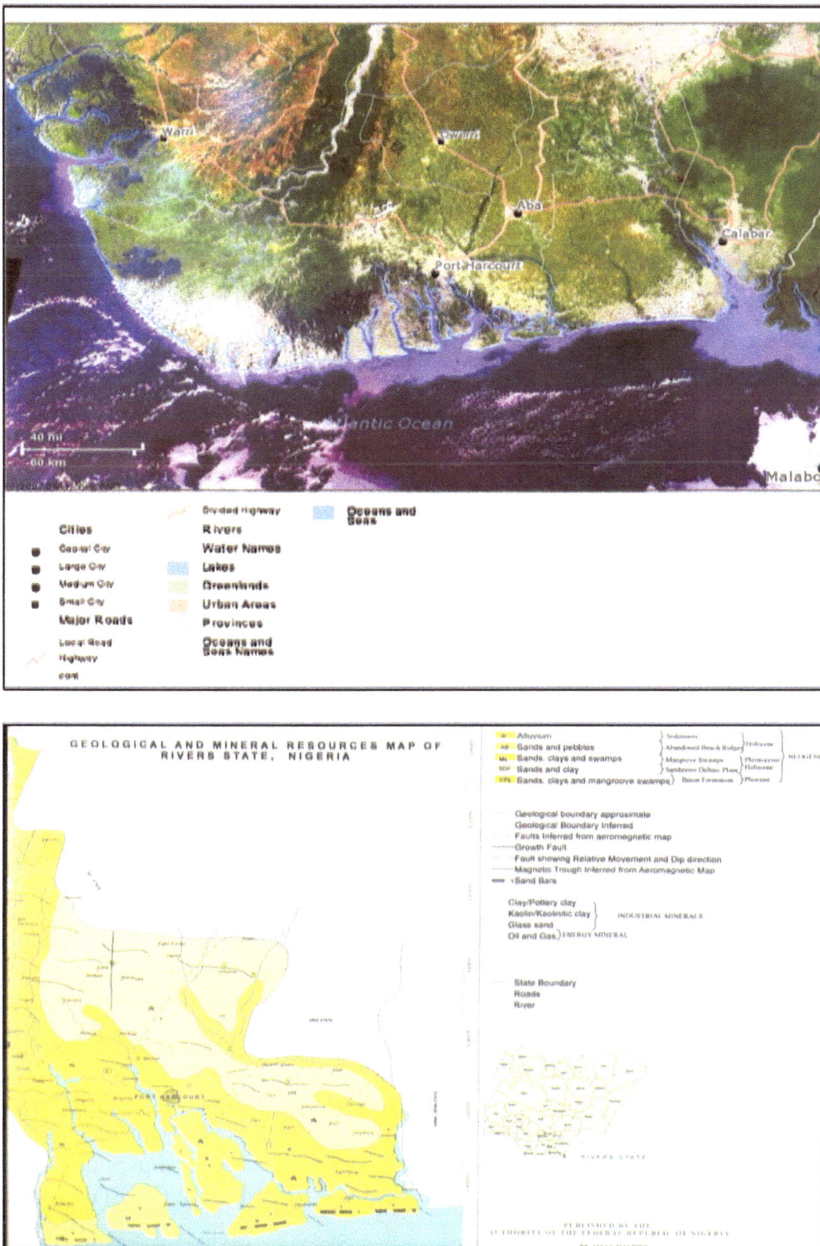

Fig. 1. (a) Location of the project area, (b) Local Geology of the Project Area

Fig. 2. (a) Entry into the Tank area # 50TK01A, (b) Entry point into the empty Crude oil Tank.

diameter tank to depths of 20 m was carried out. The manual drilling was undertaken to avoid any incidences of fire outbreak that may arise from the use of petrol engines commonly used in percussion drillings in the presence of the flammable crude oil remnants within this crude-oil drained tank, during project execution.

2.1.1 Setting up of Drilling Rig Inside the Crude Oil Tank

The dark and empty Crude Oil tank at the Refinery in question had a Floating Lid. For a Rig to be set up for drilling operation inside such tank, four (4) openings were made on the roof of the tank to accommodate the set-up Rig as well as provide light for the operations (Figs. 3 and 4). For Health, Safety and Environment (HSE) considerations,

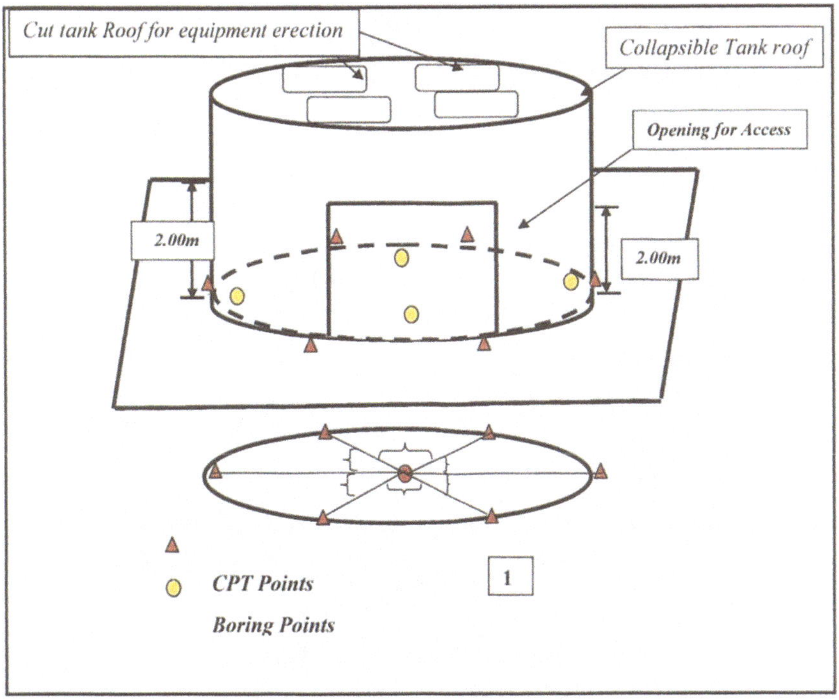

Fig. 3. Setting of the Crude Oil Tank.

an Environmental Briefing was conducted for both Clients and Consultants before commencement of soil boring operations (Fig. 5a and b). In order to avoid fire incidents, a manual portable and dismantle-able Rig was employed. (See Fig. 6a and b)

2.1.2 Standard Penetration Test (SPT) Borings

The Standard Penetration Tests (SPT) were carried out during drilling in the Crude Oil Tank where the tank roofs had been cut-off to accommodate the drilling Rig (Fig. 6a).

2.1.3 Soil Sampling and Storage of Soil Samples

Soil sampling was carried out during the SPT sounding process and retrieved (Fig. 6b), stored in wooden boxes (Fig. 6c) and transported to the Soil Mechanics and Petroleum Engineering Laboratories [Faculty of Engineering] at the Rivers State University of Science and Technology, Port Harcourt, Nigeria for testing and analysis.

Fig. 4. A view of the cut Floating Tank Roof for provision of light for drilling process.

Fig. 5. Environmental briefing for consultants (a) and discussion with clients prior to drilling inside the Crude Oil Tank at project site.

Fig. 6. (a) Cutting of the Roof of the Floating Tank for provision of light during SPT soundings, and soil, (b) Soil sampling, at site, (c) Retrieved Soil samples are stored in a core boxes prior to transportation to Laboratory.

3 Results of Field and Laboratory Testing

3.1 Results of Field Work

The field boring exercise, soil sample logging and visual soil classifications are shown in the Soil Logs below (Fig. 7a–d), while the Summary of the Laboratory test results are given on Tables 1 and 2 below.

3.2 Laboratory Test Results

These included: (i) Grain size analyses, (ii) Atterberg Limits [comprising Liquid Limits (LL); Plastic Limits (PL) and the corresponding Plasticity Indices (PI)], (iii) Shear Strength Characteristics [Triaxial tests]; (iv) Consolidation tests and (v) Extracted Crude Oil tests.

(i) Grain Size or Particle Size Distribution (PSD) Results.

The results of this test are given as distribution curves shown in Fig. 8.

Results from (ii) Atterberg Limits; (iii) Shear Strength Characteristics (iv) Consolidation tests and (v) Compactive Strength Tests are shown on Table 1; while those of (v) Extracted Crude Oil tests are given on Table 2.

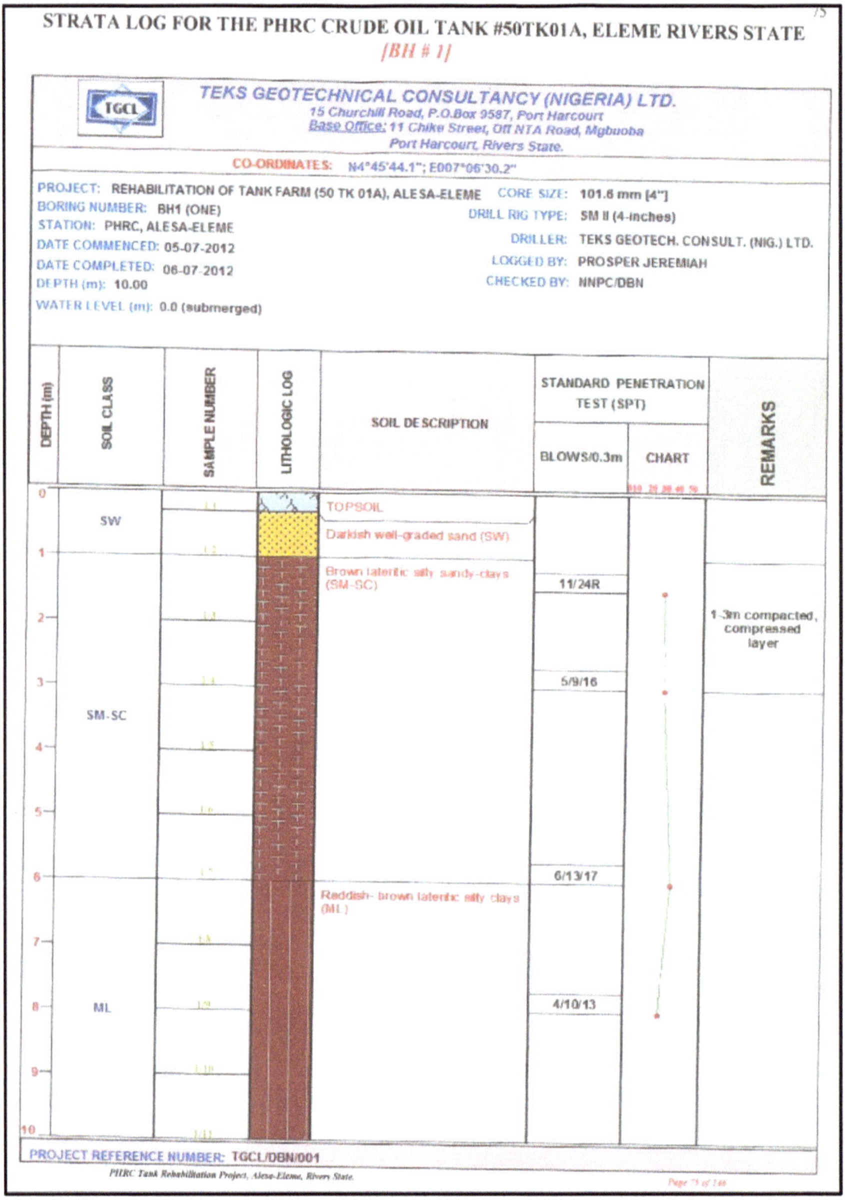

(a)

Fig. 7. (a) PHRC Crude Oil Tank BH#1, (b) PHRC Crude Oil Tank BH#2, (c) PHRC Crude Oil Tank BH# 3, (d) PHRC Crude Oil Tank BH# 4

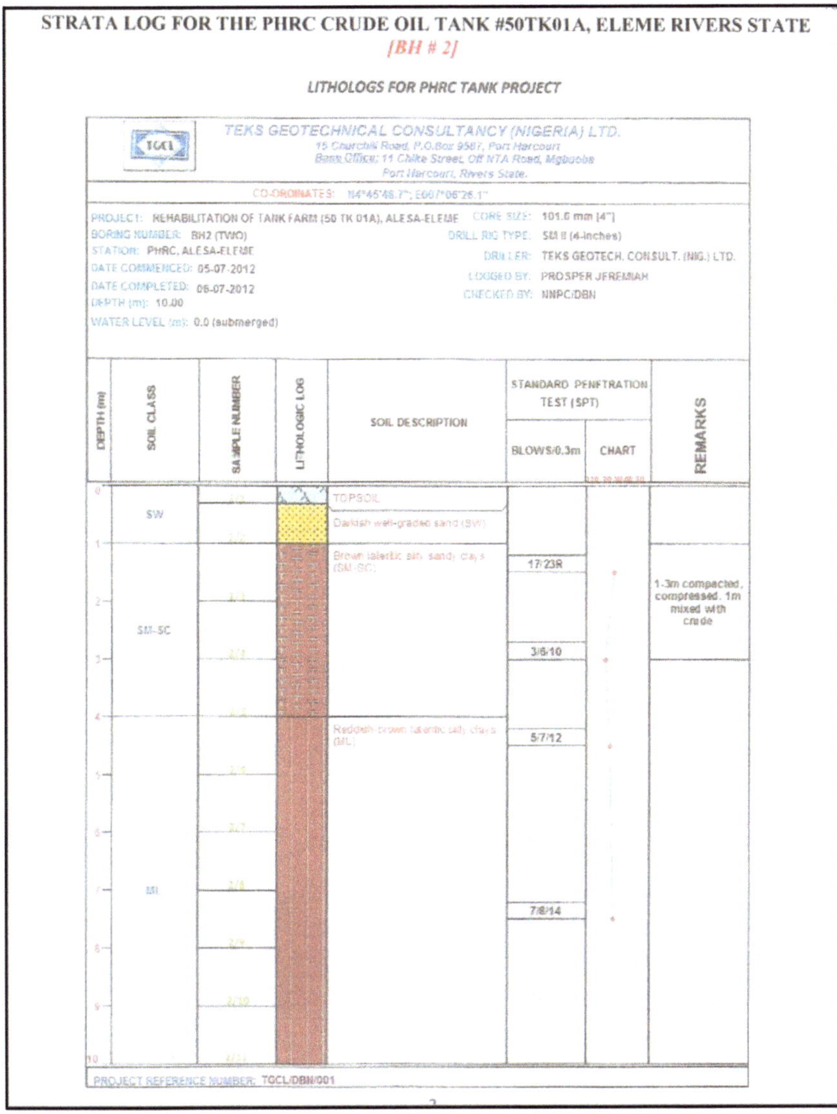

(b)

Fig. 7. (continued)

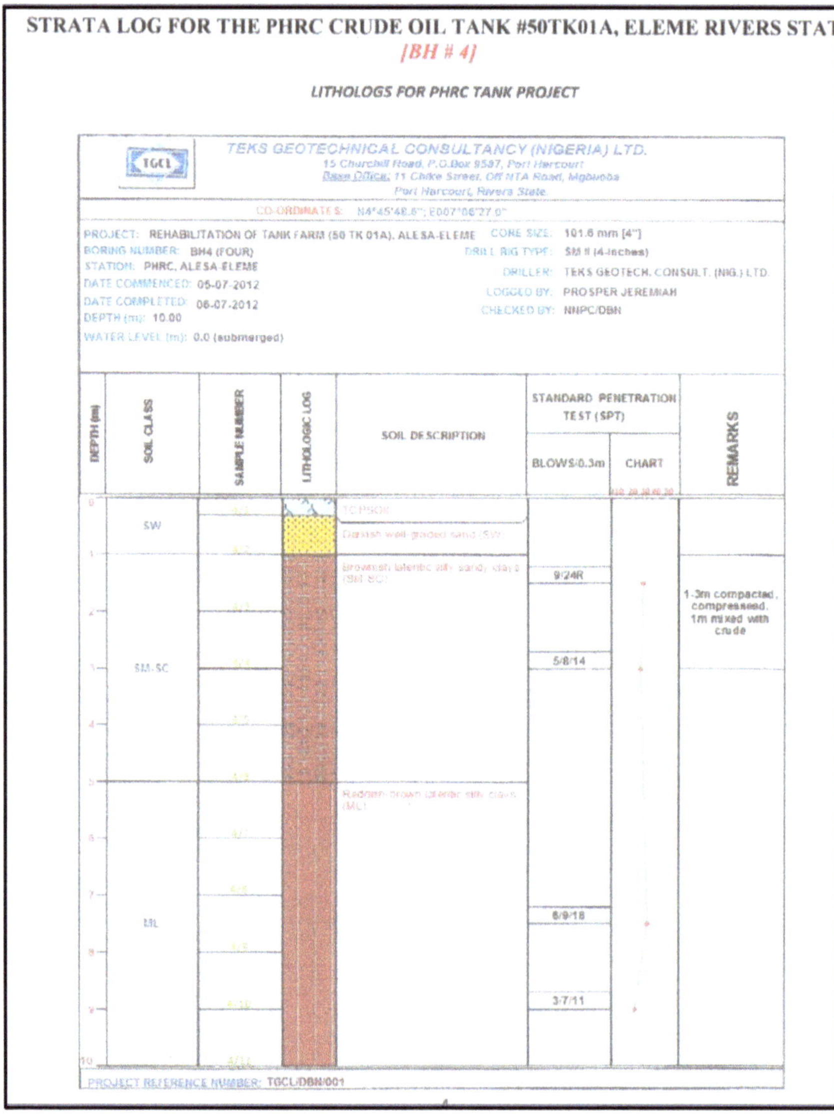

(c)

Fig. 7. (continued)

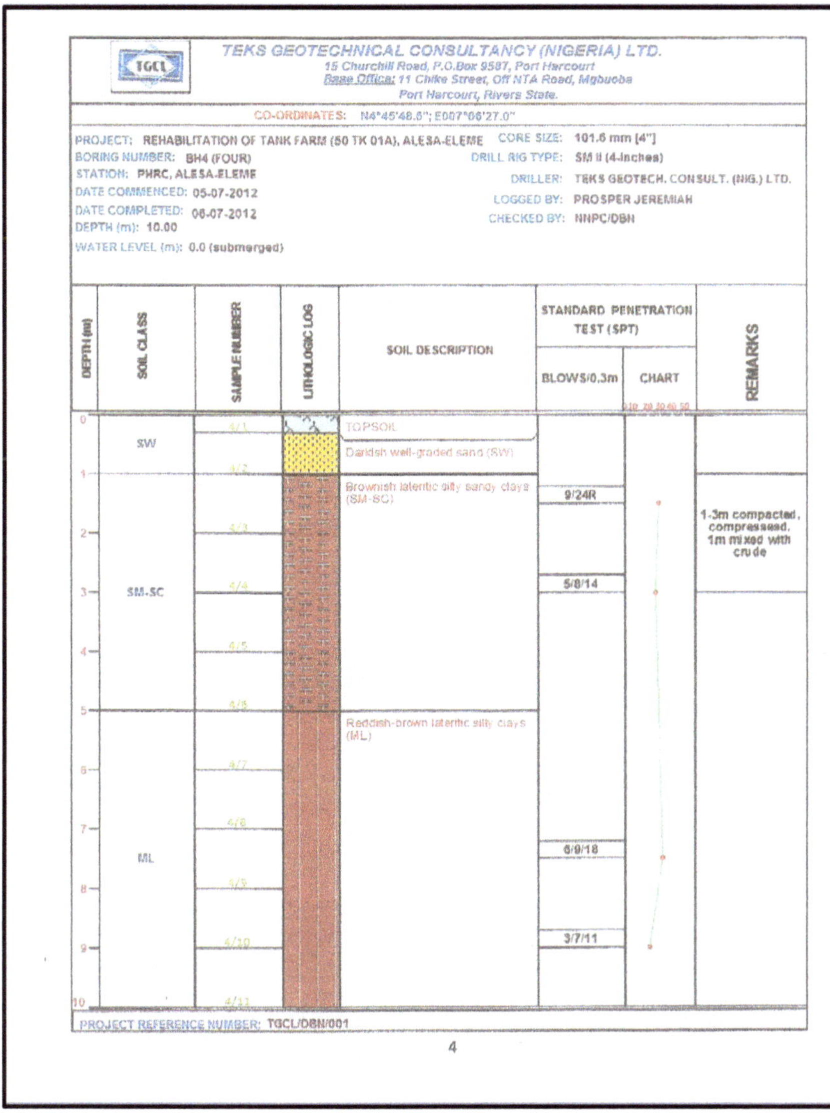

(d)

Fig. 7. (continued)

Table 1. Summary of the engineering properties of sub-surface materials obtained under the Crude Oil Tank site

Soil Type	Depth Range (m)	Soil Consistency				Saturated Unit Wt (kN/m^3	U-U Triaxial Tests (Shear Strength)		Compactive Strength		Consolidation Characteristics		Soundings		Grain Size Distribution Patterns			
		.wn (%)	LL (%)	PL (%)	PI (%)		φu (°)	c_u (kPa)	OMC (%)	MDD (Mg/m^3) x10^{-4}	Mv (m^2/MN)	Cv (m^2/yr)	SPT N_{values}	DCPT ([MPa)	<4.75 (mm)	4.75 (mm)	75μ	2μ
Brownish Lateritic Clayey Sands (SC).	0.25 – 1.00	16.9	11.3	8.7	2.6	2.78	8.75	35.2	12.18	1.97	0.002	0.60	N/A	44.87	100	96	75	5
Darkish grey Well-graded Sands (SW).	1.00 - 2.00	7.40	NP	NP	NP	2.78	34.5	0.00	8.18	2.78	N/A	N/A	51	24.89	50	30.5	27.	4
Brownish Lateritic Silty-Sandy-Clays (SM-SC).	2.00 - 5.00	16.4	13.4	9.0	2.4	2.21	10.3	26.57	13.9	2.24	0.053	0.54	31	20.51	79.29	61.71	40.	6.5
Reddish brown Lateritic Silty Clays (ML)	5.00 - 10.00	15.0	12.7	9.5	3.2	1.82	5.8	29.93	14.9	1.78	0.043	0.56	15	40.21	65.9	46.75	25.	2.9

(*) kg/m^3

Table 2. The results of the Extracted Crude Oil samples beneath the Crude Oil PHRC Tank # 50TK01A at the project site

BH #	Sample #	Sample depth (m)	Crude oil content (%)	Soils-solids contents (%)	Remarks
1	1/1	0.00	5.153	94.847	Slightly Appreciable Crude Oil
	1/2	1.00	0.092	99.908	V. Scanty Crude Oil Content
	1/3	2.00	1.537	98.463	Negligible Crude Oil Content
2	2/1	0.00	1.987	98.013	Negligible Crude Oil Content
	2/2	1.00	0.269	99.731	Scanty Crude Oil Content
	2/3	2.00	0.330	99.670	V. Scanty Crude Oil Content
3	3/1	0.00	1.046	98.954	Negligible Crude Oil Content
	3/2	1.00	0.266	99.734	Scanty Crude Oil Content
	3/3	2.00	0.621	99.379	V. Scanty Crude Oil Content
4	4/1	0.00	7.548	92.452	Appreciable Crude Oil
	4/2	1.00	0.828	99.172	Scanty Crude Oil Content
	4/3	2.00	0.407	99.593	V. Scanty Crude Oil Content

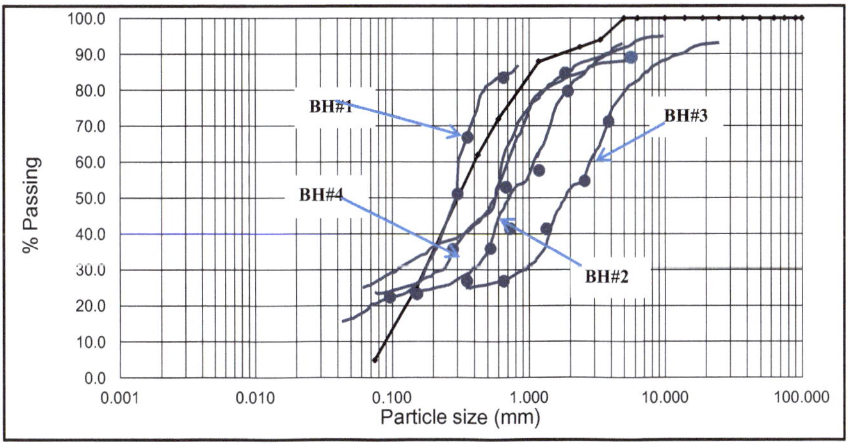

Fig. 8. Grain size distribution of soil samples from within the crude oil tank borings # 1, 2, 3 & 4 [Boring #: 1 Sample # 1; Depth: 0.25 m; Boring #: 2 Sample # 3; Depth: 3.00 m; Boring #: 3 Sample # 2; Depth: 1.50 m; Boring #: 4 Sample # 3; Depth: 3.00 m;]

3.2.1 Results of Crude Oil Content Analysis

During the SPT borings inside the Crude Oil Tank, appreciable traces of crude oil were observed coming out of the borings as shown in Fig. 9(a) and (b) below.

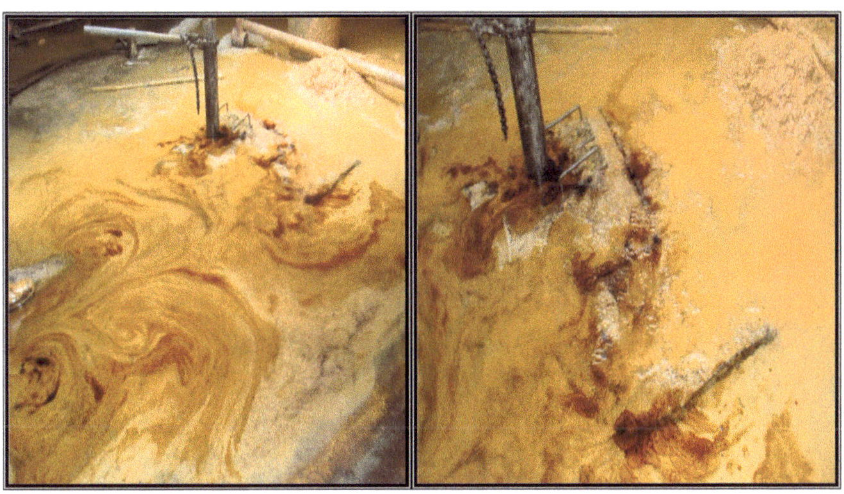

(a) (b)

Fig. 9. Crude oil observed from the soils retrieved from borings #1 & #4 at the project site.

3.2.2 Results of Determination of Subsurface Engineering Properties Around the Crude Oil Tank

A 20-Tonne Dutch Cone Penetrometer was employed in the soundings at six equidistant positions around the Crude Oil Tank (see Fig. 10 below).

Fig. 10. The 20-Ton DCPT Machine used in the subsurface sounding around the Crude Oil Tank. *[at the background is a crane for transferring the DCPT equipment from point-to-point]*

The results of the DCPT soundings at the six locations around the Crude Oil Tank are shown in Table 3 and the curves in Fig. 11(a)–(f).

Table 3. DCPT Sounding values from around the Crude Oil Tank.

Depth range (m)	Soil types	DCPT Values (MPa)					
		DCPT site No.1	DCPT site No.2	DCPT site No.3	DCPT site No.4	DCPT site No.5	DCPT site No.6
0.00–1.00	Brownish Lateritic Clayey Sands Layer (SC)	20.10	54.43	20.59	24.52	98.07	51.48
1.00–2.00	Dark grey Well-graded Sand Layer (SW)	9.81	40.21	9.81	12.26	64.97	12.26
2.00–6.00	Brownish Lateritic Silty-Clays (SM-SC) layer	12.25	25.00	14.71	9.81	46.58	14.71
6.00–20.00	Reddish brown Lateritic Silty Clays (ML)	31.87	41.68	40.21	34.32	39.23	53.84

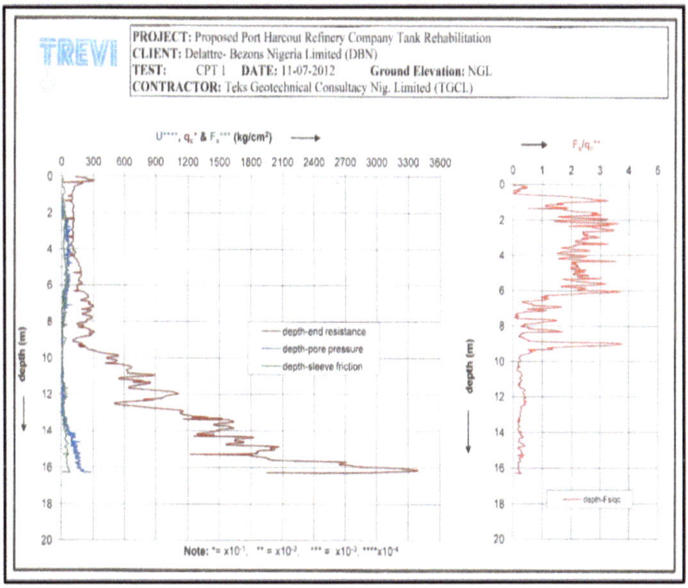

(a)

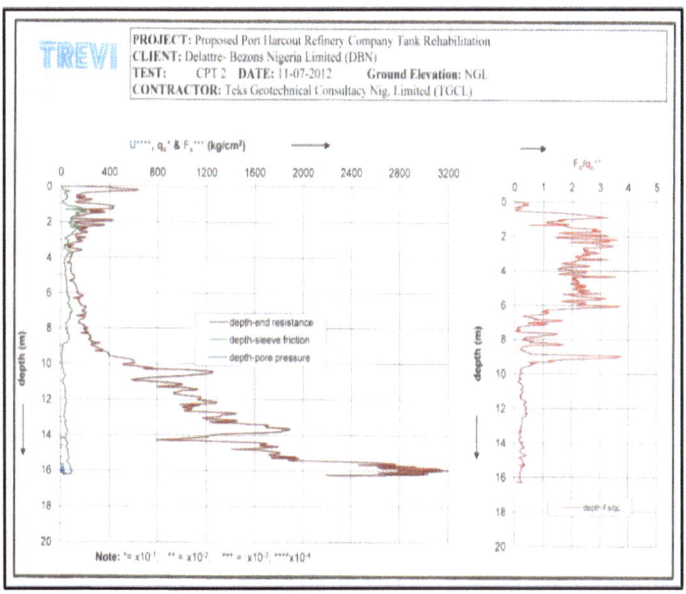

(b)

Fig. 11. (a) Results of the CPT Test at Site #1, (b) Results of the CPT Test at Site #2, (c) Results of the CPT Test at Site #3, (d) Results of the CPT Test at Site #4, (e) Results of the CPT Test at Site # 5, (f) Results of the CPT Test at Site # 6.

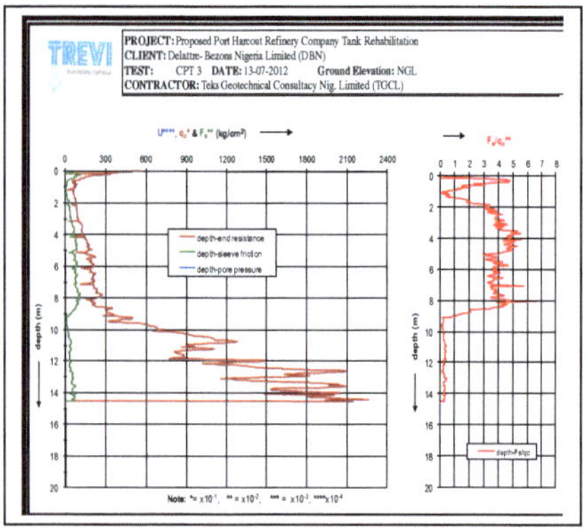

(c)

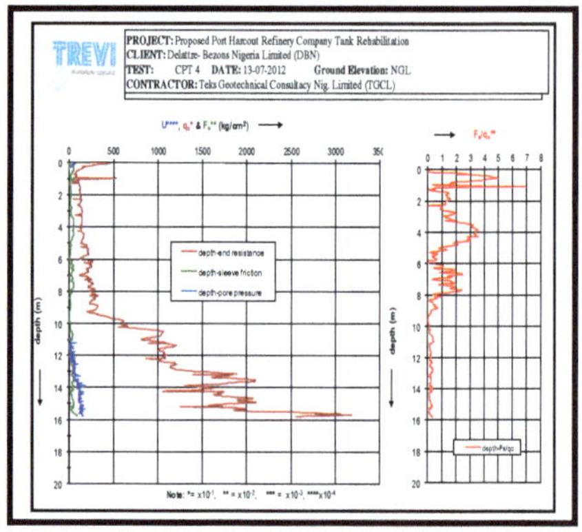

(d)

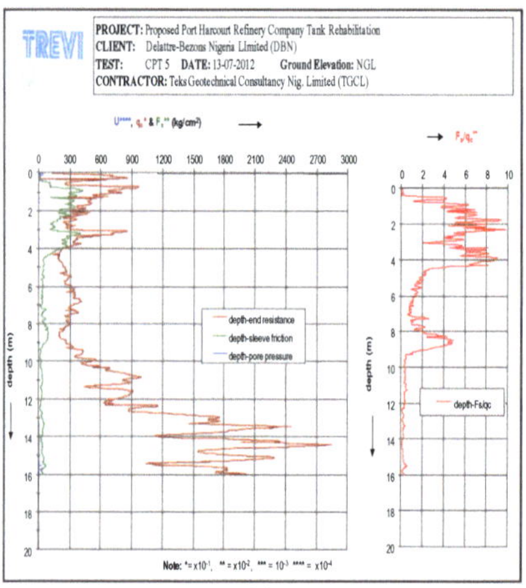

(e)

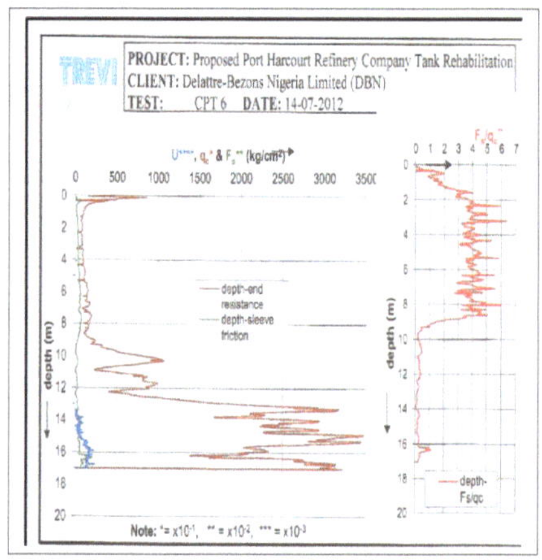

(f)

Fig. 11. (continued)

4 Discussions

The main aim of this study was to establish whether crude oil was leaked beneath the tank and if so, how much crude oil had been lost through leakage of the Tank for whatever reasons. In order to have an idea about how much Crude oil had leaked out from under the Crude Oil Tank, we had to estimate the Volume of Soils contaminated at each soil horizon beneath the tank in question to the depth of 2.00 m where crude oil was observed during the drilling.

We thereafter apply the **"4-Phase Concept of Air/Crude Oil/Water/Soils"** employing the method of Otunyo (1993) and Teme & Otunyo (1999). In summary this method involves he application of the Four (4) phases of Crude Oil, Water, Air and Solids phases as shown schematically in Fig. 12(a) and (b) below

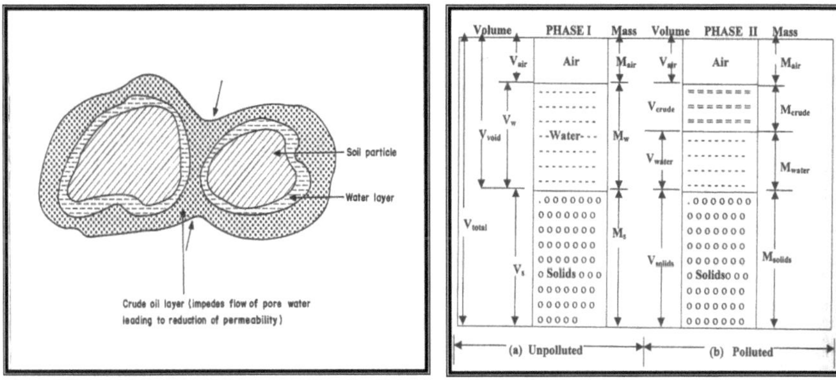

(a) **(b)**

Fig. 12. Representation of the 4-phase Air/Crude Oil/Water/Solids system in a Crude Oil contaminated Soil environment. (a) Soil particle/Water/Crude oil phase; (b) Four Phase Diagram of [Air/Crude Oil/Water/Solids] based on Volumes/Mass Relationships. (after Otunyo 1993 & Teme and Otunyo 1999)

The diameter of this tank was observed to be 62.00 m; hence the Volume (V) of a given depth (h_1) of soil beneath the Tank is taken as:

$$V = \pi R^2 = \pi.[\phi/2]^2(h_1) = \pi.\ [\phi^2]/4(h_1) \ \ldots \quad \ldots \quad \ldots \quad \ldots \quad \ldots \qquad (1)$$

where: $\pi = 3.141592654$.

Hence from a depth of 0.00 m to 1.00 m, the Volume of Soil = **(3.141592654) (62.00/2)^2 (1.00 m) = (3019.07054) m^3** of Soil. However, average % of Crude Oil Content in the Four (4) Borings carried out between 0.00 to 1.00 m = 1/8 [5.153 + 0.092 + 1.987 + 0.269 + 1.046 + 0.266 + 7.548 + 0.828]% = **2.148625%.** Therefore the "Equivalent Volume of Crude Oil" leaked into the underground for a depth of 1.00 m = 2.148625/100 [3019.07054] m^3 = 64.8685 m^3 of Crude Oil approximately equal to 17,136.31165 US gallons of Crude Oil. Also for equivalent depth range of between 1.00 m to 2.00 m beneath the Tank, the "Equivalent Volume of crude oil = [(3019.07054) m^3 of Soil (1.537 + 0.330 + 0.621 + 0.407)/4/100] m^3 equal to 21.850 m^3 of Crude Oil ~ **5,772.1145 US gallons of Crude Oil** bearing in mind that 1 m^3 is approximately 264.17 US gallons. Hence the Estimated Total Crude Oil Equivalent beneath the Crude Oil Tank was equal to 17,136.31165 US gallons +5,772.1145 US gallons of Crude Oil = 22,908.42615 US gallons or ~ **545.44 Barrels of Crude leaked.** [Note that approximately 42 US gallons = 1 Barrel of Crude Oil].

Acknowledgement. The authors are grateful to the Clients for permission to use the data for the preparation of this paper. The Managing Director and field engineers of the parent company who assisted in the field work are gratefully acknowledged.

References

Otunyo, W.A.: Effects of Crude Oil Pollution on the geotechnical Properties of Soils in Rivers State of Nigeria. An unpublished Doctoral thesis, Institute of Geosciences and Space Technology, Rivers State University of Science and Technology, Port Harcourt, 208 p. (1993)

Short, K.C., Stauble, R.J.: Outline of Geology of the Niger Delta. Bull. Am. Assoc. Pet. Geol. **54** (5), 761–779 (1967)

Tattam, C.M.: The extensive red earths and loose, ill-sorted sands underlying the Recent deposits of the Niger Delta. Nigeria Geological Survey Agency Bulletin (1943)

Teme, S.C., Otunyo, W.A.: Effect of crude oil pollution on geotechnical characteristics of subsoils in the niger delta region, Nigeria. In: Proceedings of Conference on Geotechnical Engineering Practice in the Next Millenium. Nigerian Geotechnical Association (NGA), Lagos. 6 p. (1999)

Teme, S.C.: Geotechnical engineering in environmental management – some nigerian case histories. In: Proceedings of the First Quarterly Technical Session of the Nigerian Geotechnical Association (NGA), Lagos, 14 p. (2000)

Teme, S.C.: Geotechnical site characterization in foundation engineering practice – case studies from the niger delta sub-region, Nigeria. In: Proceedings of the First International Society for Soil Mechanics and Geotechnical Engineering (ISSMGE) International Lecture Tour, Le Meridien Eko Hotel, Lagos, 19 April, 47 p. (2001)

Teme, S.C., Ubong, E.: Need for geotechnical engineering studies for foundation design: cases of collapsed buildings in Port Harcourt and Environs, Nigeria. In: (PaperNo. 8.01b) Edited. Proceedings of the Six International Conference on Case Histories in Geotechnical Engineering and Symposium in Honour of Professor James K. Mitchell, Arlington, VA (USA), 11–16 August, 6 p. (2008)

Teme, S.C., Ubong, E.: An evaluation of the geotechnical characteristics of the abutments of a proposed bridge across a 400-meter river channel in the lower niger delta, Nigeria. In: (Paper No. 8.01c) Edited. Proceedings of the Six International Conference on Case Histories in Geotechnical Engineering and Symposium in Honour of Professor James K. Mitchell, Arlington, VA (USA), 11–16 August, 6 p. (2008)

Teme, S.C., Ogbunie, P.: An assessment of the subsurface conditions beneath a collapsing Oil Rig platform at the Oil Well-6-Cellar, Ebedei, Umutu, Delta State. In: Edited Proceedings of the 45th International Conference of the Nigerian Mining and Geosciences Society (NMGS), Owerri Imo State, Nigeria, 4 p. (2009)

Teme, S.C., Ubong, E., and Falade, F.: A critical assessment of the geotechnical characteristics of the subsurface beneath the proposed AMG petro-energy oil tank farm and loading jetty at the Marina, Calabar, Cross River State, Nigeria. In: Edited Proceedings of the 45th International Conference of the Nigerian Mining and Geosciences Society (NMGS), Owerri Imo State, Nigeria, 6 p. (2009)

Teme, S.C., Ngerebara, O.D., Ubong, E.: Geotechnical and Geophysical characteristics of a proposed site for a marginal oil prospecting field in south- eastern Nigeria. In: Edited Proceedings of the 45th International Conference of the Nigerian Mining and Geosciences Society (NMGS), Owerri Imo State, Nigeria, 9 p. (2009)

Teme, S.C., Onuoha V., Otunyo, W.: The impact of crude oil and bio-solids on the niger delta environment. In: Edited Proceedings 3rd International Conference of Nigerian Union of Planetary and Radio Sciences (NUPRS) Toro, Bauchi State, Nigeria. 12 p., June 2010

Teme, S.C., David-West Jr., M., Iboroma, N.: Alignment and design of 73-km long coastal road in the south-central segment of the niger delta, Nigeria. In: Paper No. 7.06a., Edited Proceedings of the Seventeen International Conference on Case Histories in Geotechnical Engineering and Symposium in Honor of Clyde Baker, Wheeling Il, Chicago, IL AREA, 29 April–4 May, 2013, 16 p. (2013)

Teme, S.C., Adu. D., Olise, C.: Design considerations in the construction of a basement in a 7-storey hospital complex in the marginal lands of the niger delta sub-region, Nigeria. In: Edited Proceedings of the 51st Annual International Conference and Exhibitions of the Nigerian Mining and Geosciences Society (NMGS), 15–21 March, 2015, 12 p. (2015)

Teme, S.C., Olise, C., Ogunwale, D.: Design considerations for a quay-wall along a proposed oil and gas facility in the mangrove zone of the niger delta sub-region of Nigeria. In: Edited Proceedings of the 51st Annual International Conference and Exhibitions of the Nigerian Mining and Geosciences Society (NMGS), 15–21 March, 16 p. (2015)

Teme, S.C., Olise, C., Adu, D.: Geotechnical sub-soil investigation for an 18-storey high building in a coastal city, Nigeria. In: Edited Proceedings of the 51st Annual International Conference and Exhibitions of the Nigerian Mining and Geosciences Society (NMGS), 15–21 March 2015, 14 p. (2015)

The Influence of Geochemistry of Gas Hydrates on the Shear Strength and Stability of Marine Sediment Movement

J. Rajaraman[1(✉)], K. Thiruvenkatasamy[1], and S. Narasimha Rao[2]

[1] Department of Harbour and Ocean Engineering, AMET University, Kanathur, Chennai, India
rajaraman.usha44@gmail.com, swamy2667@gmail.com

[2] Dredging Corporation of India (Govt. of India), Vishakhapatnam, India
kuttyharirao@gmail.com

Abstract. Methane obtained from hydrate will become a significant energy resource in the near future. It will become a critical sustainable source of energy possibly by 2025. In this inter-disciplinary research paper the influence of Environmental Geochemistry of gas hydrates on the shear strength and stability of marine sediment movement is discussed. Methane, CH_4, the most common form of natural gas associated with petroleum reservoirs, is in gaseous state at standard temperature and pressure conditions. But under high pressure and low temperature, it combines with water molecules to form an icy – white compound called methane hydrate. Taking a continental margin (continental shelf, slope and raise) and abyssal depth the geo environment is represented as four zones namely 1. Continental shelf has high temperature and low pressure 2. Continental slope has low temperature low pressure 3. Continental rise with low temperature high pressure 4. Abyssal depth with high temperature and high pressure (it does not exist in nature in the case of gas hydrates). The same zones can be classified geo-technically for sediments as 1. Continental shelf with high angle of internal friction (φ) and low cohesion (c) 2. Continental slope with low angle of internal friction (φ) and low cohesion (c) 3. Continental rise with low angle of internal friction (φ) and high cohesion (c) 4. Abyssal depth with high angle of internal friction (φ) and high cohesion (c) (which does not exist in nature). These four zones are analyzed geotechnically for stability of sediments in the sea floor environment. The inter-disciplinary equation of shear strength and gas hydrate environment is $\tau = c + \sigma \tan \varphi$ where τ = shear strength of the sediment, c and φ are geotechnical parameters and σ = to vertical stress of sea water depth and also sediment thickness above the point of consideration with geo thermal gradient. This paper also deals with the decomposition of gas hydrates which decreases the shear strength, making the sediment move towards failure. Basically soil is a three phase system consisting of solids of the soil, water and air. Gas hydrates will behave as solid or gas depending upon the environmental conditions. Gas hydrates will (1) change pore pressure, (2) cause dilation (3) weaken sediments and (4) break down initiates major slumps in continental slopes. The application of Skempton's pore pressure parameters A and B are verified through environmental considerations. The sediment movement from coastal areas to abyssal plain should be as natural as possible. The conclusion is that, accelerated sediment movement is dangerous to marine environment.

D.N. Singh and A. Galaa (eds.), *Contemporary Issues in Geoenvironmental Engineering*, Sustainable Civil Infrastructures,
DOI 10.1007/978-3-319-61612-4_18

1 Introduction

Methane (CH_4), the most common form of natural gas associated with petroleum reservoirs, is in gaseous state at standard temperature and pressure conditions. But under high pressure and low temperature, it combines with water molecules to form an icy-white compound called methane hydrate (Carroll 2001). Other gases such as ethane, propane, butane, CO_2, H_2S etc. also form compounds with water and in general, they are called as gas hydrates. In nature such high pressure and low temperature conditions are prevalent in areas of permafrost (both offshore and onshore), on the outer continental margins and shelves, and beneath sea floors. Gas hydrates comprise more than 99 per cent of methane and in an ideally saturated gas hydrate the molar ratio of methane to water is about 1:6. Thus it contains huge volumes of gas e.g. 1 m^3 of solid hydrate (theoretically) contains up to 164 m^3 of methane gas at standard pressure and temperature. In nature, the more typical value is 40 m^3 of methane. Though the hydrates are least understood unconventional energy resource as of now, they hold the greatest volume of the global methane endowment (Carroll 2001). Methane produced from gas hydrates is truly unconventional because of its origin, trapping mechanism and production technologies. Unlike conventional traps, which are a prerequisite for the accumulation of oil and natural gas, gas hydrates can form their own trap within the pore space and expand themselves against the sediment load in the form of massive hydrates.

2 Discovery of Gas Hydrate

Davy (1811) was the first to discover gas hydrate when chlorine gas and water reacted to give rise to a solid compound at cold weather conditions in the laboratory. A variety of other molecules are later found to form hydrates called as clathrates in which they are considered as guests while the water being the host molecule. In general, a clathrate is a compound formed by the inclusion of molecules of one kind (guest) within the cavities of the crystal lattice of another (host). Clathrates, also known as "inclusion or container compounds," display no chemical bonding between the host and the guest molecules. They can form spontaneously under certain pressure, temperature conditions when a host molecule (e.g., water) crystallizes into an open lattice structure and a guest molecule of suitable size (e.g., methane) fits into the lattice voids.

Occurrence

Gas hydrates occur ubiquitously in three distinct areas: Polar (permafrost) regions, shallow offshore sediments in Arctic and sub-arctic regions, and deep ocean sediments in tropical regions. Hydrates that occur naturally exist in disseminated or dispersed, nodular, layered and massive forms. Except layered hydrates, others occur in unconsolidated sediments (Kuldeep Chandra 1997). The most favourable conditions for occurrence of gas hydrates are:

(i) Areas of hemipelagic sedimentation of Pleistocene age,
(ii) Rapid sedimentation rate (> 3 cm/1000 years),
(iii) P & T conditions, pressure (> 50 atms) and temperature (4–6 °C) envelop,
(iv) Geothermal gradient (< 6°C/100 m), and

(v) Sediment organic carbon content (> 0.5 per cent) and methane concentration in porewater (> 10 ml/1)

Crystal Structure

Methane (gas) hydrate is a crystalline ice-like solid compound. It occurs in three structures I, II and H, the most predominant being the structure I. In structure I, the cages are arranged in a body centered cubic packing, and are large enough to include methane, ethane and other gas molecules of similar molecular diameter such as CO_2 and H_2S. In structure II, some cages in the face centered (diamond) cubic packing are larger enough to include bigger molecules such as propane and isobutane. Structure H being least common in nature, contains cages still larger than structure II in a hexagonal crystallographic system (Sloan 1998).

Physical Properties

Some typical physical properties of methane hydrate are given in Table 1. The density of methane hydrate (0.91 g/cm^3) is less than water. It may vary slightly depending on the methane saturation of the hydrate lattice and with the incorporation of other molecules such as H_2S in the lattice. The heat of hydrate formation and the heat of hydrate dissociation are equal in magnitude but in opposite sign. When hydrate forms, heat is released from the exothermic system and when hydrate dissociates, heat is taken up into the endothermic system. The measured value of methane hydrate formation enthalpy at 273 °K is 54 kJ/mole (Sloan 1998).

Table 1. Some physical properties of methane hydrate

Serial No:	Property	Magnitude
1	Bulk density (gms/cm^3)	0.912
2	Thermal conductivity at 263 °K (w/m-k)	0.49 ± 0.02
3	Adiabatic compressibility at 273 °K, 10–11 pa	14
4	Heat of fusion at 273 °K (KJ/mole)	54
5	Heat of dissociation at 273 °K (KJ/mole)	–54
6	Heat capacity at constant pressure (KJ/mole)	257
7	Dielectric constant at 273 °K	58
8	Poisson's ratio	0.33

Source: Max 2000.

3 Stability of Gas Hydrates

The stability of a gas hydrate depends on pressure (P), temperature (T) and the solubility of gas as a function of P and T of the hydrate system. The stability is more susceptible to changes in temperature than to pressure. In addition to temperature, other physical properties such as thermal conductivity and thermal diffusivity are also critical which control the transfer of heat through hydrate reservoirs. Further, certain chemical and geological factors promote or inhibit hydrate stability. These include:

- Gas properties: presence of ethane, propane, CO_2/ H_2S etc.
- Pore fluid properties: dissolved ions (Na^+ K^+ Mg^{2+} Ca^{2+})
- Sediment composition: fine grained sediments (clays)
- Geological processes: erosion, rate of sedimentation, slumping, subsidence, uplift, and earthquakes.

The phase equilibria shown in Fig. 1 apply to pure methane system and structure I gas hydrate in the presence of fresh or saline water. Presence of even a fractional percentage of ethane or propane that produce structure II hydrate, is stable at higher temperatures than structure I at a given pressure, thus increasing its stability. In contrast, the presence of dissolved ions (Na^+, K+ , Ca^{2+}, Mg^{2+}) in pore fluids inhibit the stability of gas hydrate. The inhibitory effect of ionic compounds on the stability of hydrate has critical implications, for the evolution and long term stability of hydrate deposits in areas characterized by salt tectonism, evolution of seafloor brine basins, and circulation of briny liquids as evident in the Gulf of Mexico (Sassen and McDonald 1997).

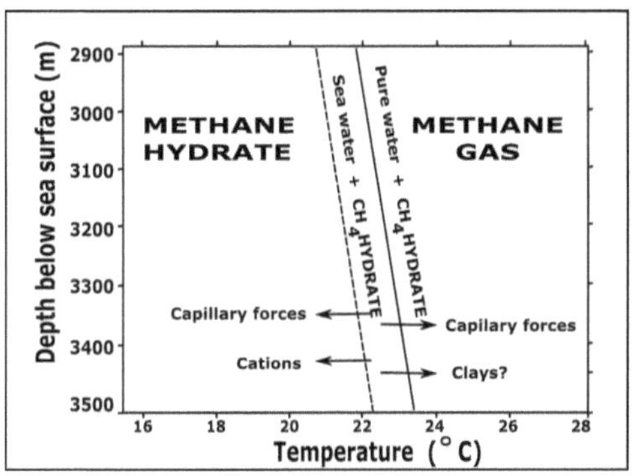

Fig. 1. Stability of marine hydrate for pure water (Solid Line) and sea water (Dashed Line) systems. Arrows schematically show various physical and chemical factors that affect stability of hydrate. (Source: Max 2000)

The composition of sediment matrix may exercise a strong influence on the temperature of hydrate dissociation in some settings. Clays appears to promote hydrate formation at higher temperatures. Physical properties of the sediment may play a critical role on the stability of hydrate as well. Large capillary forces that arise in fine grained sediments may inhibit the entry of fluids into the interstices between the grains and thus significantly inhibit hydrate stability, depressing its dissociation temperature.

Methane hydrates often form as disseminated grains and pore fillings in coarse grained sediments whereas in finer silt/clay deposits, they commonly appear as nodules and veins. In general, naturally occurring methane hydrates, form in two types of geological settings (Milkov and Sassen 2001):

(i) Structural accumulations associated with faults and mud volcanoes, where gas leaking from deeper subsurface petroleum system is rapidly transported (via faults, gas chimneys or other fluid conduits) to the gas hydrate stability zone (Cassassucem et al. 2004), and

(ii) Stratigraphically bound accumulations in relatively permeable sea floor sediments where bacterial methane generates in situ or slowly migrates from shallow depths into the gas hydrate stability zone.

4 Gas Hydrate Stability Zone (GHSZ)

Physical conditions that control the presence of methane hydrate are usually represented by a phase diagram in terms of temperature and depth fields as shown in Fig. 1.

It illustrates the phase of equilibrium among gas hydrate, free gas and aqueous solution, and physical parameters controlling the formation of gas hydrate (P, T and salinity). The phase boundary (dashed line) separates colder, higher pressure conditions where methane hydrate is stable to the left of the curve, from conditions to the right, where it is not stable (solid line) assuming a typical hydrothermal gradient in water and geothermal gradient in sediment. The figure shows the variation of temperature with depth of water and underlying sediment. The point where the solid line (representing the conditions in sediment) crosses the phase boundary represents the bottom of the zone where methane hydrate is stable (GHSZ). This phase boundary corresponds to a sediment depth of about 300–600 m below the sea floor (bsf) at a water depth of about 3,000 m along the Indian continental margins (Fig. 2). However, the hydrate zone in the water column is much shallower (about 750 m). When methane in water is of sufficient concentration (near saturation), it forms hydrate in the hydrate stability zone. However, being lighter than water, it floats upward and would dissociate when it crosses the depth where the hydrothermal gradient and the gas hydrate boundary curves intersect each other. On the other hand, if the gas hydrate forms within the sediment, it will be bound in place. The precise location of base of the GHSZ under known pressure and temperature conditions depends on several factors, the most important being the gas composition. Presences of higher hydrocarbons (ethane, propane etc.) allow the gas hydrates to form at lower pressure and higher temperature in shallow waters. The presence of salts (NaCl, KCl, $MgCl_2$ etc.), in pore waters shift the gas hydrate phase boundary to the left, low temperature and high pressure (in deep waters).

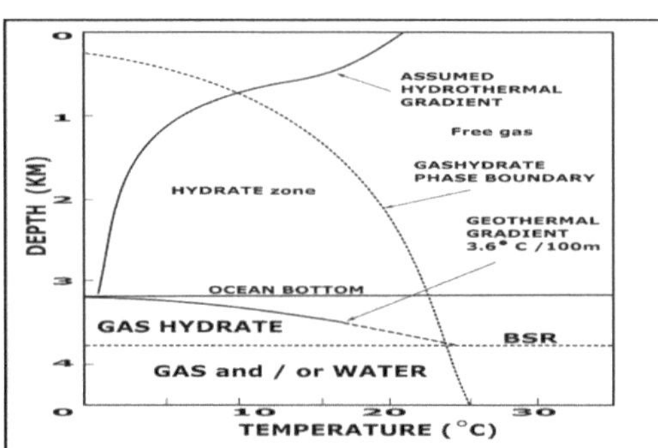

Fig. 2. Phase diagram of methane hydrate for indian continental margins. (Source: Hanumantha Rao et al. 1998)

Thickness of GHSZ

As stated above, the vertical extent of the sediment layer from sea floor to the intersection point of the phase boundary curve with the geo-thermal gradient curve in the sediment gives the thickness of GHSZ. In general, the thermal gradient tends to be uniform across broad regions where sediments do not vary. Therefore, for a given water depth, the sub-bottom depth to the base of the GHSZ will be quite constant. However, because a change in water depth causes changes in pressure, it is to be anticipated that the base of GHSZ will extend further below sea floor as water depth increases. It implies that the zone where gas hydrate exists, forms a more or less uniform layer below the sea floor, thickening towards greater depths (Fig. 3). Though this is commonly true, there are few exceptions because of disturbances of thermal structure in the sediments (Max 2000). These include:

(i) Sea floor landslides: They remove cooler near surface sediments, leaving warmer-than normal materials near the sea floor which cause local shallowing of the base of GHSZ.
(ii) Salt diapers: They produce warm spots since the salt has greater thermal conductivity than other sediments which forces the base of GHSZ to shallower regions.
(iii) Faults: Circulation of warm fluids up to shallow sub-bottom regions through faults as channel ways results in the disruption of gas hydrate formation or shift the base of GHSZ to shallow waters.

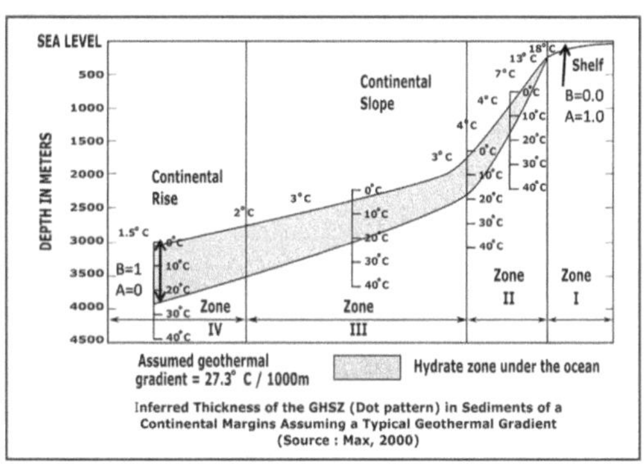

Fig. 3. A and B appearing in the figure are Skempton's pore pressure Parameters

5 Gas Hydrates, Shear Strengths and Stability of Marine Sediment Movement (Fig. 3)

In zone 1 hyper pycnal flow adds more well graded sediments and the φ is high with reasonable cohesion (c). This zone also represents high temperature and low pressure and therefore methane hydrates are not formed in sediments.

In zone 2 gas hydrates are formed in sediments but not much due to low pressure though the low temperature is favourable for the formation of gas hydrates.

Geo-technically this region has low φ (rounded and weathered grain sizes) and low cohesion (c) due to loss fines.

In zone 3 the environmental conditions are favourable for the formation of gas hydrates. The high pressure and low temperature combination is the main reason. Due to sediment movement the sediments are highly weathered with low φ and high cohesion. High cohesion indicates the production of clay.

In zone 4 as pointed out earlier in the abstract high temperature and high pressure does not exist in nature. This possibility is ruled out in marine environment. Similarly high friction φ and high cohesion © cannot exist together.

Facts About Stability of Gas Hydrates

1. Stability of gas hydrates is a function of temperature (T) and pressure (P).
2. The stability is more susceptible to changes in temperature than pressure.
3. Clays appear to promote hydrate formation at higher temperature.
4. In fine grained soils large capillary forces that arise may inhibit the entry of fluids into the interstices between the grains.

6 Interpretation of Skempton's Pore Pressure Parameters to Ocean Bottom Hydrate Sediments (Fig. 4)

The usual Skempton's pore pressure parameters are given by the following equation:

$$\Delta u = B[\Delta\sigma_3 + A(\Delta\sigma_1 - \Delta\sigma_3)] \quad (1)$$

where $\Delta\sigma_1$, $\Delta\sigma_3$ are major and minor (confined) principal stress increments and A and B are pore pressure parameters as given below:

$$B = \frac{\Delta u}{\Delta\sigma_3}, \qquad A = \frac{\Delta u}{\Delta\sigma_1 - \Delta\sigma_3}$$

← Geothermal Gradient →

Low Temperature and High Pressure Hydrate Sediments :	High Temperature and Low Pressure Hydrate Sediments:
1. Zone III and IV 2. Hydrates in solids with extra particle friction therefore φ increases 3. Hydrates are saturated and (γ_{sat}) 4. Void pore space occupied by solid hydrates. 5. τ = C+σ tanφ since Skempton's parameters B=1and A= 0 no pore pressure is involved therefore $\sigma = \sigma_{total}$ and no effective pressure, only total pressure	1. Zone I and II 2. Hydrates in gas / fluid pore pressure increases and methane escapes φ decreases 3. Unit weight of sediment ($\gamma_{Submerged}$) 4. Void space occupied by saline water 5. τ = C+(σ-u) tanφ since Skempton's Parameters B=0.0and A= 1 pore pressure is involved therefore $\sigma_{effective} = \sigma - u$

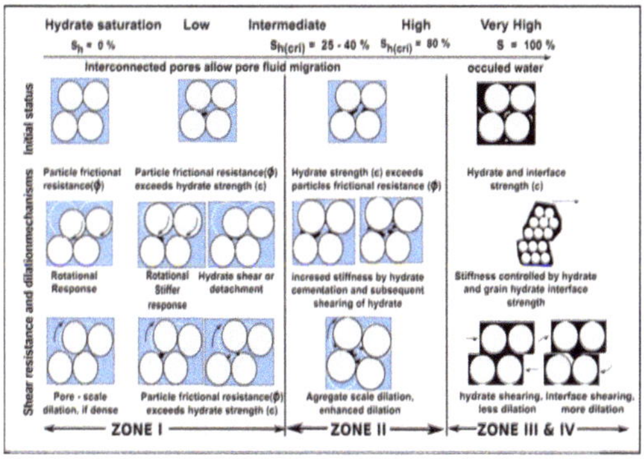

Fig. 4. A summary of the mechanisms controlling the shear strength of hydrate bearing sediments, where soil grains are shown in white, hydrate in black and water in blue (after Waite et al. 2009; modified from Yun et al. 2007). At high hydrate saturations effective stress becomes invalid and interpretations must be based on total stress.

Under global warming due to excess temperature even deeper sediments are subjected to high temperature and low pressure and therefore the solid hydrates break and the φ value decreases rapidly. The effect can be illustrated by a conceptual model example: in a ten meter high marine slope of 40° the soil has cohesion; void ratio and angle φ were found to be 2.5 tons/m^2, 0.81 and 14° respectively. When the temperature alters find the new factor of safety for the hydrate infested area for G = 2.7 and for 40° slope the stability number for different values of φ:

φ	*n*
6°	0.122
7°	0.116
14°	0.074

From the given data γ saturated = 1.94 gm/cc and γ submerged 0.94 gm/cc. The factor of safety (F) is c/γ submerged x h x n that is = 3. 59. When the temperature alters the solid gas hydrates becomes pore pressure and escapes. Therefore the new φ value is $\varphi' = 0.94/1.94 \times 14 = 6.78°$. Therefore the geothermal gradient alters the φ value from 14° to 6.78° which clearly indicates the effect of geothermal gradient on the stability of gas hydrate sediments. On the whole the sediments move towards ocean down the slope to reach the low temperature environment where the sliding sediments are stopped by the sediments with higher φ values manifested as resistive force to stop the submarine slump. As a consequence the sea level rise inundates coastal areas.

7 Conclusions

1. Climate change impacts on methane hydrates. Huge amounts of methane are stored around the world in the sea floor in the form of solid methane hydrate. These hydrates represent a large energy reserve for humanity. Global warming could cause

the hydrate to destabilize. The methane, a potential green house gas could escape unused into atmosphere and could even accelerate climatic changes.

2. Methane hydrates located directly at the boundaries of the stability zones would be primarily affected because in these locations, a temperature increase of 1 °c (Celsius) would be sufficient to release large amounts of methane from the hydrates.
3. Methane should not become silent weapon for invisible, quiet wars.
4. Methane should not become lethal to mankind.
5. Dual nature of Methane hydrates is responsible for change in shear strength. It can exist in solid form or in gaseous form depending upon the temperature, pressure and environment.
6. If a strategic area is bombed will initiate the release of methane. The released methane will take care of the rest of the destruction, leading to catastrophe.

References

Carroll, J.: Natural Gas Hydrates. Gulf Publishing, Boston (2001)

Cassassucem, F., Rector, J., Hoverstain, M.: Study of gas hydrates in the deep sea gulf of mexico from seismic data. Lead. Edge **23**, 366–372 (2004)

Davy, H.: On combination of oxymuriatic gas and oxygen gas. Philos. Trans. Royal Soc. **101**, 155 (1811)

Rao, Y.H., Reddy, S.L., Khanna, R., Rao, T.G., Thakur, N.K., Subrahmanyam, C.: Potential distribution of methane hydrates along the indian continental margins. Curr. Sci. **74**, 466–468 (1998)

Chandra, K.: Non – conventional hydro carbon resources like coal bed methane and gas hydrates: exploration imperatives for india. indian. J. Geol. **69**, 261–267 (1997)

Max, M.D.: Gas hydrate potential of the indian sector of the NE arabian sea and northern indian ocean. In: Max, M.D. (ed.) Natural Gas Hydrates. Kluwer Academic Publishers, London, pp. 213–224, 414 pp (2000)

Milkov, V.A., Sassen, R.: Economic geology of off-shore gas hydrate accumulations and provinces. Mar. Petrol. Geol. **19**, 1–11 (2001)

Sassen, R., McDonald, I.R.: Hydrocarbons of experimental and natural gas hydrates, Gulf of Mexico continental slope. Org. Geochem **26**, 289–293 (1997)

Sloan, E.D.: Clathrate Hydrates of Natural Gases, 2nd edn. Marcel- Dekker, New York (1998). 705 pp.

In-Situ Chemical Oxidation of Hydrocarbon Contaminated Groundwater (A Case Study of Baruwa Community, Lagos, Nigeria)

S.A. Ola[1,2(✉)], O.G. Fadugba[1], and M.A. Uduebor[1]

[1] Department of Civil and Environmental Engineering, The Federal University of Technology, Akure, Ondo State, Nigeria
[2] Afe Babalola University, Ado-Ekiti, Ekiti State, Nigeria

Abstract. A chemical oxidation of hydrocarbon contaminated site is attempted by using potassium permanganate ($KMnO_4$) through injection into five hand dug wells within a pilot test area one hectare. The aim is to improve the quality of groundwater by degrading the soluble organic materials for further treatment and polishing, while incorporating bioremediation. The Background characterisations of groundwater and contaminant studied have been previously carried out (Adekunte 2008; Balogun 2009). 50 g/litre of potassium permanganate ($KMnO_4$) was used as oxidant for the remediation of the Total Petroleum Hydrocarbons (TPH) in the pilot test area, the delivery of oxidant was done by gravity feed to the monitoring wells up to a height of 600 mm (2 ft) above water levels. Oxidation study was carried out by observing the TPH for and other parameters during the process of In-situ Chemical Oxidation (ISCO) remediation for 30 weeks. A reduction in Total Petroleum Hydrocarbons (TPH) ranging from 92.28% to 99.86% was observed within the period. The maximum TPH value of 512 ppm was observed at well W53, while the minimum observed TPH at the end of thirty weeks was 0.7 ppm at the same well.

Keywords: Groundwater · Chemical oxidation · Potassium permanganate · Total petroleum hydrocarbons

1 Introduction

Spills and leaks of petroleum hydrocarbons such as gasoline, diesel, motor oils, and similar materials have caused widespread contamination in the environment. Generally these contaminants are present both in NAPL form (non-aqueous phase liquid; the bulk liquid petroleum hydrocarbon) and also as dissolved contaminants in the ground water.

One of the major environmental problems today is hydrocarbon contamination resulting from the activities related to the petrochemical industry. Accidental releases of petroleum products are of particular concern in the environment.

The goal of site remediation is to restore soil and groundwater quality to precontamination conditions. Site remediation considers the limit of technical expertise, existing government regulations, environmental protection desired and economics. Numerous technologies to remediate sites exist, namely, excavation, air stripping,

D.N. Singh and A. Galaa (eds.), *Contemporary Issues in Geoenvironmental Engineering*, Sustainable Civil Infrastructures,
DOI 10.1007/978-3-319-61612-4_19

carbon bed adsorption, incineration (thermal destruction), bioremediation, soil vapour venting and natural attenuation (Ola and Ojuri 2008).

Most groundwater treatment techniques utilize a combination of technologies. Some of the biological treatment techniques include bioaugmentation, bioventing, biosparging, bioslurping and phytoremediation. Some chemical treatment techniques include ozone and oxygen gas injection, chemical precipitation, membrane separation, ion exchange, carbon absorption, aqueous chemical oxidation, and surfactant enhanced recovery physical treatment techniques including but not limited to pump and treat, air sparging, dual phase extraction and skimming.

Petroleum hydrocarbons such as gasoline are extensively utilized in various transportation and industrial operations. Improper disposal and/or accidental release of these chemicals have resulted in several large scale groundwater contamination problems (Clement et al. 2000; 2002).

Groundwater is a valuable resource both in Nigeria and throughout the world. Where surface water, such as lakes and rivers, are scarce or inaccessible, groundwater supplies many of the hydrologic needs of people everywhere. Groundwater quality impacts resulting from the leaking underground petroleum storage tanks or pipelines are of significant environmental concern due to the sheer number of such underground tanks and pipelines, also the extent of possible environmental contamination when leak. While contamination from petroleum storage tank releases can have significant impact on public health and the environment, active remediation of these contaminated sites can be difficult and expensive. Expectations of having to apply active remediation at all sites has generally not allowed the focusing of resources on those sites representing the greatest threat to public health and the environment (Steele et al. 2008).

2 Background of Study

In some decade ago, the most common technology used for remediating ground water has been to pump the water and treat it at the surface. Although still useful for certain remedial scenarios, the limitations of pump-and-treat technologies have recently been recognized, along with the need for innovative solutions to ground-water contamination (Keely 1989; National Research Council 1994). The ineffectiveness of existing remedial technologies for treatment has created the need for alternative remediation approaches. One such technology, In-situ Chemical Oxidation (ISCO), numerous laboratory column and sand tank studies have demonstrated ISCO effectiveness for the remediation of organic liquid residuals from aquifer materials (Ola et al. 2015).

In the process, called In-Situ Chemical Oxidati or ISCO, chemical oxidants are delivered in the subsurface to destroy (converted to water and carbon dioxide or to nontoxic substances) the organics molecules. The oxidants are introduced as either liquids or gasses. Oxidants include air or oxygen, ozone, and certain liquid chemicals such as hydrogen peroxide, permanganate and persulfate. Ozone and oxygen gas can be generated on site from air and electricity and directly injected into soil and groundwater contamination. The process has the potential to oxidize and/or enhance naturally occurring aerobic degradation. Chemical oxidation has proven to be an effective technique for dense non-aqueous phase liquid or DNAPL and LNAPL when it is present.

Permanganate (MnO_4^-) is available both $NaMnO_4$ and $KMnO_4$, the sodium form is distributed as a liquid available at 40% concentration by weight. Potassium permanganate is most commonly available in a crystalline form, and has a solubility of approximately 6% by weight in water at 20 °C. While permanganate is most commonly delivered as an aqueous solution, it may also be mechanically mixed into soil or introduced via pneumatic or hydraulic fracturing. Permanganate dissociates to produce the permanganate anion, an oxidizing agent with an oxidation potential of 1.7 V. The permanganate anion has an affinity for carbon-carbon double bonds (Wiberg and Saegebarth 1957).

3 Materials and Methods

3.1 Description of the Study Area

Baruwa area (Latitude 06° 35′ 12″ N, Longitude 03° 16′ 21″ E) is located in Ipaja, between the popular Iyana-Ipaja Bus-stop and the Ikotun Area of Alimosho Local Government Council of Lagos State, South West Nigeria. It is about 2.5 km from Iyana-Ipaja Bus Terminal, and is bounded by residential estates namely, Gowon Estate and Abesan Estate to the right and left respectively, while coming from Iyana-Ipaja Bus Terminal. It is accessible by a network of roads through Ipaja and Ayobo, which is a densely populated residential area with a population of over 100,000 inhabitants. The existence of Baruwa dates back to the early 20th Century, but it became prominent in

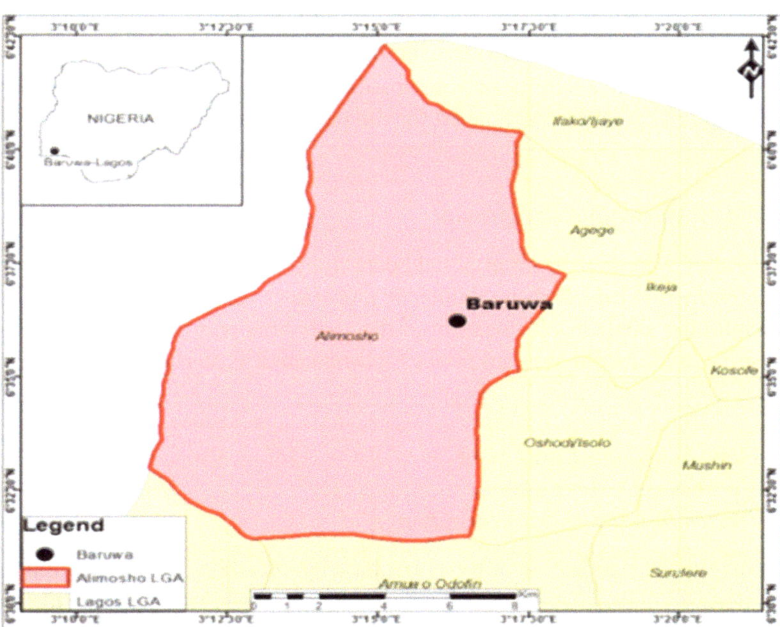

Fig. 1a. Map of Lagos State showing Baruwa Community the case study site.

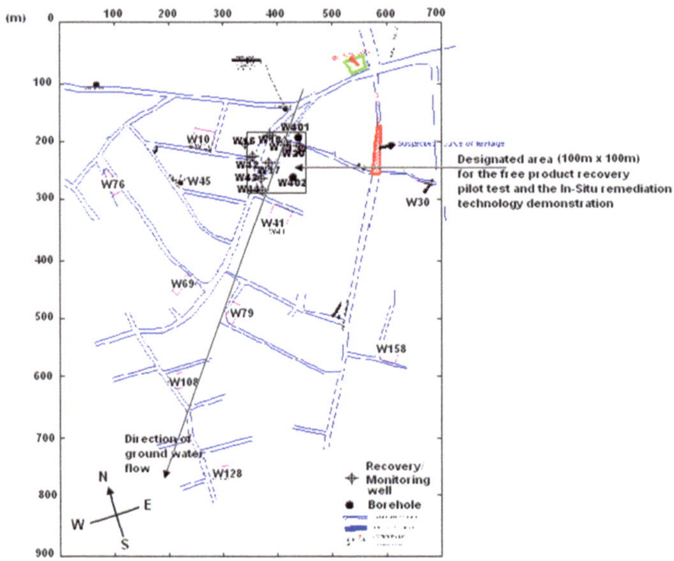

Fig. 1b. Field site (Baruwa, Nigeria) diagram showing location of *In-situ* chemical oxidation remediation

the 1970s due to population explosion witnessed in Lagos, that led to the development of satellite communities (Adekunte 2008). Figure 1a shows the map of Baruwa, in Lagos, Nigeria indicating the position of the study area (Baruwa). The area has a few cottage industries with small scale enterprises, and has a fair to good supply of electricity, with a network of earth roads without a proper drainage systems, which subsequently leads to impassability of some roads during the rainy season, and Fig. 1b shows the 100 m by 100 m designated for the pilot scheme test.

3.2 Field Experimental Programme

The study employed the collection of field data from the contaminated site in Baruwa, using equipment and methods to sufficiently characterize the hydrocarbon contamination within the area. Characterization of the study area was conducted utilizing the methodology prescribed by United States Environmental Protection Agency, US EPA (2001).

3.2.1 Field Tools and Equipment

(a) ***Oil/Water Interface Characterization***

Volumetric characterization of the pure phase hydrocarbon in the monitoring wells was monitored with the aid of water/oil interface meter (Fig. 2), the depth to the surface of liquid, thickness of the pure phase hydrocarbon and the water in the monitoring wells were determined with the equipment.

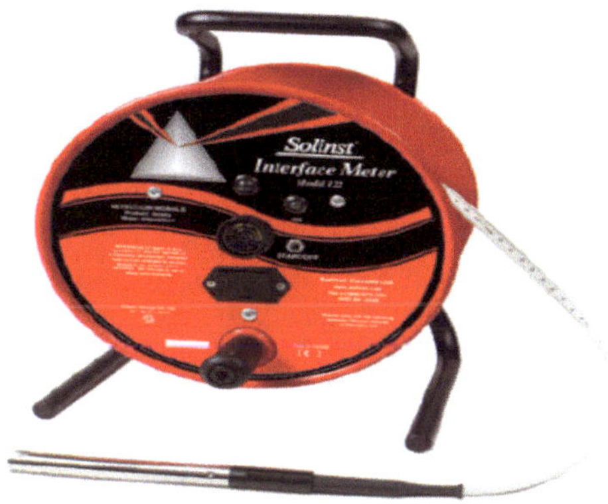

Fig. 2. Model 122 (CSA) interface meter, P8/LM3/60 m (Solinst)

(b) ***Pure Phase Hydrocarbon Skimming***

Model 8 Oil skimmer was used in skimming off the pure phase hydrocarbon from the monitoring wells within the pilot test area. The oil skimming technology was adopted to remove hydrocarbon contamination from groundwater without the pumping of groundwater with the contaminant. This is indicated in Figs. 3a and 3b.

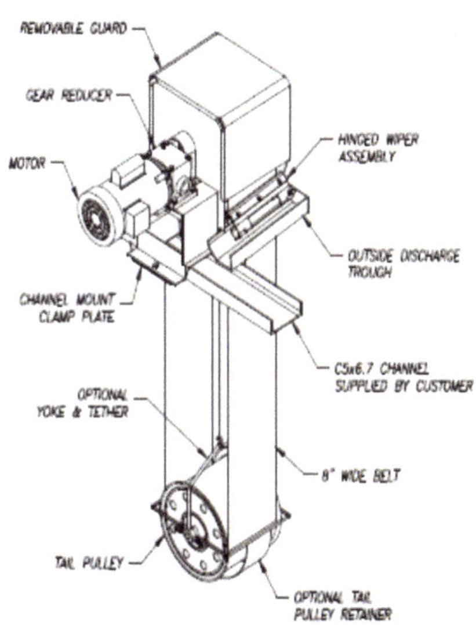

Fig. 3a. Oil grabber skimmer model 8 (Abanaki Corporation USA)

(c) ***In-situ Chemical Oxidation of the Pilot Test Area***

50 g/litre of potassium permanganate ($KMnO_4$) was used as Oxidant for the remediation of the Contaminant of Concern (COC) in the pilot test area; the delivery of

Fig. 3b. Oil skimmers with fussy belt

oxidant was done by gravity feed to the monitoring wells. Oxidation study was carried out by observing the concentration of COC for a period of 30 weeks

(i) **Determination of Concentration of Contaminant and Physico-chemical Characterization**

Pre-injection test was carried out on water samples before injecting the subsurface with oxidant so as to determine the concentration of contaminant of concern (COC); subsequently water samples were taken periodically, after the injection of oxidant (Post-injection) in order to monitor the rate of reduction of contaminant of concern

Fig. 4a. PHA-100 - Portable Hydrocarbon

Fig. 4b. Testing of water samples with PHA analyzer (FCI environmental)

per time. Unique Portable Hydrocarbon Analyzer (PHA) (Figs. 4a and 4b) was used for the purpose of monitoring the concentration of contaminant of concerns per time.

Physico-chemical characterization was determined in-situ with the aid of YSI professional multi-parameter water quality meter, also water samples were taken for other physic-chemical tests.

4 Results/Discussion

4.1 Distribution of Free LNAPL at the Study Area

The Light Non-aqueous Phase Liquid (LNAPL) thickness was monitored in all the six wells within the pilot scheme area. Figure 5a displays hydrocarbon thicknesses of the monitoring wells reported by Adekunte (2008) and Figs. 5b and 5c present the contours of free hydrocarbon thicknesses of all the six wells monitored within the pilot area and reports of all wells monitored showing the various thicknesses of free hydrocarbon product and water respectively in 2014. The free hydrocarbon thickness in the wells varies from 0.491 m to 0.001 m. W17 had the highest thickness of free hydrocarbon, and the least thickness of hydrocarbon occurred in W44 and W41.

(i) **Volume of the Skimmed-off Pure Phase Hydrocarbon**

Table 1 and Fig. 6 present the summary of the volume of free product skimmed-off from the wells within the pilot test area using the Model 8 Abanaki oil skimmer shown in Figs. 3a and 3b. About 800 litres of free hydrocarbon product was skimmed-off from the monitoring wells within the pilot scheme area and deposited at the closet depot of the national oil regulatory body in Nigeria (NNPC/PPMC depot at Ejigbo, Lagos). Thereafter, pumping of water from subsurface was carried out with the help of subsurface pumping machine. The pumping out of groundwater was done so that contaminants trapped between the monitoring wells could migrate to the wells during aggressive pumping. This was carried out for over 6 months. This operation was

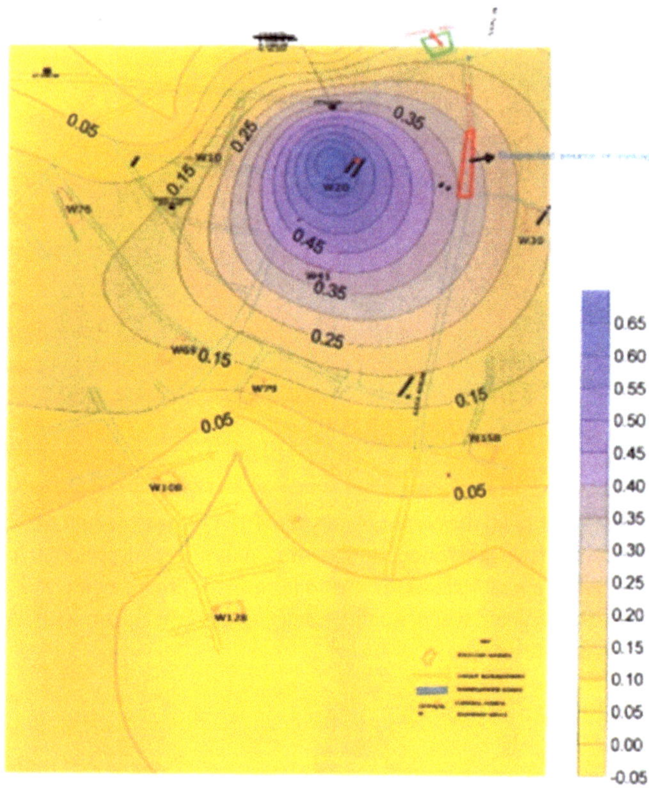

Fig. 5a. Free hydrocarbon thickness for monitoring wells within the study area (2006) (Adekunte 2008)

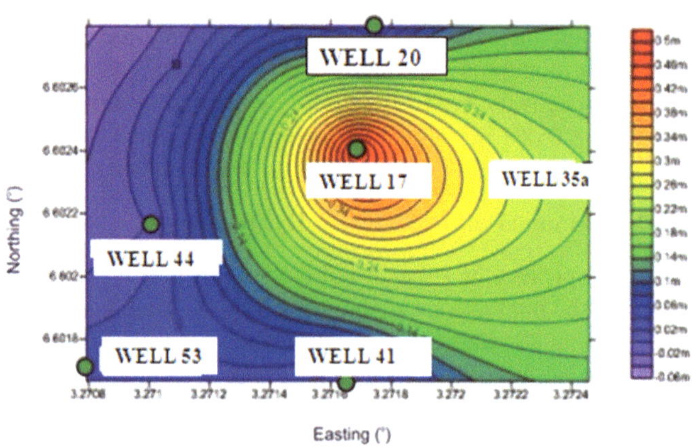

Fig. 5b. Free hydrocarbon thickness for monitoring wells within the study area

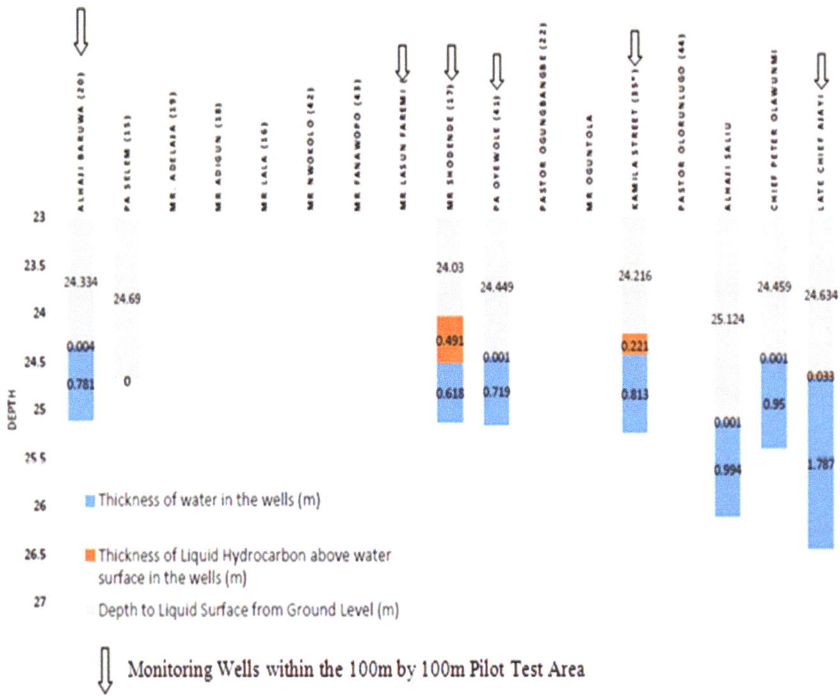

Fig. 5c. Reports of all wells monitored showing the various thicknesses of free hydrocarbon product and water

Table 1. Summary of the volume of LNAPL skimmed-off from the monitoring wells

Observation wells	W(20)	W(17)	W(41)	W(35A)	W(44)	W(53)
Thickness of LNAPL(m)	0.004	0.491	0.001	0.221	0.001	0.033
Diameter of wells (m)	1.765	0.820	1.930	1.650	1.111	0.799
Vol. of LNAPL skimmed (Litres)	9.67	259.4	2.926	472.7	0.969	16.540

successfully concluded when there was no more sheen on the surface of the groundwater in each of the monitoring wells. *In-situ* Chemical Oxidation (ISCO) process commenced only when no hydrocarbon sheen could be noticed in any of the six pilot wells.

(ii) **Dissolved Oxygen Values during ISCO**

The DO concentration usually reflects its organic contaminant load (that is, the lower the DO, the greater the contaminant concentrations). The DO increases significantly for all the observation wells dosed with the oxidant ($KMnO_4$) from the second week of injection of oxidant, except for the observation well (W17) that exhibited a contrary results at the second week of ISCO, the DO reduces from 1.8 mg/l to 1.4 mg/l, but significantly increases as ISCO progresses (Fig. 7). DO is electron acceptors and abundance of electron acceptors is helpful in the success of bioremediation.

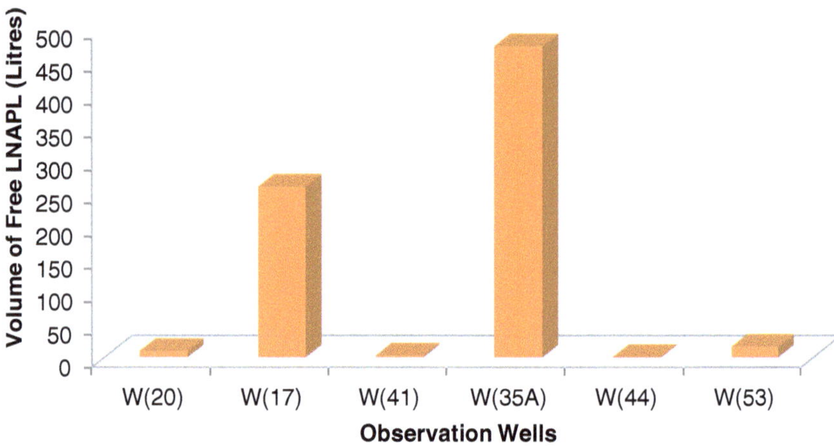

Fig. 6. Volume of LNAPL in skimmed-off prior to ISCO

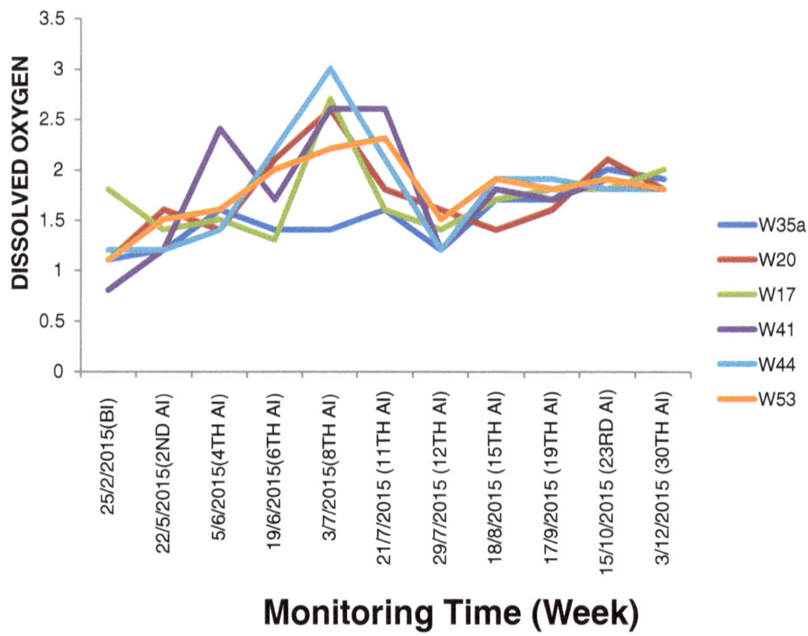

Fig. 7. Dissolved oxygen values against monitoring time

(iii) **Total Petroleum Hydrocarbon (TPH) in the Groundwater during ISCO Remediation**

Figure 8 shows the plot of TPH values for W35a, W20, W17, W41, and W53. For monitoring well W35a, the TPH value reduced from 350 ppm to 20 ppm at the end of thirty weeks of injection. The monitoring well W20 reduced from 299 ppm to 30 ppm

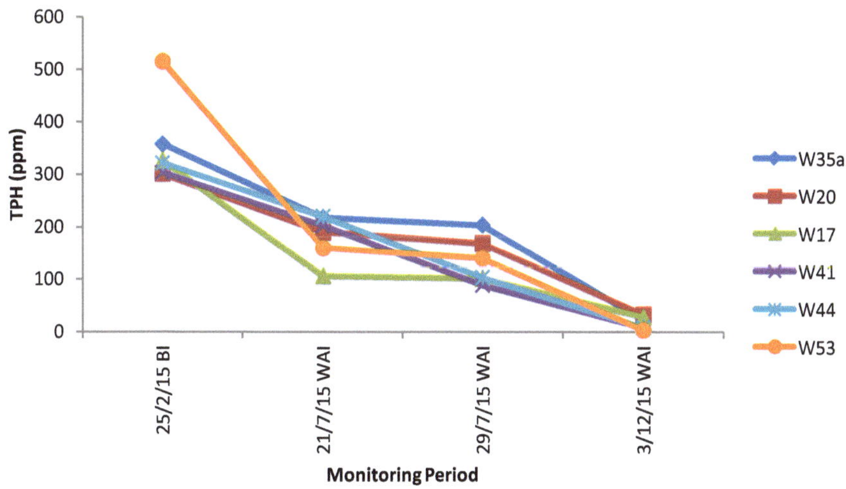

Fig. 8. Total petroleum hydrocarbon in the groundwater for monitoring wells during ISCO.

at the end of thirty weeks of injection. The TPH value of monitoring well W17 reduced from 324 ppm to 25 ppm at the end of thirty weeks of injection, also W41 reduced from 300 ppm to 6 ppm on the same time after injection.

At the monitoring well W44, TPH value before injection of oxidant was 319 ppm which reduced to 6 ppm at the end of thirty weeks after injection of oxidant; also, W53 reduced from 512 ppm to 0.7 ppm at the end of thirty weeks of injection of oxidant

5 Conclusion

The maximum record of thickness of LNAPL of 0.49 m was recorded in monitoring well W17 and 259.4 litres of the free product was skimmed off, 472.7 litres of LNAPL was skimmed off from W35a and the least product was skimmed from W20. The total volume of LNAPL skimmed off was about 800 litres. Thus, all the free products have been effectively removed in all the wells in the pilot scheme area

The DO increases significantly for all the observation wells dosed with the oxidant, the DO increases from a range of 0.8 mg/l to 8.4 mg/l.

Oxidation study was carried out by observing the concentration of COC for and other parameters during the process of ISCO remediation for 30 weeks. A reduction in Total Petroleum Hydrocarbons (TPH) ranging from 92.28% to 99.86% was observed within the period. Maximum TPH values in monitored well was 512 ppm, while the minimum observed TPH at the end of thirty weeks was 0.7 ppm (W53).

Acknowledgements. The investigators would like to acknowledge the financial support of the TETFUND National Research Fund of The Prof. S.A. Ola Research Group, Federal University of Technology, Akure referenced TETF/ES/NRF/013/Vol.I for the Research Project titled "Site Remediation in Nigeria: Proven and Innovative Technologies, Recovery of Free Hydrocarbon from Soil/Groundwater."

References

Adekunte, O.A.: Hydro-chemical analysis of shallow wells for hydrocarbon contamination in Baruwa, Lagos, Nigeria. M.Eng thesis, Civil Engineering Department, The Federal University of Technology, Akure, Nigeria (2008)

Balogun, S.: Geoenvironmental assessment and stratigraphic characterization of a contaminated site in Baruwa, Lagos State. M.Eng thesis, Civil Engineering Department, The Federal University of Technology, Akure, Nigeria (2009)

Clement, P., Kim, Y., Gautam, R., Lee, K.: Experimental and numerical investigation of DNAPL dissolution process in a laboratory aquifer model. Groundwater Monit. Rem. **24**(4), 88–96 (2004)

Environmental Protection Agency (EPA): How to evaluate alternative cleanup technologies for underground storage tank sites. EPA/510-B-95-007.5 (2005) "Historical site assessment" (1995)

Fadugba, O.G.: In situ pilot scale remediation of hydrocarbon contaminated groundwater (A case study of Baruwa Community Lagos, Nigeria). A Ph.D. progress seminar presented at School of Engineering and Engineering Technology, Department of Civil and Environmental Engineering, The Federal University of Technology, Akure, Nigeria (2016)

Jegele, T.O., Ola, S.A.: Deep groundwater remediation by means of oil skimmer in Baruwa Community, Lagos State. Int. J. Environ. Eng. Sci. Technol. Res, **4**(1), 1–11 (2016). ISSN 2326-3113 (Online)

Keely, J.F.: Performance evaluations of pump-and-treat remediations. U.S. environmental protection agency, Superfund ground water issue, EPA/540/4-89/005. Robert S. Kerr Environmental Research Laboratory (1989)

National Research Council: Alternatives for Ground Water Cleanup. National Academy Press, Washington, D.C. (1994)

Ola, S.A., Ojuri, O.O.: Remediation of hydrocarbon contaminated sites. In: Paper Presented at the 2nd ISSMGE International Lecture Tour in Environmental Geotechnics (NSE-NGA/ISSMGE), Nigeria, 14–16 January 2008 (2008)

Ola, S.A., Fadugba, O.G., Uduebor, M.A.: Slug tests for determination of hydraulic conductivity of contaminated wells. Environ. Nat. Resour. Res. **6**(2), 156–165 (2016)

Ola, S.A., Fadugba, O.G., Ojuri, O.O.: Comparison of effectiveness of two remediating agents on hydrocarbon contaminated soil/ groundwater in the laboratory. Environ. Nat. Resour. Res. (2015). ISSN 1927-0496, doi:http://dx.doi.org/10.5539/enrr.v5n1p1. Accessed 12 Dec 2015

Ojuri, O.O., Ola, S.A., Fadugba, O.G., Uduebor, M.A.: Site remediation in nigeria: proven and innovative technologies (Recovery of free hydrocarbon from soil/groundwater), GEOMATE-Brisbane, 19–21 November 2014, pp. 585–590 (2014)

Ojuri O.O., Ola, S.A.: Estimation of contaminant transport parameters for tropical sand in a sand tank model. Int. J. Environ. Sci. Tech. **7**(2), 385–394 (2010). ISSN 1735-1472

Ojuri O.O.: Geo-environmental assessment and remediation of contaminated sites: a case study of Baruwa Community, Lagos, Nigeria, Ph.D. thesis (2009)

Steele, E.A., Johnson, H.M., Sandstrom, M.W., Capel, P.D., Barbash, J.E.: Occurrence and fate of pesticides in four contrasting agricultural settings in the United States. J. Environ. Qual. **37** (3), 116–1132 (2008)

Uduebor, M.A., Ola, S.A.: Hydrocarbon remediation by natural attenuation at Baruwa, Lagos, Nigeria. Electron. J. Geotech. Eng. **2016**(21), 501–511 (2016). ejge.com

United States Environmental Protection Agency: In situ remediation technology: in situ chemical oxidation, EPA-510-B-95-007 (1995)

US EPA: The State-of-the-practice of characterisation and remediation of contaminated ground water at fractured rock sites, Publication: 542-R-01-010, Washington DC, USA (2001)

Wiberg, K.B., Saegebarth, K.A.: The mechanisms of permanganate oxidation: IV. hydroxylation of olefins and related reactions. J. Am. Chem. Soc. **79**, 2822–2824 (1957)

Analysis of Infiltration, Colapsividade and Absorption of Surface Layer in Area of Deployment in the Setor Habitacional Jardins Mangueiral in Federal District - Brazil

Rideci Farias[1(✉)], Haroldo Paranhos[1], João Marcelo Lopes Siqueira[2], Elson Oliveira de Almeida[3], and Roberto Pimentel[4]

[1] UCB/Reforsolo Engenharia/UniCEUB/IesPlan, Brasília, Brazil
rideci.reforsolo@gmail.com, reforsolo@gmail.com
[2] Construsaneo, Brasília, Brazil
construsaneo@gmail.com
[3] UniCEUB,B, Brasília, Brazil
elsul0@hotmail.com
[4] UnB, Brasília, Brazil
eng.robertopimentel@gmail.com

Abstract. The Setor Habitacional Jardins Mangueiral is the result of a public-private partnership in which the Government of the Federal District gave the area to build a certain group and financing held by the Caixa Economica Federal. Therefore, beneficiaries who meet the requirements of the Minha Casa, Minha Vida program, the Federal Government, have access to special terms of payment in real estate. This is a neighborhood with installed infrastructure, equipped with squares, accessibility, water supply and sewage, gated communities, paved and lighted streets, parking spaces, among other devices. Thus, with the installation of the housing area, made necessary the allocation of areas with a view to infiltration of rainwater consistent with the neighborhood in question. In this context, this article seeks to present the results and discussions of the geotechnical investigations carried out in the area and collected soil at the site of the Setor Habitacional deployment. The study consisted in carrying infiltration assays "in situ" removal of soil samples granulometry tests, natural density of the soil, density of the soil grains, direct shear and densification in order to verify the possible behavior of soils located in the areas that were intended for infiltration of rainwater. Among other findings, it was increased densification with the depth of contrasting material with the decrease of the permeability as the depth increases. This fact highlights the influence of weathering in the upper layers of existing soil strata in the area. The specific weights show that in terms of the superficial genesis and underlying soils are similar and typical compositions of clays and sands. The porosity of the material then decreases with depth as a result of densification (increased bulk density). It is found that the permeability values obtained are of the order of magnitude of 10–3 cm/s in the depths of 1.50 m and 10–4 cm/s to the depths of 3.0 m. For laboratory tests were collected three undisturbed soil samples. These tests comprised grain size, natural density of the soil, density of soil grains, direct shear and densification. Abstracts are presented the results of tests carried out

D.N. Singh and A. Galaa (eds.), *Contemporary Issues in Geoenvironmental Engineering*, Sustainable Civil Infrastructures,
DOI 10.1007/978-3-319-61612-4_20

geotechnical and certain parameters. In terms of densification were performed a total of six (6) tests in order to verify a possible ground behavior in the natural state and soil collapse in predetermined voltages. These tests were conducted on the three soil samples collected along the area.

1 Introduction

This article presents the results of geotechnical investigations carried out in the area and soil collected at the site of implantation of the Housing Sector. The study consisted in carrying infiltration assays "in situ" removal of soil samples granulometry tests, natural density of the soil, density of the soil grains, direct shear and densification in order to verify the possible behavior of soils located in the areas that were intended for infiltration of rainwater. The criteria used to define the points was based on the verification of possible areas designed for infiltration of rainwater.

2 Location of Studies Area

The study area was located in DF 463, DF 001 and the Botanical Garden of Brasilia, according to Figs. 1, 2 and 3, taken from Google Earth .

Fig. 1. Global view of implementation area of Setor Habitacional Jardins Mangueiral.

Fig. 2. Global view of implementation area of Setor Habitacional Jardins Mangueiral.

Fig. 3. Micro view of implementation area of Setor Habitacional Jardins Mangueiral.

2.1 Executed Studies

The test campaign involved field trials and laboratory tests. Field trials understood infiltration and tactile-visual identification of soils. Already laboratory tests comprised the grain size tests, natural density of the soil, density of soil grains, direct shear and densification. The following presents the test run.

2.2 Field Tests: Infiltration Tests

For infiltration tests "in situ" were defined twenty (20) points along the deployment area of the Housing Sector Jardins Mangueiral. The locations of the 20 points were defined with the designer and also field checks. The criteria used to define the points was based on the verification of possible areas designed for infiltration of rainwater. These tests were aimed at checking the coefficients of permeability of the study area. The test consisted of running the depth of 1.5 m and 3.0 m, i.e., every section were performed two infiltration tests in order to evaluate an average over the soil profile study. From the test results it was possible to generate iso permeability curves estimated to depths of 1.5 and 3.0 m, a result can also determine the curves of average iso permeability estimated for the housing industry as a whole. Table 1 permeability obtained for the points studied in depths 1.5 m and 3.0 m for each point studied, but also the average over the study profile. It is found that the permeability values obtained are predominantly of the order of magnitude 10–3 cm/s and the depth of 1.50 m 10–4 cm/s to depths of 3.0 m.

2.3 Lab Tests

In total we were collected three undisturbed soil samples for carrying out the tests in the laboratory. These tests comprised grain size, natural density of the soil, density of soil grains, direct shear and densification. Tables 2, 3 and 4 below present summary of the results of tests carried out geotechnical and certain parameters.

Table 1. Permeabilities obtained for the studied holes.

Furo	Permeabilidade (cm/s)		
	Prof. 1,5 m	Prof. 3,0 m	Média
CA 01	2,49E-04	1,37E-05	1,31 E-04
CA 02	1,91E-03	1,97E-04	1,05E-03
CA 03	2,04E-03	2,81E-04	1,16E-03
CA 04	2,49E-03	5,92E-05	1,28E-03
CA 05	1,36E-03	2,73E-05	6,95E-04
CA 06	1,96E-03	5,13E-04	1,24E-03
CA 07	1,88E-03	3,09E-04	1,09E-03
CA 08	1,46E-03	1,02E-03	1,24E-03
CA 09	2,66E-03	7,05E-04	1,68E-03
CA 10	1,66E-03	4,57E-04	1,06E-03
CA 11	1,15E 03	3,85E-04	7,69E-04
CA 12	1,15E-03	5,35E-04	8,43E-04
CA 13	2,83E-03	6,18E-04	1,73E-03
CA 14	4,40E-03	5,97E-04	2,50E-03
CA 15	3,00E-03	6,36E-04	1,82E-03
CA 16	2,05E-03	1,09E-03	1,57E-03
CA 17	7,03E-03	9,27E-04	3,98E-03
CA 18	3,31 E-03	4,65E-04	1,89E-03
CA 19	6,43E-03	2,87E-04	3,36E-03
CA 20	1,19E-03	3,61 E-04	7,75E-04

Table 2. Summary of Geotechnical tests results made and determined parameters for sample 01.

Sample 01 (Depth 2,5 m)		
Natural specific weight	kN/m^3	17,00
Natural humidity	%	39,75
Specific gravity apparent dry (kN/m^3)	kN/m^3	12,16
Specific weight of solids (or grains)	g/cm^3	2,73
Specific weight saturated (kN/m^3)	kN/m^3	17,51
Empty index	–	1,20
Porosity	%	54,57
Degree of saturation (%)	%	90,35

In terms of densification were performed a total of six (6) tests in order to verify a possible ground behavior in the natural state and soil collapse in predetermined voltages. These tests were conducted on the three soil samples collected along the deployment area of the Housing Sector Jardins Mangueiral. Two samples referred to as "Sample 1" and "Sample 3" were collected in areas of local infiltration and intended for another sample named "Sample 2" was collected between points 19 and 20 (points called for permeability). Table 5 summarizes the results of the compaction testing for pre-determined tensions.

Table 3. Summary of results of geotechnical tests realized and parameters of sample 02.

AMOSTRA 02 (Profundidade de 1,5 m)		
Natural specific weight	kN/m^3	11,04
Natural humidity	%	23,21
Effective cohesion (Flooded condition)	kPa	3
Effective friction angle (Flooded condition)	graus	16
Effective cohesion (Natural condition)	kPa	11
Effective friction angle (Natural condition)	graus	17
Specific gravity apparent dry (kN/m^3)	kN/m^3	8,96
Specific weight of solids (or grains)	g/cm^3	2,71
Specific weight saturated (kN/m^3)	kN/m^3	15,46
Empty index	–	1,97
Porosity	%	66,29
Degree of saturation (%)	%	31,98

Table 4. Summary of results of the geotechnical tests realized and parameters of sample 03.

Amostra 03 (Profundidade de 1,5 m)		
Natural specific weight	kN/m^3	12,59
Natural humidity	%	39,91
Specific gravity apparent dry (kN/m^3)	kN/m^3	9,00
Specific weight of solids (or grains)	g/cm^3	2,71
Specific weight saturated (kN/m^3)	kN/m^3	15,48
Empty index	–	1,95
Porosity	%	66, 14
Degree of saturation (%)	%	55,37

Table 5. Summary of results for consolidation tests.

Amostra	Tipo de Ensaio	Tensao de Colapso (kPa)	Índice de Vazios		Coefic. de Colapso
			Inicial	Final	
01	Collapse	25	1,168	1,127	0,019
02	Natural (Until 480 kPa)	–	2,0765	1,6021	–
	Collapse	24	2,0100	1,920	0,030
	Collapse	106	1,8388	1,2700	0,200
03	Natural (Until 805 kPa)	–	2,0781	1,1861	–
	Collapse	423	1,6857	1,1364	0,2045

Figures 4, 5, 6 and 7 show the preparation and execution of infiltration tests "in situ". Then, in Figs. 8, 9 and 10, excavation and removal of soil samples for laboratory tests, and in Figs. 11, 12, 13 and 14 shows views of the Housing Sector Jardins Mangueiral, already consolidated with areas for infiltration of rainwater.

Fig. 4. Hole execution for infiltration test.

Fig. 5. Hole saturation for infiltration test.

Fig. 6. Measurement of water infiltration in the hole.

Fig. 7. Measurement of water infiltration in the hole.

Fig. 8. Hole for withdrawal undisturbed sample.

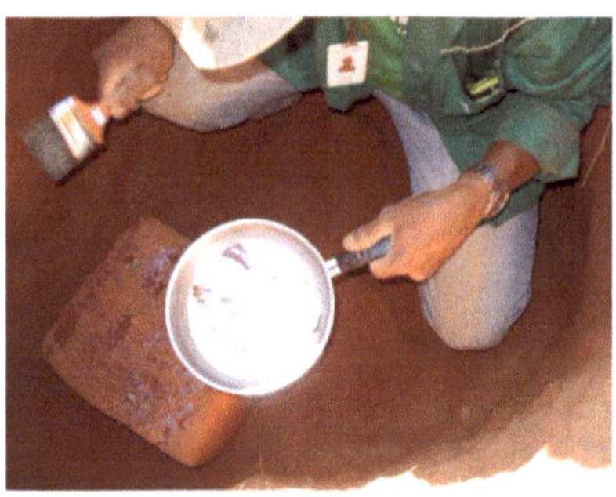

Fig. 9. Parafining of underturbed sample of soil.

Fig. 10. Undisturbed soil sample parafinn for lab tests.

Fig. 11. Global view of implementation area of Setor Habitacional Jardins Mangueiral.

Fig. 12. Micro view of one of the housing blocks.

Fig. 13. Micro view of one of the housing blocks.

Fig. 14. View of one infiltration basin of stormwaters.

3 Results and Considerations of Executed Geotechnical Studies

This article aims to present the results of geotechnical investigations carried out in the area and collected soil at the site of implantation of the Housing Sector Jardins Mangueiral. On the studies we found that in stratigraphic terms there is a predominance of clay and sandy soils and/or clay-silty in the surface layer of the study area. Based on the conditions of densification, permeability and density of the investigated surface strata it was concluded that:

(1) The densification depth in the material under study with the decrease of the permeability as the depth increases. This fact highlights the influence of weathering in the upper layers of existing soil strata in the area. The specific weights of the particles show that in terms of the superficial genesis and underlying soils are similar and typical compositions of clays and sands. The porosity of the material so decrease depth as a result of densification (increased bulk density). The conducted field tests show the permeability soil permeability decreases with depth. It is found that the permeability values obtained are predominantly of the order of 10–3 cm/s in the depths of 1.50 m and 10–4 cm/s to the depths of 3.0 m;

(2) The porosity and permeability values in the geotechnical field investigations and laboratory have recognized the important influence of weathering the storm water infiltration properties in the superficial layers of the study area. The combination of stratigraphic data of geotechnical analyzes performed here confirm densification with increasing depth of the surface layers of area. This fact reflects the lower permeability of the lower strata which still maintains a more consistent structure. It should be noted, however, that such consistency certainly would not hold if the permeability tests were carried out on the possible preferred plans "fracturing" of land;
(3) The surface layers have adequate permeability to certain liquids storage capacity of which decreases with depth as shown with decreasing porosity (densification) with increasing depth. Allied to this fact can be verified that there is spatial variability of hydraulic parameters (porosity and permeability) resulting from the degree of weathering. Based on these results it can be estimated that surface water that sometimes reach the existing cover soil will seep into relatively short time depending on the surface of the ground retention capacity in terms of topographic basins;
(4) In the alternative that clay landfills to be executed and (or) silty with a certain degree of compression and relatively lower than that of natural soil either existing permeability, then there would be to predict the performance of a drainage system and lead the water the surface around the compacted area then evaluating whether to infiltration of surface water based on the permeability values now presented in this study;
(5) The studies presented here demonstrate adequate permeability of the surface soils of the study area. Notably, also checks to compatibility of the area to the accumulation of surface water in topographic basins small. However waterproofing a given area due to the construction of an improvement raises concerns that are controversial in nature. One concern relates to the influence of runoff water to be directed to adjacent areas with potential erodibility to be due to the drainage project; This project should match the energy in the water discharge points with soil erodibility where these discharges occur;
(6) In terms of collapsivity it is reported that some non-saturated soils suffer a sharp reduction of the void ratio when saturate and (or) receive some sort of request. This behavior, for example, is typical of porous clays of Brasilia. The cause is associated with macro pores which are formed from a leaching process by water, where the fine particles are carried into deeper layers and allowed to contact one binder materials (usually calcium carbonate or iron oxide). In saturation, the cementitious material is dissolved and the collapse of the clay structure may occur due to some request. The role of cementitious material can be done by capillary forces which arise in the unsaturated soil and also lose the saturation. Thus, the degree of collapsibility of a clay can be set from an index called collapsing ratio "i" obtained in densification assays wherein the sample is flooded at some voltage greater than or equal to that operates in the field. Generally, the collapse coefficient (i) is >0.02 is considered collapsible soil. For the samples tested, there was a low value of collapse coefficient (0.019 for Sample 01 and Sample 02 to 0,030), voltages of 25 and 24 kPa at flooding, respectively. For higher voltages, we have

what might be expected, higher values collapse coefficient. However, as the analysis to hold small bowls of infiltration, it is expected that the disposal of rainwater, in adequate amounts, not cause damage in terms of collapse in these topographic basins;

Acknowledgments. The Reforsolo Engenharia Ltda., Catholic University of Brasilia (UCB), University Center of Brasilia (UniCEUB) and the Institute of Higher Education Plateau with important contributions that made possible the realization of this work.

References

ABNT/NBR 9603 – Sondagem a trado
ABNT/NBR 6122 – Projeto e execução de fundações
ABNT/NBR 13441 – Rochas e solos
ABNT/NBR 13969 – Procedimento para estimar a capacidade de percolação do solo
Associação Brasileira de Geologia de Engenharia (ABGE). Ensaios de Permeabilidade em Solos – Boletim nº 4. Orientações para sua execução no campo, 3a. edição, 35 p. (1996)

Environmental Aging Effect and Geotechnical Stabilization of Organic Waste on Soil

Nirali Bhaskar Hasilkar(✉), M. Shweta, Mehul Parmar, Pradeep Rathod, and Ravi Patel

Civil Engineering Department, Parul University, Vadodara 391760, Gujarat, India
nironearth@gmail.com

Abstract. This study analyzes the aging effect of organic waste like sugarcane bagasse on soil and its geotechnical stabilization. Our aim of this research is to make economical use of sugarcane bagasse and to maintain environmental balance by avoiding random disposal. Sugarcane bagasse is the major by-product of the sugar industry. It has a great potential for the production of bio-fuels and chemicals due to its considerable amount of cellulose and hemicelluloses. After crushing and milling the stalks, a fibrous residue called bagasse remains. Bagasse is universally used as a fuel source for the milling and refining plants and typically releases enough energy to power other parts of the power grid as well.

Soil taken for this research was CL type as per Indian standard classification system (ISCS) by performing basic tests in laboratory. The organic waste material like sugarcane bagasses were added in different percentage (1 & 3%) in black cotton soil and yellow soil. Also, geotechnical parameters were checked on soil + organic waste material with and without vermi composting. The performance of sugarcane bagasse stabilized soil was evaluated using strength performance tests namely; California bearing ratio (CBR) and Unconfined compressive strength test (UCS). From the results, it was observed that there was significant increase in the strength after the addition of sugarcane bagasse also the effect of organic waste on the yellow soil was quite more, as compared to black cotton soil after vermi composting. It was noticed that the changes in the environmental properties of soil with respect to time was of better quality in yellow soil.

1 Introduction

Billions of dollars in damages are attributed to expansive soils in many countries each year. Geotechnical design and analyses in/on/with expansive soils may involve additional complications that otherwise would not have to be dealt with if expansive soils were not present. Traditional methods for chemical stabilization of soils include the addition of lime, class-C or class-F fly ash, portland cement, or other industrial waste by-products such as cement kiln dust, steel or copper slag. In building systems, every displacement can be led to generate internal stresses which have not been predicted in analysis and design of structures which should be anticipated. Generally, clayey soils are very weak soils. So, they definitely need stabilization for the better usage.

D.N. Singh and A. Galaa (eds.), *Contemporary Issues in Geoenvironmental Engineering*, Sustainable Civil Infrastructures,
DOI 10.1007/978-3-319-61612-4_21

Sugarcane is the source of food for billion peoples across the world. Sugarcane bagasse being agricultural waste dumped near the mills or burnt in open fields. Numerous problems arises from corn straw disposal such as, methane generation due to fermentation microorganisms, being light and fine particles causing breathing problems, bad effect on the health are reported such as acute and chronic effect affecting eyes, skin and upper respiratory tract and allegoric response such as nasal catarrh, asthma. Long periods required for decomposition are not appropriate for composting of manure. Thus, proper and safe disposal of sugarcane straw is again a big problem.

Soils with low-bearing capacity can be strengthened economically for building purposes through the process of "soil stabilization" using different types of stabilizers. Soil bed should bear all generated stresses transmitted by shallows foundation or piles. Also, the soil often is weak and has no enough stability in heavy loading. In this regard, it is necessary to reinforce and or stabilize the soil.

Vermicompost is the product or process of composting using various worms, usually red wigglers, white worms, and other earthworms to create a heterogeneous mixture of decomposing vegetable or food waste, bedding materials, and vermicast. Vermicast, also called worm castings, worm humus or worm manure, is the end-product of the breakdown of organic matter by an earthworm. These castings have been shown to contain reduced levels of contaminants and a higher saturation of nutrients than do organic materials before vermi composting.

2 Literature Review

M. Chittaranjan, M. Vijay, D. Keerthi Studied the agricultural wastes as soil stabilizers. In this study agricultural wastes such as sugarcane bagasse ash, rice husk ash and groundnut shell ash are used to stabilize the weak sub grade soil. The weak sub grade soil was treated with the above three wastes separately at 0%, 3%, 6%, 9%, 12% and 15% and California Bearing Ratio (CBR) test is carried out for each percent. The results of these tests showed improvement in CBR value with the increase in percentage of waste.

Kiran R.G., Kiran Lhad studied "the analysis of strength characteristics of black cotton soil using bagasse ash and additives as stabilizer". In this study the black cotton soil was taken from Harihara, Davanagere district, Karnataka. Under this study laboratory experiments are carried out for different percentages (4%, 8% and 12%) of bagasse ash and additive mix proportions. The strength parameters like CBR, Unconfined Compression Strength (UCS) are determined. It was observed that, the blend results of bagasse ash with different percentage of cement for black cotton soil gave change in density, CBR and UCS values. The density values got increased from 15.16 KN/m^3 to 16.5 KN/m^3 for addition of 8% bagasse ash with 8% cement.

Moses G., K.J. Osinubi studied the 'influence of compactive efforts on cement-bagasse ash treatment on expansive black cotton soil'. The dark grey soil used in this study was obtained along Gombe-Biu road in Yamatu Deba Local Government Area of Gombe State using the method of disturbed sampling. The index properties were determined on the natural and treated soils with stepped percentages of cement (i.e., 0, 2, 4, 6 and 8%) were admixed with 0, 2, 4, 6 and 8% of bagasse ash by dry weight of soil. All the compactions involving moisture-density relationships, CBR and UCS tests were carried out by using energies derived from the standard Proctor (SP), West African Standard (WAS) and modified Proctor (MP) energies. Finally an optimal blend of 8% OPC/4% BA was recommended for treatment of expansive black cotton soil for use as a sub-base material.

3 Objectives

This paper mainly emphasizes on stabilization of expansive soil using locally available sugarcane bagasse and to classify the soil according to Indian standard classification system (ISCS). Also this research was aimed to study the aging effect of soil both yellow soil as well as black cotton soil by adding sugarcane bagasse. The study was to compare strength properties of soil + waste materials combinations i.e. soil + sugarcane bagasse. In addition to all above aims, soil + waste materials combinations were compared for strength properties with and without vermi composting.

4 Material Used

In this study, both the soil (Yellow and Black Cotton Soil) under scrutiny is gathered from "Vanadra Gaun" near to waghodia village, Vadodara. The black cotton soil has an expansive surface zone because of level and lengthened molecule shapes that stick

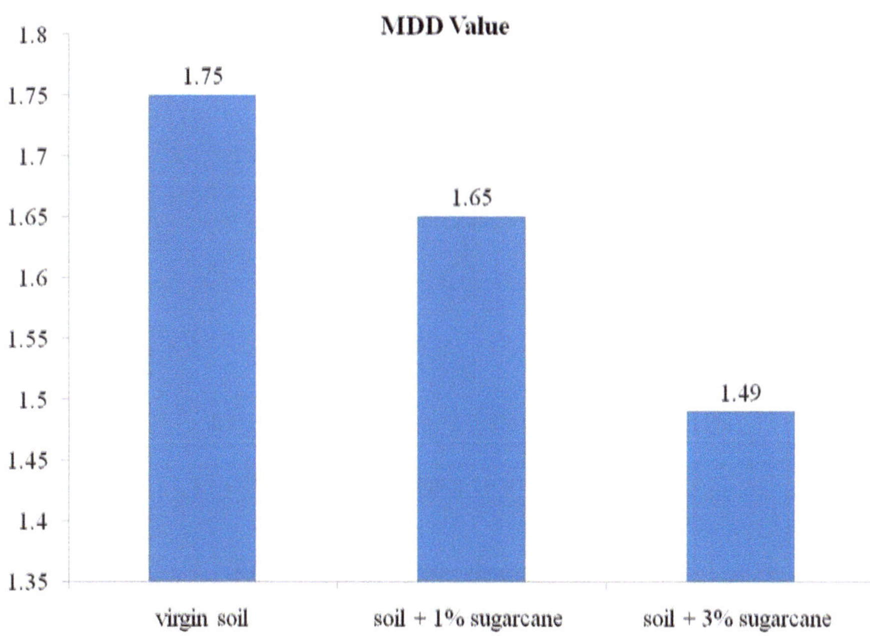

Fig. 1. Comparative graph of MDD value (gm/cc)

together when wet, avoiding typical waste procedures. When it is wet it doesn't get to be dry soon. In like way, when completely dry, it is not soon wetted and shrinks causing breaks. Sugarcane bagasse were used as a soil stabilizer and that sugarcane bagasse were collected from one vender making sugarcane juice at Vadodara city.

5 Experimental Work

Virgin black cotton soil & Yellow soil were tested for its geotechnical classification and test results shows that yellow soil was falling in CL type of soil, where black soil taken for this research was having effect of impurities and therefore our results were nearer to CL type of soil according to Indian standard classification system (ISCS). Soil was replaced by sugarcane bagasse in 1 & 3 percentages of weight and then tested for the comparative strength analysis as per IS code procedure. The test results of the virgin soil, 1% & 3% of sugarcane with soil are as follows:

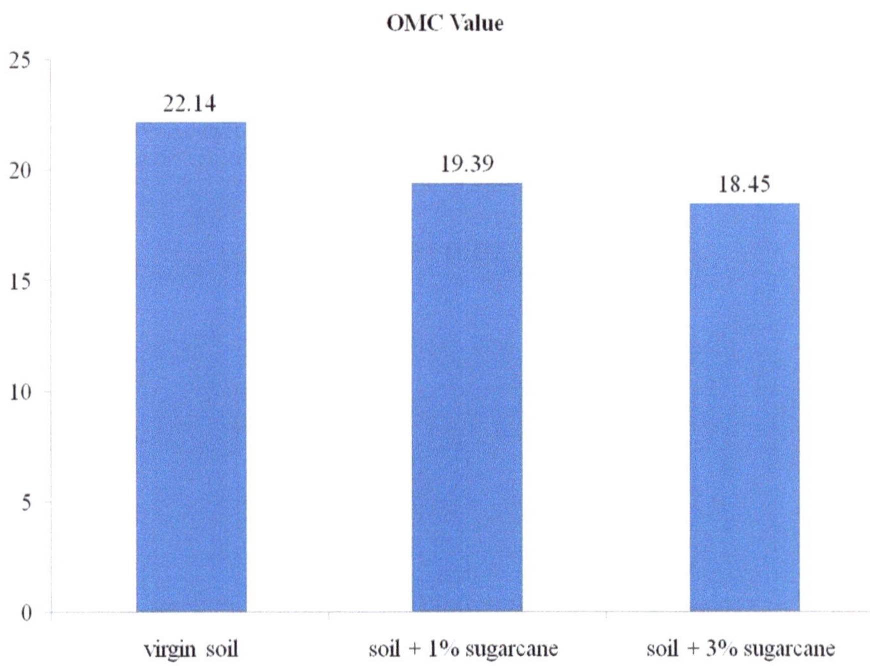

Fig. 2. Comparative graph of OMC value (%)

Table 1. Comparative data for the proctor test

	Virgin soil	Soil + 1% sugarcane	Soil + 3% sugarcane
MDD (gm/cc)	1.75	1.65	1.49
OMC	22.14	19.39	18.45

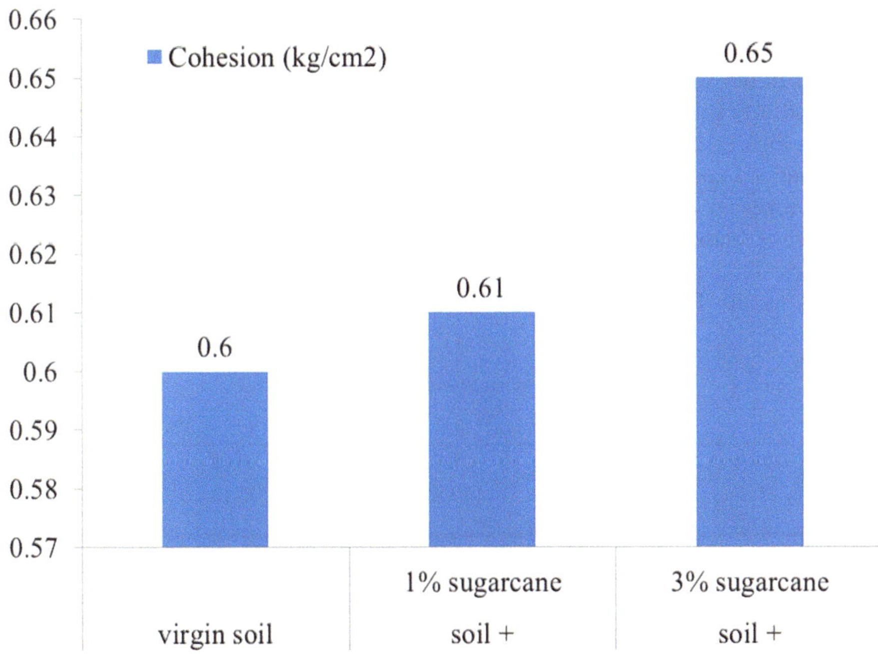

Fig. 3. Comparative graph of cohesion value (kg/cm^2)

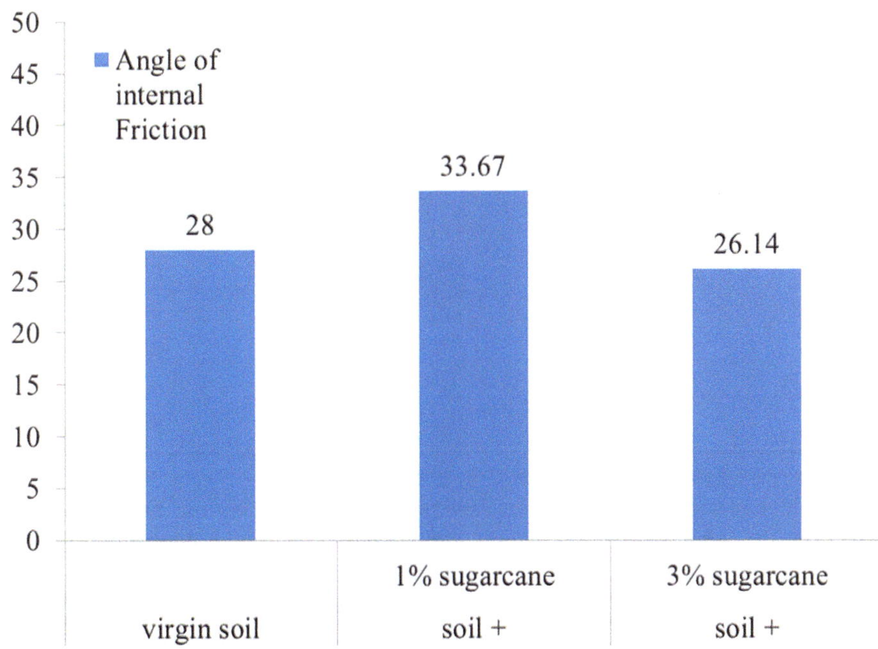

Fig. 4. Comparative graph of angle of internal friction value

Compaction is a procedure by which the air in the pores of the dirt uprooted by mechanical intends to accomplish the thickness necessities. Soil thickness is generally measured in dry unit weight. Optimum moisture content (OMC) and maximum dry density (MDD) values got decreased as the percentage of waste material increases as shown in Figs. 1 and 2 and Table 1.

Unconfined compression strength (UCS) test give the strength parameters in the form of cohesion and angle of internal friction. Figures 3 and 4 show that as the percentage of waste material increases the cohesion value increases but for the angle of internal friction the value increases for the one percentage of waste material and then it decreases for the three percentage of waste material as given in Table 2.

Table 2. Comparative data for the UCS test

	Virgin soil	Soil + 1% sugarcane	Soil + 3% sugarcane
Cohesion (kg/cm^2)	0.6	0.61	0.65
Angle of internal Friction	28	33.67	26.14

California bearing ratio (CBR) test is led to focus the CBR estimation of the specimens, this test did in light of the standard system given in (IS: 2720 Part-16) (Bureau of Indian Standard 1979). The CBR values got increased as the percentage of waste material increased as shown in Fig. 5 and Table 3.

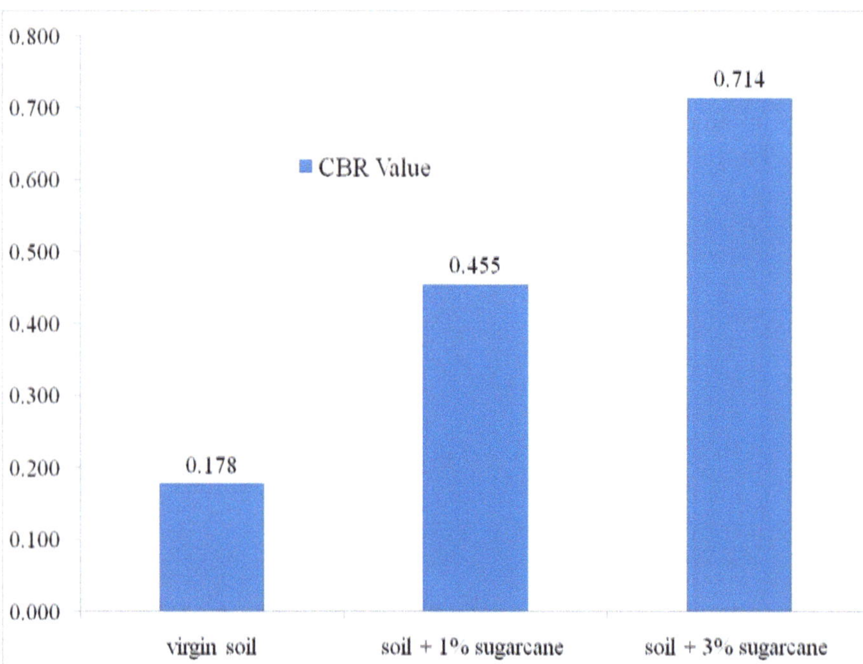

Fig. 5. Comparative graph of CBR value

Table 3. Comparative data for the CBR test

	Virgin soil	Soil + 1% sugarcane	Soil + 3% sugarcane
CBR VALUE	0.178	0.455	0.714

Both the soils were kept for vermi composting for the maturing time periods of one week and one month as shown in Fig. 6. The reading of nitrogen, phosphorous & potash (N P K) values checked at these time periods are as given in Tables 4 and 5. Figures 7 and 8 give the comparative graph of cohesion and angle of internal friction value, with and without vermi composting.

Fig. 6. Soil placed for Vermi composting

Table 4. Comparative data for the N P K value for Black cotton soil

Type of soil: Black cotton soil		9/3/2016		15/3/2016		14/4/2016	
Sr no	Test	Result	Quality	Result	Quality	Result	Quality
1	Total Nitrogen/Organic Carbon (OC)%	1.07	More	1.17	More	1.43	Excess
2	Phosphorous/(P_2O_5,Kg/Ac)	50.00	Excess	60.00	Excess	210.00	Excess
3	Potash/acre (K_2O,Kg/Ac)	503.00	Excess	553.00	Excess	640.00	Excess
4	PH	6.49	Acidic	7.04	Neutral	7.05	Neutral
5	Conductivity ds/m (EC 1:2)	1.4	Soluble salt more	1.08	Soluble salt more	1.68	Soluble salt more
6	S,ppm	26.9	More	Nil	Nil	31.00	More
7	Zn	3.48	More	Nil	Nil	6.04	More
8	Fe	30.60	More	Nil	Nil	36.40	More

(continued)

Table 4. (*continued*)

Type of soil: Black cotton soil		9/3/2016		15/3/2016		14/4/2016	
Sr no	Test	Result	Quality	Result	Quality	Result	Quality
9	Mn	33.10	More	Nil	Nil	33.40	More
10	Cu	2.78	More	Nil	Nil	6.10	More

Table 5. Comparative data for the N P K value for Black cotton soil

Type of soil: Yellow soil		9/3/2016		15/3/2016		14/4/2016	
Sr no	Test	Result	Quality	Result	Quality	Result	Quality
1	Total Nitrogen/Organic Carbon (OC)%	1.17	More	1.17	More	1.38	Excess
2	Phosphorous/(P_2O_5,Kg/Ac)	83.00	Excess	55.00	Excess	216.00	Excess
3	Potash/acre (K_2O,Kg/Ac)	558.00	Excess	520.00	Excess	565.00	Excess
4	PH	7.46	Neutral	7.54	Neutral	6.97	Neutral
5	Conductivity ds/m (EC 1:2)	1.36	Soluble salt more	1.07	Soluble salt more	2.30	Soluble salt more
6	S, ppm	20.10	More	Nil	Nil	29.70	More
7	Zn	3.32	More	Nil	Nil	5.42	More
8	Fe	11.20	More	Nil	Nil	24.60	More
9	Mn	30.30	More	Nil	Nil	34.20	More
10	Cu	1.16	More	Nil	Nil	3.86	More

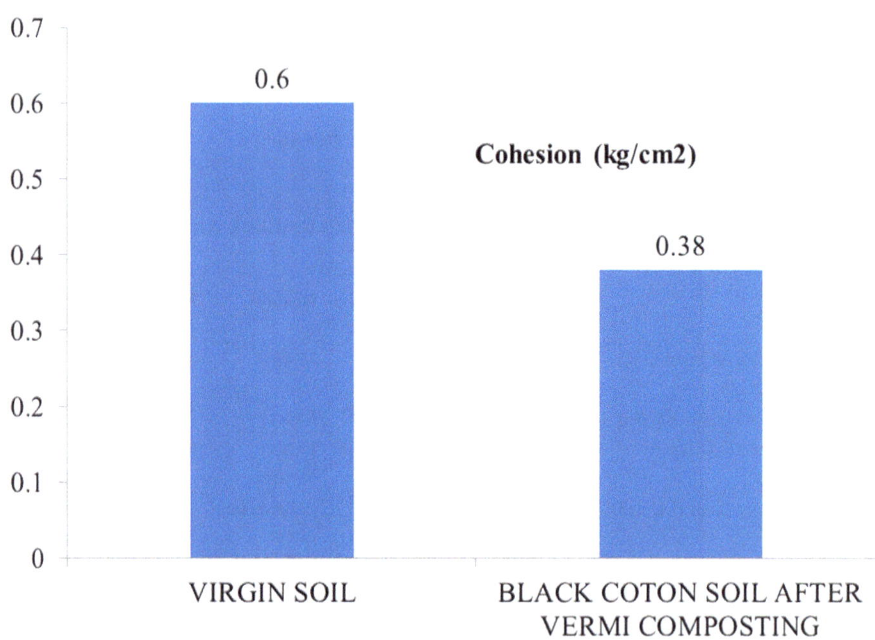

Fig. 7. Comparative graph of cohesion value with and without vermi composting

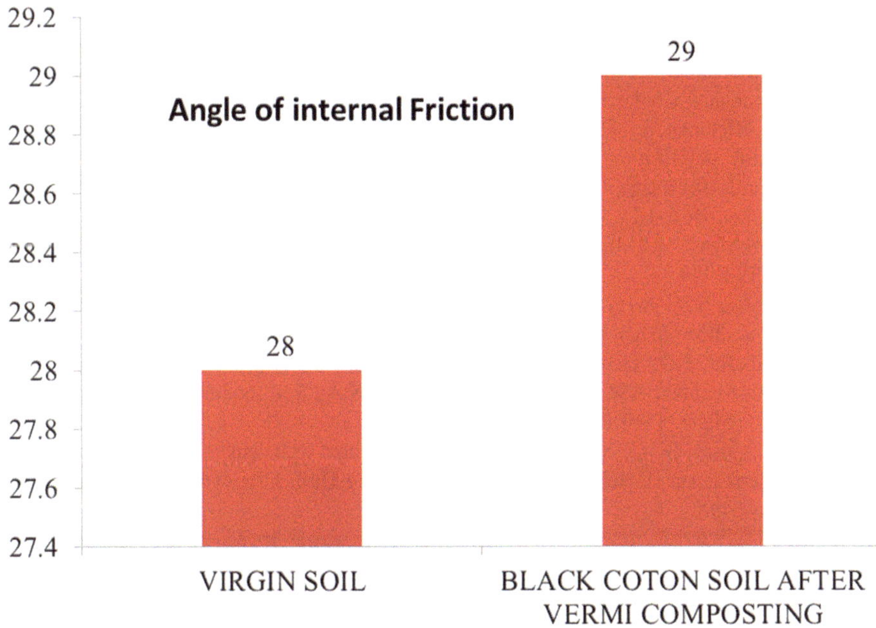

Fig. 8. Comparative graph of angle of internal friction value with and without vermi composting

6 Conclusions

- After adding sugarcane bagasse in the soil, cohesion value get increases with increase in the percentages of waste material.
- Angle of internal friction reading shows increased value for the one percentage of waste material but it got decreased for the three percentage of waste material.
- The CBR value get increased with increase in the percentage waste in the soil.
- One can effectively utilized the organic waste material for the soil stabilization.
- After vermi composting cohesion value get decreased compared to original soil approximately half the value.
- From the N P K value one can see that environmental effect of such a organic waste on soil increases with increase in time.

References

Kiran, R.G., Kiran, L.: Analysis of strength characteristics of black cotton soil using bagasse ash and additives as stabilizer. Int. J. Eng. Res. Technol. (7) (2013)

Moses, G., Osinubi, K.J.: Influence of compactive efforts on cement-bagasse ash treatment of expansive black cotton soil. World Acad. Sci. Eng. Technol. **7**(7), 1559–1566 (2013)

Khandaker, M., Hossain, A.: Stabilized soils incorporating combinations of rice husk ash and cement kiln dust. 1320/J. Mater. Civil Eng. ASCE (2011)

Ramaji, A.E.: A review on the soil stabilization using low-cost methods. J. Appl. Sci. Res. **8**(4), 2193–2196 (2012). ISSN: 1819-544X

Adaska, W.S.: Beneficial uses of cement kiln dust. In: Presented at 2008 IEEE/PCA 50th Cement Industry Technical Conference, Miami, FL, 19–22 May 2008

Rahman, M.K., Rehman, S., Al-Amoudi, O.S.B.: Literature review on cement kiln dust usage in soil and waste stabilization and experimental investigation. IJRRAS **7**(1) (2011)

Sen, T., Mishra, U.: Usage of industrial waste products in village road construction. Int. J. Environ. Sci. Dev. **1**(2), (2010). ISSN: 2010-0264

Elmashad, M.E.M.A.: Soil improvement using cement dust mixture. World Acad. Sci. Eng. Technol. **58** (2011)

Patel, M.A., Patel, H.S.: A review on effects of stabilizing agents for stabilization of soil. Civil Environ. Res. **2**(6) (2012). ISSN 2222-1719 (Paper) ISSN 2222-2863 (Online)

Ken, C.O., Okoafor, F.O.: Geochemistry of soil stabilization. **1**(1) (2012). ISSN 2305-493X

Baghdadi, Z.A., Member, ASCE, Fatani, M.N., Sabban, N.A.: Soil modification by cement kiln dust. 2181 J. Mater. Civil Eng. (1995)

Oriola, F.O.P., Moses, G.: Compacted black cotton soil treated with cement kiln dust as hydraulic barrier material. Am. J. Sci. Ind. Res. (2011). Science Huβ. http://www.scihub.org/AJSIR. ISSN: 2153-649X

Laboratory Study on the Permeability of Fresh Municipal Solid Waste

Zhang Zhenying[1(✉)], Zhang Lingfeng[1], Wu Dazhi[1], Ding Zhengkun[1], Wang Yingfeng[1], and Yuan Qiuyan[2]

[1] School of Civil Engineering and Architecture, Zhejiang Sci-Tech University, Hangzhou, China
zhangzhenyinga@163.com
[2] Department of Civil Engineering, University of Manitoba, Winnipeg, Canada

Abstract. In this study, the permeability of fresh municipal solid waste (MSW) was investigated to understand the relationship between the permeation rate and hydraulic gradients, and the relationship between the permeability coefficient and the initial density. A large-scale permeability test apparatus was used and 15 fresh MSW samples were examined. Three mixture proportions were adopted and five initial densities of 0.7, 0.8, 0.9, 1.0, and 1.1 g/cm^3 were selected for each mixture proportion. It was found that the seepage did not occur until the hydraulic gradient reached a critical value and the critical hydraulic gradient increased with the initial density. The permeation rate increased with the hydraulic gradient. A linear relationship was found between permeation rate and hydraulic gradient. It was also found that the permeability coefficient decreased gradually with the initial density of fresh MSW and they followed a linear relationship with the fitted coefficients greater than 0.99. In addition, the prediction model of the permeability coefficient was proposed. Furthermore, the permeability coefficient of fresh MSW was found to have some interrelationship with the sample's mixture proportions and the organic content. Under the same condition, the permeability coefficient decreased with the organic content. The range of the permeability coefficient obtained for different mixture proportions along with other findings from this study could provide a reference guide for the analysis of the permeability of a landfill.

Keywords: Fresh municipal solid waste · Large-scale permeability test · Critical hydraulic gradient · Initial density · Darcy's law

1 Introduction

The hydraulic characteristics of a landfill are key elements in the design of an effective leachate collection, drainage and recirculation system. Amount all the factors that affect the hydraulic characteristics of a landfill, the permeability of MSW is one the most important factors. Research on the permeability characteristics of waste is often divided into field and laboratory experiments. Outdoor experiments on a landfill often included water-injection, water-pumping, well-drilling and prediction based on the field measurement data. The disadvantages associated with field permeability coefficient test are large discreteness and high cost. Therefore, the laboratory experiment has been

D.N. Singh and A. Galaa (eds.), *Contemporary Issues in Geoenvironmental Engineering*, Sustainable Civil Infrastructures,
DOI 10.1007/978-3-319-61612-4_22

commonly employed to study the permeability of MSW. Laboratory experiments included model-tank method and permeation column method. The best advantage of laboratory experiment was the effective control on the sample size, temperature, initial density, mixture ratio, and other test parameters. Jang et al. (2002) carried out laboratory permeability experiment on the waste samples from Kimpo metropolitan landfill, Korea, and found the permeability coefficient ranging from 2.91×10^{-4} to 2.95×10^{-3} cm/s. Korfiatis et al. (1984) adopted the constant head method to measure the permeability of waste samples from New Jersey landfill, USA, and gained the permeability coefficient ranging from 3×10^{-3} to 5×10^{-3} cm/s. Powrie and Beaven (1999) also used the constant head method to study the permeability of waste beneath 50 m of a landfill, and got permeability coefficient ranging from 10^{-5} to 10^{-2} cm/s. Durmusoglu et al. (2006) studied the permeability of a landfill by the varying head method, and obtained the permeability coefficient ranging from 4.7×10^{-4} to 1.2×10^{-2}cm/s. Reddy et al. (2009) used the constant head method to investigate the permeability of Orchard Hills landfill, USA, and achieved the permeability coefficient ranging from 1.0×10^{-6} to 1.0×10^{-2} cm/s. Zhan et al. (2014) carried out laboratory permeability experiment on the waste from different depths of Qizishan landfill, Suzhou, China. He found the range of permeability coefficient of the landfill from 1.4×10^{-5} to 5.5×10^{-4} cm/s. Chen et al. (2014) carried out constant head permeability test on the waste from a landfill, and gained the permeability coefficient range of 1.8×10^{-6} to 5×10^{-4}cm/s. Jie et al. (2005) collected samples from 8 locations in a landfill from Beijing and found the permeability coefficient ranging from 2.67×10^{-7} to 1.07×10^{-3} cm/s using a self-made apparatus. Zhang et al. (2009) carried out constant head permeability experiment by using a self-made permeability model apparatus, and found the permeability coefficients of the deep layer, the middle layer and the shallow layer as 3.56×10^{-4} cm/s, 3.50×10^{-4} cm/s, 4.81×10^{-2} cm/s, respectively. Qu et al. (2005) carried out permeability experiment by using a permeation column, and the permeability coefficient ranging from 1.35×10^{-4} to 1.26×10^{-3} cm/s was reported. Ke et al. (2013) carried out constant head experiment on the self-made waste samples by using a large-diameter triaxial test apparatus and the permeability coefficients obtained under different degradation ages and different vertical loads ranged from 2.06×10^{-4} to 5.91×10^{-2} cm/s. Liu and Shi (2012) carried out the Soil-Water Characteristic Curve (SWCC) experiment on self-made fresh waste sample and reported the permeability coefficient ranging from 10^{-5} to 10^{-2} cm/s.

In the present study, a large-scale permeability test apparatus was used to carry out the experiment on 15 self-made fresh MSW samples. Three mixture proportions were adopted and five initial densities were selected for each mixture proportion. The permeability of fresh MSW was systematically studied to provide a reference guide for analysis on the permeability of a landfill.

2 Test Apparatus and Materials

2.1 Test Apparatus

The large-scale permeability test apparatus (Fig. 1) constituted an operating floor, a permeation column, a water level controller, a water bucket, and connection tubes.

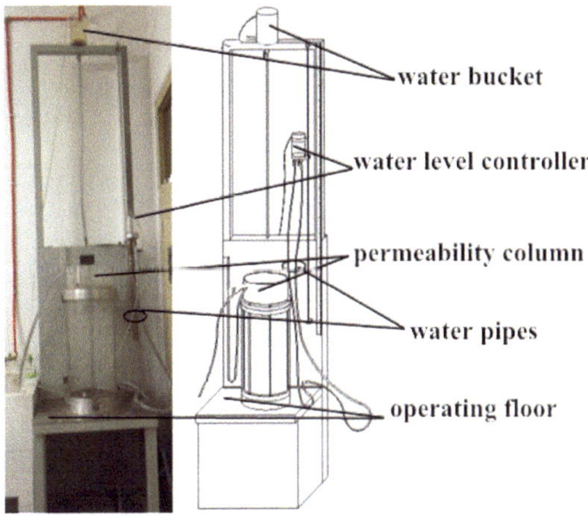

Fig. 1. Large-scale permeability test apparatus

The operating floor had a stainless steel countertop, and the length, width, and height were 600, 500, and 650 mm, respectively. The height, internal diameter, and the wall thickness of the permeation column were 600, 300, and 10 mm, respectively. The thickness and the diameter of the draining stone were 15 and 298 mm, respectively. The top of the permeation column was connected to a seal cover. A draining stone was inset into the seal cover. The water level controller consisted of a cylindrical bucket and a sliding rod. The height and the diameter of the cylindrical bucket were 150 and 100 mm. The length of the sliding rod was 1800 mm, which was fixed to the operating floor. The bucket can move up and down along the sliding rod, and the hydraulic gradient was adjustable. The materials used in apparatus were fabricated from high strength PVC material and No. 45 stainless steel, and all screw joints were embedded with sealing rings.

2.2 Test Materials

To simulate the different economic development levels, three mixture proportions (MPs) were selected (Zhang et al. 2014). The composition of fresh MSW and the mixture proportions were shown in Table 1. Figure 2 shows every component of the

Table 1. Composition of fresh MSW (wet weight)

Material classification	Easily degradable materials				Hardly degradable materials				Incompressible solid materials			
MP1 (%)	80				5				15			
MP2 (%)	65				10				25			
MP3 (%)	50				15				35			
Ingredients	Waste paper	Cabbage, greens, celery	Fruit peel	Wood	Textile	Plastic	Rubber	Bone	Metal	Gravel	Ceramic	Glass
Proportion	4%	70%	20%	6%	30%	20%	20%	30%	5%	85%	5%	5%

Fig. 2. Components of fresh MSW sample

fresh MSW. The fresh MSW was prepared according to the *Technical specification for soil test of landfill municipal solid waste* (CJJ/T204-2013).

3 Test Process

The permeability test was performed based on the technical specification of Soil Test (CJJ/T204-2013). The maximum size of the material was less than a tenth of the column's diameter. First of all, according to the selected mixture proportion of every component, the volume of permeation column and the initial densities, the required amount of every material was calculated, then weighed, and mixed evenly. Figure 3 presents the picture of mixed fresh MSW samples. The test materials were mixed on a geomembrane to avoid the loss of water, then divided into five equal parts, and added into the permeation column in sequence. To keep the initial density at the desired value, every part was compacted to a fifth of the column. The first initial density was selected as 0.7 g/cm^3. After all the test materials were loaded, the sealing cover was locked.

Fig. 3. Mixed fresh MSW sample

Pounding and filling of the fresh MSW sample were shown in Fig. 4. The permeability experiment was carried out two hours after locking the sealing cover. During the test process, the water flow was kept at a stable seepage rate. The hydraulic gradient was set from low to high with the following values of 0.5, 0.75, 1.0, 1.25, 1.5, 1.75, and 2.0. The quantity of discharge water was recorded for every experiment. After the experiment was over, the waste was poured out and the permeation column was cleaned for the next experiment. Then the initial densities were changed to 0.8, 0.9, 1.0, and 1.1 g/cm^3, respectively. After the experiments on the first mixture proportion were finished, the experiments on the second and the third mixture proportions were carried out similarly.

Fig. 4. Pounding and filling of the fresh MSW sample

4 Test Results and Analysis

4.1 Relationship Between the Permeation Rate and the Hydraulic Gradient

Figure 5 shows the relationship between the permeation rate and the hydraulic gradient. It was shown that the permeation did not occur till the hydraulic gradient reached a critical value, and the permeation rate increased with the hydraulic gradient. Besides, the linear relationship had a high correlation (>0.99) and can be expressed as:

$$v = k(i - i_{cr}) \tag{1}$$

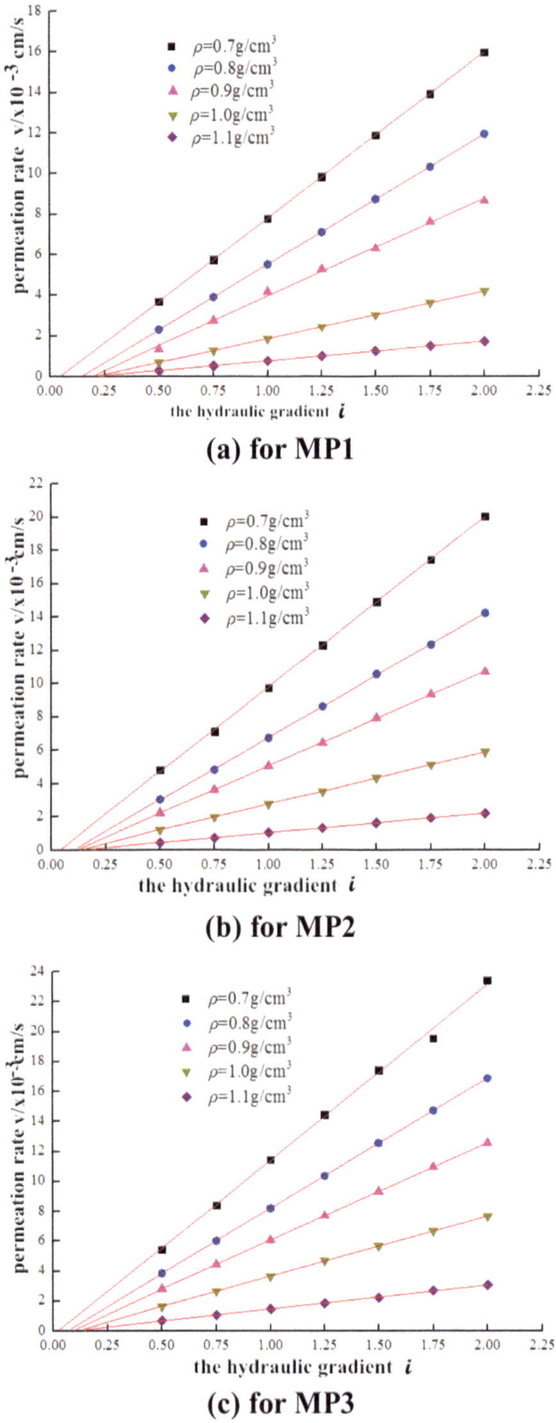

Fig. 5. Relationship between the permeation rate and the hydraulic gradient at various initial densities

Table 2. Values of parameters (k and i_{cr})

Density, ρ (g/cm^3)	0.7		0.8		0.9		1.0		1.1	
Coefficient	k	i_{cr}	k	i_{cr}	k	i_{cr}	k	i_{cr}	k	i_{cr}
MP1	8.2	0.05	6.42	0.14	4.83	0.18	2.34	0.21	0.97	0.22
MP2	10.2	0.04	7.49	0.10	5.67	0.11	3.13	0.12	1.18	0.14
MP3	11.7	0.03	8.69	0.06	6.5	0.07	4	0.09	1.59	0.09

where v was the permeation rate, k was the permeability coefficient, i was the hydraulic gradient, and i_{cr} was the critical hydraulic gradient. The values of k and i_{cr} are shown in Table 2.

Moreover, Fig. 5 showed that the permeation rates under same hydraulic gradient and various initial densities had the same trend, and the permeation rate decreased with the initial density. The reason might be that the pore size of different waste samples was different and as a result, the porosity of waste samples decreased with the increase in initial density. Consequently, the seepage section also decreased, which would decrease the permeation rate of waste samples.

According to the data presented in Fig. 5, the critical hydraulic gradient could also be obtained. The relationship between the critical hydraulic gradient and the initial density was shown in Fig. 6.

Figure 6 showed that the critical hydraulic gradient increased with the initial density, and the relationship could also be formulated as a linear line, the fitted coefficients are greater than 0.8. The linear relationship can be represented by:

$$i_{cr} = f\rho + j \tag{2}$$

where ρ is the initial density and j is a undetermined coefficient. Table 3 presents the values of both these parameters.

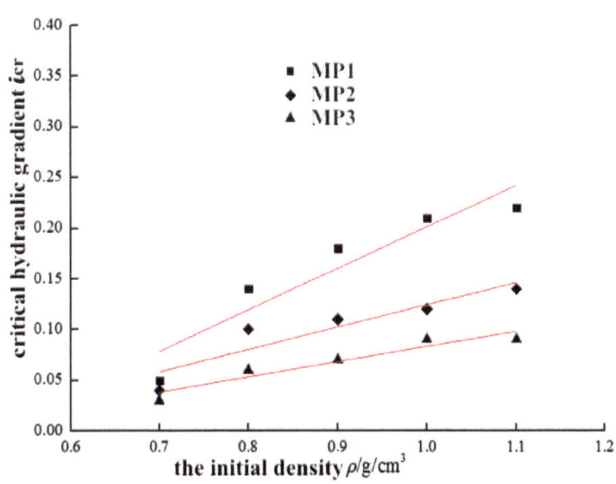

Fig. 6. Relationship between the critical hydraulic gradient and the initial density

Table 3. Values of parameters (f and j)

Parameters	MP1	MP2	MP3
f	0.41	0.22	0.15
j	−0.21	−0.1	−0.07

Figure 6 shows that the critical hydraulic gradient of MP1 was the largest, and that of MP3 was the smallest at the same initial density. This implied that the organic content had an effect on the critical hydraulic gradient, and the critical hydraulic gradient of waste samples with high organic content was large whereas the critical hydraulic gradient of waste samples with low organic content was small.

4.2 Relationship Between the Permeability Coefficient and the Initial Density

Based on Fig. 5, it was found that the relationship between the permeation rate and the hydraulic gradient can also be represented by a straight line, and it conformed to Darcy's law (Matsuoka et al. 2001). The data allowed for calculation of the permeability coefficient, and the relationship between the permeability coefficient and the initial density was shown in Fig. 7.

Figure 7 showed that the permeability coefficient gradually decreased with the initial density of fresh MSW. The relationship could also be formulated as a straight line and the fitted coefficients had a correlation coefficient greater than 0.99. The relationship can be expressed by the Eq. (3) as follows:

$$k = m\rho + n \tag{3}$$

where k is the permeability coefficient, and m and n are the fitted coefficient. The values of all these parameters were shown in Table 4.

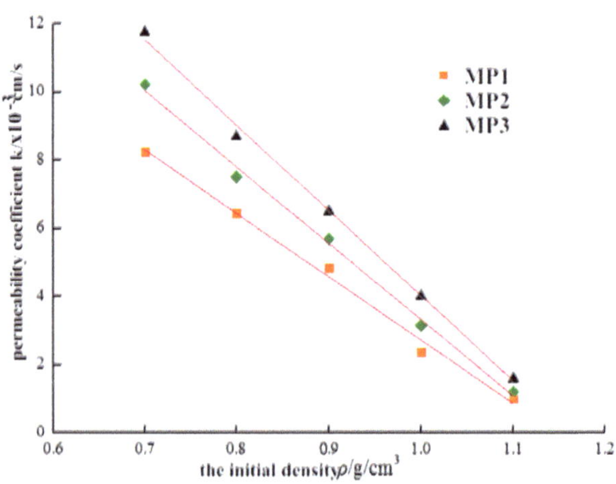

Fig. 7. Relationship between the permeability coefficient and the initial density

Table 4. Values of parameters (m and n)

Parameters	MP1	MP2	MP3
m	−24.7	−22.4	−18.6
n	28.46	25.7	21.28

Figure 7 shows that the permeability coefficient of MP1 was the smallest, and that of MP3 was the largest at the same initial density. The reason might be that the water permeability of organics in waste was lower than the permeability of inorganics. The waste samples with high organics content had poor permeability, which resulted in a small permeability coefficient. With the increase in initial density, the effect of the organic content on the permeability coefficient decreased. For the waste sample with lower initial density, the effect of the organic content on the permeability coefficient was higher. Therefore, to improve the permeability of a landfill, the organic content in the waste should be reduced, on the other hand, treatment method that degrades organic waste should be adopted.

Based on the test results, the range of permeability coefficient for three mixture proportions were obtained from Fig. 7. The permeability coefficient range of MP1 was from 9.70×10^{-4} to 8.22×10^{-3} cm/s. The permeability coefficient range of MP2 was from 1.19×10^{-3} to 1.02×10^{-2} cm/s. And the permeability coefficient range of MP3 was from 1.59×10^{-3} to 1.17×10^{-2} cm/s. Overall, the permeability coefficient of fresh MSW ranged from 10^{-4} to 10^{-2} cm/s.

5 Conclusions

In this study, a large-scale permeability test apparatus was used to carry out the experiment on 15 fresh MSW samples. Three mixture proportions were adopted and five initial densities were selected for each mixture proportion. Based on the experimental results, following conclusions were drawn.

(1) The seepage did not occur until the hydraulic gradient reached a critical value, and the permeation rate increased with the hydraulic gradient. The relationship between hydraulic gradient and permeation rate could be formulated as a linear line with the fitted coefficients greater than 0.99, and it conformed to Darcy's law. In addition, the critical hydraulic gradient increase with the initial density and they followed a linear relationship.

(2) The permeability coefficient gradually decreased with the initial density of fresh MSW, showing a linear relationship with a high correlation coefficient of greater than >0.99. Thereby, a model to predict the permeability coefficient was proposed.

(3) The permeability coefficient of fresh MSW was found to have some interrelationship with the sample's mixture proportions and the organic content. Under the same condition, the permeability coefficient decreased with the organic content. Furthermore, the range of the permeability coefficient was obtained for different mixture proportions which could provide a reference guide for the analysis of the permeability of a landfill.

Acknowledgments. This research was funded by the National Natural Science Foundation of China (Contract Nos. 51478436, and 51678532), the Zhejiang Provincial Natural Science Foundation of China (No. LY14E080021), and the Foundation of Key Laboratory of Soft Soils and Geoenvironmental Engineering (Zhejiang University), Ministry of Education.

References

Chen, Y.M., et al.: Development and control of leachate mound in MSW landfills. Chin. J. Rock Mechan. Eng. **33**(1), 154–163 (2014)

Durmusoglu, E., Sanchez, I.M., Corapcioglu, M.Y.: Permeability and compression characteristics of municipal solid waste samples. Environ. Geol. **50**(6), 773–786 (2006)

Jang, Y.S., Kim, Y.W., Lee, S.I.: Hydraulic properties and leachate level analysis of kimpo metropolitan landfill, Korea. Waste Manage. **22**(3), 261–267 (2002)

Jie, Y.X., Danzeng, D.Z., Wei, Y.F.: Study on the permeability of municipal solid waste. Geotechn. Eng. Techn. **19**(6), 307–310 (2005)

Ke, H., et al.: Experimental study on saturated hydraulic conductivity of MSW under different conditions. J. Zhejiang Univ.: Eng. Sci. **47**(7), 1164–1171 (2013)

Korfiatis, G.P., Demetracopoulos, A.C., Nawy, E.G.: Moisture transport in a solid waste column. J. Environ. Eng. **110**(4), 780–796 (1984)

Liu, X.D., Shi, J.Y.: Laboratory tests on hydraulic properties of municipal solid waste. Adv. Sci. Technol. Water Res. **32**(1), 35–38 (2012)

Matsuoka, L.T., Yao, Y.P.: Soil mechanics. China. Water Power Press, Beijing (2001)

Powrie, W., Beaven, R.P.: Hydraulic properties of household waste and implications for landfills. Proc. Inst. Civ. Eng. **137**(4), 235–247 (1999)

Qu, X., et al.: Research on hydraulic conductivity test of compacted municipal solid waste. Techniq. Equip. Environ. Pollut. Control **6**(12), 13–17 (2005)

Reddy, K.R., et al.: Hydraulic conductivity of MSW in landfills. J. Environ. Eng. **135**(8), 677–683 (2009)

Technical specification for soil test of landfill municipal solid waste (CJJ/T204-2013). Chinese Building Industry Press, Beijing (2013)

Wang, W.F.: Laboratory Research on Saturated Hydraulic Conductivity of Municipal Solid Waste under Different Degradation age. Zhejiang University, Hang-zhou (2010)

Zhan, L.T., et al.: Field and laboratory study on hydraulic characteristic of MSWs. J. Zhejiang Univ.: Eng. Sci. **48**(3), 478–486 (2014)

Zhang, W.J., Chen, Y.M., Qiu, Z.H.: Laboratory and field tests on hydraulic properties of landfill waste. Rock Soil Mech. **30**(11), 3313–3323 (2009)

Zhang, Z.Y., et al.: Study on the permeability and the permeable mechanism of the municipal solid waste. Energy Educ. Sci. Technol. Part A: Energy Sci. Res. **32**(3), 1517–1524 (2014)

Compaction Properties of Municipal Solid Waste

F.N. Okonta[1,2(✉)], N. Ngcobo[1,2], M. Mtsweni[1,2], and D. Harris[1,2]

[1] Head of Department of Civil Engineering Science, University of Johannesburg, Johannesburg, South Africa
fnokonta@uj.ac.za
[2] Johannesburg Waste Agency (PIKITUP), Johannesburg, South Africa

Abstract. Waste compaction decrease insitu waste void, improve landfill stability and extend landfill life. Recently buried waste was excavated from two large pits in Johannesburg Robinson Deep Landfill site. The compaction characteristics such as optimum moisture content (w_{omc}), compaction effort (Ce) and maximum dry density (γ_{mdd}) of the bulk and reconstituted waste mass were investigated in the laboratory. Compaction effort ranging from 5 layers and 35 blows per layer to 5 layers and 175 blow per layer of as received and reconstituted samples was investigated. The result of series of modified proctor effort revealed that the density of waste mass is dependent on the relative constitution of organics and plastics. Waste mass constituted of 50% of organics, the maximum. Maximum density of waste mass decreased with percentage of plastic and waste mass constituted of 20% plastics was not compactible. The maximum dry unit weight of as received waste was dependent on the pre-compaction moisture content. The maximum dry unit weight of initially soaked was greater than that of the initially dried waste mass. The reproducibility of specific gravity of the compacted waste mass was dependent on specimen mass and test results which were determined by the use of waste specimens that were are less than 200 g was not reproducible. The relationship between density and compaction effort (KN.m/m^3) was nonlinear with $R^2 = 0.9196$, for the range of 35 blow/layer [44 KN.m/m^3] and 175 blows/layer [222.04 KN.m/m^3], and can provide good estimate of field parameters of compaction plant in relation to number of passes.

Keywords: Domestic waste · Compaction · Dry density · Insitu bulk unit weight · Dry unit weight · Compaction effort · Reconstituted waste mass · Specific gravity · Maximum dry density · Optimum moisture content · Degree of saturation · Moisture content

1 Introduction

Gauteng is the smallest province in South Africa; it has the highest population density and contributes about 25% of the country's municipal waste at 142 million tons a year. The Johannesburg waste agency PIKITUP dispose municipal waste in four landfills; Goudkoppies (Devland), Mare Loiuse (Roodeport), Ennerdale (Lawley) and the oldest and closest to the CBD, Robinson Deep in Turffontien.

D.N. Singh and A. Galaa (eds.), *Contemporary Issues in Geoenvironmental Engineering*, Sustainable Civil Infrastructures,
DOI 10.1007/978-3-319-61612-4_23

Johannesburg is landlocked and semi-arid. Robinson Deep Landfill which opened in 1930, is the oldest PIKITUP landfill with a remaining lifespan of approximately 7 years. With increasing population and changes in land use, Robinson Deep is now situated and trapped in the centre of Johannesburg metropolis, with encroaching low cost infrastructure development and no land space for lateral expansion. It has an average base area of 50000 m^2 and average height of 80 m.

MSW collected from different communities and industries by trucks were disposed in demarcated cells by rear and front loaders and compacted by sheep foot compactors in cells of approximately 5 to 10 m. The dozers were used to rework the cells and external berns to estimated slope of 1:3. The waste materials are compacted to 300 mm and covered with 150 mm layer of soil (PIKITUP Johannesburg Ltd 2009). The sheep foot compactor can impose high pressure ranging from 5.2 to 30 kg/cm^2 on disposed waste. Although high quality and strict supervision of waste compaction was regularly implemented, it is important to evaluate the compaction properties of the waste mass under controlled condition in the laboratory so as to provide some estimate of the optimum water content for maximum insitu density, as well as the number of passes of a compaction plant to achieve maximum insitu density.

Knowledge of unit weight is required for all aspects of landfill design. Unit weight values vary significantly both between sites and within a single site. MSW has highly variable components, and the types and amounts of cover soil differ between sites. In addition placement procedures and environmental conditions play important role. Unit weight can be estimated and measured using several techniques. Methods based on direct field measurement are considered to be the most reliable as the influence of waste placement and overburden stress could be accounted for.

For laboratory methods of determining unit weight, Powrie and Beaven (1999) recommended the use of large-size samples. They noted that large size of disturbed sample can be used to represent density over a wide range of vertical stresses, while the use of small-size samples require series of tests to give reliable results. There are no internationally accepted standard sampling and testing procedures for waste materials and there is presently no accepted guidance on selection of appropriate values of the engineering parameters for use in design, or agreed approaches for assessing waste behaviour as part of the design process (Dixon and Langer 2006). The existing classification systems do not fulfil the requirements of a rigorous classification framework. A new classification framework for MSW that fulfils broad requirements proposed by Langer and Dixon (2004); Dixon and Langer (2006) requires data from different parts of the world for validation.

Data on systematic studies of field compaction of wastes that reflect unit weight-moisture content values immediately after compaction generally are not available on a field scale. (Hanson et al. 2010). In addition variations in unit weight due to changes in waste moisture content with seasons are also limited.

Soil compaction theory was found to be applicable to wastes with the exception that the Gs of waste solids increased with compactive effort resulting in steep degree of saturation curves and low change in wopt between efforts (Hanson et al. 2010). Moisture addition to wastes during compaction increased the workability, the unit weight, and the amount of incoming wastes disposed, and reduced the compaction time. The combined effects have significant environmental and economic implications for landfill operations (Kavazanjian 2001; Zekkos et al. 2006).

Evaluating the engineering properties and hence behaviour of MSW bodies in the field is challenging due to the variety of materials present. The unit weight of compacted waste depends upon the waste components, thickness of layer, weight and type of compaction plant and the number of times equipment passes over the waste (Powrie and Beaven 1999). Due to the difficulties and costs involved there are few field measurements of unit weight variation with depth. Thus studies that relate laboratory compaction properties to field values are critical for landfill operation and management.

2 Materials and Test Methods

The waste samples were excavated from two large pits in Robinson Deep landfill. 12 bags of samples were collected. Some of the samples were sealed, and others were dried slowly at temperature of 60 C and humidity of 60% in a drying chamber in the laboratory. The dried samples were reconstituted and the major constituents were shown in Fig. 1.

Fig. 1. Major constituent of domestic waste mass

2.1 Compaction Test

The aim of this experiment is to determine the unit weight of the characterized MSW using the Modified Proctor Test. The tests conform to the ASTM D1557 and AASHTO T-180 standards. For compaction tests, Hammer weight of 4.54 kg) and cylindrical steel moulds of 152.4 mm diameter and 450 mm height were used. Compaction effort of 55 blows per waste layer and total of five layers were delivered by the hammer from a drop height of 457 mm.

Large pieces of plastics, textiles, papers and organics were cut into smaller pieces. Generally the largest permissible pieces were 20 mm diameter for textiles and plastics and approximately 60 mm square for papers and plastics.

Series of tests were conduction on the following set of waste samples.

As received or excavated waste mass.

Reconstituted waste mass containing 25% each of organics, textiles, plastics and paper.

Reconstituted waste mass containing 50% organics and 50% (textiles, plastics and paper).

Reconstituted waste mass containing 50% paper and 50% (textiles, plastics and organics).

Reconstituted waste mass containing 50% textiles and 50% (organics, plastics and paper).

Reconstituted waste mass containing 50% plastics and 50% (textiles, organics and paper).

Waste mass containing 25% each of organics, textiles, plastics and paper covered with capping soil material.

The waste samples from pit 1 were used for compaction tests and the major constituents considered were papers, plastics, organics and textiles. Waste sample such as large stones, glasses and metals were not considered due to their small percentage size and uneasiness to compact.

2.2 Moisture Content and Specific Gravity (Gs)

The specific gravity (Gs) of municipal solid waste (MSW) was determined with a large glass pycnometer, with capacity of 1800 ml. Different masses of as received waste and reconstituted waste were tested in both loose state and after compaction. In particular the effect sample size on the reproducibility of results was investigated and samples between 50 g and 400 g were tests.

The moisture content of MSW was determined by the test method used in the laboratory for soils and organic soils. The test conforms to the ASTM D2216 standards. In relation to compaction characteristics, sets of samples that were gradually dried after 24 h wetting and samples that were gradually wetted from dried states were investigated.

3 Results and Analysis

The components of waste mass that were excavated from two large pits in Robinson Deep landfill were presented in Tables 1 and 2. The MSW had been covered two months prior to excavation. The materials from the two pits strongly reflect the waste constitution of landfills sites that are either close to or receive high percentage of industry and commercial waste. It is common practice in landfill operations to use large

Table 1. Waste mass classification

Bag number	Pit no 1	Pit no 2
Measurement	%	%
Paper	3.8	10
Plastic	5	11.9
Textile	1.6	10.3
Soil	39.8	18.1
Steel	0.7	4
Building debris	40.3	12.2
Glasses	0.9	1.8
Organic	4.8	9.6
Others	3.2	22.2
Sum	**100**	**100**

quantities of available construction debris as recycled cover materials. Pit 2 had almost the same percentages by mass for plastics, papers and textiles and also indicate that Robinson Deep receive large quantities of construction debris.

3.1 Grading Analysis

The result of sieve analysis of construction debris from the two pits and exported soil materials used for waste capping after insitu compaction was presented in Fig. 2. The waste materials from the two pits had 0% mass percentage passing sieve 0.001 mm and 0.01 mm. Pit 1 had 42% particles passing the 1 mm sieve since it was mostly soils and construction debris but Pit 2 had 7% particles passing the 1 mm sieve and uniformly graded with most particles between 1 mm and 10 mm. The capping soil and debris and

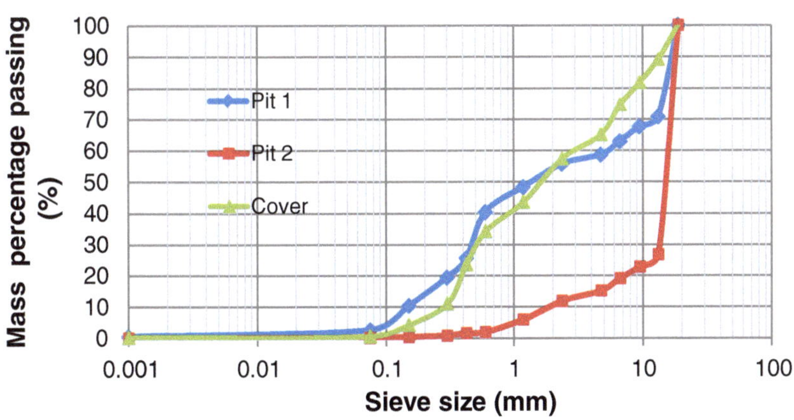

Fig. 2. Sieve analysis of construction debris from the two pits and exported soil materials used for waste capping

soil from Pit 1 are well graded with smooth curve. The instu waste profile of capping soil overlying materials in pit 1 is more impermeable to percolation of surface waste and could be easily compacted to dense insitu waste mass than materials from pit 2.

3.2 Specific Gravity of Municipal Solid Waste

The major factors that influence the specific gravity of MSW are particle size, compaction, decomposition and overburden stress. The specific gravity (Gs) of municipal solid waste (MSW) was determined with a large glass pycnometer, and the effect of compaction as well as the effect of sample size on the reproducibility of results was investigated.

The measured average specific gravities shown in Table 3, were 1.522 and 1.355 for compacted and loose and uncompacted wastes, respectively. For reconstituted materials, the waste mass with 50% plastics has the lowest specific gravity in both compacted and loose state and the highest specific gravity in both loose and compacted state was reflected by the waste mass with 50% organics. There is no significant difference in the specific gravity of waste containing either 50% plastic, 50% textile or 50% paper in both compacted and loose state. In general Gs increased with compaction effort and decreased with reconstitution of waste mass. For fresh wastes, reductions in sample and particle size due to compaction caused occluded particle pores to be exposed and waste particles to be deformed resulting in increases in specific gravity. The reproducibility of Gs was also evaluated in the laboratory. The result shown in Fig. 3 indicated that Gs values are dependent on test method and larger samples yielded more reproducible values. Tests conducted with 20 g and 50 g samples produced Gs results with wide variation. Increasing the sample mass to 100 g and 200 g yielded results with narrow variation in Gs values. It is thus re commended that Specific gravity test be conducted using representative waste composition and on waste that reflect the field compaction state.

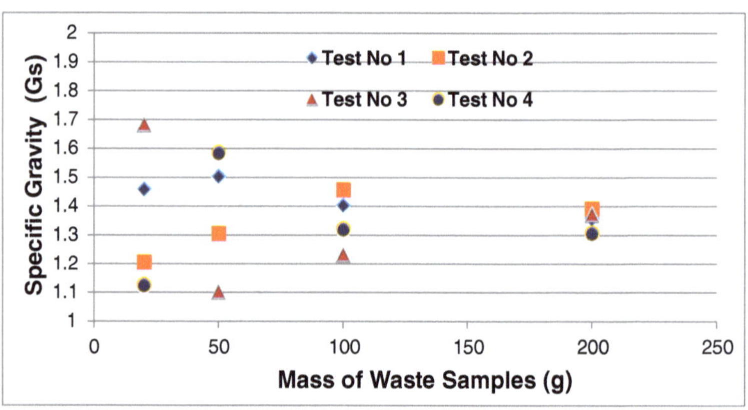

Fig. 3. Effect of sample mass in the reproducibility of Gs

3.3 Laboratory Compaction Test Results

Series of laboratory compaction tests were conducted to determine the maximum dry density and optimum moisture contents of the as received and reconstituted MSW. The effect of cover soil and compaction effort on the dry density of MSW was also investigated. These tests were conducted to evaluate the different compaction scenarios in typical landfill site.

Figure 4 present the results of the compaction curve for waste samples with equal masses. The density of as received samples which were soaked for 24 h at high humidity chamber and compacted after step wise drying in the chamber (the wetting path) than the density of samples that were initially dried to low moisture and compacted after stepwise wetting (drying path). Thus for relatively fresh waste that were buried for 2 months, stepwise drying of initially wet samples resulted in high maximum dry density. The density of reconstituted samples was also lower because the process of reconstitution required that samples be initially dried before separation. Relatively fresh waste mass are easily compressed because of the closure of intra particle voids into denser particle parking. The optimum moisture content are which the maximum density was mobilized was presented in Fig. 5. Relatively fresh samples require the lowest moisture content for dense parting. Drying and reconstitution reduced the plasticity of the waste due to reduced specific surface and thus such samples require higher moisture content for inter particle lubrication and dense packing.

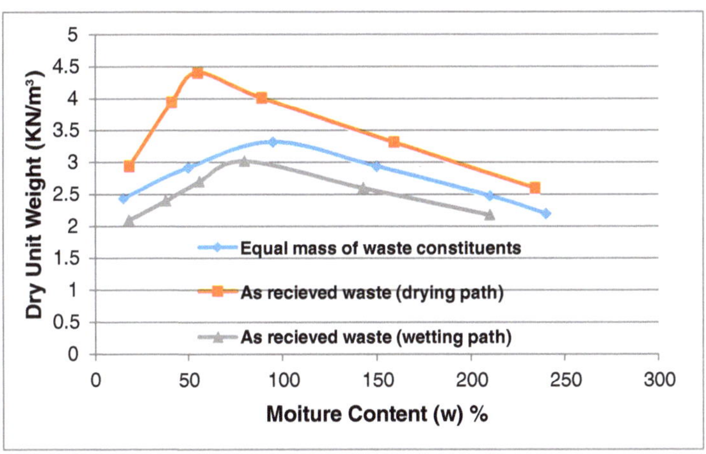

Fig. 4. Effect of wetting path on density of as received waste and reconstituted waste mass.

The results of the compaction tests of reconstituted waste samples was presented in varying masses was presented in Figs. 6 and 7. For specimens consisting of 50% papers, 16.7% organics, 16.7% textiles and 16.7% plastics, the maximum dry unit weight of 3.07 KN/m^3 and optimum moisture content of 85.7% were obtained. For specimens consisting of 50% organics, 16.7% papers, 16.7% textiles and 16.7% plastics, the maximum dry unit weight of 4.1 KN/m^3 and optimum moisture content of

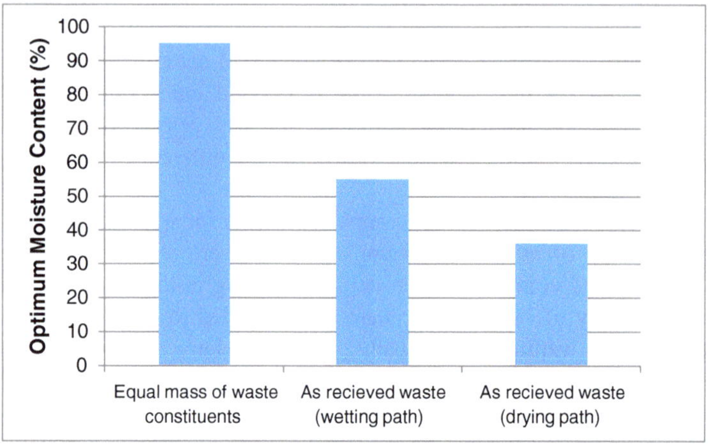

Fig. 5. Compaction for equal mass percentages of waste samples

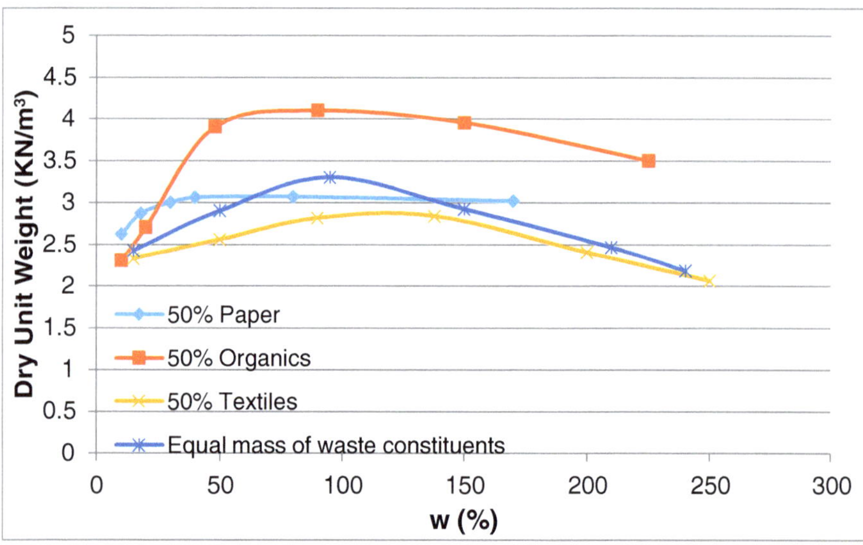

Fig. 6. Compaction parameters of reconstituted waste samples

91.3% were obtained. For specimens consisting of 50% plastics, 16.7% organics, 16.7% textiles and 16.7% papers, the maximum dry unit weight of 2.18 KN/m^3 and optimum moisture content of 16.7% were obtained. For specimens consisting of 50% textiles, 16.7% organics, 16.7% papers and 16.7% plastics the maximum dry unit weight of 2.83 KN/m^3 and optimum moisture content of 133.3% were obtained.

Pulat and Yukselen-Aksoy (2013) investigated the effects of composition, degradation and energy on the compaction behavior of artificially-prepared and natural fresh and aged MSW samples from Europe, Turkey and the USA. It was noted that with

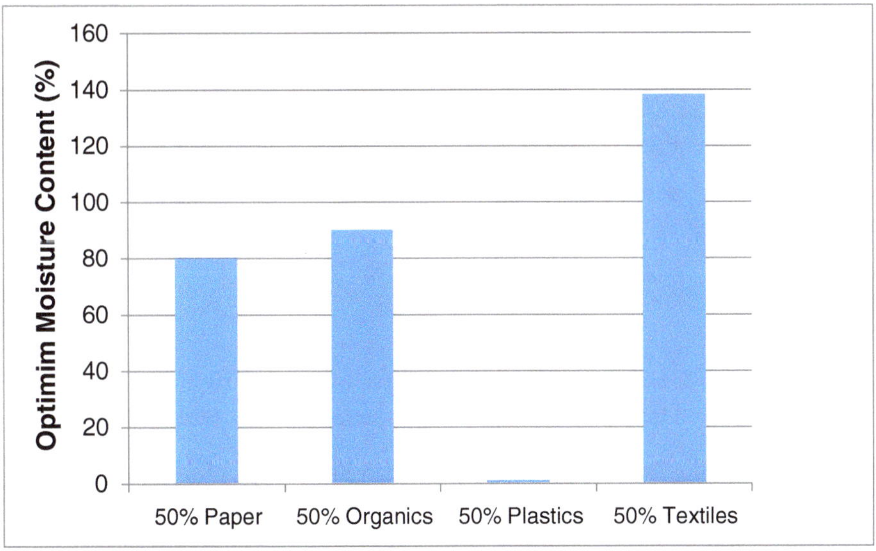

Fig. 7. Optimum moisture content of reconstituted waste samples

respect to the composition effect, the percentage of organic waste was the most important factor on the compaction behavior of MSWs. As paper, organic and plastic contents increase in the MSW composition the maximum dry density value decreases and optimum moisture content increases. The result presented in the current study indicate that the densities of reconstituted waste materials consisting of 50% organics was slighter lower than the density of the fresh waste mass presented in Fig. 5. Figure 6 also indicated that while waste consisting of 16 and 25 percentage of buried plastic can be compacted, waste consisting of 50% cannot be compacted. Figure 6 also indicated that waste mass with 50% textiles require large amount of water for remoulding into maximum particle packing.

3.4 High Compaction Effort

The effect of high compaction effort on the compaction parameters were shown in Fig. 10. MSW samples were compacted with blows per layer varying from 35 to 175 blows per layer. For 35 blows per layer the maximum dry unit weight of MSW obtained was 2.84 KN/m^3 with optimum moisture content of 54.5%. The other results obtained are γ_d = 3.30 KN/m^3 with w = 87.5% for 55 blows per layer and γ_d = 4.61 KN/m^3 with w = 54.5% for 75 blows per layer and γ_d = 5.83 KN/m^3 with w = 82.2% for 95 blows per layer. The maximum dry unit weight of 6.055 KN/m^3 and optimum moisture content of 78.6% for 150 blows per layer and γ_d = 6.096 KN/m^3, w = 77.8% for 175 blows per layer. Figures 8 and 9, indicated that quite unlike soils, when the compaction effort was increased, both the maximum dry unit weight and the optimum moisture content of the waste mass also increased. High compaction effort reworked the waste mass into finer particles with increased specific surface, plasticity and thus

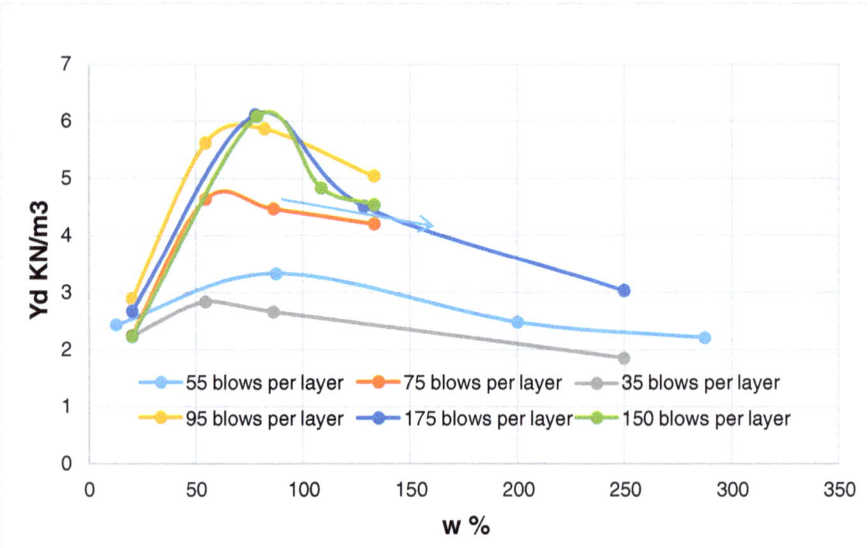

Fig. 8. The effect of high compaction effort on the dry unit weight

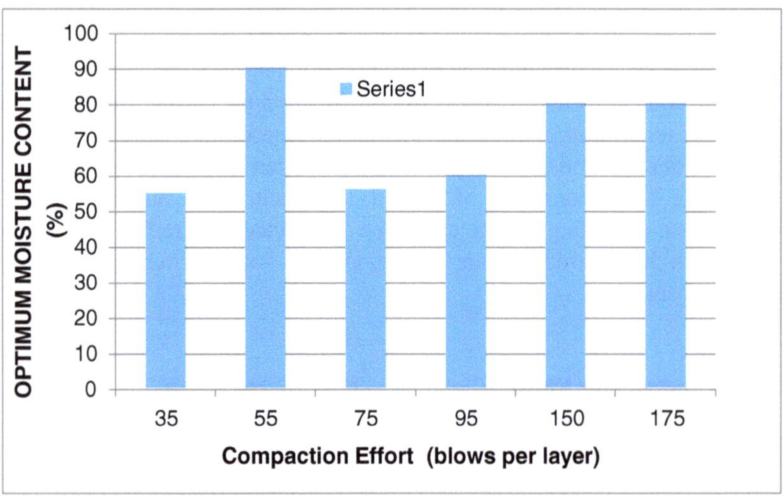

Fig. 9. The effect of high compaction effort on the optimum moisture content

moisture content. The high compaction energy was used in the laboratory to simulate the compaction energy range used in most municipal solid waste landfill site. The increase in optimum moisture content with compaction effort was erratic but reached a constant value at 150 blows per layer and maximum unit weight of 6.10 KN/m^3.

3.5 Compaction Effort in the Laboratory and the Field

Das (2015) noted that the contact pressure under the spikes of a sheep foot compacter can vary from 1400 KN/m^2 to 7000 KN/m^2. The spiked drums are either mounted singly or are in pairs. The number of spikes in a 5000 kg drum varies from 64 to 88 and the shape are clubbed shaped of 75 × 100 mm or tapered 57 mm × 57 mm (Smith and Smith 2013)

Different models of sheep foot compaction plant are used for waste mass compaction in Robinson Deep landfill. Different models have different dead weights and number of spikes (sheep foot), but most models have four (4) wheels. The general features of the most common BORMAG sheep foot used in the landfill are

- Landfill uses a sheep foot roller with 4 wheels and 48 spikes per wheel.
- The effective number of spikes supporting a wheel and bearing down on the waste mass at any instantaneous time per wheel is 12spikes
- The area of 1 spike is 100 mm × 100 mm.
- BOMAG compactor weighs 42 tons (38101.8 kg)
- Each wheel carries (1/4) 25% of the total weight of the compactor (95254.5 N).

Since there are 24 spikes per wheel supporting that weight, thus 1 spike carries a total force of 3968 N/ over an area of 0.06 m × 0.06 m and average compact pressure 1102.47 kPa.

The Equivalent compaction effort over a waste mass of 0.06 m × 0.06 m × 0.1 m applied from a drop height equal the height of a spike (0.1 m) = 1102.47 KNm/m^3 or (90 blows per layer).

The subgrade coefficient of a stiff plate on a soft heterogeneous formation is 0.90 (Das 2015).

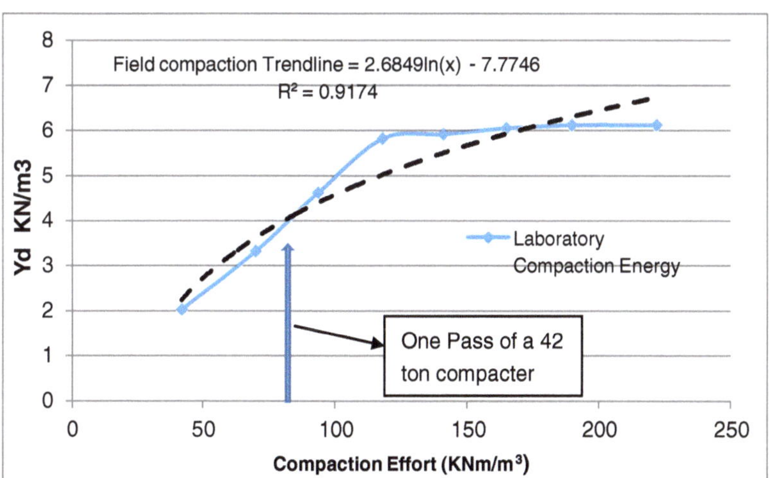

Fig. 10. Compaction effort equivalents in the laboratory and the field in relation to maximum dry unit weight

Thus the equivalent blows per layer of the 42 ton compacter is 81 blows per layer or 992 KNm/m^3.

150 blows per layer from large mould of 450 mm height and 154 mm diameter is 1815 KNm/m^3.

The effect of higher compaction effort from the laboratory in KNm/m^3 and the equivalent compaction effort from one pass of the 42 ton sheep foot compacter. The general trend reflected in Fig. 10 is that a minimum of two passes was needed to compact the waste mass to dry unit weight value close to the maximum unit weight.

4 Conclusions

The compaction properties of recently buried (six months) municipal solid waste was investigated. The following conclusion was drawn from the result of series of tests on as received waste mass, reconstituted was mass and the effects of high compaction effort.

The maximum dry unit weight of as received waste was dependent on the pre-compaction moisture content. Initially wet or soaked waste exhibited greater maximum dry unit weight than initially dried waste mass.

The reproducibility of the specific gravity of domestic waste mass was dependent on the mass of sample tested. In addition, the specific gravity of as received waste mass was greater than that of the reconstituted waste mass.

The dry unit weight of as received waste mass was greater than the maximum dry unit weight of reconstituted waste mass. Drying induced reduction in specific surface and plasticity implied that reconstituted waste mass required greater moisture content for lubrication and particle parking than as received waste mass.

Waste consisting of 16.7%–25% plastics was compactable but waste with more than 25% plastics is not compactable.

Maximum dry unit weight increased with compaction effort and up to 95% of maximum dry unit weight of waste mass is achieved by compaction effort of 150 blows per layer on five layers. The optimum moisture content also increased marginally because of the increase in specific surface of the reduced particle size of the waste mass.

The laboratory compaction effect – dry density relationship can be used to estimate the number of passes of a field compaction plant in relation to insitu dry unit weight.

References

Coduto, D.: Geotechnical Engineering: Principles and Practices, 5th edn. Prentice Hall Inc, Upper Saddle River (2011)

Dixon, N., Jones, D.R.V.: Stress states in, and stiffness of, landfill waste. In: Geotechnical Engineering of Landfills, pp. 19–34. Thomas Telford, London (1998)

Dixon, H., Ng'ambi, S.C., Jones, D.R.V., Connell, A.K.: The role of waste deformations on landfill steep side wall lining stability. In: Proceedings of Wastecon 2000, September, Cape Town, vol. 2, pp. 379–388 (2000)

Dixon, N., Ng'ambi, S., Jones, D.R.V.: Structural performance of a steep slope landfill lining system. Proc. Inst. Civil Eng. Geotechn. Eng. **157**, 115–125 (2004)
Fassett, J.B., Leonardo, G.A., Repetto, P.C.: Geotechnical properties of municipal solid waste and their use in landfill design. In: Waste Tech 1994, Landfill Technology Technical Proceedings, Charleston, SC (USA), 13–14 January (1994)
Gourc, J.P., Thomas, S., Vuillemin, M.: Proposal of a waste settlement survey methodology. In: Pinto, S. (ed.) Proceedings of Environmental Geotechnics, pp. 195–200. Balkema (1998)
Gourc, J.P., Olivier, F., Thomas, S., Chatelet, L., Denecheau, P., Munoz, M.L.: Monitoring of waste settlements on five landfills: comparison of the efficiency of different devices. In: Proceedings of Sardinia 2001, Eighth International Waste Management and Landfill Symposium, Cagliari, Italy, pp. 515–524 (2001)
Grisolia, M., Napoleoni, Q., Tancredi, G.: Contribution to a technical classification of MSW. In: Proceedings of Sardinia 1995, Fifth International Landfill Symposium, Cagliari, Italy, pp. 761–767 (1995)
Grisolia, M., Napoleoni, Q., Tancredi, G.: The use of triaxial tests for the mechanical characterization of MSW. In: Proceedings of Sardinia 1995, Fifth International Landfill Symposium, Cagliari, Italy, pp. 703–710 (1995)
Kavazanjian, E.: Mechanical properties of municipal solid waste. In: Proceedings of Sardinia 2001, Eighth International Waste Management and Landfill Symposium, Cagliari, Italy, pp. 415–424 (2001)
Kavazanjian, E.: Evaluation of MSW properties using field measurements. In: Koerner, R.M., Koerner, G.R., Hsuan, Y., Ashley, M.V. (eds.) Proceedings of 17th Geosynthetic Research Institute Conference, Hot Topics in Geosynthetics—IV, pp. 74–113 (2003)
Kavazanjian, E., Matasovic, R., Bachus, R.C.: Large-diameter static and cyclic laboratory testing of municipal solid waste. In: Proceedings of Sardinia 1999, Seventh International Waste Management and Landfill Symposium, Cagliari, Italy, pp. 437–444 (1999)
Kavazanjian, N., Matasovic, R., Bonaparte, G.R., Schmertmazin, E.: Evaluation of MSW properties for seismic analysis. In: Geoenvironment 2000, Geotechnical Special Publication No. 46, pp. 1126–1141. ASCE (1995)
Kolsch, F.: Material values for some mechanical properties of domestic waste. In: Proceedings of Sardinia 1995, Fifth International Landfill Symposium, Sardinia, vol. 2, p. 20 (1995)
Landva, A., Clark, J.I.: Geotechnical testing of waste fill. In: Proceedings of Canadian Geotechnical Conference, Ottawa, Ontario, pp. 371–385 (1986)
Landva, A., Clark, J.I.: Geotechnics of waste fills. Geotechnics of Waste Fills—Theory and Practice. ASTM STP **1070**, 86–106 (1990)
Landva, A.O., Valsangkar, A.J., Pelkey, S.G.: Lateral earth pressure at rest and compressibility of municipal solid waste. Can. Geotech. J. **37**, 1157–1165 (2000)
Langer, U., Dixon, N.: Mechanical properties of MSW: development of a classification system. In: Yong, R.N., Thomas, H.R. (eds.) Proceedings Fourth British Geotechnical Association Geoenvironmental Engineering Conference. Integrated Management of Groundwater and Contaminated Land, pp. 267–274 (2004)
Manassero, M., Van Impe, W.F., Bouazza, A.: Waste disposal and containment. In: Proceedings of Second International Congress on Environmental Geotechnics, Osaka, vol. 3, pp. 193–242 (1996)
Powrie, W., Beaven, R.P.: Hydraulic properties of household waste and implications for landfills. Inst. Civil Eng. Geotechn. Eng. J. **137**, 235–247 (1999)
Powrie, W., Richards, D.J., Beaven, R.P.: Compression of waste and implications for practice. In: Geotechnical Engineering of Landfills, pp. 3–18. Thomas Telford (1998)

Siegel, R.A., Robertson, R.J., Anderson, D.G.: Slope Stability Investigations at a Landfill in Southern California. In: Geotechnics of Waste Fill—Theory and Practice, ASTM STP, p. 1070 (1990)

Turczynski, U.: Geotechnische Aspekte beim Aufbau von Mehrkomponentendeponien. Ph.D. Thesis, Fakultät der technischen Wissenschaften, Bergakademie Freiberg, Germany (1988)

Dixon, N., Russel, D., Jones, V.: Engineering properties of municipal solid waste (2005)

Zekkos, D., et al.: Unit weight of municipal solid waste. J. Geotechn. Geoenviron. Eng. (2006)

Wong, W.: Investigation of geotechnical properties of municipal solid waste as a function of placement conditions. Faculty of California Polytechnic State University, San Luis Obispo (2009)

Sowers, G.: Settlement of waste disposal fills. In: Proceedings of the 8th International Conference on Soil Mechanics and Foundation Engineering, Moscow, Russia (1973)

Cowland, J., Tang, K., Gabay, J.: Density and strength properties of Hong Kong Refuse. In: Christensen, T.H., Cossu, R., Stegmann, R. (eds.) Proceedings of Sardinia 1993, 4th International Landfill Symposium, CISA, S. Margherita di Puls, Cagliari, Italy, pp. 1433–1446 (1993)

Jamal, S.: garbage-we-dump-in-landfills-is-worth-billions. Times Live, 25 March 2015. http://www.timeslive.co.za

Yesiller, N., Hanson, J.L., Cox, J.T., Noce, D.E.: Determination of specific gravity of municipal solid waste. Waste Manage. **34**(5), 848–858 (2014). doi:10.1016/j.wasman.2014.02.002

Pulat, H.F., Yukselen-Aksoy, Y.: Compaction behaviour of synthetic and natural MSW samples in different compositions. Waste Manage. Res. **31**(12), 1255–1261 (2013). doi:10.1177/0734242X13507967

Strength Characteristics of Dispersive Soil by Using Industrial By-Products

Samaptika Mohanty(✉), N. Roy, and S.P. Singh

Civil Engineering, NIT Rourkela, Rourkela, India
mohantysamaptika4@gmail.com,
{nroy, spsingh}@nitrkl.ac.in

Abstract. Soils of low salt concentration that dislodge easily and quickly, when comes in contact with water are called dispersive soils. These are unstable and disintegrate or erodible. These soils are present in many parts of the world such as India, United States, Australia, Greece, America, South Africa, Thailand, and others. They pose serious problems in stability of earthen structures, road fills, and other engineering structures. Even if there are simple methods to identify the dispersivity of the soils but it is more difficult to quantify the dispersivity. Visual classification such that Atterberg's limits and particle size distribution is not sufficient to differentiate between ordinary erosion resistant clays and dispersive clays. In the present work, the dispersive soil is identified by conducting double hydrometer test to find the percentage of dispersion. The CBR results are presented and unconfined compressive strength tests are carried out at different curing period such as 0, 7, 14, 28 days. After stabilization using cement clinker, the dispersive soil can be prevented from collapse, settlement, pipe failures and slope failures in earth dams, embankments and may become suitable for other engineering purposes.

1 Introduction

The dispersive soils are extensively found in various parts of the world such as India, Australia, United States, South Africa, Greece, Thailand and Latin America. These soils may constitute extensive distress if not taken care of effectively. One of the simple ways to prevent the problems encountered in dispersive soils is to replace them with suitable materials.

Many researchers have studied the fundamental aspects of the interaction between dispersive soil, aluminum sulfate, fly ash, cement, and lime to bind the soil clay particles and reduce the dispersivity. A.H. Vakili observed that adding pozzolan followed by curing reduces dispersivity potential of the samples. Hamza Gullu and Kenan Hazirbaba studied relatively new non-traditional stabilizer (synthetic fluid) in conjunction with fiber to improve the strength characteristics of a low-plastic fine-grain soil. Hamza Gullu and Ali Shahabaldeen Khudir observed the effect of freeze–thaw cycles on the unconfined compressive strength of fine-grained soil treated with steel fiber, jute fiber, and lime. In the present study, an attempt has been made to improve the strength by mixing the dispersive soil with clinker.

D.N. Singh and A. Galaa (eds.), *Contemporary Issues in Geoenvironmental Engineering*, Sustainable Civil Infrastructures,
DOI 10.1007/978-3-319-61612-4_24

2 Materials

In the present study, the sodium bentonite and cement clinker are used for the mixture material. Sodium bentonite is collected from Cuttack, Odisha, India, which contains montmorillonite and sodium as the predominant clay mineral. Cement clinker is made by heating clay and limestone in a rotary kiln at high temperature. It is collected from Rourkela, Odisha, India. The cement clinker used in the experiment is first crushed in crusher machine and then fined in a ball mill.

3 Methodology

The Sodium bentonite and cement clinker are oven dried. After that the geotechnical properties (Table 1) like grain size analysis, specific gravity, double hydrometer test, CBR test, compaction characteristics, and UCS test are performed to characterize the strength of dispersive soil. Following these XRD test is performed for sodium bentonite and cement clinker. Different proportion of cement clinker is mixed with bentonite to find the optimum strength.

Table 1. Physical properties of sodium bentonite and cement clinker

Sl. No.	Properties	Sodium bentonite	Cement clinker
1	Liquid limit (%) Plastic limit (%) Plasticity index (%)	318 62 256	–
2	Specific gravity	2.74	3.24
3	IS classification	CH	–
4	Free swell index, FSI (%)	610	–
5	Linear shrinkage index, LSI (%)	45	14.36 18.01
6	Optimum Moisture Content (%), Maximum dry density (kN/m^3)	28.6 13.98	14.6 18.05
7	Unconfined compressive strength, σ (kPa)	175.47	110.38

4 Results and Discussion

4.1 Dispersive Test

The double hydrometer test (ASTM D4221-99) is performed to identify the dispersiveness of soil. Primarily the particle size distribution of the soil is determined using the standard hydrometer test where the soil is dispersed in water with the chemical dispersing agent (Sodium hexametaphosphate). A parallel hydrometer test is then made on a soil specimen but without a chemical dispersing agent, the percent dispersion is the ratio of the dry mass of particles smaller than 0.005 mm diameter of the second test to the first expressed as a percentage. The value of greater than 50 is highly dispersive.

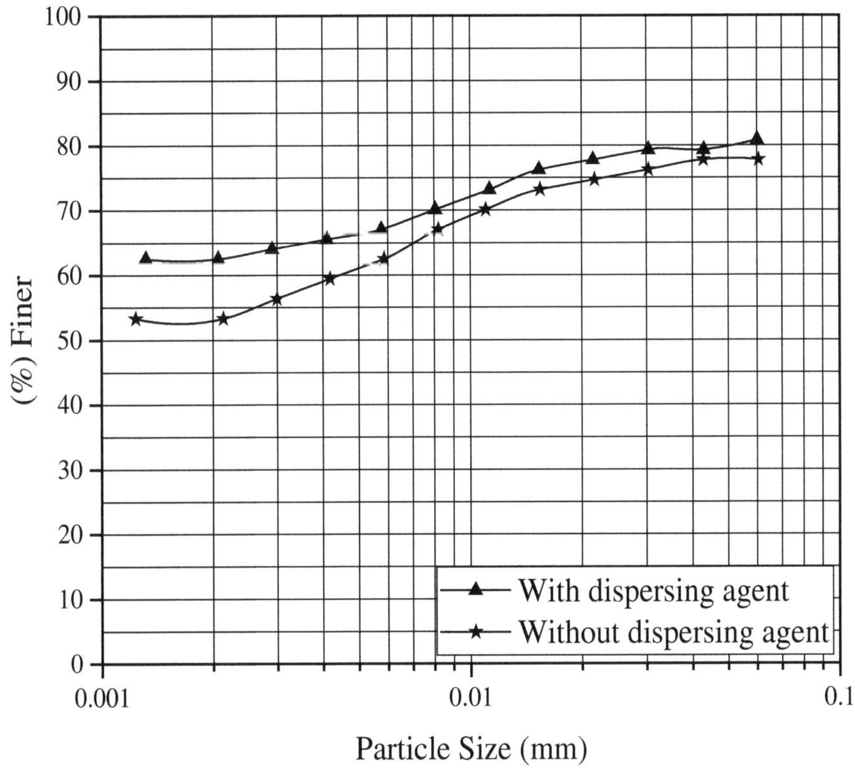

Fig. 1. Double hydrometer test for sodium bentonite

Figure 1 shows that from double hydrometer test the percentage of sodium bentonite is found to be 90.66% and classified under highly dispersive soil. Table 2 shows the results of double hydrometer test of bentonite with different % of cement clinker.

Table 2. Results of double hydrometer test of bentonite-cement clinker mixture

Sl. No.	Degree of dispersion and classification of dispersive soil	Specimen code	% of dispersion by double hydrometer test	Classification
1	<30% - Non-dispersive 30% to 50% - Intermediate dispersive >50% - Highly dispersive	Bentonite	90.66	Highly dispersive
2		BC9505	78.41	Highly dispersive
3		BC9010	49.94	Intermediate-dispersive
4		BC8515	29.96	Non-dispersive
5		BC8020	23.62	Non-dispersive
6		BC7525	16.80	Non-dispersive

4.2 Grain Size Distribution

Figure 2 shows that cement clinker used in the present study is found to be coarse grained and poorly graded.

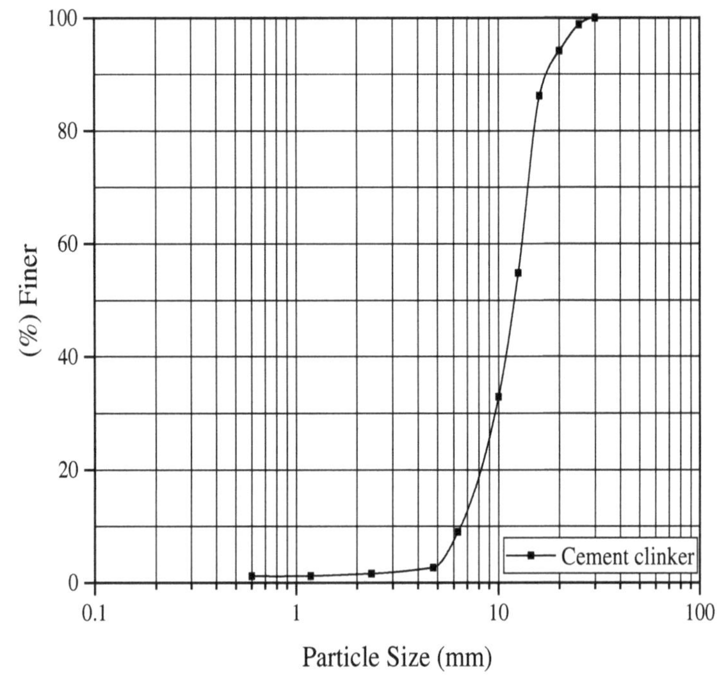

Fig. 2. Grain size analysis curves cement clinker

4.3 XRD (X-Ray Diffraction)

The XRD pattern in Fig. 3(a) shows that the bentonite used in the present work contains quartz, kaolinite, montmorillonite, and phillipsite & Fig. 3(b) shows that cement clinker contains calcium silicate oxide barium strontium tellurite and strontium tungsten nitride.

4.4 Compaction Characteristics

Standard Proctor compaction tests are conducted on specimens of sodium bentonite, cement clinker and sodium bentonite with 5,10,15,20 and 25 percentage of cement clinker respectively (IS: 2720-7) (1980). The maximum dry density of sodium bentonite is found to be 13.98 kN/m^3 at an optimum moisture content of 28.6% as

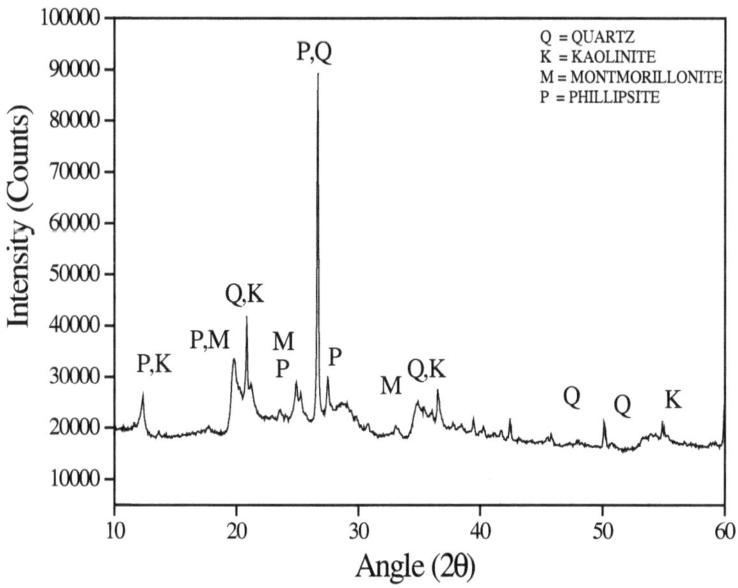

(a)

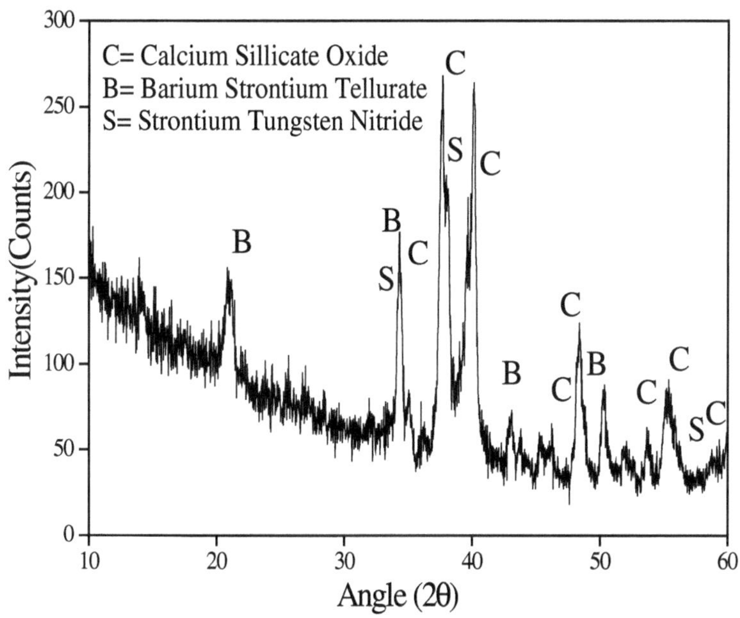

(b)

Fig. 3. XRD analysis of (a) Sodium bentonite (b) Cement clinker

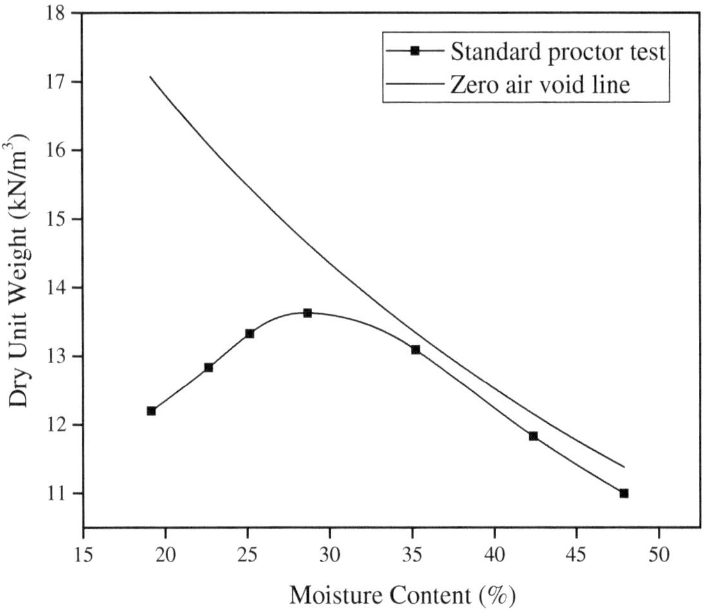

(a)

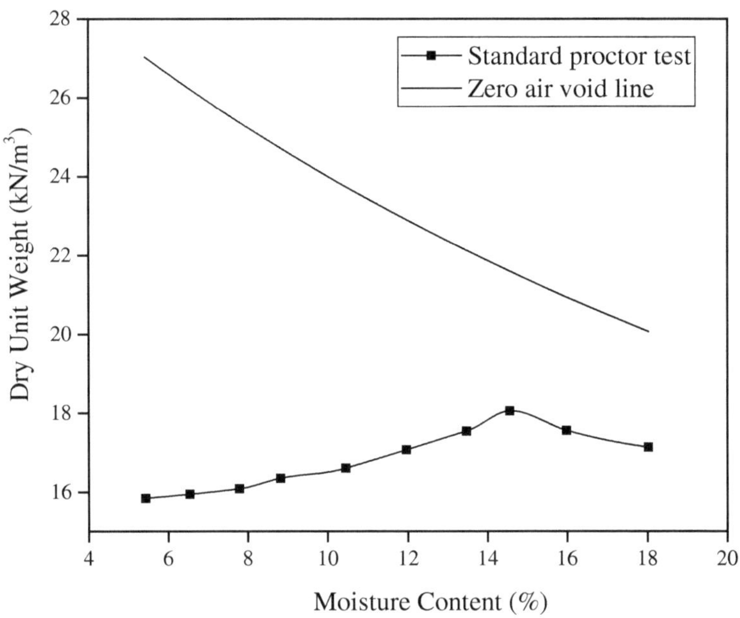

(b)

Fig. 4. Compaction curve of (a) Sodium bentonite, (b) Cement clinker, (c) Bentonite-cement clinker mix

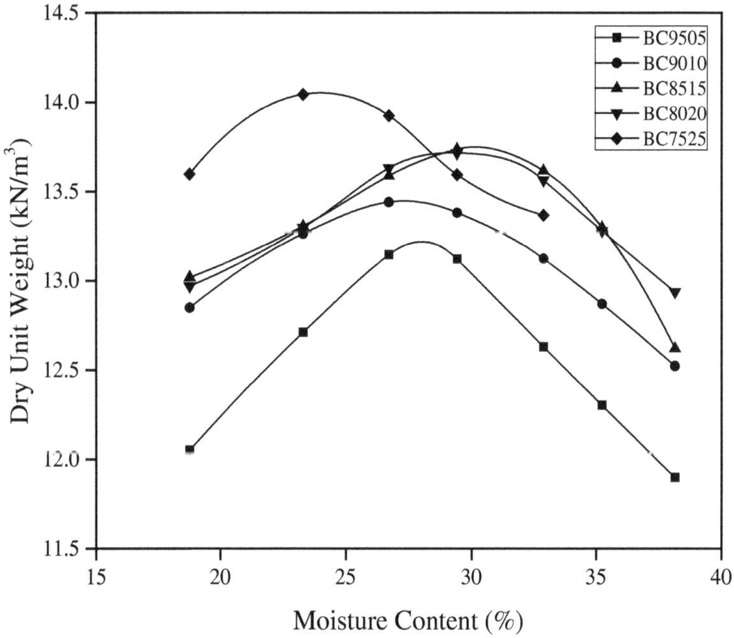

(c)

Fig. 4. (continued)

illustrated in Fig. 4(a). The maximum dry density of cement clinker is found to be 18.05 kN/m^3 at an optimum moisture content of 14.6% as illustrated in Fig. 4(b). Similarly, the compaction curve for bentonite-cement clinker mixture is presented in Fig. 4(c). Table 3 shows the OMC and MDD values of bentonite–cement clinker mixture.

Table 3. OMC and MDD values of bentonite–cement clinker mixture

Sl. No.	Specimen code	Specific gravity	Optimum moisture content (%)	Maximum dry density (kN/m^3)
1	BC9055	2.73	26.71	13.15
2	BC9010	2.61	27.59	13.44
3	BC8515	2.62	27.13	13.73
4	BC8020	2.64	29.63	13.64
5	BC7525	2.68	27.64	14.03

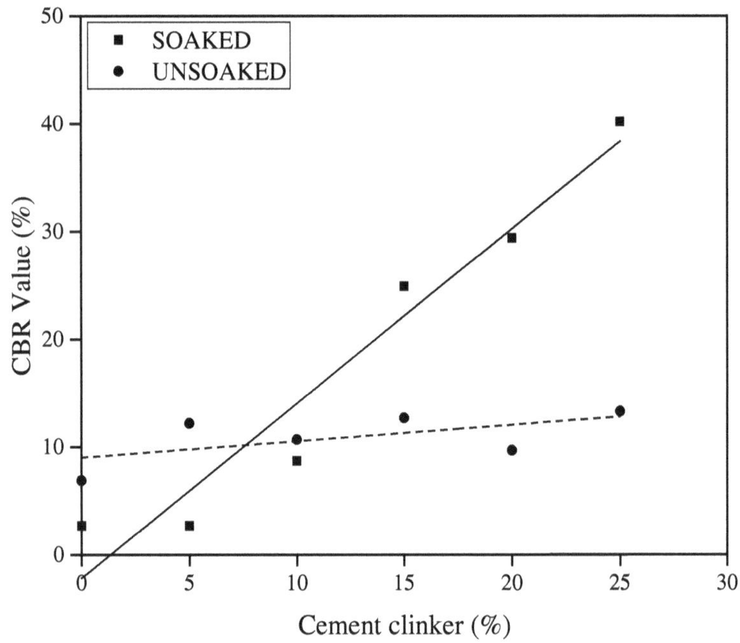

Fig. 5. Variation of CBR with % of Cement clinker content

4.5 CBR Test

CBR tests are conducted for both soaked and unsoaked condition on the samples as per IS Code IS: 2720-16 (1987). The soaked CBR test is performed on soil sample with a surcharge for 96 h. Both soaked and unsoaked condition CBR value of bentonite-cement clinker mixture is presented in Fig. 5. It is found that with the addition of cement clinker of 15%, 20%, and 25%, CBR values are higher in soaked condition than the unsoaked condition.

4.6 UCS Test

Unconfined compressive strength tests are conducted on soil sample using IS Code IS: 2720-10 (1991), to find the strength of Bentonite-cement clinker mixture. Figure 6 shows the results obtained from UCS test of sodium bentonite mixed with different percentages of 5%, 10%, 15%, 20% and 25% cement clinker at curing periods 0,7,14 and 28 days in normal room temperature. Results show that in addition of 25% cement clinker, the strength of bentonite-cement clinker mix gives higher strength as compared to other percentages of cement clinker.

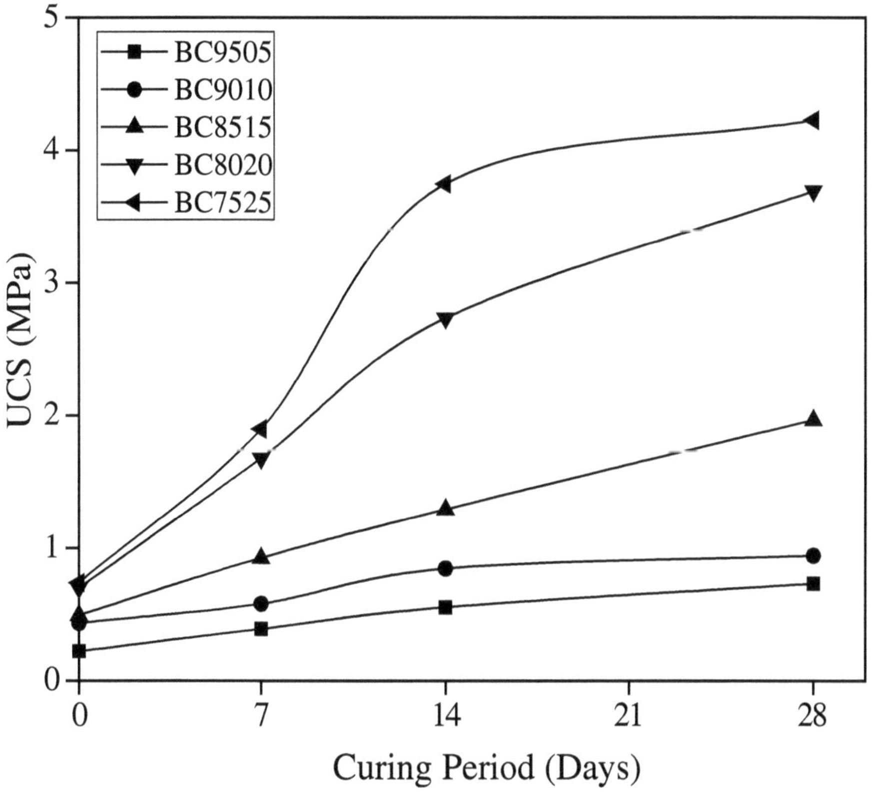

Fig. 6. UCS vs. Curing Period for of Bentonite- cement clinker mix

5 Conclusions

Sodium bentonite is a highly dispersive soil and needs to be stabilized to avoid failure of structures resting on such type of soil. As observed from the double hydrometer test results, the addition of 25% cement clinker to bentonite reduces dispersion from 90.66% to 16.80% and thus the bentonite and cement clinker mixture become non-dispersive. In XRD test it was observed that bentonite mainly consists of montmorillonite, kaolinite, quartz, and phillipsite minerals. Philipsite minerals contain sodium, which is mainly responsible for dispersive nature of the soil. With the addition of cement clinker of 15%, 20%, and 25% CBR values are found to be higher in soaked condition than the unsoaked condition. The unconfined compressive strength is found to be higher (0.34 to 4.23 MPa) when 25% cement clinker is added to bentonite at 28 days curing period. So, based on the experimental results presented here in, it is observed that when the cement clinker is mixed with bentonite, the dispersive nature of sodium bentonite decreases and its strength increases.

References

Vakili, A.H., Selmat, M.R., Moayedi, H., Amani, H.: Stabilization of dispersive soils by pozzolan. In: Proceedings of Forensic Engineering, pp. 726–735 (2013). doi:10.1061/9780784412640.077

ASTM D4221-99 – Standard Test method for dispersive characteristics of clay soil by Double Hydrometer. doi:10.1520/D4221-99

Gullu, H., Hazirbaba, K.: Unconfined compressive strength and post-freeze–thaw behavior of fine-grained soils treated with geofiber and synthetic fluid. Cold Reg. Sci. Technol. **62**, 142–150 (2010). doi:10.1016/j.coldregions.2010.04.001

Gullu, H., khudir, A.S.: Effect of freeze–thaw cycles on unconfined compressive strength of fine-grained soil treated with jute fiber, steel fiber and lime. Cold Reg. Sci. Technol. **106–107**, 55–65 (2014). doi:10.1016/j.coldregions.2014.06.008

IS: 2720-Part 7: Methods of test for soils-determination of water content dry density relationship using light compactionBureau of Indian StandardsNew Delhi, India (1980)

IS: 2720-Part 10: Methods of test for soils-determination of unconfined compressive strength Bureau of Indian StandardsNew Delhi, India (1991)

IS: 2720-Part 16: Methods of test for soils-laboratory determination of CBRBureau of Indian StandardsNew Delhi, India (1987)

Abbasi, N., Nazif, M.H.: Assessment and modification of sherard chemical method for evaluation of dispersion potential of soils. Geotech. Geol. Eng. **31**, 337–346 (2013). doi:10.1007/s10706-012-9573-7

Umesha, T.S., Dinesh, S.V., Sivapullaiah, P.V.: Control of dispersivity of soil using lime and cement. Mater. Sci. Appl. **3**, 8–15 (2009)

Ouhadi, V.R., Goodarzi, A.R.: Assessment of the stability of a dispersive soil treated by alum. Eng. Geol. **85**, 91–101 (2006). doi:10.1016/j.enggeo.2005.09.042

The Sustainable Use of Fine Marble Waste Powder for the Stabilization of Desert Sand in Oman

Mohsin U. Qureshi(✉), Bushra Al-Sawafi, Maryam Al-Washahi, Moza Al-Saidi, and Salima Al-Badi

Civil Engineering, Sohar University, Sohar, Oman
mohsinqureshi81@gmail.com

Abstract. The Sultanate of Oman has gone through a tremendous development over the last four decades and envisions more environment friendly and sustainable infrastructure. As a result of this infrastructure development, quantities of several types of waste have dramatically increased. Recently, the use of waste materials as a soil additive has become widespread. Nevertheless, in Oman marble waste generally ends up in municipality dump sites. Recent practice has demonstrated partial replacement of marble waste for soil stabilization. This paper explored possibilities for the sustainable invigoration of marble waste as an environment friendly recycled material in Oman. In this study, marble waste powder was mixed with desert sand at various proportions and the strength, deformation, compaction and permeability characteristics of sand-marble waste mix were elucidated. Triaxial test results indicated an increase in strength and deformation response by the addition of marble waste in sand. Further, a decrease in permeability was observed by the addition of marble waste. The strength parameters obtained were utilized to calculate the bearing capacity for shallow foundations. The geotechnical properties of sand-marble waste mixes elucidated in this study reveals that the waste marble if mixed with sand, not only increase the bearing capacity for shallow foundations in desert but also eliminates its disposal at the landfill sites.

1 Introduction

Waste management is going through a transition stage in Oman. It started to learn from technologically of advanced countries like Germany, Japan and Canada, that having waste recycling rates in excess of 62, 27 and 21%, respectively. Ministry of Environment in Oman is working together with the municipalities to manage solid waste from both: industries and household sectors. Presently, industrial waste disposal comes under the responsibility of industries and huge quantities of waste are dumped to the landfill sites with a very little recovery. Likewise, all the wastes of marble industry are dumped into the landfill sites. Moreover, landfills occupy large space and can be a considerable source of pollution. To date only few engineered landfill sites exist, among more than 300 dump sites over the Sultanate (Al-Kindi 2010). In Oman, the studies on recycling of wastes in civil engineering applications is very limited, such as the partial utilization of reclaimed asphalt pavement aggregates, demolition concrete,

D.N. Singh and A. Galaa (eds.), *Contemporary Issues in Geoenvironmental Engineering*, Sustainable Civil Infrastructures,
DOI 10.1007/978-3-319-61612-4_25

cement by-pass dust, copper slag; petroleum contaminated soils and discarded tires (Taha et al. 2004). Soil Improvement is considered wide field for using wastes materials such as stabilization of dune sand by using the ash (Mohamedzein and Al-Aghbari 2012) and using FCC spent catalyst in mortar (Al-Jabri et al. 2013).

The production of fine particles (<2 mm) while cutting marble is one of the major problems for the marble industry. Around 50% of marble quarries are turned to waste during the cutting and polishing process (Gurer et al. 2004). Rizzo et al. (2008) highlighted the problems related to soil and water pollution due to the disposal of marble waste slurry. They recommended that the slurry should be subjected to inactivation treatment before disposal or, alternatively recycled as secondary raw material. Luodes et al. (2012) assessed the suitability of various rock wastes for environmental friendly industrial applications. The application of waste from stone industry having particle size of 100 microns as recycled materials is very well-practiced worldwide. A number of studies detailed by Sivrikaya et al. (2014), Agarwal and Gupta (2011), Khan (2006), Ramadas et al. (2010) and Acchar et al. (2006) were devoted to stabilize the problematic soils by using marble waste. In further application, Khan (2006) utilized the marble waste as a partial replacement for clay core of a dam structure. According to his optimized results, partial replacement of marble dust helps not only in reducing the permeability of core but also increased its durability.

The enhancement of clay permeability by the addition of marble waste powder can also be utilized for the construction of engineered landfills which require clay in huge quantities. Zorluer and Taspolat (2009) utilized marble waste dust to build liner for an engineered landfill site. Taspolat et al. (2006) concluded that the durability of clay liner can also be enhanced by the addition of marble waste. They also reported an increase in freeze-thaw resistance of marble waste mixed clay liner. The promising studies in Oman concentrated on proposing the suitable material for landfill liner included the mixture of desert sand and local clay (Al-Rawas et al. 2006). The utilization of marble waste for the treatment of problematic soils, as well as an additive in soil to improve its physical, strength and permeability characteristics are a worldwide practice. However, such practice is not common in Oman because marble needs valorization.

In addition, Misra et al. (2010) proposed that dust and slurry of marble can be used for constructing road layers. There are various applications of marble waste in concrete industry among them are applications of marble dust addition to improve the mechanical behavior of concrete against freeze-thaw weathering (Unal and Uygunoglu 2003) and improve the workability and strength of self-compacting concrete (Uysal 2010). Moreover, the utilization of marble waste in manufacturing of construction bricks and tiles have also been implemented. Recent studies reported the manufacturing of bricks by the addition of marble dust in clay (Bilgin et al. 2012; Dhanapandian et al. 2009) and eco-blocks or cement bricks made of marble sludge (Aukour 2009).

However, to the best of knowledge no studies was conducted in Oman for the feasibility of marble waste to utilize in various applications. Therefore, the present study will be a foremost study for proposing marble waste as a recyclable material in Oman. The authors aimed to elucidate the improvements of strength, compaction and permeability characteristics of desert sand mixed with marble waste powder at various propositions. The adopted methodology is presented in the next section.

2 Methodology

2.1 Materials

Sand employed in this study was sampled from Al-Bidiya in northern part of Wahiba sand dunes (Fig. 1), in Al-Sharqia Region which is about 200 km to the south of Capital Muscat. The Wahiba Sand formed in Quatenary period (Glennie 1998) is composed of two physiographic units that can be roughly divided into northern and southern regions. The Northern Wahiba is predominantly a large megaridge system, whereas the Southern Wahiba mostly comprises linear dunes, sand sheets, and sabkha fields (Pease and Tchakerian 2003). Carbonate sands present throughout the Wahiba sand area with some quartz-rich sands in the Southern region. The Al-Bidiya sand, BS (Fig. 2) used in this study is sampled from the Northern region. The grain size distribution (ASTM D6913) of the sand is shown in Fig. 4. The sand can be classified as poorly-graded sand (SP) according to the Unified Soil Classification System (USCS). The sand has a specific gravity (ASTM D854) of 2.62 and a fine content of 1.1%. The standard Proctor compaction test (ASTM D558) gave a maximum dry density of

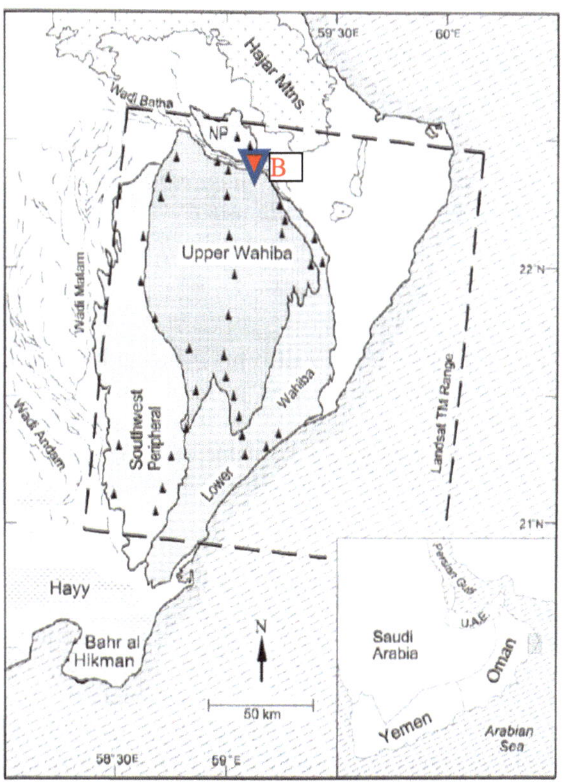

Fig. 1. Location of sampling site at Al-Bidiya, Wahiba sand Dunes, Al-Sharqia, Oman (Pease et al. 1999)

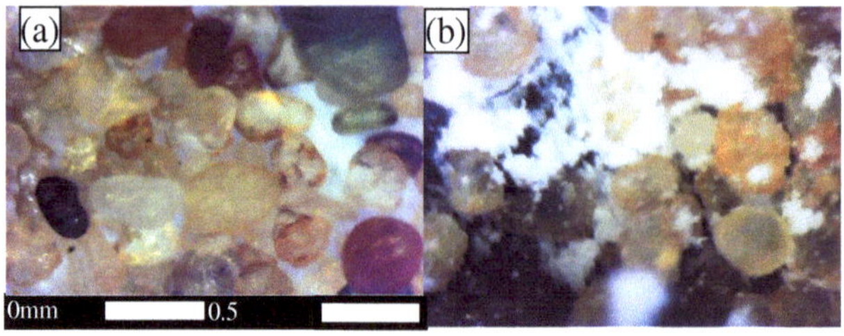

Fig. 2. (a) Al-Bidiya sand (BS), (b) Al-Bidiya sand mixed with marble waste powder (BSM)

1.64 g/cc and an optimum water content of 19.5%. The micrograph of BS indicates sub-rounded particles and lack of fines as shown in Fig. 2(a).

Marble is a metamorphic rock that consists predominantly of calcite and/or dolomite (Dietrich and Skinner 1979). Marble may be considered as metamorphosed limestone (i.e. limestone which has been fully re-crystallized and hardened under hydrothermal conditions). In this study waste calcite marble dust was used. Marble waste was sampled from Oman Marble, Al-Turki Enterprices L.L.C., Sohar Industrial Estate, Sohar, Sultanate of Oman. Marble waste slurry is produced as a result of sawing and polishing of marble for construction applications. Water is then removed from slurry at the factory and fine marble waste is dumped at the landfill sites. The authors collected waste marble powder from the factory and performed particle size distribution test which indicated a fine content of 82.4%. The micrograph of sand-marble mix (BSM) presented in Fig. 2(b) indicates the potential of fine marble to fill in the voids between poorly graded sand particles, thereby improving its gradation. Waste marble powder was passed through # 40 sieve before usage.

2.2 Equipment

The standard equipment manufactured by ELE International Inc. UK was employed to perform compaction tests and constant head permeability tests in accordance with ASTM D558 (ASTM 2011) and ASTM D2434 (ASTM 2000) respectively. A state-of-the-art automatic triaxial test machine (Fig. 3) manufactured by Utest, Turkey was employed to perform the drained triaxial tests in accordance with ASTM D7181 (ASTM 2011a).

2.3 Test Specimens

Marble waste was added in percentages of 5, 10, 15 and 50% by dry weight of the sand. The required amount of marble waste powder was added to a dry sand sample passing No. 4 sieve (4.75 mm). The sand and marble waste were mixed thoroughly to a uniform color. The summary of test specimens prepared for gradation analysis and compaction tests is shown in Table 1. The grain size distribution for all the mixes as well as the virgin samples are plotted in Fig. 4. The gradation analysis clearly indicates that the addition of marble waste powder has improved the gradation of the sand-marble mix to a

Fig. 3. A state-of-the-art automated triaxial test equipment at Sohar University

certain extent. To ensure reproducibility of the results, for each marble content large amount of the mixture was prepared for each test. Identical samples were tested and the average value was obtained. The variability of the measured quantities was assessed based on published guidelines. For example ASTM D558 (ASTM 2011) recommends that the variability be evaluated in terms of the average difference between high and low values. Average differences of less than 10% were considered acceptable in this study. The cylindrical specimens (10 cm height and 5 cm diameter) were prepared for conducting the drained triaxial tests as shown in Fig. 5. All the specimens were compacted at their respective optimum moisture contents to achieve maximum dry densities. Thereafter preparation, the specimens were dried in oven at 110 °C for 72 h to be at the same condition of moisture content at the time of triaxial tests.

Table 1. Summary of test specimens

Specimen	Marble by wt. (%)	Fine content (%)	Gs
BS	0	1.1	2.62
BSM5	5	1.9	–
BSM10	10	4.5	–
BSM15	15	12.8	–
BSM50	50	37.2	–
M	100	82.4	2.52

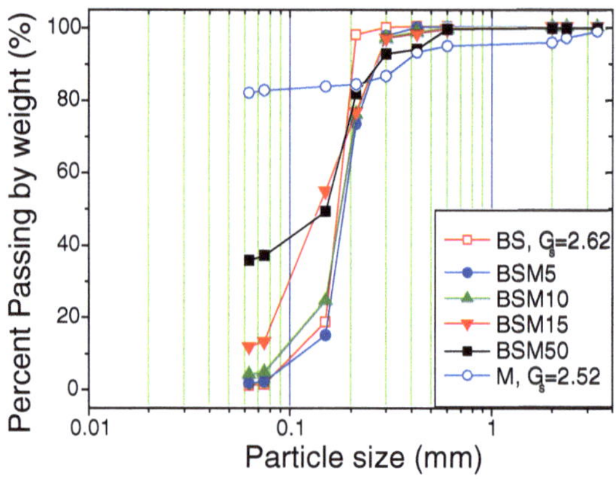

Fig. 4. Grain size distribution of the specimens

Fig. 5. Cylindrical specimens of marble-sand mixes

3 Test Results and Discussions

3.1 Compaction

Standard Proctor compaction tests were performed on the mixtures in accordance with ASTM D558 (ASTM 2011). The results of compaction tests for various mixtures are shown in Fig. 6. The data obtained for pure sand (BS) is also shown in the figure for comparison purposes. The maximum dry density is increased upto marble content of 15% and then decreased with the increase in marble content. Initially, the fine marble powder filled the voids between the sand particles and once the voids are filled with marble waste powder (at marble content of about 15%), further increase in the marble waste content decreases the maximum dry density as shown in Fig. 7. However, the optimum water content reduces up to 15% of marble and then it starts to increase with the increase in marble content as shown in Fig. 7.

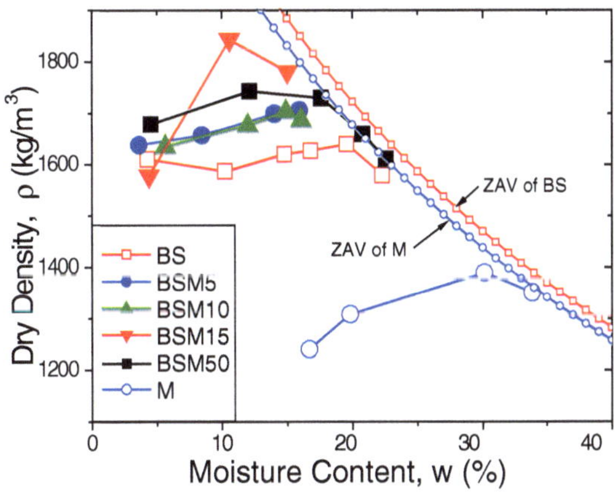

Fig. 6. Compaction Curves for sand-marble mixes

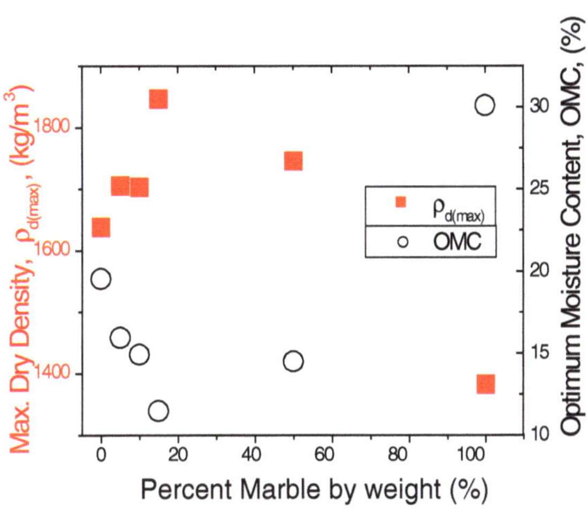

Fig. 7. Variations of maximum dry density and optimum moisture content with marble content in sand

3.2 Permeability

Constant head permeability tests were performed in accordance with ASTM D2434 (ASTM 2000) on virgin sand and the marble-sand mixes. At least five reading were noted and an average value of permeability was computed. The average hydraulic conductivity was corrected for temperature effects. The results are shown in Fig. 8. The permeability decreased substantially with the increase of marble content. The permeability of BSM50 which has marble content of 50% is in range of 10–5 cm/s which is typically for silts. The marble waste powder fills the voids between the sand particles and provides a more tortuous flow path for water to follow and hence reduces the permeability.

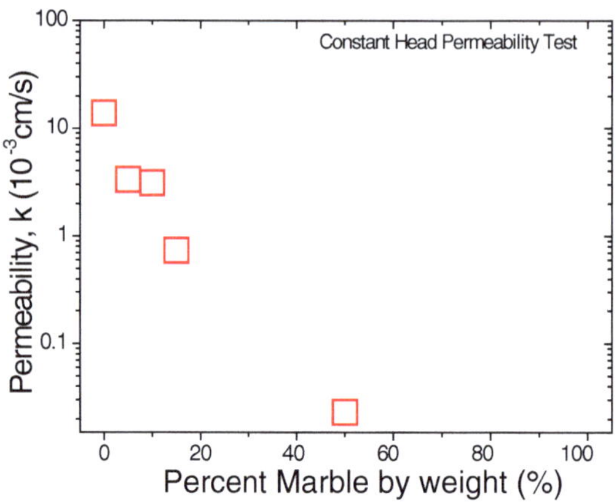

Fig. 8. Decrease in permeability with increase of marble content in sand

3.3 Triaxial Compression

Drained triaxial compression tests were performed on BS, BSM5 and BSM15 specimens. The tests were performed at different levels of lateral stresses (σ_r) for the computation of shear strength parameters, i.e., cohesion, c and angle of internal friction, ϕ at peak. All the tests were performed at a constant axial strain rate of 0.5 mm/min. The typical results of triaxial tests performed on BSM5 are shown in Fig. 9. Thereafter, shear strength parameters were determined by the Mohr Coulomb failure criterion.

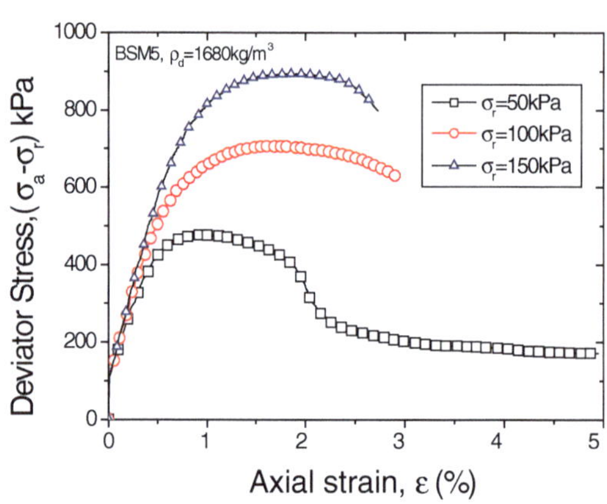

Fig. 9. Triaxial test results on BSM5

The change in cohesion and angle of internal friction of marble sand mix specimens is plotted against the marble content in Fig. 10. It can be seen that there is no significant change in angle of internal friction, however the cohesion drastically increased by the increase in the percentage of fine marble waste powder in sand. The cohesion is around 60 kPa for BSM15 which is considered a substantial improvement in the shear strength of the desert sand which transferred from cohesionless state to cemented sand.

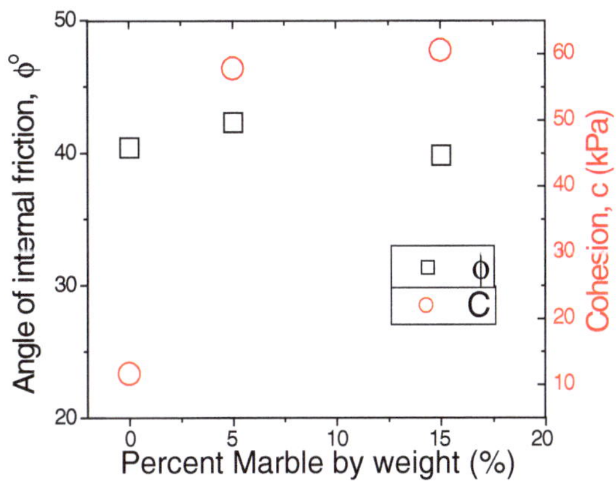

Fig. 10. Change in cohesion and internal friction angle with the addition of marble in sand

3.4 Field Application

A remarkable increase in inter-particle cohesion of the marble mixed sand is expected to provide significant ground improvement in practical geotechnical applications. To address considerations for foundation engineering, the ultimate bearing capacities for square footing with unit dimensions and unit depth resting on virgin sand and marble mixed sand (BSM15) compacted to respective optimum water content are computed in accordance with Terzaghi's formula on ultimate bearing capacity (Coduto 2001). The ultimate bearing capacities at the aforementioned conditions were determined to be 3050 kPa for virgin sand and 6780 kPa for BSM15. Thus, it is vigilant to conclude that the improvement efficiency of the mixing marble waste in sand provides effective ground improvement for various geotechnical engineering applications including foundation design. The improvement of sand is noticeable, but the effect of water on the internal structure of soil should be studied.

4 Conclusions

The research was carried out to improve the geotechnical properties of desert sand by the partial addition of marble waste powder. The gradation of desert sand which is classified as fine-grained, poorly-graded sand (SP) was improved by the addition of fine marble waste powder. Based on the results presented in this paper the authors have drawn the following conclusions.

The results of compaction tests showed that maximum dry density of sand is increased by the addition of marble waste powder up to 15% and further addition of marble waste powder decreased the maximum dry density.

Sand sampled from the desert was lacking of fines. The addition of fine marble waste has also reduced the permeability of sand from 0.013 cm/s to 2.27 × 10–5 cm/s at a mix ratio of 50%.

The results of triaxial tests showed that the sand-marble mixtures possess both cohesion (c) and angle of internal friction (ϕ), indicating that the fine marble waste acted as a cementing agent between the sand particles.

An increase in bearing capacity is observed due to the addition of marble in sand so marble waste can be a locally available additive to desert sand for the improvement of strength and quality of shallow foundations. Furthermore, the marble waste has no monetary value since it is a waste and if employed in ground improvement will eliminate the need for its disposal in landfills. Therefore, the mixing of fine marble waste powder to sand can be employed in improving the properties of desert sand for shallow foundations.

Acknowledgements. The present research was carried out under the financial support provided by the Research Council of Oman under the FURAP grant awarded in 2015. The authors are also indebted to the technicians at Sohar University for their support in the performance of experiments. The authors are thankful to Al-Turki Enterprices L.L.C., Sohar Industrial Estate, Sohar, Sultanate of Oman, for allowing the visit to the facility and permission for the collection of marble waste powder.

References

Acchar, W., et al.: Effect of marble and granite sludge in clay materials. Mater. Sci. Eng., A **419**, 306–309 (2006)

Agrawal, V., Gupta, M.: Expansive soils stabilization using marble dust. Int. J. Earth Sci. Eng. **4**, 59–66 (2011)

Al-Jabri, K.S., et al.: Potential use of FCC spent catalyst as partial replacement of cement or sand in cement mortars. Constr. Build. Mater. **39**, 77–81 (2013)

Al-Kindi, A.H.: Oman waste management sector workshop by Oman Environment Services Holding Company, Muscat, Oman, 5 February 2010

Almeida, N., et al.: Recycling of stone slurry in industrial activities: application to concrete mixtures. Build. Environ. **42**(2), 810–819 (2007)

Al-Rawas, A.A., et al.: Sand-attapulgite clay mixtures as a landfill liner. Geotech. Geol. Eng. **24**, 1365–1383 (2006)

ASTM. ASTM designation ASTM D2434: Standard Test Method for Permeability of Granular Soils (Constant Head) (2000)

ASTM. ASTM designation ASTM D6913: Standard Test Methods for Particle-Size Distribution (Gradation) of Soils Using Sieve Analysis (2009)

ASTM. ASTM designation D 558: standard test methods for moisture-density (unit weight) relations of soil-cement mixtures (2011)

ASTM. ASTM designation ASTM D7181: Method for Consolidated Drained Triaxial Compression Test for Soils (2011a)

ASTM. ASTM designation ASTM D854: Standard Test Methods for Specific Gravity of Soil Solids by Water Pycnometer (2014)

Aukour, F.J.: Incorporation of marble sludge in industrial building eco-blocks or cement bricks formulation. Jordan J. Civ. Eng. **3**(1), 58–65 (2009)

Bilgin, N., et al.: Use of waste marble powder in brick industry. Constr. Build. Mater. **29**, 449–457 (2012)

Coduto, D.P.: Foundation Design, Principles and Practices, 2nd edn. Prentice Hall, New Jersey (2001)

Dhanapandian, S., et al.: Utilization of granite and marble sawing powder wastes as brick materials. Carpathian J. Earth Environ. Sci. **4**(2), 147–160 (2009)

Dietrich, R.V., Skinner, B.J.: Rocks and Rock Minerals, 6th edn. John Wiley and Sons, New York (1979)

Glennie, K.W.: The desert of SE Arabia: a product of climatic change. In: Alsharhan, A.S., Glennie, K.W., Whittle, G.L., Kendall, C.G.S.T.C. (eds.) Quaternary Deserts and Climatic Change, pp. 279–291. Balkema, Rotterdam (1998)

Gurer, C., et al.: Recycling and re-evaluation of different building materials as a source of raw material in the construction industry. In: 5th Industrial Raw Materials Symposium, pp. 28–36, Izmir, Turkey, 13–14 May 2004

Khan, S.A.: Physical characteristics of fine soil stabilized with marble industry waste. In: 7th International Congress on Civil Engineering, Tehran, Iran, 7–10 May 2006

Luodes, H., et al.: Characteristics and the environmental acceptability of the natural stone quarrying waste rocks. Bull. Eng. Geol. Environ. **71**, 257–261 (2012)

Misra, A.K., et al.: A new technology of marble slurry waste utilization in road. J. Sci. Ind. Res. **69**, 67–72 (2010)

Mohamedzein, Y.E.A., Al-Aghbari, M.Y.: The use of municipal solid waste incinerator ash to stabilize dune sands. Geotech. Geol. Eng. **30**, 1335–1344 (2012)

Pease, P., et al.: Mineralogical characterization and transport pathways of dune sand using Landsat TM data, Wahiba Sand Sea, Sultanate of Oman. J. Geomorphol. **29**(3–4), 235–249 (1999)

Pease, P.P., Tchakerian, V.P.: Composition and sources of sand in the Wahiba Sand Sea, Sultanate of Oman. Can. Geotech. J. **40**(5), 416–434 (2003)

Ramadas, T.L., et al.: Swelling and strength characteristics of expansive soil treated with stone dust and fly Ash. In: Indian Geotechnical Conference 2010, GEOtrendz, Bombay, pp. 557–560, 16–18 December 2010

Rizzo, G., et al.: Problems of soil and groundwater pollution in the disposal of "marble" slurries in NW Sicily. Environ. Geol. **55**, 929–935 (2008)

Sivrikaya, O., et al.: Recycling of waste from natural stone processing plants to stabilize clayey soil. Environ. Earth Sci. **71**, 4397–4407 (2014)

Taha, R., et al.: An overview of waste materials recycling in Sultanate of Oman. Resour. Conserv. Recycl. **41**(4), 293–306 (2004)

Taspolat, L.T., et al.: The effect of marble waste powder on the freeze—thaw in impermeable clay layers. Electron. J. Constr. Technol. **2**, 11–16 (2006)

Unal, O., Uygunoglu, T.: Investigation of mechanical properties of concretes with waste marble powder under the effect of freeze-thaw. In: Turkish 4th Marble Symposium, Afyon, pp. 147–157, 18–19 December 2003

Uysal, M.: The effect of marble powder on the workability and strength properties of self-compacting concrete. In: 2nd International Congress on the Marble and Natural Stones, Izmir, Turkey, pp. 326–334, 2–4 February 2010

Zorluer, I., Taspolat, L.T.: Reuse of waste marble dust in the landfill layer. In: 1st International Symposium on Sustainable Development, Sarajevo, Bosnia and Herzegovina, 9–10 June 2009

Assessing the Suitability of Compacted Bentonite-Pond Ash Mixes as Landfill Liner

Suryaleen Rout[1(✉)] and S.P. Singh[2]

[1] National Institute of Technology, Rourkela, Odisha, India
suryaleen1512@gmail.com
[2] Department of Civil Engineering, National Institute of Technology, Rourkela, Odisha, India
spsingh@nitrkl.ac.in

Abstract. The use of compacted bentonite-sand mixes as landfill liner has been accepted worldwide. However, the non-availability of sand at all landfill sites triggers a search for an alternate material. Pond ash is an industrial waste and has the distinct advantage of a geotechnical material as it resembles the natural earth material. Moreover, it has the similar mineralogical composition to conventional earth materials and needs special attention for its effective utilization rather than disposing of it by consuming large land area leading to the loss of agricultural production. With the addition of bentonite to the pond ash, the plasticity, and strength is expected to increase whereas the permeability is expected to decrease. Further, as pond ash is non-plastic and possesses very low shrinkage, it is expected to reduce the swelling and shrinkage, preventing the formation of any cracks in the compacted liner material. The present study examines the suitability of compacted bentonite-pond ash mixes as landfill liner material. The proportion of bentonite in the pond ash-bentonite mix is varied from 5 to 30% by weight at 5% intervals. The plasticity characteristics, compaction properties, strength and permeability properties of pond ash-bentonite mixes are evaluated. It is observed that with the addition of bentonite the plasticity index, maximum dry density, differential free swell index, cohesion, and compressive strength increases gradual and on the other hand the optimum moisture content, hydraulic conductivity, and frictional angle decreases. It is also found that an addition of 5% bentonite has no effect on the geotechnical properties of pond ash. The designed mixture with 20% bentonite attained its plasticity index, strength and permeability values as per the requirements of a suitable landfill liner material as per the EPA regulatory act.

Keywords: Pond ash · Bentonite · Landfill liner · Geoengineering properties

1 Introduction

With the growing industrialization and population, wastes are generated more and more. Based on the safety level, these wastes can be hazardous or non-hazardous. These wastes can be controlled by recycling, incineration, transformation and environmentally sustainable disposal on land. Despite all effort, the most frequently used option is disposing of waste in the landfill due to its efficiency and low cost. The liner is

D.N. Singh and A. Galaa (eds.), *Contemporary Issues in Geoenvironmental Engineering*, Sustainable Civil Infrastructures,
DOI 10.1007/978-3-319-61612-4_26

the essential component of a landfill. According to EPA regulatory, liner must have adequate strength with minimum hydraulic conductivity and plasticity index should vary between 10 to 30%. Previously, clay has been used as a liner material due to its low cost, large leachate attenuation capacity and resistance to damage and puncture. Nowadays many more researchers have found out that compacted clay can be effectively replaced by wastes and any other material. Kayabali (1997) conducted tests on seven different ratios of bentonite to zeolite to obtain a mix ideal as landfill liner material. Owing to its swelling ability, bentonite content serves as pore-sealant at the saturated condition. The optimum water content and the corresponding dry density ranged from 33 to 42% and 1.16 to 1.26 g/cc, respectively. At the smallest ratio of bentonite to zeolite (i.e., 0.05), the water content of the bentonite component was as high as 850%, and full saturation was reached. The average range of hydraulic conductivity obtained was 2×10^{-8} to 4×10^{-8} cm/s. Mohamedzein et al. (2005) investigated the use of two types of crushed shales as a liner material by conducting different laboratory tests. It was concluded that both the shales satisfied the basic properties of the liner having low compressibility and hydraulic conductivity for all level of compaction and the shear strengths were with the acceptable range of earthen material. Tanit and Surapon (2005) conducted direct shear and hydraulic conductivity tests to assess the shear strength parameters and hydraulic conductivity of compacted sand-bentonite mixtures. Test results indicated that the hydraulic conductivity of mixtures decreased about four orders of magnitude when mixed with 5% bentonite or more. Based on test results, a suitable sand-bentonite mixture that yields low hydraulic conductivity, while maintaining relatively high shear strength, was recommended for use as a liner in hydraulic containment applications. Shankar and Phanikumar (2012) studied the variation of index properties with the concentration of permeating fluids. This paper proposed an innovative clay liner material in the form of fly ash-stabilised expansive clay and presented the experimental findings on the variation of index properties and free swell index (FSI) values of the blend with varying fly ash content in the blend and concentration of salt solutions. The various permeating liquids used were deionized water and salt solutions such as $CaCl_2$, NaCl, and KCl. It was observed that with an increase in the fly ash content for all concentrations of salt solution, the liquid limit (LL), FSI and hydraulic conductivity decreased. A correlation between LL and FSI has been established for the liner material. Akcanca and Aytekin (2012) studied the influence of wetting–drying cycles on swelling pressures of sand–bentonite mixtures used in the construction of sanitary landfills before and after lime treatment of the mixtures. The specimens were prepared by mixing bentonite in various proportions with their OMC and also lime was added to the sand bentonite mixture. Both the specimens were compacted with standard Proctor energy. The test results indicated that the swelling pressures of the specimens made of sand–bentonite mixtures stabilized by lime were lower than the swelling pressures of the specimens made of only sand–bentonite mixtures. Also, the swelling pressure is found to reduce by wetting–drying cycles. The present study examines the suitability of compacted bentonite-pond ash mixes as landfill liner material. The proportion of bentonite in the pond ash-bentonite mix is varied from 5 to 30% by weight at 5% intervals. The plasticity characteristics, compaction properties, strength and permeability properties of pond ash-bentonite mixes are evaluated.

2 Materials and Methodology

2.1 Materials

The basic raw materials used in this experimental program are pond ash and commercial bentonite. Pond ash (PA) used in this study was collected from the ash pond of the national thermal power plant (NTPC), Angul (Odisha). These samples were collected from the outlet points of slurry pipes. The samples were mixed thoroughly to bring homogeneity and dried at an oven temperature of 105–1100 C. The geotechnical properties of the pond ash (PA) and bentonite (B) are given in Table 1. Commercially available sodium bentonite of Neelkanth Minechem, Jodhpur, (India) is used in this experimental program. According to the IS soil classification system, it is classified as high plasticity clay (CH) as plasticity index of bentonite is more than 50%. The grain size distribution curves for both the materials are shown in Fig. 1. Synthetic mixes are prepared by mixing bentonite with pond ash in different proportions. The percentage of bentonite in the mix is varied from 5 to 30% by weight at 5% intervals, and these mixes are designated as B5, B10, B15, B20, B25, and B30 respectively.

Table 1. Index and engineering property of pond ash and bentonite

Parameters	Bentonite	Pond ash
Colour	Cream	Light grey
Particle shape	Flaky type	Sub-rounded to irregular
Fines content, <75 μm (%)	96.93	1.99
Fine sand (%)	3.07	64.28
Medium sand (%)	0	33.73
Median diameter, D_{50} (mm)	0.04	0.28
Uniformity coefficient, C_u	–	2.91
Coefficient of curvature, C_c	–	1.15
Specific Gravity, G	2.7	2.23
Liquid limit (%)	301	Non-plastic
Plastic limit (%)	65	
Shrinkage limit (%)	9.42	
Plasticity index (%)	236	
OMC (%)	32	27
MDD (kN/m^3)	13.53	11.63

2.2 Methodology

In this present study, synthetic mixes are prepared by adding bentonite of 5 to 30% by weight at 5% intervals with the pond ash. In this manner, six different mixes were synthesized. The basic properties like plasticity characteristics, specific gravity, and differential free swell index values are evaluated to check whether it meet the liner requirement. The OMC and MDD of these mixes corresponding to compactive efforts of 595 kJ/m^3 and 2674 kJ/m^3 are determined as per IS: 2720 (Part 7) 1980 and

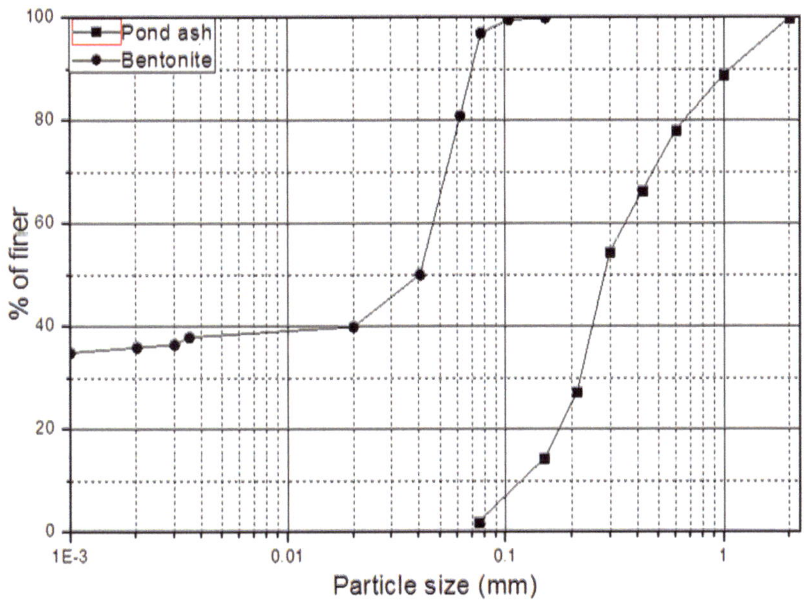

Fig. 1. Grain size distribution curve of pond ash and bentonite

IS: 2720 (Part 8) 1983 respectively. Before the compaction tests, these synthetic materials were mixed with initial water and were kept in an airtight container for 48 h for proper distribution of water in the mix. To find out the shear strength parameters synthetic mixes were compacted in standard Proctor mould to their respective MDD at OMC using compactive efforts of 595 kJ/m^3, and 2674 kJ/m^3. Undisturbed specimens of size 60 mm × 60 mm × 25 mm were recovered from the mould and were sheared at a rate of 0.2 mm/min. The shear parameters (i.e. c and Φ values) were found out from the plots between normal stress and shear stress. These tests were performed as per IS: 2720 (Part 13) 1986. Unconfined compressive strength tests on specimens compacted to their corresponding MDD at OMC was performed according to IS: 2720 (Part 10) 1991. The cylindrical test specimens were of size 50 mm in diameter and 100 mm in height, and these were sheared at an axial strain rate of 1.25 mm/min. Hydraulic conductivity of different pond ash-bentonite mixes, compacted to their respective MDD at OMC are evaluated as per IS: 2720(Part 17) 1987. Pressure permeameter is used to determine the hydraulic conductivity of specimens. A constant fluid pressure of 4 kg/cm^2 is used during the experiment. The length of the specimens used is 100 mm. Portable water with pH of 7.2 is used in this experiment.

3 Result and Discussion

A well designed compacted clay liner must have required plasticity, minimum fines content and a low permeability to prevent or minimize leachate leakage, adequate shear strength for stability and minimal shrinkage potential to prevent desiccation cracking.

The design objective for a compacted clay liner is to determine the range of water content and dry unit weight within which compacted test specimens will have low hydraulic conductivity, adequate shear strength and minimal shrinkage upon drying. Some index and engineering properties are scrutinized based on the addition of the bentonite to pond ash.

3.1 Index Property

The index properties of the pond ash, bentonite, and the mixes i.e. specific gravity, grain size distribution and plasticity characteristics are evaluated and presented. The specific gravity of pond ash and bentonite was found to be 2.23 and 2.70 respectively. The specific gravity of pond ash is comparatively lower than the similar graded conventional earth material. The specific gravity of resulting flyash is affected by the presence of foreign materials in the coal fissures, and it also depends on the source of coal, degree of pulverization and firing temperature. Moreover the pond ash is subjected to mixing with other earth materials during its transportation and depositions, which influences its specific gravity (Singh et al. 2014 and Asrith et al. 2015). Though the chemical composition of pond ash is very much similar to earth material but as the particles are cenospheres, it results in a lower specific gravity. The pond ash consists of grains mostly of medium to fine sand size with fines content (<75 μm) of 1.99% only. The coefficient of uniformity and coefficient of curvature are 1.15 & 2.91 respectively, indicating uniform gradation of samples. The sample is classified as 'SP' as for unified soil classification system. The fines content of the commercial bentonite is 96.93%, and as per unified soil classification system it is classified as 'CH'. The grain size distribution curves of pond ash and bentonite is presented in Fig. 1. The test results show that the pond ash is a non-plastic material whereas bentonite is highly plastic in nature. Both the liquid limit (LL) and plastic limit (PL) of the pond ash-bentonite mixture could not be determined up to bentonite content of 10% in the mix and these mixes are reported as to be non-plastic. However, with further increase of bentonite content of the mix, it behaved as plastic material and both the LL and PL is found to increase significantly with increase in bentonite content of the mixture. When small amount of bentonite is added to the pond ash sample, these finer particles of bentonite get entrapped in the intra-particle and inter-particle void space of pond ash. As the percentage of fines in the mixture is increased these particles tries to coat the surface of the ash particles thus imparting plasticity to the mixture. As the amount of water needed to satisfy the double layer of bentonite is very high an addition of bentonite above threshold amount to pond ash significantly increases both the liquid limit and plastic limit of the mix (Fig. 2). An increase in the bentonite content in the mix, results a marginal decrease of shrinkage limit (SL) value. This is due to the development of high suction force in the capillary pores of the desiccating pond ash–bentonite specimens with higher bentonite content. On the other hand, the differential swelling index value increases with the increase in bentonite content due to the presence of montmorillonite mineral in bentonite, which even allowed more water to enter into unit layers. The variation of the DFS value of the mix with bentonite content is shown in Fig. 3.

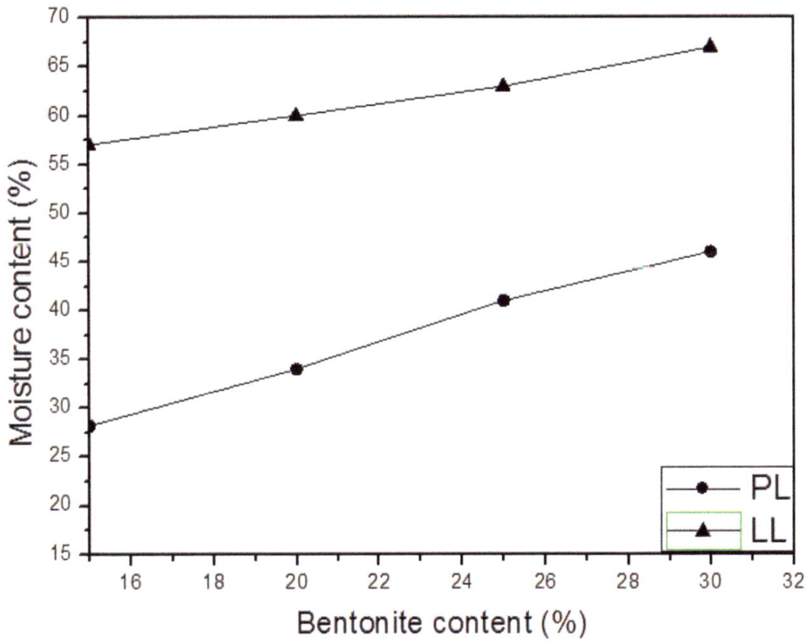

Fig. 2. Variation of liquid limit (LL) and plastic limit (PL) with bentonite content

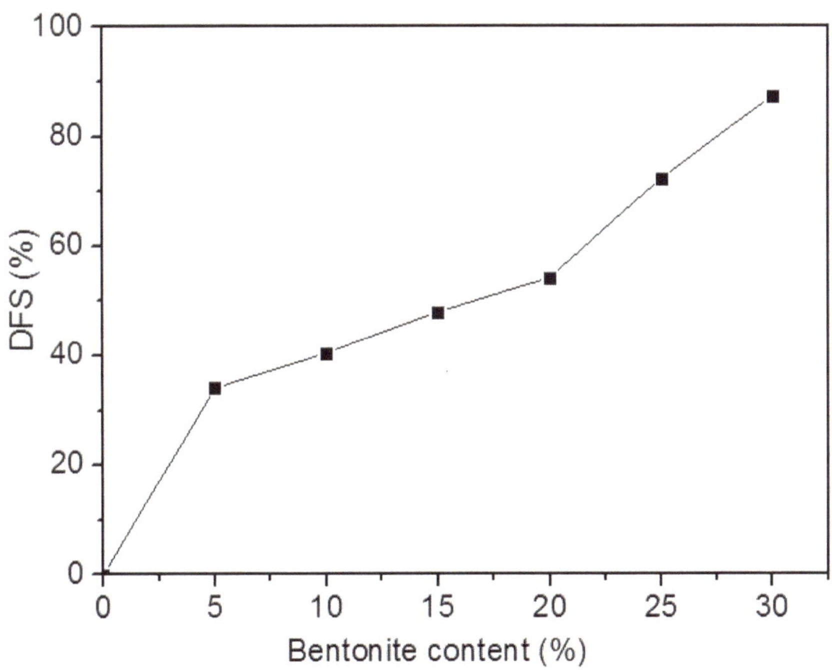

Fig. 3. Variation of DFS value with bentonite content

3.2 Compaction Characteristics

Figures 4(a) and (b) show the compaction curves of pond ash bentonite mixes compacted with standard and modified Proctor energies respectively. It is observed that as the bentonite content increases the compaction curves shifts towards the left that is an addition of bentonite to the pond ash results in a decrease of the OMC value up to optimum bentonite content after that, it remains almost unchanged. This behaviour is observed for samples compacted with either standard or modified Proctor compaction energies. As seen from the gradation curves of pond ash and bentonite both the materials are uniformly graded, and the particle of pond ash ranges from fine to medium sand size while for the bentonite it is mostly clay size particles. An addition of bentonite to pond ash changes the gradation making the mixture better graded from uniformly graded material. The change in gradation results in a well packing of particles during compaction thus increases the MDD value (Fig. 5) and reducing the OMC value of the mix. Figure 6 shows the variation of MDD with bentonite content. It is found that with the addition of bentonite to pond ash, MDD value increases up to an optimum bentonite content after that it remains almost constant. This trend is observed for samples compacted either with 595 kJ/m^3 or 2674 kJ/m^3 energy. The optimum content of bentonite for both the cases is found to be 20%. This may be due to the fact that when bentonite percentage is small, the finer particles of bentonite mostly try to enter into the intra-void spaces present in the conglomerated ash particles. As the percentage of bentonite is further increased, these particles filled-up the inter-particular void spaces available in the pond ash sample. This increases the MDD value. Further increase of bentonite creates a space for itself forcing the ash particles to move apart, thus stabilizing the MDD value.

3.3 Hydraulic Conductivity

Hydraulic conductivity values of bentonite-pond ash mixes compacted with 595 kJ/m^3 and 2674 kJ/m^3 energy are found to be in the range of 8.83×10^{-4} to 9.23×10^{-9} cm/s and 3.42×10^{-4} to 5.15×10^{-9} cm/sec respectively. The variation of hydraulic conductivity with bentonite content is shown in Fig. 7. It is obvious that an increase of bentonite content in the compacted mixtures lowers the hydraulic conductivity values. This is verified by many studies. Since bentonite alone has characteristics of the impervious material and has extremely low hydraulic conductivity values in the order of 10^{-10} to 10^{-12} cm/s, it is expected to have lower hydraulic conductivities as its percentage in the mixture increases. The experimental data shows that the reduction in hydraulic conductivity with bentonite content is not linear. Initially, the rate of reduction is mild followed by a sharp decrease in the hydraulic conductivity value. At low bentonite content, the finer bentonite particles get adjusted in the intra-particle void space of pond ash thus there is an appreciable reduction in capillary void space. However, as its content increases it reduces the capillary void thus a sharp reduction in the hydraulic conductivity value.

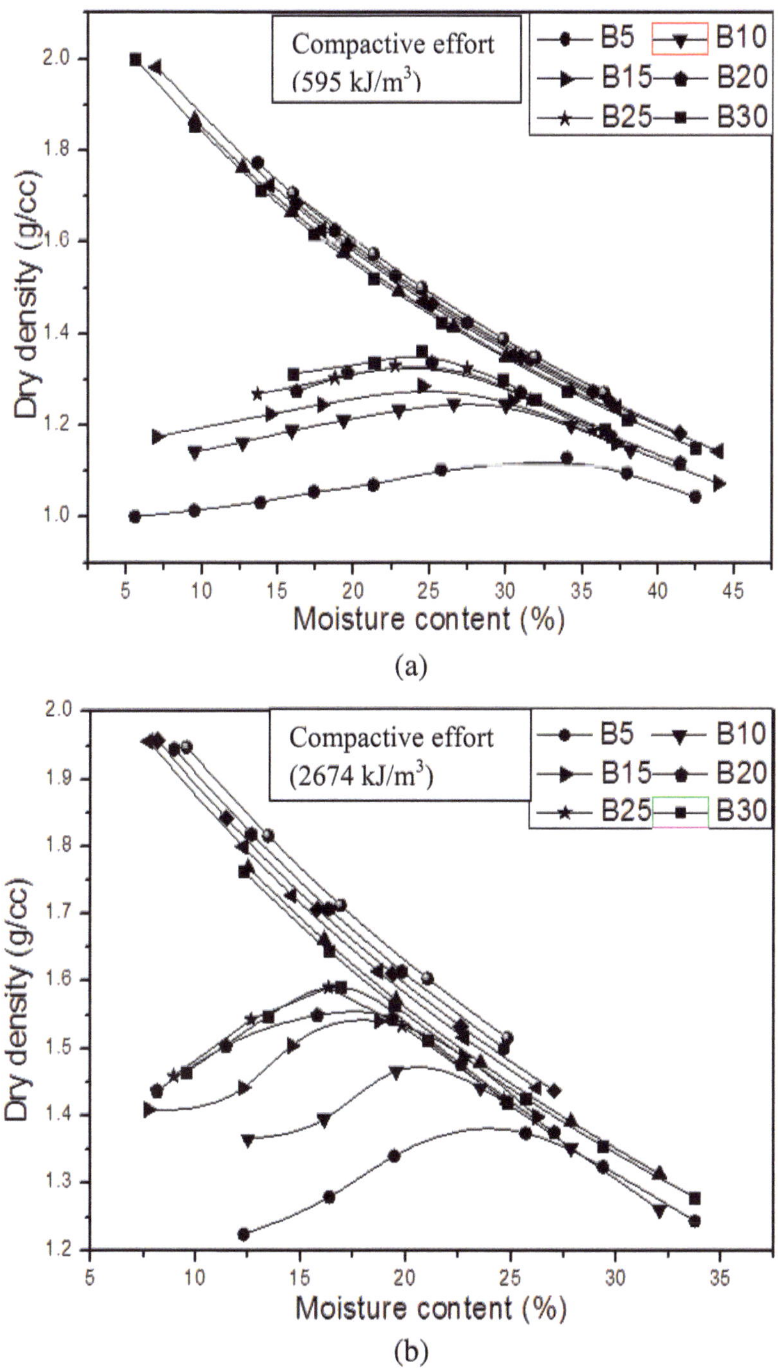

Fig. 4. (a). Compaction curves of pond ash-bentonite mixes at standard Proctor energy (b). Compaction curves of pond ash-bentonite mixes at modified Proctor energy

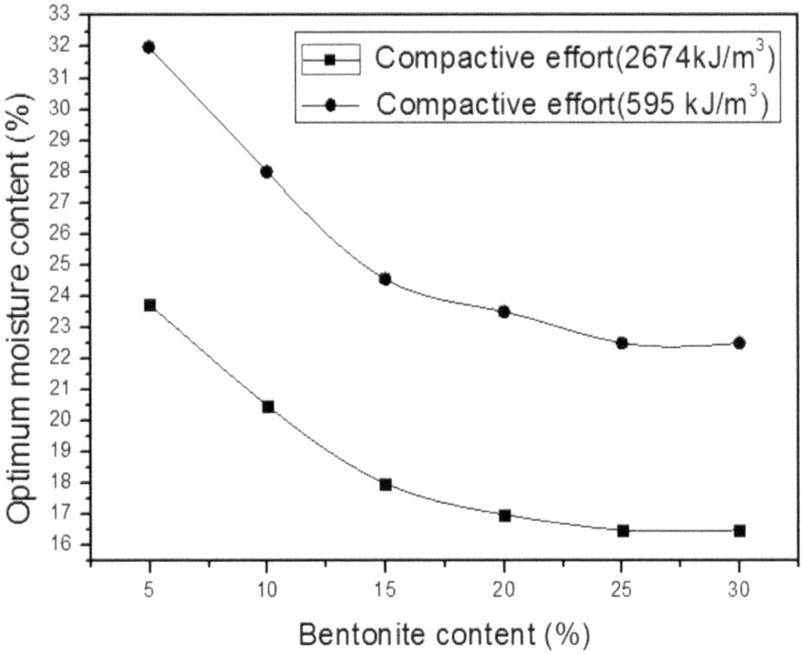

Fig. 5. Variation of OMC with bentonite content

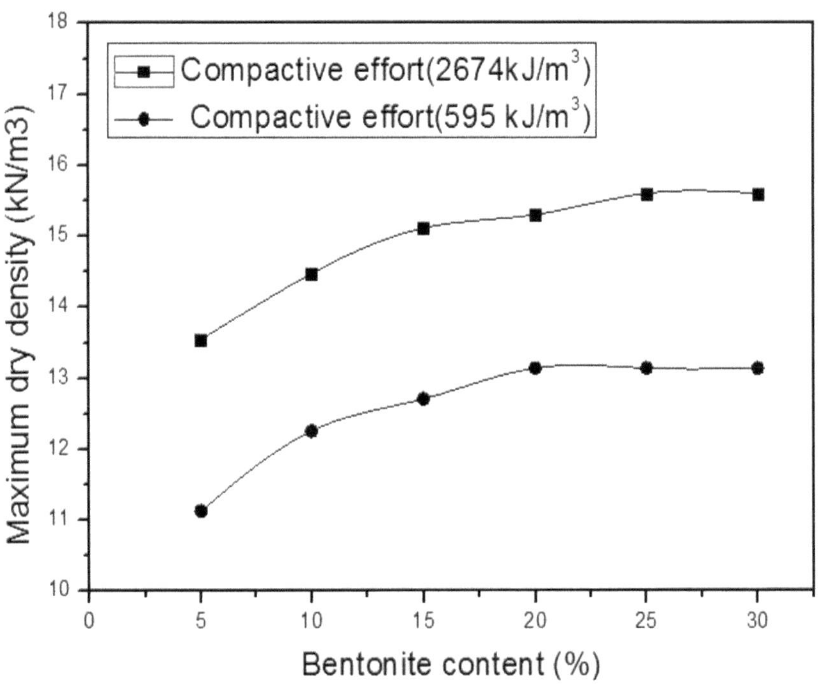

Fig. 6. Variation of MDD with bentonite content

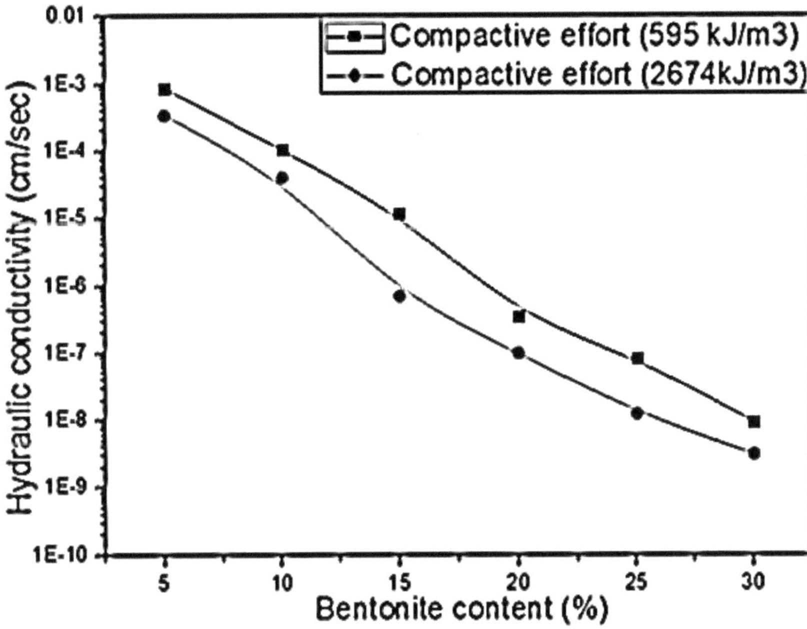

Fig. 7. Variation of hydraulic conductivity with bentonite content

3.4 Unconfined Compressive Strength

Unconfined compressive strength test is performed for sample compacted with both 595 kJ/m^3 and 2674 kJ/m^3 energy. Compressive strength of pond ash-bentonite mixture compacted for both the energy are found be in the range of 52 to 197 kPa and 99 to 691 kPa respectively as shown in Fig. 8. As per the general requirement of isolation material such as liner, the compressive strength should have a value more than 200 kPa. The above strength is accomplished for the samples containing 15% or more bentonite and compacted with 2674 kJ/m^3 energy.

3.5 Shear Strength Parameter

Shear strength parameters such as unit cohesion (c) and internal friction angle (ϕ) are found to be in the range of 16.27 to 63.81 kPa and 42.5° to 28.4° for mixtures compacted with 595 kJ/m^3 energy, 12.67 to 89.59 kPa and 44.6° to 30.9° for the mixtures compacted with 2674 kJ/m^{3}energy when the bentonite content of the mixture is changed from 5 to 30%. Figures 9 and 10 present the variation of unit cohesion and angle of internal friction with bentonite content. It is seen that as the bentonite content increases, the cohesion increases and internal friction angle decreases. This may be due to the fact that as pond ash particles are irregular in shape, there exist a good interlocking between the particles hence the angle of internal friction is high. When a smaller amount of bentonite is added to the mixture, the finer particle of bentonite tries

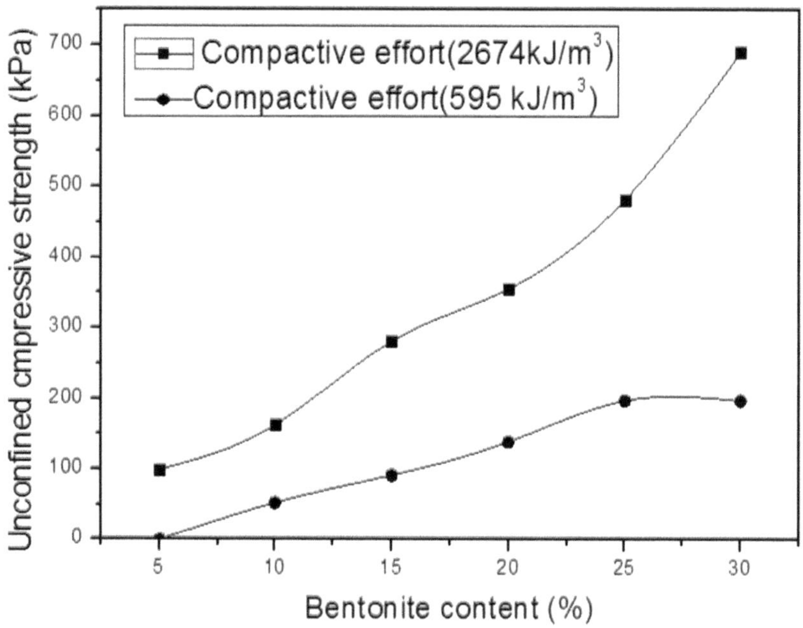

Fig. 8. Variations of unconfined compressive strength with bentonite content

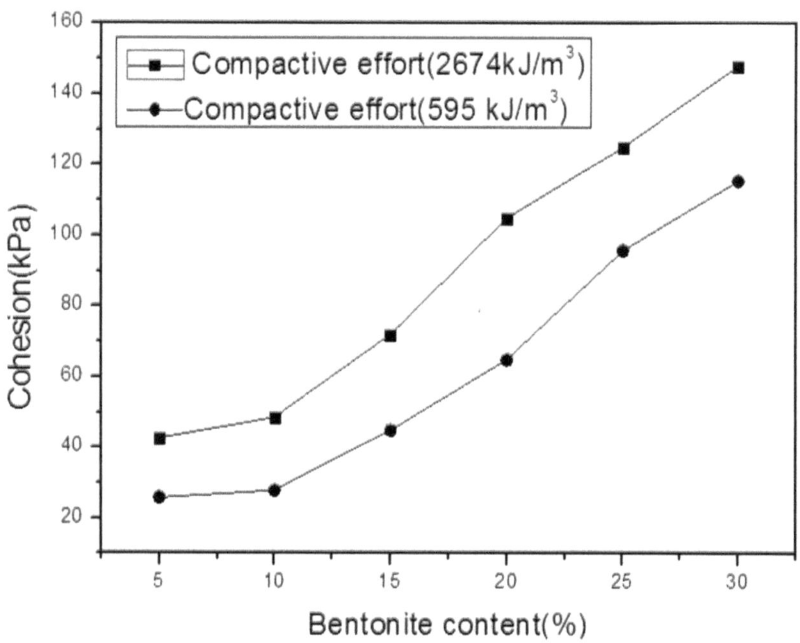

Fig. 9. Variations of cohesion with bentonite content

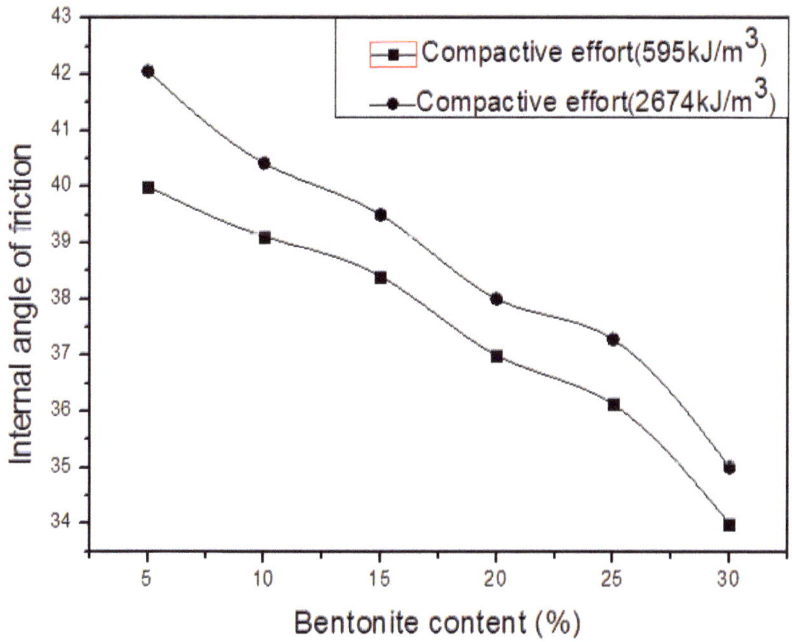

Fig. 10. Variations of internal angle of friction with bentonite content

to fill the intra-void spaces present in the conglomerated ash particles. But as the bentonite content increases, the excess bentonite coats the surfaces of the ash particle. As bentonite particles are lubricative in nature, the bentonite coated pond ash particles losses their contact and slips over each other and hence frictional angle decreases. However, due to the presence of strong inter-particular attractive forces between the bentonite particles the mixture gradually develops cohesion with an increase in bentonite content.

4 Conclusion

The suitability of pond ash-bentonite mixes as landfill liner material was investigated in this work. The compaction, strength, and hydraulic conductivity characteristics are evaluated. Based on the experimental investigation, the following conclusions are drawn:

- From the grain size distribution curve, it was inferred that the pond ash is a non-plastic material whereas bentonite is highly plastic in nature. It was also inferred that the percentages of fines content (<75 μm) present in pond ash and bentonite are 1.99 and 96.93 respectively. Both the LL and PL was found to increase significantly and on the other hand; differential swelling index value was increased from 34 to 87% with the increase in bentonite content.

- When the samples were compacted with compactive effort of 595 kJ/m^3 the maximum dry unit weight increases from 11.13 to13.14 kN/m^3 and the corresponding optimum water content of the compacted mixtures decreases from 32 to 22.5% as the bentonite content varies from 5 to 30%. However with the compactive effort of 2674 kJ/m^3, the maximum dry unit weight increases from 13.54 to 15.6 kN/m^3 and the corresponding optimum water content decreases from 23.75 to 16.5%
- The compressive strength of pond ash-bentonite mixture compacted for both the energy were found vary from 52 to 197 kPa and 99 to 691 kPa respectively.
- Shear strength parameters such as cohesion (c) and internal friction angle (ϕ) was found to be in the range of 16.27 to 63.81 kPa and 42.54° to 28.4° for mixtures compacted with compactive effort of 595 kJ/m^3. These values are 12.672 to 89.59 kPa and 44.55° to 30.90° for the mixtures compacted with compactive effort of 2674 kJ/m^3.

From this study, it can be concluded that sand bentonite mixture can effectively be replaced by pond ash bentonite mixture when sand is not easily available and wastes are to be utilized.

References

Akcanca, F., Aytekin, M.: Effect of wetting–drying cycles on swelling behavior of lime stabilized sand–bentonite mixtures. Environ. Earth Sci. **66**(1), 67–74 (2012). doi:10.1007/s12665-011-1207-5. Springer

Asrith, V., Viswanath, G.: Geotechnical strength of randomly oriented fiber reinforced compacted pond ashes. Int. J. Comput. Eng. Res. Trends **2**(12), 1056–1061 (2015)

Bose, B.: Geo engineering properties of expansive soil stabilized with fly ash. Electron. J. Geotech. Eng. **17**, 1339–1353 (2012)

Chalermyanont, T., Arrykul, S.: Compacted sand-bentonite mixtures for hydraulic containment liners. Songklanakarin J. Sci. Technol. **27**(2), 313–323 (2005)

Cho, W.J., Lee, J.O., Kang, C.H.: Hydraulic conductivity of compacted soil-bentonite mixture for a liner material in landfill facilities. Environmental engineering research-Seoul, Korean Society of Environmental Engineering (2002). doi:10.4491/eer.2002.7.3.121

Kaya, A., Durukan, S.: Utilization of bentonite-embedded zeolite as clay liner. Appl. Clay Sci. **25** (1), 83–91 (2004). doi:10.1016/j.clay.2004.10.001. Elsevier

Kayabali, K.: Engineering aspects of a novel landfill liner material: bentonite-amended natural zeolite. Eng. Geol. **46**, 105–114 (1997). doi:10.1016/S0013-7952(96)00102-0. Elsevier

Mir, B.A.: Effect of fly ash on swelling potential of BC soil. In: Proceedings of Indian Geotechnical Conference on Geotechnical Advances and Noval Geomechanical Applications, Roorkee, 1–9 December 2013

Mishra, A.K., Ohtsubo, M., Li, L.Y., Higashi, T.: Influence of the bentonite on the consolidation behaviour of soil–bentonite mixtures. Carbonates Evaporites **25**(1), 43–49 (2010). doi:10.1007/s13146-010-0006-525(1). Springer

Mollamahmutoğlu, M., Yilmaz, Y.: Potential use of fly ash and bentonite mixture as liner or cover at waste disposal areas. Environ. Geol. **40**(11–12), 1316–1324 (2001)

Satyanarayana, P.V.V.: Utilization of red soil bentonite mixes as clay liner materials. Int. J. Sci. Eng. Res. **4**(5), 876–882 (2013)

Shankar, M.U., Phanikumar, B.R.: Correlation studies on index properties of fly ash-stabilised expansive clay liners. Geomechanics and Geoengineering. Taylor & Francis (2012). doi:10.1080/17486025.2011.631036

Singh, S.P., Sharan, A.: Strength characteristics of compacted pond ash. Geomechanics and Geoengineering, Taylor & Francis (2014). doi:10.1080/17486025.2013.772661

U.S. Environmental Protection Agency. Requirements for Hazardous Waste Landfill Design, Construction, and Closure, Publication No. EPA-625/4–89-022, Cincinnati, Ohio (1989)

Feasibility Study on Usage of Metalized Plastic Waste in Concrete

Ankur Bhogayata[1(✉)] and N.K. Arora[2]

[1] Department of Civil Engineering, Marwadi Education Foundation Group of Institutions, Rajkot, Gujarat, India
ankur.bhogayata@gmail.com
[2] Lukdhirji Engineering College, Morbi, Gujarat, India
narendrakarora@gmail.com

Abstract. Polypropylene (PP) and polyethylene (PE) are the leading plastics used for food packaging with surface metallization to protect the foodstuff. Unlike the polyethylene terephthalate (PET) bottles, metalized plastic wastes (MPW) produced by discarded food packaging articles are not suitable for recycling and reuse owing to the metalized layer. MPW litter in waste streams or sent to the landfill as a part of municipal solid waste (MSW) resulting in hazardous environmental impacts. Efforts are made in this work to utilize MPW in concrete as a sustainable usage of the specific waste. In present work, MPW is shredded into fibers of varying sizes and added to concrete ranging 0% to 2% by volume of concrete. The response of addition of MPW in concrete was evaluated by assessing workability and strength properties namely slump, compaction factor, compressive strength, splitting tensile strength and flexure strength. The results show significantly improved crack resistance and ductility in concrete reinforced with MPW. The increase in tensile resistance of concrete containing MPW up to 1% of volume fraction was observed. Flexure test results exhibited improved ductility of concrete containing MPW. However, reduction in the fresh properties and compressive and flexure strength of concrete was noticed with the increased fraction of MPW beyond 1% dosage. The experimental study establishes the potential of the sustainable use of MPW in concrete to reduce adverse environmental impacts and to improve the tensile resistance of the conventional concrete.

Keywords: Metalized plastic waste · Concrete · Workability · Compressive strength · Splitting tensile strength

1 Introduction

The annual generation of plastic waste has reached beyond 300 million tons by the year 2015 (Plastics-the facts 2015). Plastic being the most durable material it has gained popularity in past few decades with changing modern lifestyles. Most of the plastics are recyclable and being reused in many forms except a few namely metalized plastics. Polyethylene (PE) and Polypropylene (PP) are the plastics used in food packaging with additional metalized layer called metalized plastic (A report: American Plastic Council 1997) making them unfit for recycling. Metalized plastic waste, therefore, has become

D.N. Singh and A. Galaa (eds.), *Contemporary Issues in Geoenvironmental Engineering*, Sustainable Civil Infrastructures,
DOI 10.1007/978-3-319-61612-4_27

one of the challenges raised by the ever-increasing post-consumer activities for a safe disposal. Major concerns of metalized plastic wastes are the littering issues and prolonged existence in the environments.

The construction industry has initiated the usage of post-consumer plastic waste in past few years. The area of research has attracted the attention of researchers and plastic wastes generated by post consumer activities namely polyethylene terephthalate (PET) bottle wastes and plastic carry bags have been used in concrete as aggregates and fibers. Partial replacement of natural aggregates by recycled aggregates has shown a balanced influence on concrete properties. Properties namely density, workability are influenced due to the addition of plastic aggregates (Rahmani et al. 2013). Hardened properties of concrete are tested for obtaining the response of addition of aggregates. It is observed that modulus of elasticity; density and strengths properties namely compressive strength, splitting tensile strength are reduced for a higher dosage of replacement of natural aggregates by PET aggregates (Saiki and de Brito 2012). However, the addition of PET aggregates improved the flexural response of concrete (Ferreira et al. 2012).

Recycled plastics are incorporated as fibers in concrete by other researchers. Most of the researchers have used PET bottles to retrieve fibers. The addition of plastic fibers influenced workability of concrete and showed a reduction in higher dosage (Pacheco-Torgal et al. 2012). Recycled plastic fibers affected the properties namely shrinkage, drying shrinkage, slump up to an extent and showed improvement for a given dosage (Yin et al. 2015; Borg et al. 2016). Hardened properties namely compressive strength, splitting tensile strength, flexural strength was improved due to the addition of recycled PET waste fibers (Foti 2011, 2013; Yin et al. 2016).

The deformation characteristics of concrete were improved and the addition of PET fibers improved the ductility of concrete (Kim et al. 2010; Fantilli et al. 2016). PET fibers are utilized in self-compacting concrete (SCC) also. It is studied that the fresh properties of SCC at a given dosage of PET fibers exhibits the acceptable response of concrete. Hardened properties of SCC were improved by the use of recycled PET fibers (Mazaheripour et al. 2011).

The literature review provided extended information on the usage of recycled plastic waste in concrete. However, metalized plastic wastes have not been utilized in the desired quantum of research. The paper, therefore, focuses on a feasibility study of metalized waste plastic (MPW) in concrete as a constituent. Fresh and strength properties of concrete are investigated to obtain the influence of MPW.

2 Experimental Program

Materials for Concrete Mixes

Ordinary Portland Cement (OPC grade 53) manufactured by the Ultratech cement company from Gujarat having a specific gravity of 3.15 g/cm^3 was used. Locally available aggregates around the Rajkot city area of the average sizes of 20 mm and 10 mm were used. River sand from the banks of Aaji River near Rajkot was used as fine aggregates. The specific gravity of coarse and fine aggregates was 2.71 and 2.63

with 0.5% and 0.78% water absorption respectively. The fineness modulus of coarse aggregates was 4.76 and fine aggregates were 2.99.

Metalized Plastic Waste

Metalized plastic used by food packaging industries was obtained from Shri Umiya Plastics – a plastic packaging industrial unit at Shapar-Veraval industrial area near Rajkot city. Metalized plastic film was shredded into fibers and flake from as shown in Fig. 1. Two types of plastics namely type A and Type B were designated as per the size of 1 mm and 5 mm of average dimensions respectively. The fibers were mixed in varying fractions from 0% to 2% by volume of concrete mix. Concrete containing 0% of MPWF was considered as reference concrete. General properties of metalized plastics are shown in Table 1.

(a) Type A fibers (b) Type B fibers

Fig. 1. MPW fibers in shredded form

Table 1. General properties of MPW

Property	Values	Unit
Resin category	Polythene	–
Plastic type	LDPE	–
Recycling code	4	–
Density range	0.94-1.4	gm/cm^3
Thickness	0.08	mm
Water vapor resistance	Good	–
Oxygen permeability	High	–
Elongation	10–30	%
Co efficient of friction	0.45 – 0.55	–

Mixture Proportions and Specimens

Total 10 mixes were prepared including reference concrete mix at 0.45 water cement ratio. Detailed mixture proportions and quantities of MPW are shown in Tables 2 and 3 respectively. Concrete mixes were prepared from references suggesting the inclusion of macro recycled plastic fibers. All ingredients were first mixed in a mixer in dry conditions first. At last, the MPW were mixed in the dry mix. Water was added after few

Table 2. Mix proportions

Sr. no.	Mixture	W/C ratio	Cement	Aggregates 20 mm	Aggregates 10 mm	Sand	Water
			kg	kg	kg	kg	kg
1	Mix-1	0.45	438	669	446	638	197

Table 3. Mixes and MPW proportions

Sr. no.	MPW type	Mix designation	MPW details			
		Mix-1 w/c: 0.45	Volume fraction	Weight kg	Size	Thickness mm
1	Type-A	B1	0%	–	–	–
		B2	0.5%	4.5	1 mm average (W*L)	0.08
		B3	1%	9.2		
		B4	1.5%	14		
		B5	2%	18.5		
2	Type-B	B6	0%	–	–	–
		B7	0.5%	4.5	5 mm average (W*L)	0.08
		B8	1%	9.2		
		B9	1.5%	14		
		B10	2%	18.5		

initial revolutions of the mixer to allow MPW settlement in the mix. The MPW was utilized as one of the constituents in the conventional mix. It was observed that they readily get settled in the mixture.

Specimens namely cubes (150 mm*150 mm*150 mm), and cylinders (150 mm*300 mm) were prepared to obtain compressive strength and splitting tensile strength respectively.

3 Tests on Concrete

Workability Tests

Slump and compaction factor tests were employed on the fresh mix to obtain the workability of concrete. The tests are commonly used to determine the consistency of fresh concrete. As shown in Fig. 2, standard apparatus for slump and compaction factor were used. The results obtained were compared with the standard values suggested in the literature for comparative study.

Strength Tests

Cubes and cylinder specimens were tested to determine compression and splitting tensile strength respectively. All specimens were cured for 28 days. The experimental arrangements are shown in Fig. 3. Average of three specimens test results were regarded as the strength of the concrete.

(a) Slump test (b) Compaction factor test

Fig. 2. Workability tests

The tests were carried out with CTM machine of 2000kN loading capacity and the rate of loading was maintained at 0.5 MPa/s cubes were subjected to axial compression until a final failure of the specimen. cylinders were conditioned for the splitting tensile strength by mounting in CTM machine with necessary attachments like timber planks

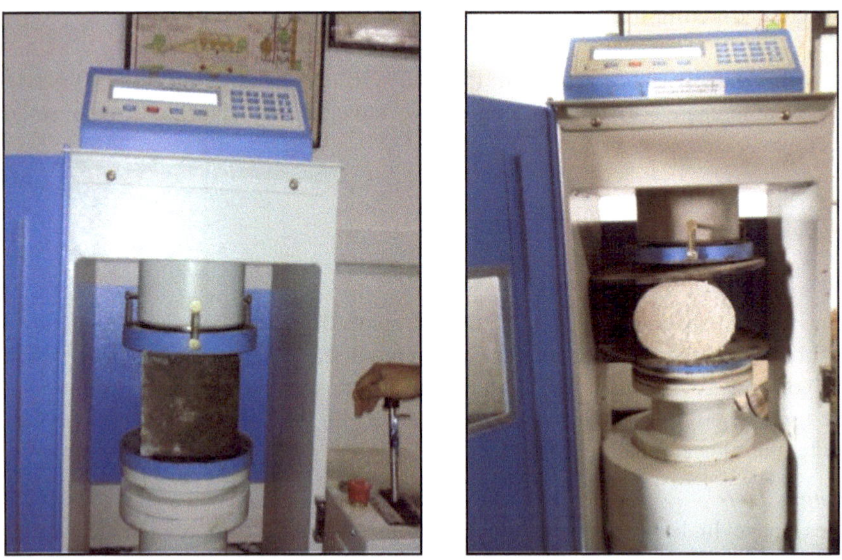

(a) Compression test (b) Split tension test

Fig. 3. Strength test apparatus

and metal bars to achieve effective splitting action in the specimens. The tests results were compared with the results of reference concrete for analysis purpose.

4 Results and Discussion

Workability of Concrete

Slump and compaction factor test results are shown in Figs. 4 and 5 respectively. It was observed that addition of MPW in concrete reduced workability. Different works carried out on the use of waste plastic fibers have shown a similar response (Yin et al. 2015). The inclusion of plastic fibers tends to reduce the segregation required to maintain the flow of concrete. The presence of MPW improved the viscosity of mix and obstructed the relative movements of sand and aggregate particles as observed during the mixing and testing of concrete.

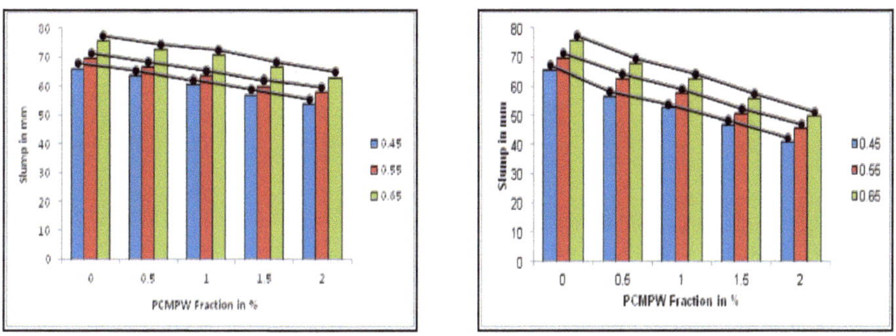

(a) Test results for type A fibers (b) Test results for type B fibers

Fig. 4. Slump test results

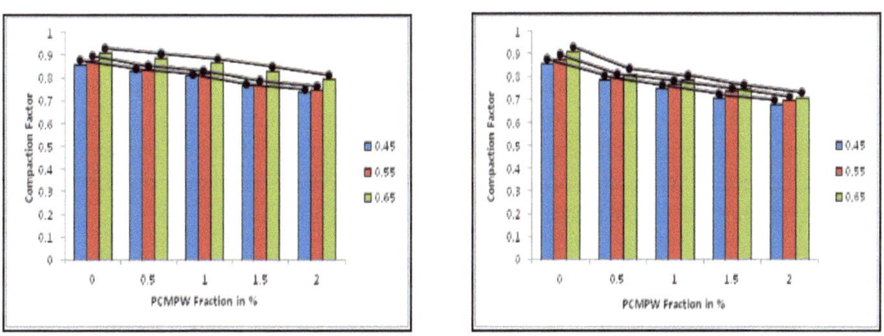

(a) Test results Type A fibers (b) Test results Type B fibers

Fig. 5. Compaction factor test

Type A fibers reduced the slump in the range of 3%, 7%, 12% and 18% for the range of dosage of 0.5% to 2% respectively. Type B fibers with larger dimensions showed a more significant reduction in workability. The fresh properties namely slump were reduced in the range of 5% to 25% and compaction factor reduced from 3% to 15% for the dosage range of 0.5% to 2% for Type B fibers of MPW. It was noticed that the response of freshly mixed concrete was influenced by fraction and type of MPW simultaneously.

Compressive Strength

Figure 6 shows test results of compressive strength of concrete reinforced with varying size and fractions of MPW fibers. It is reported by different researchers that compressive strength is less influenced by the addition of macro plastic fibers (Kim et al. 2010). However, the test results in this study exhibited a reduction in strength at a higher dosage of MPW.

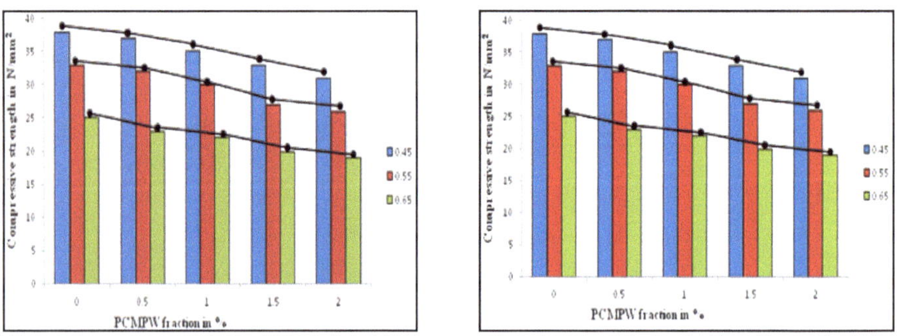

(a) Test results Type A

(b) Test results Type B

Fig. 6. Compressive strength results

Type A and Type B showed a nearly similar response in compressive strength and showed a negligible reduction for a small dosage of MPW till 1% by volume of a mix. However, the reduction was dominant beyond the 1% dosage and the strength reduced up to 15% and 25% at 1.5% and 2% dosage of MPW. Irrespective of the type of MPW, compressive strength reduced with increased dosage of MPW. However, the failure patterns showed a uniform and controlled crack patterns at the surface at initial loading stages. Brittle failure of concrete was observed in the case of cubes prepared by the reference concrete, while the presence of fibers of MPW provided ductility to the material and showed a less brittle failure when mixed in concrete.

Splitting Tensile Strength

The indirect tensile strength of concrete was obtained by performing splitting tensile strength tests on cylinders of concrete containing varying types and fractions of MPW fibers. The addition of MPW improved the splitting tensile strength of concrete for all test conditions. Type A showed a moderate rate of increase of strength compared to Type B fibers for a given fraction.

Tensile strength was improved from 5%, 9%, 14% and 18% for fractions values from 0.5% to 2% in case of Type A fibers. Similarly, improvement of strength from 8%, 15%, 22% and 27% was observed in the case of Type B fibers for the full range of fractions of MPW. The test results are shown in Fig. 7.

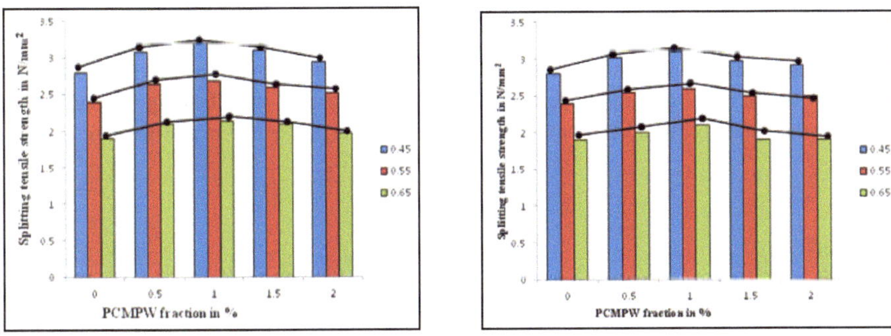

(a) Test results Type A

(b) Test results Type B

Fig. 7. Splitting tensile strength results

During the tests on cylinders, it was observed that the cracks developed at initial loading were similar for both types of fibers. However, progressively at higher loading Type B fibers showed better resistance against the splitting action in concrete. Type B fibers tend to hold the cracked concrete with better grip and interrupted the development of macrocracks with higher fractions. The randomly distributed fibers were capable of providing the ductility to the cement paste by getting anchored effectively in the hardened cement paste. Researchers have confirmed similar response by macro-sized plastic fibers (Yin et al. 2015). The availability of macro sized fibers locally entrapped the development of micro-cracks those may result in large sized cracks eventually. A similar response was observed at the surface of the specimens subjected to splitting action.

5 Conclusion

Experimental investigations on concrete containing MPW fibers provided following conclusions,

- The addition of MPW fibers in concrete improves splitting tensile strength of concrete significantly.
- Workability of fresh concrete was adversely affected by MPW fibers and slump and compaction factors reduced with increased fractions of MPW.
- Compressive strength was less influenced by MPW at a lower dosage up to 1% by volume.

- The experimental work ensured the potential of usage of metalized plastic waste generated by different post consumer activities with 1% volume fraction as optimum dosage.
- The present work encouraged the possible usage of metalized plastic waste as an alternate way of disposal and dual advantage of improvement of effective crack resistance of concrete subjected to tensile forces.

Acknowledgements. The authors are thankful to Umiya plastics, Rajkot for providing a friendly supply of metalized plastic and to Marwadi Education foundation for extending the facilities of laboratory and funding for all the raw materials required for the experimental work.

References

PLASTICS-THE FACTS 2015: An Analysis of European Plastics Production, Demand and Waste Data. A report by plastics Europe, Association of plastic manufacturers. www.plasticseurope.org

Understanding Plastic Films: It's Uses, Benefits and Waste Management Options. A report prepared by Headly Pratt Consulting for American Plastic Council (1997)

Rahmani, E., Dehestani, M., Beygi, M.H.A., Allahyari, H., Nikbin, I.M.: On the mechanical properties of concrete containing waste PET particles. Constr. Buil. Mater. **47**, 1302–1308 (2013). doi:10.1016/j.conbuildmat.2013.06.041

Saikia, N., de Brito, J.: Use of plastic waste as aggregate in cement mortar and concrete preparation: a review. Constr. Build. Mater. **34**, 385–401 (2012). doi:10.1016/j.conbuildmat.2012.02.066

Ferreira, L., de Brito, J., Saikia, N.: Influence of curing conditions on the mechanical performance of concrete containing recycled plastic aggregate. Constr. Build. Mater. **36**, 196–204 (2012). doi:10.1016/j.conbuildmat.2012.02.098

Pacheco-Torgal, F., Ding, Y., Jalali, S.: Properties and durability of concrete containing polymeric wastes (tire rubber and polyethylene terephthalate bottles): an overview. Constr. Build. Mater. **30**, 714–724 (2012). doi:10.1016/j.conbuildmat.2011.11.047

Yin, S., Tuladhar, R., Shi, F., Combe, M., Collister, T., Sivakugan, N.: Use of macro plastic fibers in concrete: a review. Constr. Build. Mater. **93**, 180–188 (2015). doi:10.1016/j.conbuildmat.2015.05.105

Borg, R.P., Baldacchino, O., Ferrara, L.: Early age performance and mechanical characteristics of recycled PET fiber reinforced concrete. Constr. Build. Mater. **108**, 29–47 (2016). doi:10.1016/j.conbuildmat.2016.01.029

Foti, D.: Preliminary analysis of concrete reinforced with waste bottles PET fibers. Constr. Build. Mater. **25**(4), 1906–1915 (2011). doi:10.1016/j.conbuildmat.2010.11.066

Foti, D.: Use of recycled waste pet bottles fibers for the reinforcement of concrete. Compos. Struct. **96**, 396–404 (2013). doi:10.1016/j.compstruct.2012.09.019

Yin, S., Tuladhar, R., Riella, J., Chung, D., Collister, T., Combe, M., Sivakugan, N.: Comparative evaluation of virgin and recycled polypropylene fiber reinforced concrete. Constr. Build. Mater. **114**, 134–141 (2016). doi:10.1016/j.conbuildmat.2016.03.162

Kim, S.B., Yi, N.H., Kim, H.Y., Kim, J.H.J., Song, Y.C.: Material and structural performance evaluation of recycled PET fiber reinforced concrete. Cement Concr. Compos. **32**(3), 232–240 (2010). doi:10.1016/j.cemconcomp.2009.11.002

Fantilli, A.P., Chiaia, B., Gorino, A.: Fiber volume fraction and ductility index of concrete beams. Cement Concr. Compos. **65**, 139–149 (2016). doi:10.1016/j.cemconcomp.2015.10.019

Mazaheripour, H., Ghanbarpour, S., Mirmoradi, S.H., Hosseinpour, I.: The effect of polypropylene fibers on the properties of fresh and hardened lightweight self-compacting concrete. Constr. Build. Mater. **25**(1), 351–358 (2011). doi:10.1016/j.conbuildmat.2010.06.018

Stabilization of Red Mud Using Ground Granulated Blast Furnace Slag by Geopolymerization Technique

S.P. Singh(✉), S. Samantasinghar, and D. Sindhuja

Department of Civil Engineering, National Institute of Technology Rourkala, Rourkela, India
spsingh@nitrkl.ac.in, shree.singhar@gmail.com, sindhuja.dl16@gmail.com

Abstract. Red mud is an alkaline by-product which is generated during the processing of bauxite. Generally red mud is considered as a waste material which is disposed into large lagoons. Red mud contains large amount of Fe_2O_3, Al_2O_3 and small amounts of CaO, SiO_2 and some valuable metals such as titanium however it devoid of reactive alumino-silicates. As red mud is highly alkaline, it is possible to stabilize red mud by adding industrial waste materials rich in alumino-silicates by geopolymerization. The main principle involved in geopolymerization is the chemical reaction between alumino-silicate materials with highly alkaline solution to form amorphous to semi crystalline inorganic polymers. In the present research work an attempt has been made to stabilize red mud using ground granulated blast furnace slag. The amount of slag is varied as 0%, 10%, 20%, 40%, 60% and 80% by weight of the red mud-slag mixture. For each combination of red mud-slag the Atterberg's limits, pH value, differential free swell value, compaction characteristics corresponding to light compaction and heavy compaction energies and unconfined compressive strength (UCS) are determined. Further additional doses of sodium hydroxide (NaOH) as an alkali activator varying as 2%, 4%, 8%, 12% of dry mass of the above red mud-slag mixes were given to accelerate the geopolymerization process. The UCS values of specimens were determined after curing periods of 0, 3, 15, 30 and 60 days. Also, an addition 4% NaOH with 40% slag gave a compressive strength of 14 MPa after a curing period of 60 days. Hence it can be concluded that red mud can be stabilized effectively by the addition of little amount of alkali and ground granulated blast furnace slag.

1 Introduction

Rapid industrialization and infrastructure development produces enormous waste materials which in turn require large area and safe disposal. During past decades more concern is shown towards the utilization of industrial wastes and by-products in Civil engineering works. Utilizing the waste is a partial solution to ecological and environmental problems. If the industrial wastes are used in production of cement, concrete and in some other construction materials as a replacement to conventional materials then the manufacturing/project cost will be reduced and reduces the land fill area.

D.N. Singh and A. Galaa (eds.), *Contemporary Issues in Geoenvironmental Engineering*, Sustainable Civil Infrastructures,
DOI 10.1007/978-3-319-61612-4_28

Geopolymerization technique is an alternative in which an alkali activated alumino-silicate material is used as the binder. The major processes involve the reaction between an alumino-silicate source such as fly ash, metakaolin or blast furnace slag and an alkaline solution which leads to final hardening of the matrix by exclusion of excess water and the growth of an inorganic polymer. India has huge reserve of bauxite ore. Red mud is a solid waste generated from aluminium industry in the process of alumina production by Bayer's process. For every ton of alumina one tonne of red mud produced. Due to its alkaline nature, it may cause serious environmental issues. Red mud contains large amount of oxides of iron and aluminium; and small amount of oxides of calcium and silicon; and some valuable metals such as titanium however it devoid of reactive alumino-silicate materials. The bulk utilization of red mud includes neutralization of acidic soils, construction of road/railway embankments, brick and cement manufacturing etc. However, its use in construction industries is very limited. Blast furnace slag is formed from the manufacture of pig iron. It mainly contains silicates, aluminosilicates and calcium. So, addition of blast furnace slag to red mud may have substantial influence on its index and engineering properties. Furthermore, addition of small amount of alkali may accelerate the geopolymerization process thus improves the engineering properties.

Researches have been made to stabilize red mud either by addition of admixtures or by suitable chemicals. He et al. (2013) has developed a new kind of geopolymer from red mud with rice husk ash and compressive strength ranging from 3.2 to 20.5 MPa was obtained. Hajjaji et al. (2013) has developed a new geopolymer from metakaolin and red mud by sodium silicate solution as alkali activator and concluded that lesser amount of red mud with longer curing period gave more compressive strength. Rai et al. (2013) studied feasibility of reducing the alkaline nature of red mud by sintering using fly ash as an additive via Taguchi methodology. A pH of 8.9 was obtained at 25–50% of red mud and 50–75% fly ash with water and temperature of 1100 °C. Deelwal et al. (2014) studied the effectiveness of red mud stabilized by lime with gypsum. The results showed that 12% lime and 1% gypsum gave more UCS value at 7 days curing. Singh et al. (2014) stated addition of cement kiln dust (CKD) up to 8% has shown higher strength further addition of CKD does not play any vital role in increasing the strength of red mud-CKD mix. Zhang et al. (2014) derived geopolymers from red mud and class F fly ash and found 28 days unconfined compressive strength of about 11.3 to 21.3 MPa. Kaya and Uzun (2016) investigated the behavior of red mud and metakaolin based geopolymers. Red mud is added in the range of 0 to 40% by weight of red mud-metakaolin mixture. 51.5 MPa was the maximum compressive strength for metakaolin based geopolymer.

2 Materials and Methods

2.1 Materials

The raw materials used in this investigation are red mud (RM), granulated blast furnace slag (GBFS) and sodium hydroxide. Red mud was collected from NALCO, Damanjodi, India. The red mud was air dried, homogenized and pulverized to powder in order to provide better geopolymerization and minimizes the compositional error in the final product. The granulated blast furnace slag was collected from slag granulation plant of the Rourkela Steel Plant (RSP), Sundargarh, India. The waste materials were dried in oven to remove the water present in raw material. A NaOH flake with 98% purity was obtained from Loba Chemie, Mumbai, India.

The grain size distribution curves from Fig. 1 shows that the red mud is well graded whereas the granulated blast furnace slag contains sand sized particles, so it was ground in a ball mill to a Blaine's fineness of 410 m^2/kg to obtain ground granulated blast-furnace slag (GGBFS) for better reactivity. The index and engineering properties for both RM and GGBFS are tabulated in Table 1. The chemical compositions of raw materials were examined by X-ray fluorescence (XRF) analysis and presented in Table 2. The microstructure of the raw materials was studied by Scanning electron microscopy (SEM) and the SEM images are shown in Fig. 2. The SEM image for red mud shows the particles are angular in shape. The microstructure for GGBFS reveals that most of the particles are rough and angular shaped. The chemical compounds present in the raw materials are investigated by X-ray diffraction (XRD) method and the XRD patterns are shown in Fig. 3. The red mud majorly composed of iron oxide, alumina oxide, and silicon dioxide. From the XRD test result of slag it is observed that the slag is glassy material.

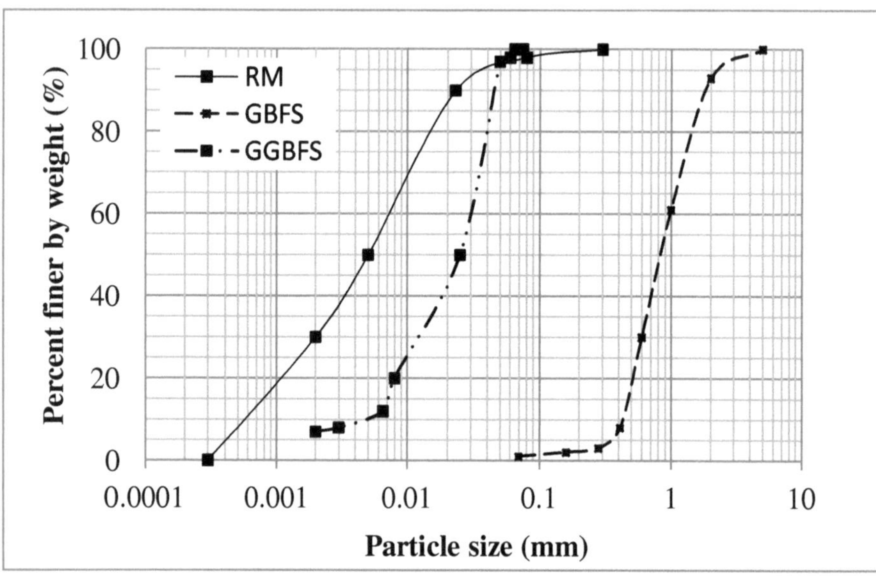

Fig. 1. Grain size distribution for raw materials

2.2 Methods

The present work is divided into two phases. In the first phase, red mud is stabilized with GGBFS of 10, 20, 40, 60 and 80% by dry weight of the total mixture. For all these

Table 1. Index and engineering properties of raw materials

Properties	RM	GGBFS
pH	10.43	10.24
Specific gravity	3.3	2.84
Liquid limit (%)	29	Non-plastic
Plastic limit (%)	23	
Coefficient of uniformity	2.45	1
Coefficient of curvature	20	6
Optimum moisture content (%)	23	19
Maximum dry density (kN/m^3)	17.88	16.9
Unconfined compressive strength (MPa)	0.22	0.09

Table 2. Chemical compositions of raw materials

Composition (%)	RM	GGBFS
MgO	0.020	9.52
Al_2O_3	16.07	21.06
SiO_2	8.25	30.82
K_2O	–	1.04
CaO	1.48	32.02
Fe_2O_3	53.75	1.37
Na_2O	3.82	0.088
MnO	0.157	0.14
TiO_2	4.24	1.04
SO_3	–	0.66
V_2O_5	0.148	–
Loss on ignition	11.83	1.81

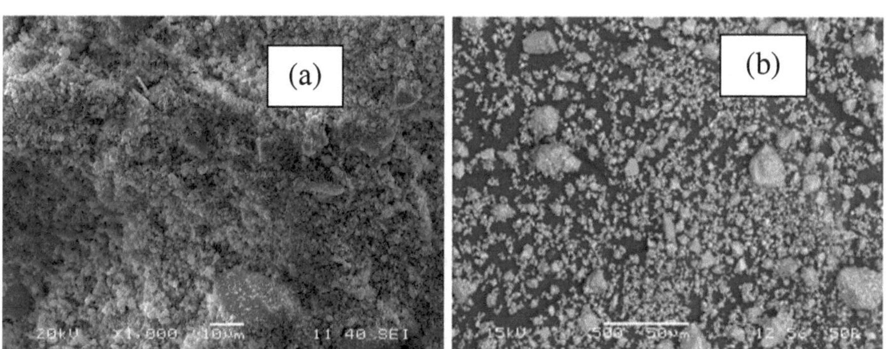

Fig. 2. SEM images for (a) RM and (b) GGBFS

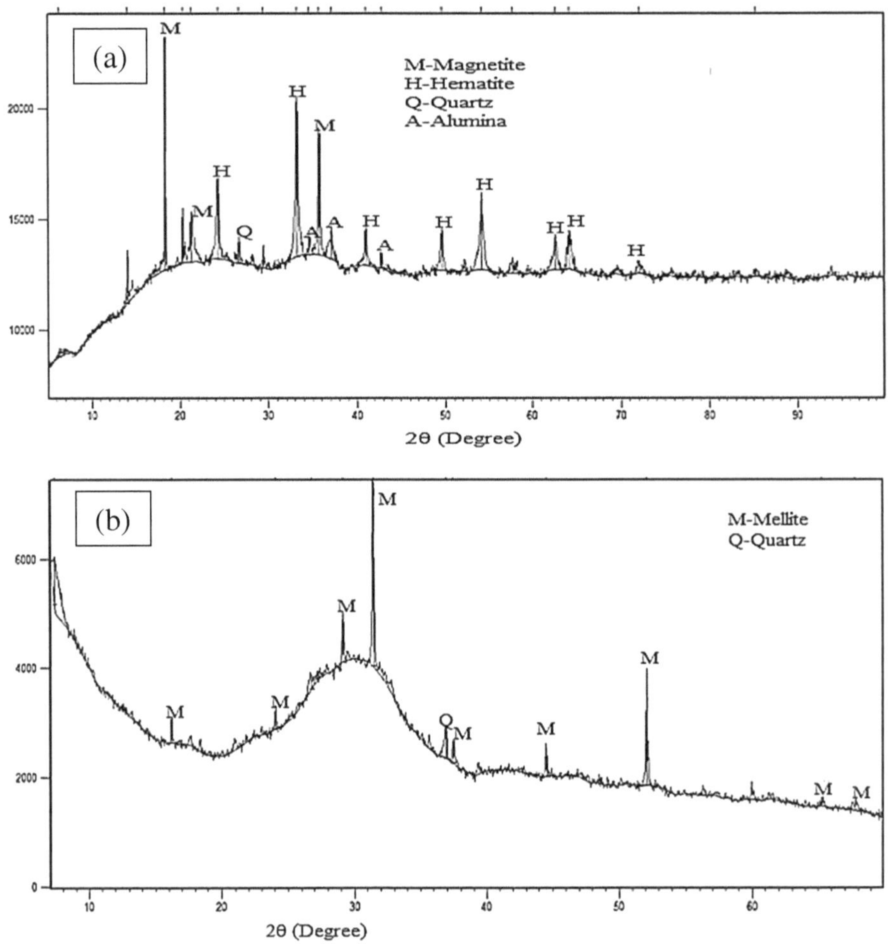

Fig. 3. XRD patterns for (a) RM and (b) GGBFS

proportions index properties such as Atterberg's limits (liquid limit, plastic limit and shrinkage limit), differential free swell (DFS) value and pH were determined. Furthermore, engineering properties like optimum moisture content (OMC) and maximum dry density (MDD) values both for heavy and light compaction energy (compaction effort of 2674 kJ/m^3 and 595 kJ/m^3 respectively) and unconfined compressive strength for different curing periods of 0, 3, 15, 30 and 60 days were determined. The second phase of work includes additional alkali to accelerate the geopolymerization process. To study this, the unconfined compressive strength of red mud-slag mixtures activated with different dosage of sodium hydroxide was determined. NaOH is taken in the percentages of 2, 4, 8, and 12% by dry mass of red mud-slag mixture. NaOH solutions were prepared by adding the required amount of NaOH to the water obtained as OMC for the red mud-slag mixture. The NaOH solutions were prepared 1 day prior to the sample preparation. Liquid limit test is done using Cassagrande apparatus, whereas the plastic limit is obtained by the thread rolling method as per IS: 2720 (Part 5) - 1985. The shrinkage limit has been determined conforming IS: 2720 (Part 6) - 1992. The free swell index is determined following IS: 2720(part XL) - 1977. The Optimum moisture content and dry density were determined by performing the as per IS: 2720 (part VII) - 1965. The unconfined compressive strength is determined as per IS: 2720 (part-10) - 1991. Also, the tests for stabilized material were done referring IS: 4332.

3 Results and Discussion

3.1 pH Value

From Fig. 4, it is observed that pH values are decreasing with the replacement of red mud with slag in red mud-slag mixes. The pH value decreased from 10.43 to 10.24 with the increase of slag content. Red mud contains residual sodium hydroxide and slag contains more calcium oxide, but red mud is more alkaline compared to slag which is responsible for decreasing pattern of pH values with the increase of slag percentage. Also, the fusion of free soda in red mud with silica and alumina in slag may cause the decreased alkalinity. Similar observation has been stated by Rai et al. (2013) where red mud is stabilized with fly ash.

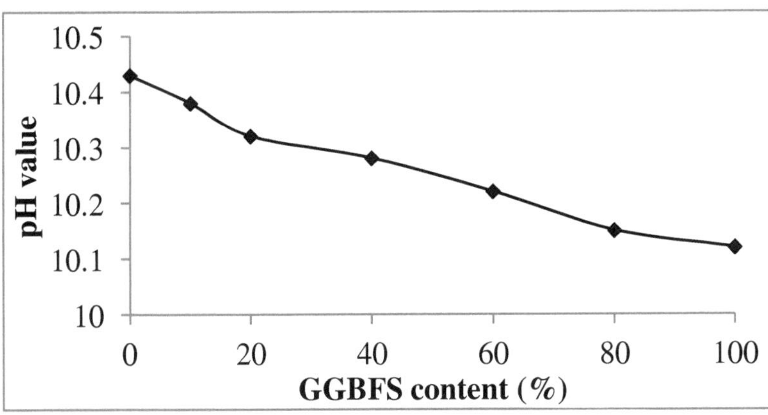

Fig. 4. Variation of pH value with slag content

3.2 Atterberg's Limits

From the plasticity characteristics of red mud and GGBFS (Table 1), it is seen that red mud possesses low plasticity with its liquid limit of 29% (<35%), whereas slag is a non-plastic material. For each combination of red mud-slag, liquid limit, plastic limit and shrinkage limit are illustrated in Fig. 5. As the amount of slag in red mud is increased the liquid limit, plastic limit and shrinkage limit reduces. This is due to increase in the percentage of non-plastic material in the mixture. Deelwal et al. (2014)

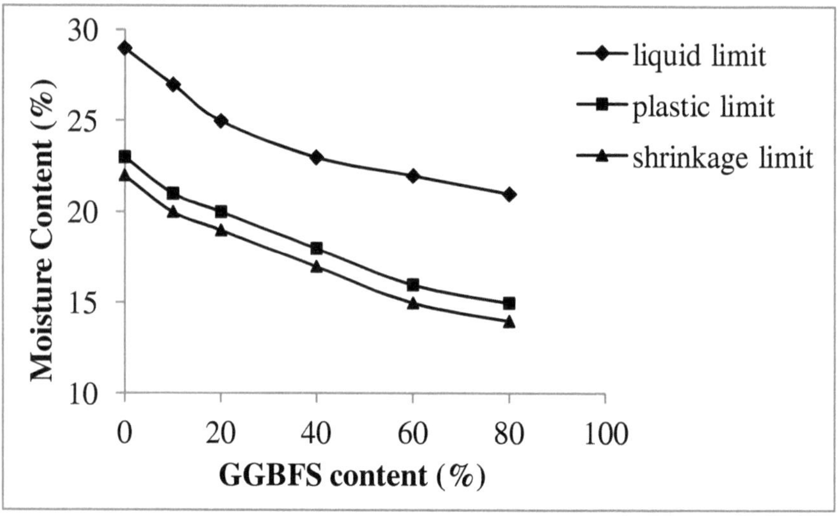

Fig. 5. Variation of Atterberg's limits with slag content

has reported that the liquid limit and plasticity index of red mud is decreased consistently when lime is added.

3.3 Differential Free Swell

Figure 6 shows the variation of DFS values with the slag content. Red mud, slag, and their blends exhibit negative values of differential free swelling. This type of negative DFS values are reported for materials containing pure kaolinite minerals, rock powders,

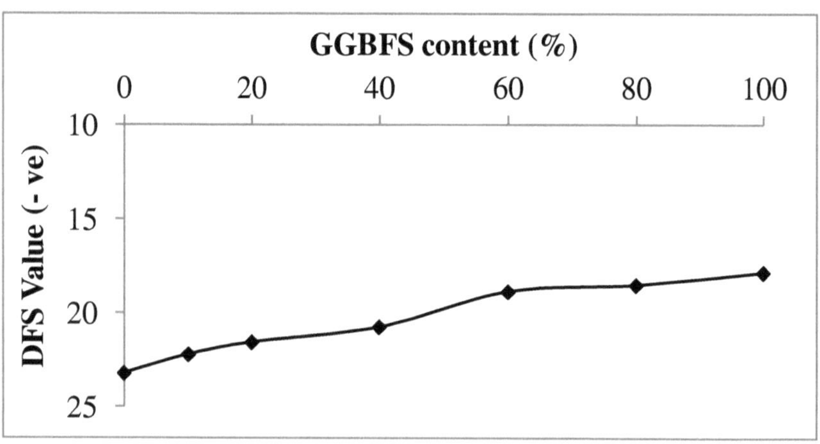

Fig. 6. Variation of DFS value with slag content

fly ash as these materials occupy high sediment volume in non-polar solvents like kerosene than in water. In the present case it is observed that the addition of slag to red mud reduced the negative DFS value. This indicates the inter particle attraction in the materials is non-existent and the material is dispersive in nature.

3.4 Compaction Characteristics

The values OMC and MDD were determined for each combination of red mud and slag compacted with light and heavy compaction energies. The variations of MDD and OMC values with slag content are shown in Fig. 7. In red mud-slag mixture, as the percentage of slag added to red mud is increased, the MDD increases up to 40% slag

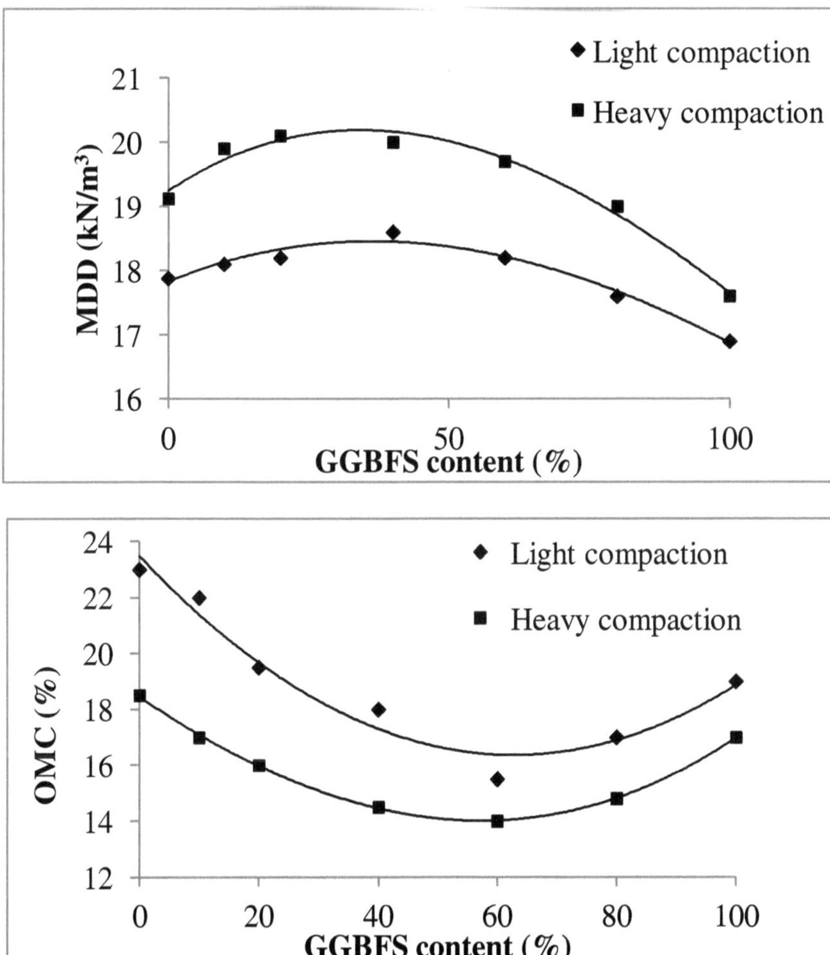

Fig. 7. (a). Variation in MDD with slag content, (b). Variation in OMC with slag content

and further addition of slag resulted in decreased MDD value. Both red mud and slag possess higher specific gravity than the conventional earth material. Change in MDD value of the mixture mainly depends on the specific gravity of the constituent materials and gradation of the mix. As a well graded material possess a high MDD value compared to poorly graded material, observed changes in MDD values may be attributed to the above reason. In red mud-slag mixture (40% slag content) the MDD value obtained in heavy compaction test is 1.07 times than that of light compaction test. The OMC for red mud is higher than that of slag. When slag is added upto 60% to red mud, the OMC reduced but thereafter with increased slag content, the OMC value increased again. The reduction in OMC is may be due to the reduction in adsorbed water by red mud when replaced by slag.

3.5 Unconfined Compressive Strength

The unconfined compressive strength (UCS) for all the RM-GGBFS mixtures was obtained for different curing periods without and with addition of the chemical activator that is the NaOH solution.

Stabilization Without NaOH Activation

The variation in compressive strength of specimens with different GGBFS contents and cured for different periods are shown in Fig. 8. For 60% red mud and 40% slag, maximum strength of 5.2 MPa is obtained. The strength increment is due to the formation of calcium silicate hydrate (CSH) gel as the GGBFS has a high amount of calcium which gets activated by the alkali content of the red mud. However, a decreased strength is observed with further addition of slag which might be explained by the fact that the amount of sodium hydroxide present in red mud became insufficient to react with alumino silicates present in slag. For all red mud-slag mixes, compressive

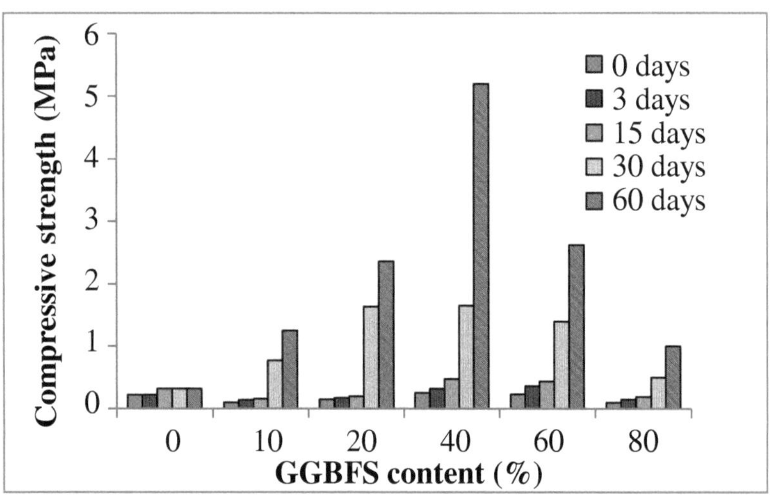

Fig. 8. Variation in UCS value with slag content for different curing periods

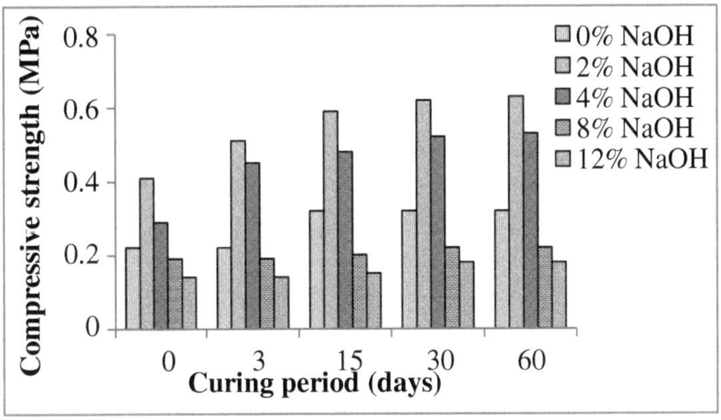

(a) 100% RM + 0% GGBFS

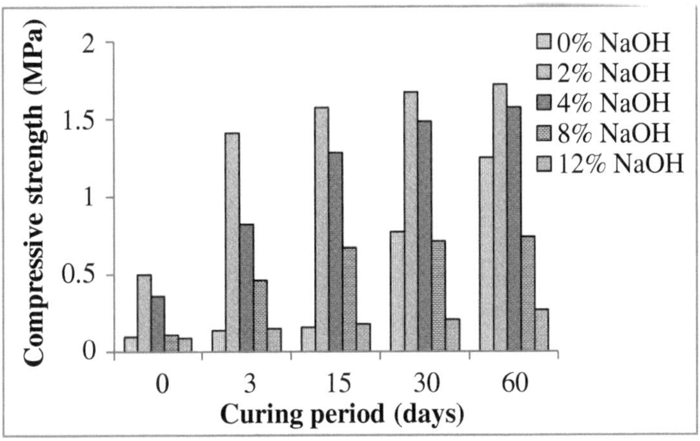

(b) 90% RM + 10% GGBFS

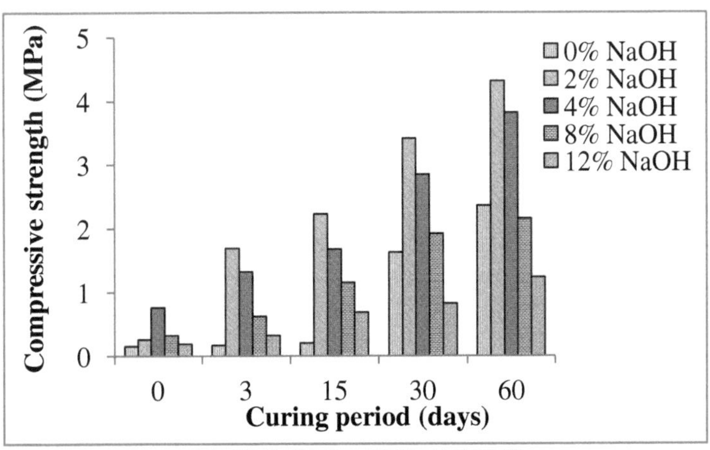

(c) 80% RM + 20% GGBFS

Fig. 9. Variation in UCS value with alkali content for different curing periods (a) 100% RM + 0% GGBFS, (b) 90% RM + 10% GGBFS, (c) 80% RM + 20% GGBFS, (d) 60% RM + 40% GGBFS, (e) 40% RM + 60% GGBFS, (f) 20% RM + 80% GGBFS

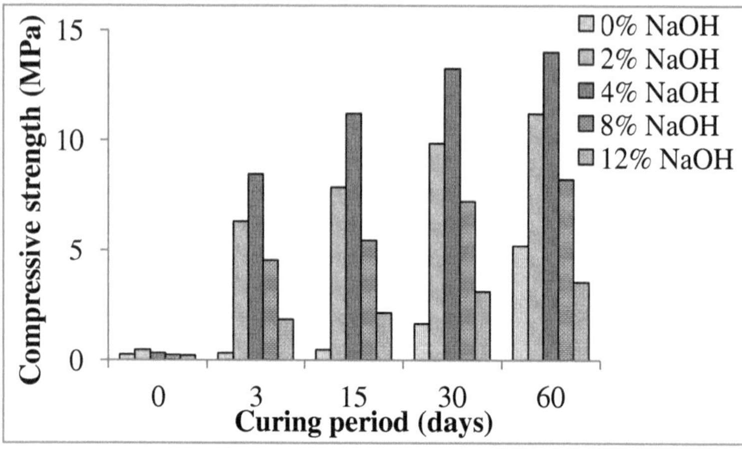

(d) 60% RM + 40% GGBFS

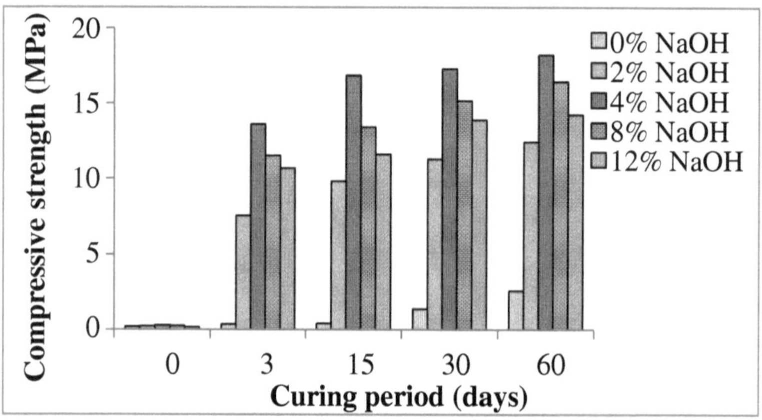

(e) 40% RM + 60% GGBFS

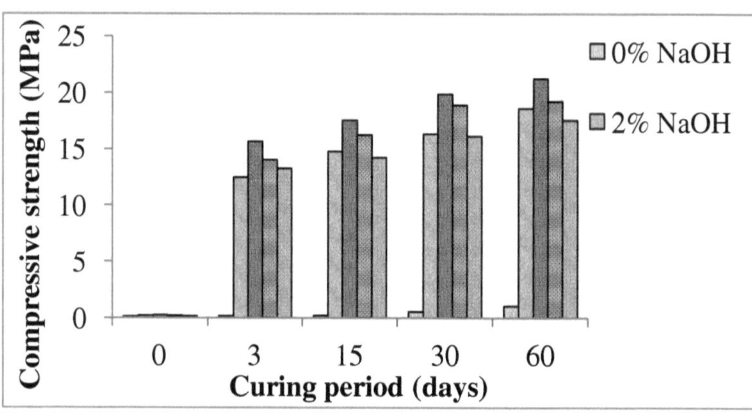

(f) 20% RM + 80% GGBFS

Fig. 9. (continued)

strength at 0 days curing is very less which increases from 0.22 MPa to 0.26 MPa with the addition of 40% slag. For the mix 60% red mud and 40% slag, strength increased about 20 times from 0 days to 60 days curing period.

Stabilization with NaOH Activation
The results from Fig. 9 show that compressive strength was significantly influenced by the increase of sodium hydroxide content. For all red mud-slag mixtures, the compressive strength increment is achieved for an optimum NaOH content. In case of red mud, maximum compressive strength of 0.63 MPa was obtained when activated with 2% NaOH for 60 days curing period. Further increment in NaOH content resulted in a reduction of the compressive strength. This might be due to the lubricating effect of excess NaOH. Whereas, 60% RM + 40% GGBFS with 4% NaOH achieved the highest compressive strength of 14.02 MPa at 60 days curing period and other red mud-slag mixtures with higher slag content showed relatively less significant strength gain.

4 Conclusions

Experiments are carried out to investigate the geo-engineering properties of red mud treated with slag. The effects of slag, NaOH and curing period on the strength were investigated. Based on the experimental finding the following conclusions are arrived.

1. Plastic limit and shrinkage limit reduces with increased slag content but the liquid limit is found to decrease up to 40% slag content thereafter with further increase in slag content the liquid limit increases marginally.
2. Red mud-slag mixes exhibits negative differential free swell values indicating that these materials are dispersive in nature.
3. Red mud shows moderate decrease in pH values with increase in slag content.
4. The maximum dry density increases upto 40% slag content in red mud-slag mixture whereas the optimum moisture content decreases upto 60% slag content Addition of slag to red mud without any additional alkali increases the UCS value up to slag content of 40%. Further addition of slag beyond 40% results in reduction in unconfined compressive strength value. Virgin red mud possesses UCS value of 0.22 MPa which is increased to 5.2 MPa with the addition of 40% slag at curing period of 60 days.
5. Addition of alkali i.e. NaOH solution enhances the strength remarkably for red mud-slag mixes. With the addition of 40% slag and 4% NaOH, red mud attains an unconfined compressive strength of 14.02 MPa at 60 days of curing
6. Red mud is a highly alkaline material which can be stabilized effectively by addition of slag through geopolymerization process. It reduces the environmental pollution and it can be used in the construction works such as embankments, structural fills, road base and sub-base courses etc.

References

Deelwal, K., Dharavath, K., Kulshreshtha, M.: Stabilization of red mud by lime, gypsum and investigating its possible use as a geotechnical material in the civil engineering construction. Int. J. Adv. Eng. Technol. (2014). doi:10.3390/ma5101800

Hajjaji, W., Andrejkovic, S., Zanelli, C., Alshaaer, M., Dondi, M., Labrincha, J.A., Rocha, F.: Composition and technological properties of Geopolymers based metakaolin and red mud. Mater. Des. (2013). doi:10.1016/j.matdes.2013.05.058

He, J., Zhang, J., Yu, Y., Zhang, G.: The strength and microstructure of two Geopolymers derived from metakaolin and red mud-fly ash admixture. Constr. Build. Mater. (2012). doi:10.1016/j.conbuildmat.2011.12.011

He, J., Zhang, J., Yu, Y., Zhang, G.: Synthesis and characterization of red mud and rice husk ash based geopolymercomposites. Cement Concr. Compos. (2013). doi:10.1016/j.cemconcomp.2012.11.010

Kaya, K., Uzun, S.: Evolution of structural characteristics and compressive strength in red mud-metakaolin based geopolymersystems. Ceram. Int. (2016). doi:10.1016/j.ceramint.2016.01.144

Rai, S., Lataye, D.H., Chaddha, M.J., Mishra, R.S., Mahendiran, P., Mukhopadhyay, J., Yoo, C., Wasewar, K.L.: An alternative to clay in building materials: red mud sintering using fly ash via taguchi's methodology. Adv. Mater. Sci. Eng. (2013). doi:10.1155/2013/757923

Singh, K., Pandey, R.K., Mishra, C.S., Rai, A.K., Bind, Y.K.: Analysis on utilization of cement kiln dust stabilized red mud for road construction. Int. J. Civil Eng. Technol. **5**, 56–61 (2014)

Zhang, M., Korchi, T., Zhang, G., Liang, J., Tao, M.: Synthesis factors affecting mechanical properties, micro structure and chemical composition of red mud-fly as based geopolymers. J. Fuel. (2014). doi:10.1016/j.fuel.2014.05.058

Effect of Physico-Chemical Treatment on the Mechanical Properties of Dredged Sediment

Maghnia Asmahane Bourabah[1(✉)] and Said Taibi[2]

[1] Laboratoire Eau et Ouvrages dans leur Environnement EOLE, Université de Tlemcen, Tlemcen, Algérie
mbourabah2000@yahoo.fr

[2] Laboratoire Ondes et Milieux Complexes LOMC, UMR 6294 CNRS, Université du Havre, Le Havre, France
Said.Taibi@univ-lehavre.fr

Abstract. In Algeria, the siltation phenomenon affects all dams. Million cubic meters of sediments are deposited annually. A research program is developed for a rational use of dredged sediments.

This study aims to develop sediment management methodology and valorization of dredged materials in western region of Algeria. Experimental results are presented for the exploitation of dredged sediments and its potential uses as a new material resource for road engineering. The goal is to provide formulations economically competitive and easy to implement in situ.

The paper is composed of two parts. The first one presents the geotechnical characteristics of the dredged sediment. The second provides an analysis of the experimental results and discusses the influence of physico-chemical treatment composed of a size particle correction and treatment with binders on the compaction properties and bearing capacity of the material as well as its durability.

The results analysis show that the addition of lime influences the plasticity of material which is clearly decreases (55%). An improvement of the dry density is also observed (25%), which indicating an acceptable level of compactness.

The bearing capacity indices are significantly improved after treatment and continue to increase with lower water contents. The values of bearing capacity are ranging from 34–42%, acceptable values according to the technical guide (I_{IBCR} > 25%), and sufficient to use in the pavement layers.

The compressive strength increases with the curing time or a significant increase appears for the lime treated sediment from a cure period of 28 days and the minimal strength criterion (1 MPa) set by the technical guide to allow trafficability on the treated layers is reached before seven days of curing time. This increasing of the curing time also improves the elastic modulus which indicated that the curing time affected the compressive strength and elastic modulus.

1 Introduction

Mediterranean countries are commonly reported to be severely affected by soil erosion. In north Africa, the high degree of climatic variability, the torrential nature of storm runoff, the low vegetation cover, and the excessive soil exploitation, have a great

D.N. Singh and A. Galaa (eds.), *Contemporary Issues in Geoenvironmental Engineering*, Sustainable Civil Infrastructures,
DOI 10.1007/978-3-319-61612-4_29

impact on erosion, sediment yield, and fluvial transport, on dam sedimentation and, consequently, on socio-economic conditions (Lahlou 1996).

The increased development of water resources in North African countries and Algeria in particular, increasingly demands an improved knowledge and understanding of the sediment generation, transport and deposition which seriously threatens the life of many reservoirs.

According to the ICOLD World Register of dams, there are 114 large dams in Algeria where 35% of dams are represented by the average sediment yield rate (Basson 2010).

The sediment deposits in Algerian reservoirs generate a siltation rate of about 50 million m^3/year (Bensafia and Remini 2014).

Faced with this difficult situation, Algeria has taken a number of measures in order to limit the negative impact of erosion and suspended solids inflow, and to preserve the water resources stored in the dams. Dredging is therefore of crucial importance for the sustainable development of these natural resources. A significant number of dams were recovered whose lifetime is extended by several years.

However, the quantities of sediment removed by dredging and deposited downstream of the dam may lead to long-term pollution in rural areas. The management of such sediments is a worldwide problem at present.

For that, the recovery and re-use of this sediment should allow in the medium-term, the valorization of these materials in the field of Civil Engineering and Public Works (road engineering, bricks manufacturing and production of cementitious materials). This study focuses on the road sector because the road construction companies consume large quantities of natural aggregates; consequently, their natural resources are being depleted as it has become increasingly difficult to open new quarries. The dredged sediments can be viewed as a new source of materials and can be proposed, what will reduce the consumption of natural aggregates coming from quarries.

Previous research on the use of dredged sediments in road construction has shown that different processes of treatment have been developed to improve the geotechnical parameters of these sediments. The solidification-stabilization (S/S) treatment is frequently used to improve the mechanical properties of such dredged sediments. In this treatment process, the improvement of mechanical behavior of sediment by adding physico-chemicals such as, granular material, cement, fly ash, geofiber, lime, or a combination of these often alters the physical and chemical properties of the treated material. Two primary mechanisms are considered (Boutouil and Levacher 2001; Asgari et al. 2015):

1. Increase in particle size due to cementation, increase in strength parameters, change in the plasticity properties, and reduced deformation potential.
2. Absorption and chemical binding of moisture that will facilitate compaction.

According to Levacher et al. (2012), the addition of lime to sediments is suggested in a stabilization process for the following effects: immediate decreasing of water content of sediments, improvement of workability for the lime-sediment composite, obtaining a possible long-term mechanical performance, and attenuation of the effects of organic matter. Indeed, Song et al. (2003) investigated solid waste soils for use as sub-base materials in road construction. It was shown that the geotechnical properties are clearly affected by the magnitude of organic matter content. As it increases, the maximum dry unit weight, shear strength and bearing capacity of ground decrease, while the void ratio and compressibility increase. Song et al. (2003) noted that if the organic matter content is more than about 8% in solid waste soils, it is not suitable for use as a sub-base material in road construction. This is due to the significant decrease of shear strength and bearing capacity.

Le Guern et al. (2014) studied the mechanical performances of three marine dredged sediments treated with hydraulic binders from three different French harbors, showing that the organic matter content and optimal dry density have a major influence on unconfined compressive strength, and a moderate one on the indirect tensile strength.

According to Wang et al. (2013), the addition of lime results in a significant improvement in the unconfined compressive strength (UCS) and the shear strength. The failure mode of sediments treated with lime develops gradually from plastic to brittle failure. In the meantime, failure strength increases, whereas failure strain decreases with lime content. The shear strength parameters of the solidified sediments are increased owing to the pozzolanic reaction between lime and clay minerals, compared to those of untreated sediments. In addition, the preconsolidation pressure, coefficient of consolidation and the compression index are increased, while the swelling index decreased after lime treatment.

Dermatas et al. (2003) studied the geotechnical properties of cement treated dredged sediment. They found that the compressibility decreased with decreasing water content and increasing cement content. The higher the coefficient of consolidation value and the lower the volumetric strain, the higher the unconfined compressive strength will be. Moreover, the hydraulic conductivity increased with the cement content.

Nor Zurairahetty et al. (2015) investigated the effectiveness of lime stabilization of marine clay based on their strength and compressibility characteristics. An increase in strength was found with increasing lime content. In addition, strength also increases significantly as early as seven curing days and continues for 28 curing days. The coefficient of consolidation with the addition of lime decreases as the lime content increases compared to the untreated marine clay. The compression index value decreases as the lime content increases.

In conclusion, the research in the literature agrees generally on the increase in strength parameters (UCS, shear strength and preconsolidation pressure) of treated sediments with binders; however, for the compressibility characteristics, results in the literature differ somewhat. These variations depend on several parameters: mineralogy, binder content, binder type, curing time, organic matter content, etc.

This study is part of a research program on the dredged sediments management of dams in the western Algeria for reuse in road engineering. Several dam study sites have been selected, depending on the accessibility of the site, their siltation rates and the dredged sediments extraction area. The Cheurfas dam site is part of this selection list.

In this paper, an experimental study based on a treatment protocol, and aiming at the valorization of Cheurfas dredged sediments as a new material resource for road embankments. The treatment proposed is based on the formulations of mixtures with different percentages of dredged sediment, granular materials and binders according to the technical guide (GTS 2000). Several physico-chemical and mechanical tests are performed such as compactness, bearing capacity, elastic modulus and the unconfined compressive strength. The aim is to propose a treated material that is economically competitive and easy to implement in situ.

2 Presentation of Site

The studied sediment comes from the Cheurfas dam, which is located in the north-west of Algeria. Its exploitation began in 1992 and was intended to irrigate surrounding agricultural lands and supply drinking water for the town of Sig (west of Algeria).

This dam, which crosses the Wadi El Mebtouh, regulates 45 10^6 m^3 of water, and is built on the ruins of an old dam swept away by a flood in 1927 (Labiod et al. 2001). Its major river basin has a total estimated area of 4190 km^2 with a maximum altitude of 1754 m and an average altitude of 855 m.

According to the USCS/LCPC classification and the road technical guide (GTR 2000), which are based on the values of geotechnical parameters presented in Table 1, the Cheurfas sediment is classified as an organic and highly plastic soil OH (Ot) and a F1-A2 class material, respectively.

Table 1. Geotechnical properties of Cheurfas sediment

	Clay (%)	Silt (%)	BV	LL (%)	PI (%)	SI (%)	OM (%)
Cheurfas sediment (CS)	53	43	3.22	62	26	53	6.5

This sediment cannot be used in its natural state in road engineering, and therefore needs to be treated to improve its mechanical performance.

3 Use of Cheurfas Sediment in Road Engineering

The sediments have a complex structural composition and can be considered as an evolutionary material. The presence of organic matter constitutes a plastic and moderately compressible behavior. Recent studies revealed that the organic matter has a significant effect on the geotechnical properties of soils (Thiyyakkandi and Annex 2011; Adejumo 2012).

Indeed, the physico-chemical and geotechnical testing performed on natural and calcined Cheurfas sediments have demonstrate that the organic activity is highly influential on the sedimentary matrix. The plasticity decreases with the organic matter content. The relative difference observed is higher than 50%, with a significant difference found in the size fraction and swelling-shrinkage potential (Bourabah 2012).

This difference is probably attributed to the elimination of organic matter, leading to deterioration in the internal mineralogical complex of the sediment (clay humic complex). Consequently, a new internal structure material is created, having a non evolutive internal structure in time.

Thus, as long as no particular treatment, simple or complex, will be made on these sediments to destroy organic matter, they can always act in the material structure.

4 Physico-Chemical Behavior of Sediment After Treatment

To be used as pavement layer structure, a well-graded granular distribution of material is recommended (Dubois 2006; Predis 2007).

For this purpose, the particle size and organic matter content are the most important factors in choosing the valorization of dredged sediment in road engineering. It is necessary to manage their impact on the achievement of durable material on the mechanical and environmental plan, based on the general valorization methodology approach taken in this study, which is mainly not to use treatments heavy and expensive.

The studied sediment is mixed with other granular components in interesting proportions from the economic point of view, which allow an easy application on the industrial scale and an interesting reduction of the water content and the proportion of organic matter. The addition of hydraulic binders, for obtaining the final road material, will allow, reducing the water content, by the phenomena of hydration and dry matter contribution (Sfar Felfoul et al. 2003).

The added granular fraction is a quarry sand (0/4) mm with 3% of elements smaller than 80 μm, and 47% coarse sand. It is a well-graded clean sand SW (Sb) according to the USCS/LCPC classification. The lime used is quicklime (0/2) mm.

The treatment consists of adding a granular fraction of sand (0/4) mm and secondly 3% of quicklime (0/2) mm. This binder content corresponds to what is usually used in road construction of pavement layers for soil treatment.

Table 2 shows the types, physico-chemical and geotechnical characteristics of the two studied formulations. The proportion of granular addition with fine sediment has given a mainly clayey sand mixture with a well-graded granular distribution.

The addition of lime in treatment process influences the plasticity of material which is clearly decreases (55%). In general, adding lime causes the rapid flocculation of clays. This action results in an improvement of the consistency, which causes a

Table 2. Effects of treatment on the physico-chemical and geotechnical characteristics of studied formulations

Formulations	For 1 = 30% CS + 70% Sand	For 2 = 30% CS + 67% Sand + 3% Lime
Passing 80 μm (%)	45	25
% >63 μm	57	77
Silt (%)	18	11
Clay (%)	25	12
LL (%)	41	32
PI (%)	20	9
SI (%)	27	7

displacement of the plasticity domain to larger water contents. The shift of the plasticity index is related to the lime content, the nature of the clay minerals and the water content (Ola 1977; Bell 1996; Lasledj and Al-Mukhtar 2008).

5 Mechanical Behavior of Sediment After Treatment

A series of mechanical tests were conducted on the studied formulations, which allow knowing the changes in the parameters of the compactness (γ_{dMPO}), bearing capacity (I_{IBCR}, CBR), unconfined compressive strength (UCS) and the elastic modulus (E).

5.1 Compaction and Bearing Capacity

The compaction-bearing capacity test was performed according to the modified Proctor method (MPO). Samples of about 6 kg were spray-moisturized in order to reach different water contents. Following the method, amounts of soil from these homogenized wet samples were compacted into five equal layers in a cylindrical compaction mold of 152 mm in diameter and height. Each layer received 56 blows of a 4.53 kg falling hammer of 5.1 cm diameter and 45.7 cm height. The energy density was 2.41 kJ dm^{-3}. The weight of the wet compacted soil in the mold was determined, and the samples were dried and weighed again to estimate the moisture content and dry density.

This process is repeated for various moisture contents, and the dry densities are determined for each. Two graphical relationships of the dry density and bearing capacity to moisture content are then plotted to establish the compaction and bearing capacity curves. The maximum dry density (γ_{dMPO}) is finally obtained from the peak point of the compaction curve, and its corresponding moisture content, also known as the optimum moisture content (w_{MPO}). Two indices are determined for the bearing capacity curve:

- The Immediate Bearing Capacity Ratio index (I_{IBCR}): This index is carried out to evaluate the ability of the soil to support the movement of construction machinery. The principle is to measure the force corresponding to the driving of a cylindrical punch (with a diameter equal to 49.6 mm) into the sample at a constant rate of 1.27 mm/min (NF P 94-078). The forces for two conventional displacements (2.5 and 5 mm) are compared to the corresponding values for a reference material. The Immediate Bearing Capacity Ratio index (I_{IBCR}): is determined by the following equation:

$$I_{IBCR} = max(I_{IBCR25};\ I_{IBCR50}) \tag{1}$$

$$\mathrm{IIBCR25} = \frac{F25}{13.35} \times 100 \quad \mathrm{IIBCR50} = \frac{F50}{20} \times 100 \tag{2}$$

F_{25} = load corresponding to 2.5 mm of penetration (kN)
F_{50} = load corresponding to 5 mm of penetration (kN)

- The California Bearing Ratio index (CBR): This index is performed to estimate the bearing capacity of a finished pavement. The principle of the test for this index is the same as that of the I_{IBCR} test, the difference being that the specimen is overloaded at the top by annular discs of 150 mm diameter and 2.3 kg mass representing the equivalent of imposed stress by the pavement on the platform (NF P 94-078).

Table 3 indicates the values of the Modified Proctor Optimum (MPO) corresponding for each formulation.

Table 3. Effects of treatment on the mechanical parameters of studied formulations

	Cheurfas sediment (CS)	For 1	For 2
w_{MPO} (%)	23.3	12.8	13.35
γ_{dMPO} (kN/m^3)	15.54	19.41	18.82
A_{MPO} (%)	5.57	2.37	5.17
$I_{IBCRMPO}$	14	20	34
$I_{IBCR95\ \%of\ MPO}$	27	28	47
CBR	–	32	42
UCS 7 days (MPa)	–	1.73	1.94
E 7 days (MPa)	–	2679.03	2225.79
UCS 28 days (MPa)	–	1.81	2.01
E 28 days (MPa)	–	2273.58	2460.84

For these mixtures, a decrease in water content at the MPO is clearly observed compared to that of the Cheurfas sediment (CS).

It is also observed that, more a mixture contains sediment, more the optimum water content is high, and therefore the dry density and bearing capacity corresponding are less important. The use of this sediment without treatment cannot be considered in the road construction.

The results presented in Table 3 show that the maximum dry density achieved for Cheurfas sediment is 15.54 kN/m^3 and the optimum water content is 23.3% and 14% for immediate bearing capacity ratio index. The material will be considered as usable for a pavement layer if the water content allows a dry density greater than 95% of γ_{dMPO} and a I_{IBCR} higher than 25%. The Cheurfas sediment studied in its natural state does not check for this condition for use in pavement layer.

The treatment of Cheurfas sediment (CS) by 70% of sand (0/4) mm (For 1), indicates the following:

- The water content was clearly decreased by the addition of sand (0/4) mm compared to that of Cheurfas sediment (CS), showing that the addition of sand as a granulometric corrector significantly affects the optimal water content of the mixture.
- An improvement of the maximum dry density is also observed, where an average increase of γ_{dMPO} is of around 25%. However, we note a little variation of the immediate bearing capacity ratio index (I_{IBCR}) with this decrease in the optimum water content, compared to the Cheurfas sediment. This is probably due to occluded

air bubbles generating a reversible compressibility of the interstitial fluid and thus occurrence of the phenomenon known as "quilting". This phenomenon describes the behavior of a too humid soil during compaction. This operation results in the formation of pads due to the instability of the engine and finally an impossibility of compaction.

The sediments, such as mixtures are sensitive to the water and their use without binders may be detrimental to the durability of the pavement (Boutouil and Saussaye 2011). This sensitivity is related to the presence of organic matter, which by the water retention increases the phenomena.

The addition of lime as binder type in the composition of the final material can change the values of mechanical parameters (Wang et al. 2013). Indeed, the analysis of the results, in Table 3, indicates the following:

- Adding lime to sediments increases their optimum moisture content and slightly reduces their maximum dry density;
- The bearing capacity indices are significantly improved after treatment. According to Fig. 1, the I_{IBCR} and CBR indices increase immediately after the addition of lime and continue to increase with lower water contents. The treated sediment gives values of I_{IBCR} and CBR ranging from 34–42%.

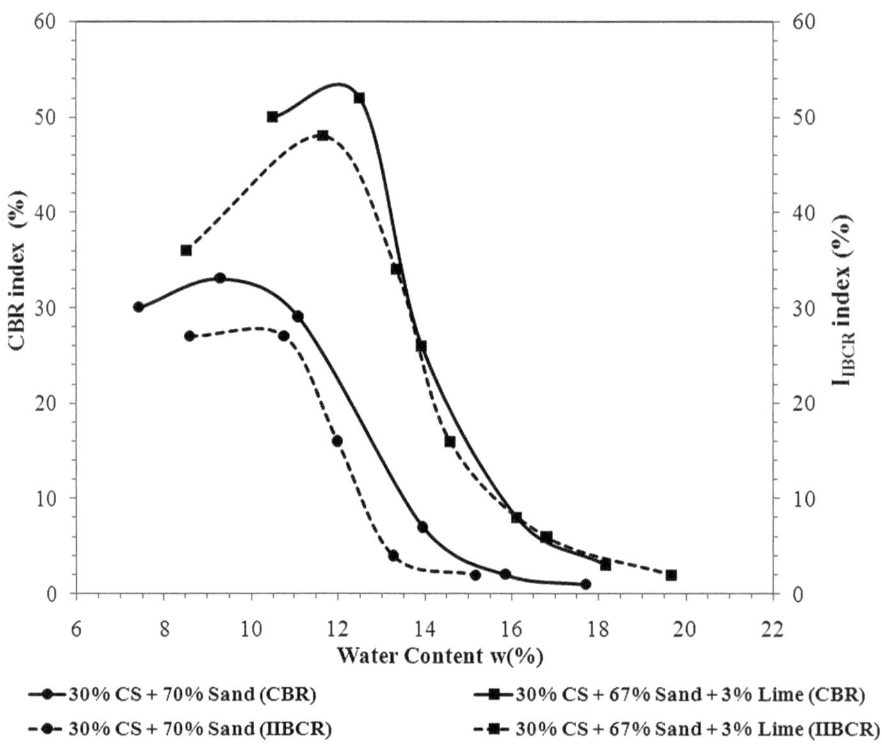

Fig. 1. Variation of bearing capacity indices versus the water content

According to the French standard GTR (2000), these parameters are sufficient to use this treated material in a pavement layers ($I_{IBCR} > 25\%$). However, the bearing capacity curves show a rapid decrease when the water content is higher.

5.2 Compressive Strength and Elasticity

The compressive strength determines the trafficability criterion of the soil according to NF P 98-232-1 standard that allows one, in the short term, to estimate the necessary waiting period in order to allow the harmless passage of construction equipment on the soil layer.

Based on the results of Proctor Modified test, the compression strength test is performed on cylindrical specimens of 35 mm diameter and 70 mm height according to the NF P 98 114-3 standard, compacted statically at w_{MPO} and 98.5% γ_{dMPO}, with respect to the NF P 98-230-2 standard. Then, the specimens of treated materials are stored for 24 h at constant water content in an airtight container for homogenizing the water content, and then placed in the open air in a room at controlled humidity for even drying during the curing time, respectively of 7 to 180 days. When the targeted curing times are achieved, the elastic modulus (E) and the unconfined compressive strength (UCS) are measured.

The elastic modulus (E) is determined by the ultrasonic test according to the NF P 18-418 standard. This test consists of emitting a wave in a sample (treated sediment) and measuring the time and the propagation velocity of this wave by traveling a known distance. The elastic modulus (E) is determined by the following equation:

$$E(\text{MPa}) = \frac{V^2\,\rho(1+\nu)(1-2\nu)}{g(1-\nu)} \qquad V = \frac{d}{t} \tag{3}$$

V = Propagation velocity (m/s)
ρ = Solid density (MN/m^3)
g = Acceleration of gravity (m/s^2)
ν = Poisson's ratio
d = Distance between the transmitter and receiver transducers (m)
t = Propagation time of the wave (s)

The unconfined compression strength (UCS) is determined by applying increasing pressures up to the rupture of the sample. The results of the UCS test in Fig. 2 show that:

- The UCS increases with the curing time or a significant increase of this strength appears for the formulations from a cure period of 28 days.
- The minimal mechanical strength criterion (1 MPa) set by the technical guide (GTS 2000) is reached before seven days of curing time, which allows them to be used as a road layer.

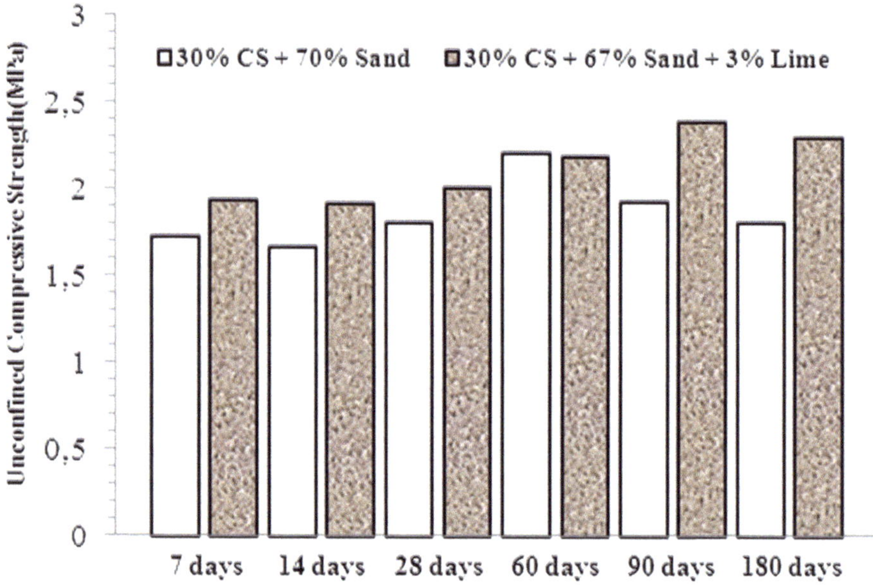

Fig. 2. Variation of unconfined compressive strength (UCS) versus the curing time

- After a curing time of 28 days, there is a significant change in the mechanical strength for the formulation treated with quicklime which stabilizes at a curing time of 90 days. This sharp increase of the mechanical strength after 28 days is justified by the lime hydration reaction.

According to Bell (1996), when lime is added to a clay soil, it has an immediate effect on the properties of the soil, as cation exchange begins to take place between the metallic ions associated with the surfaces of the clay particles and the calcium ions of the lime. The clay particles are surrounded by a diffuse hydrous double layer that is modified by the ion exchange of calcium. This alters the density of the electrical charge around the clay particles, which leads them to be attracted closer to each other to form flocs, the process being termed flocculation, which is responsible for the modification of the engineering properties of soils.

Figure 3 shows the effect of lime and curing time on the elastic modulus of stabilized sediments. It can be seen that the addition of lime, leads to an increase in the stabilized samples' stiffness. Consequently, the strength and elastic modulus of the treated sediment increased after lime treatment.

As can be seen from Fig. 3, increasing the curing time leads to an increase in the elastic modulus of the treated specimens. It can be concluded that the curing time affected the compressive strength and elastic modulus.

This may be due to the fact that the hydration process still occurs between the binder and fine sediment and, in continuation; the pozzolanic reactions are responsible for increasing the strength and stiffness.

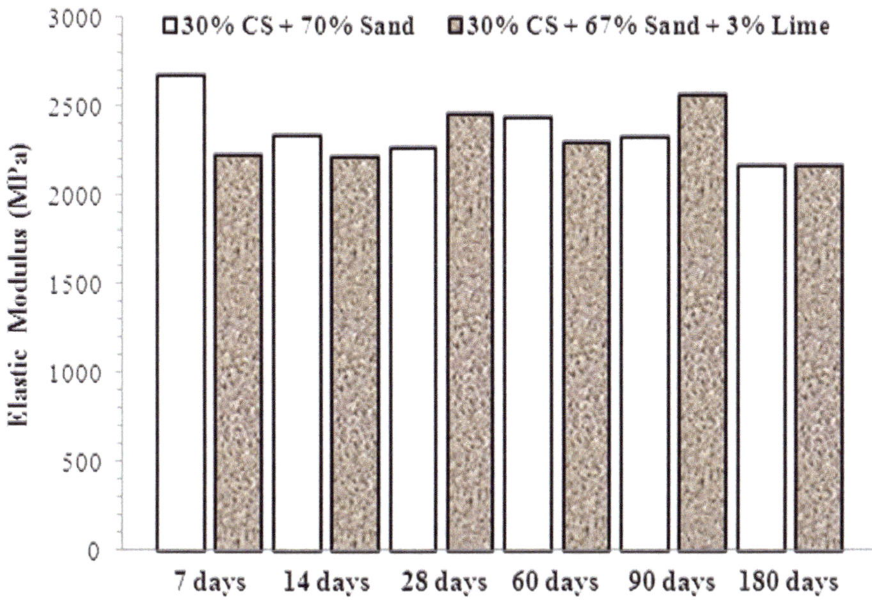

Fig. 3. Variation of elastic modulus versus the curing time

6 Conclusions

The valorization approach proposed in this study is based on choice of preparation and simple formulation, facilitating the use of Cheurfas sediment in road engineering. The main objectives were to identify the influence of granular material and lime, on the strength behavior and durability of sediments under extreme conditions. The improvement in mechanical properties such as dry density, bearing capacity, unconfined compressive strength and elastic modulus, were investigated.

The studied fine sediment presents a high value of the organic matter content, which can influence the physical and mechanical properties of this material.

The analysis of physico-chemical and mechanical results on studied formulations is encouraging for the management of this dredging sediment where a particle size correction has been well observed, followed by an improvement of the maximum dry density obtained by substitution of this sediment by a granular material. The optimum water content is relatively small and the organic matter content does not affect the compactness of the material, it acts adversely against the bearing capacity.

The addition of lime as a type of hydraulic binders in these sediment mixtures allowed us to notice that their influence is on the plasticity of the material which clearly decreases and largely on these mechanical parameters which are clearly improved, which indicates a good hydraulic jack.

This research has shown that, the addition of granular material and binder may be considered as an interesting alternative for stabilization of Cheurfas sediments and its beneficial use as a new material for the construction of foundation and base layers for roads.

References

Adejumo, T.E.: Effect of organic content on compaction and consolidation characteristics of Lagos organic clay. Electron. J. Geotechn. Eng. (EJGE) **17**, 2201–2211 (2012)

Asgari, M.R., et al.: Experimental study on stabilization of a low plasticity clayey soil with cement/lime. Arab. J. Geosci. (2015). doi:10.1007/s12517-013-1173-1. Springer

Basson, G.: Sedimentation and sustainable use of reservoirs and river systems. In: International Commission on Large Dams (ICOLD), Bulletin (2010)

Bell, F.G.: Lime stabilization of clay minerals and soils. Eng. Geol. (1996). doi:10.1016/0013-7952(96)00028-2. Elsevier

Bensafia, D., Remini, B.: Le rôle de la vase dans l'accélération de l'eutrophisation des eaux de barrages. Etude expérimentale. Larhyss J. **19**, 161–181 (2014)

Bourabah, M.A.: Comportement mécanique des sols fins. Application à la valorisation des sédiments de barrages en technique routière. Thèse de Doctorat, Université de Tlemcen, Algérie (2012)

Boutouil, M., Levacher, D.: Etude expérimentale de la solidification d'une vase de dragage à base de ciments: Evaluation de la porosité et de la structure poreuse après solidification (Experimental study of cement-based solidification of dredged sludge: assessment of the porosity and pore structure after solidification). Mater. Struct./Matériaux et Const. (2001). doi:10.1007/BF02480510. Springer

Boutouil, M., Saussaye, L.: Influence of granulometric corrector on the properties of the sediments treated with hydraulic binders. Revue Paralia, Editions Paralia CFL (2011). doi:10.5150/revue-paralia.2011.008

Dermatas, D. et al.: Geotechnical properties of cement treated dredged sediment to be used as transportation fill. In: Garbaciak Jr., S. (ed.) Proceedings of Dredging 2002, Key Technologies for Global Prosperity. ASCE, New York (2003)

Dubois, V.: Etude du comportement physico-mécanique et caractérisation environnementale des sédiments marins-Valorisation en technique routière. Thèse de doctorat, Ecole Nationale Supérieure des Mines de Douai, France (2006)

GTR: Guide technique pour la réalisation des remblais et des couches de forme, fascicules 1, principes généraux (Technical Guide for the realization of embankments and subgrades, fascicles 1 General Principles), LCPC-SETRA (2000)

GTS: Traitement des sols à la chaux et/ou aux liants hydrauliques en remblai et en couche de forme, Guide technique (Treatment of soils with lime and/or hydraulic binders embankment and subgrade, Technical Guide), LCPC-SETRA (2000)

Labiod, Z., et al.: Auscultation des barrages en Algérie (Auscultation of dams in Algeria). In: Séminaire National de Génie Civil SNGC 2001 (2001)

Lahlou, A.: Environmental and socio-economic impacts of erosion and sedimentation in North Africa. In: Erosion and Sediment Yield: Global and Regional Perspectives, Proceedings of the Exeter Symposium. IAHS Publ. No. 236 (1996)

Lasledj, A., Al-Mukhtar, M.: Effect of hydrated lime on the engineering behaviour and the microstructure of highly expansive clay. In: The 12th International Conference of International Association for Computer Methods and Advances in Geomechanics (IACMAG), Goa, India (2008)

Le Guern, M., et al.: Geotechnical and mechanical characterisation of three marine dredged sediments treated with hydraulic binders. In: Proceedings of the South Baltic Conference on Dredged Materials in Dike Construction, Rostock, Germany (2014)

Levacher, D., et al.: A civil engineering application in sediment management: A confined embankment in Cherbourg harbour. XIIèmes Journées Nationales Génie Côtier-Génie Civil, Cherbourg, France, Editions Paralia CFL (2012). doi:10.5150/jngcgc.2012.114-L

NF P 18-418: Concrete - Sonic auscultation - Measurement of the sonic wave transmission time in concrete. Association Française de Normalisation, France (1989)

NF P 94-078: Soils: investigation and tests - CBR after immersion - Immediate CBR - Immediate bearing ratio - Measurement on sample compacted in CBR mould. Association Française de Normalisation, France (1997)

NF P 98 114-3: Roadway foundations - Methodology for laboratory study of materials treated with hydraulic binders - Part 3: Soils treated with hydraulic binders possibly combined with lime. Association Française de Normalisation, France (2001)

NF P 98-230-2: Test relating to pavements - Preparation of materials bound with cementitious binders or aggregates mixes - Part 2: Preparation by static compression of sands or fine grained soils specimens. Association Française de Normalisation, France (1993)

NF P 98-232-1: Tests relating to pavements - Determination of the mechanical characteristics material bound with hydraulic binders - Part 1: Unconfined compression test on aggregate. Association Française de Normalisation, France (1991)

Nor Zurairahetty, M.Y. *et al.:* Performance of lime-treated marine clay on strength and compressibility characteristics. Int. J. Geomate Jpn. Geotech. Const. Mat. Env., 2186–2990 (2015). ISSN: 2186-2982(P)

Ola, S.A.: The potentials of lime stabilization of lateritic soils. Eng. Geol. (1977). doi:10.1016/0013-7952(77)90036-9. Elsevier

Predis: Améliorer la valorisation des déchets industriels en BTP. Guide Technique Régional Relatif à la méthodologie de gestion des sédiments de dragage portuaire (Improve the recycling of industrial waste in construction. Regional Technical Guide Related to the methodology of port dredging sediments management). Démarche PREDIS Nord Pas de Calais, Groupe de travail no. 5, Ecole des Mines de Douai, Université Sherbrooke, Université D'artois (2007)

Sfar Felfoul, H., et al.: Traitement des sols à la chaux et au ciment en technique routière tunisienne: présentation d'un exemple. Séminaire International de Géomatériaux, M'sila, Algérie (2003)

Song, Y.S., et al.: Investigation of solid waste soil as road construction material. Environ. Geol. (2003). doi:10.1007/s00254-002-0746-1. Springer

Thiyyakkandi, S., Annex, S.: Effect of organic content on geotechnical properties of Kuttanad clay. Electron. J. Geotechn. Eng. (EJGE) **16**, 1653–1663 (2011)

Wang, D., et al.: Effect of lime treatment on geotechnical properties of Dunkirk sediments in France. Road Mater. Pavement Des. (2013). doi:10.1080/14680629.2012.755935. Taylor & Francis Online

Characterization of Dredged Sand from Chilka Lake as a Geo Material With and Without Biopolymer

Harishankar Jeevan Dash(✉) and Sarat Kumar Das

Civil Engineering Department, National Institute of Technology, Rourkela, India
harishankar51.bbsr@gmail.com, saratdas@rediffmail.com

Abstract. Chilika lake is a brackish water lagoon as well as the largest coastal lagoon in India and the second largest lagoon in the world, of Odisha state on the east coast of India covering an area of over 1,160 km^2. The size of the lagoon is decreasing over the years due to large siltation of the rivers joining Chilika lake and dredging is required to deal with siltation. Utilization of dredging sand from Chilika for construction purposes is a sustainable method to deal with siltation. In the present study, the dredged sand was collected from Chilika lake (Satapada site) and is characterized as an alternate geo material. The chemical, mineralogical, morphology are also studied. The particle shape parameters analysis was done using optical microscope. The results indicate that Chilika sand is poorly graded sandy soil with of 99.9% sand with a specific gravity of 2.64, optimum moisture content of 10.3% and pH 7.72. An effort was also made to use biopolymer for stabilization of sand while using this as barrier in the lake. It was found that biopolymer, guar gum is effective in imparting shear strength of sand. The present study indicates that the Chilika sand can be used in various construction works as well as foundation and fill materials. The use of biopolymer is a sustainable method for stablilization. The present study will help in effective utilization of the dredged sand; thereby it will not only help in reclamation of the Chilka lake, but also save the natural resources.

Keywords: Chilika lake · Dredged sand · Physical · Chemical · Morphological characterization · Biopolymer

1 Introduction

Chilika lake is a brackish water lagoon of Odisha state on the east coast of India. It is the largest coastal lagoon in India and the second largest lagoon in the world covering an area of over 1,160 km^2 (Sahu et al. 2014). The Chilika lake receives freshwater from 52 rivers and rivulets opening into it from different directions. The rivers on the north-east (e.g. Bhargavi, Daya, luna, Makara) contribute about 60–80% of total freshwater input, while the western rivers (e.g. Kansari, Kusumi, Janjira, Tarimi etc.) and rivulets contribute the rest with seawater intrusion from two sources. The lake is broadly divided into four different zones based on the water quality, distribution of biotic components and fishery yield, as (i) northern sector, (ii) central sector, (iii) southern sector and (iv) outer channel area. Ecological degradation of Chilika lake

D.N. Singh and A. Galaa (eds.), *Contemporary Issues in Geoenvironmental Engineering*, Sustainable Civil Infrastructures,
DOI 10.1007/978-3-319-61612-4_30

had begun since 1970s and then increased rapidly during 1980s and 1990s. Sahu et al. (2014) studied the environmental conditions of Chilika lake during pre and post hydrological intervention in which they studied that the geomorphology, water quality and biological productivity of the lake had undergone significant changes over the years under the influence of natural events and anthropogenic interventions. Depletion of fisheries and loss of biodiversity in 1980s and 1990s have been linked to the northward shifting of lakes inlet and silting up of the outer channel. The lake has emerged as one of the highly degraded and hypersensitive coastal marine ecosystem and recognized as Ramsar site in 1981 (Sahu et al. 2014). Siltation of the lagoon basin was identified as the major threat to the ecology of the lake. The amount of silt brought into the lake by the distributaries of the Mahanadi river system is about 1.5 million tons per year, while that from the western catchment stands at 0.3 million tons per year. Land use pattern promoting erosion in the upstream areas has increased the sediment load considerably. Siltation of outer channel restricting the outflux of silt laden freshwater had favoured silting up of the lagoon. In order to restore the biodiversity and to improve the health of the Chilika lagoon, a new mouth at Magarmukha (shark face) is locally known as the inner Channel was dredged in September 2000 and it serves to connect the Outer Channel to the main body of Chilika lake. Opening of the new mouth has resulted in some improvements with substantial increase in capture fishery, reduction in weed infestation, growth of seagrasses, appearance of dolphins and increase in the population of migratory birds.

Now, efforts have been made to maintain the size of the lagoon through dredging at different location on the inlets of the lagoon. The major problem is effective utilization of the dredged soil. The dredged soil consists of sand, silt and clay materials depending upon the dredging point. The properties of the dredged soil depend upon physical properties of parent rock, hydrogeological process and temperature and tidal variations (www.chilika.com).

The investigation related dredged marine sand is limited. Wang et al. (2014) investigated that the Dunkirk marine sediments can be used as an alternative roadbed material. There is also restriction for excavation of river sand as a construction material. Effective utilization of the dredged soil will not only help in reclamation of the Chilka lake, but also the dredged soil can be used as alternate construction material thereby saving the natural resources. Subudhi et al. (2014) studied about the geotechnical characterization of dredged mud from Chilika lake and studied the geotechnical parameters such as void ratio, porosity, particle size analysis, specific gravity, Atterberg's limits, differential free swell index, compaction characteristics.

The dredged sand of the Chilika lake is also being also stored at site to make small (dumps) islands. But, these small dumps are sometimes becoming unstable due to wind erosion and small waves. The stabilization of these mounds is also one of the important issues. Keeping in mind the aqua and fauna sustainability of the lake, chemical stabilization of the sand using cement and others may not be suitable.

Chen et al. (2013) investigated about the feasibility of using two natural and renewable biopolymers, xanthan gum and guar gum, to stabilize Mine tailings for dust control. The results indicate that both xanthan gum and guar gum are effective in enhancing the moisture retention capacity, improving the dust resistance, and increasing the surface strength of Mine tailings beyond that of water wetting. Shashank et al. (2016) investigated about the 'bio-mediated soil improvement techniques' such as bio-cementation, bio-clogging, bio-remediation and phytoremediation on dredged sediments. Hence, in the present study an attempt has been made to characterize dredged sand as an alternate construction material and also to stabilize the dredged sand with biopolymer guar gum, to stabilize the mound made of dredged sand.

2 Materials and Methods

2.1 Materials

The dredged Chilika sand collected from Satapada site, which is close to sea (Bay of Bengal) and is shown in Fig. 1. The commercially available guar gum is used.

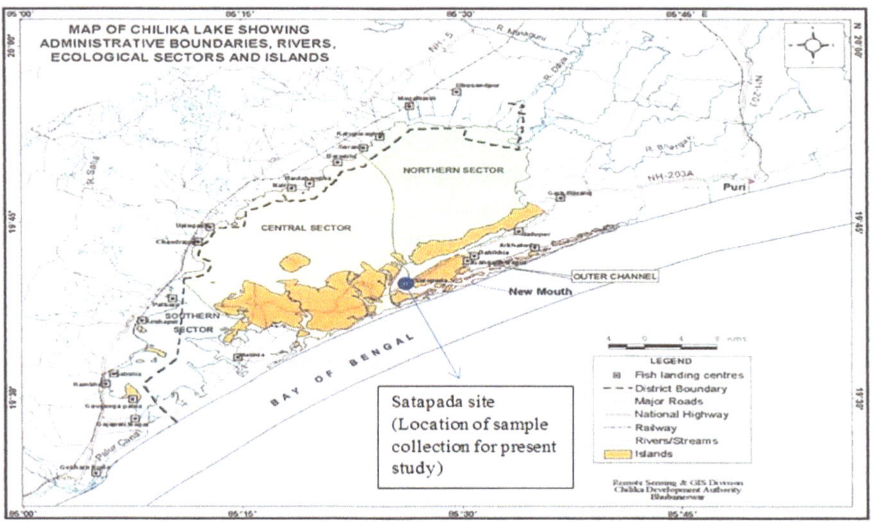

Fig. 1. Map of Chilika lake showing satapada site, location of sample collection (Ref: www.chilika.com)

2.2 Methods

The present study includes the characterization and use of dredged sand of Chilka lake through physical, chemical and bioremediation. The physical and geotechnical characteristics, such as grain size analysis (ASTM D 422-63 2007), specific gravity (ASTM D854-14 2014), compaction characteristics (ASTM D1557-12e1 2012), maximum and minimum void ratios (ASTM D4253) and direct shear tests were conducted as per (ASTM D3080/D3080M). The morphological and mineralogical characteristics of dredged sand also determined by scanning electron microscope (SEM) and x-ray diffraction (XRD) analysis. Jeol-840-A model SEM, fitted with electron dispersive X-ray (EDX) analyser was used. The mineralogy of the materials is studied using XRD analysis on a Rich-Siefert X-ray diffractometer using copper target and Ni filter. The pH tests were conducted by digital pH meter (Hach, HQ 40 D). To study the particle shape parameter optical microscope (Axio lab A1, Zeiss) was used. The permeability test according to ASTM D 5084 was conducted using flexible permeameter of Humboldt, USA.

3 Results and Discussion

In this present study the physical characterization includes determination of moisture content, specific gravity, grain size distribution characteristics, compaction characteristics, permeability and shear strength of dredged Chilika sand.

3.1 Grain Size Distribution, Specific Gravity, Compaction Characteristics and Permeability Tests Without Biopolymer

The grain size distribution curves for Chilika sand was compared with dredged marine sand collected from different sources and shown in Fig. 2. The Chilika sand is with 80.1% of fine sand and 19.7% of medium sand and it is poorly graded sand (SP) as compared to other dredged sands. The physical properties of the dredged Chilika sand are shown in Table 1. The specific gravity of sand is 2.64. The standard Proctor compaction curve for Chilika sand is shown in Fig. 3. The optimum moisture content (OMC) was found to be 10.3% and maximum dry density (MDD) as 17.03 kN/m^3. The maximum and minimum void ratio values of Chilika sand were found to be 0.91 and 0.66, respectively. Three different direct shear test (DST) tests were conducted at OMC, e_{max} and e_{min} conditions. The angles of friction values (ϕ) were found to be 34.95°,

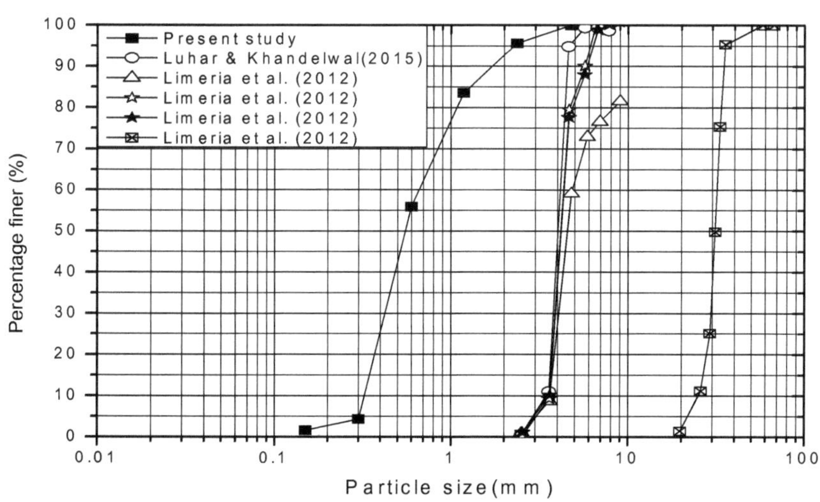

Fig. 2. Grain size Distribution curves for Chilika sand compared with dredged marine sand collected from different sources

Table 1. Physical characteristics of Chilika sand

Parameter	Values
Dredged Chilika sand collected from depth (m)	0.6
Specific gravity (G_s)	2.64
Natural moisture content (%)	10.39
e_{max}	0.91
e_{min}	0.66
γ_{dmin} (kN/m^3)	13.53
γ_{dmax} (kN/m^3)	15.59
pH Value	7.72
D_{10} (mm)	0.22
D_{30} (mm)	0.31
D_{60} (mm)	0.37
Uniformity coefficient (C_u)	1.70
Coefficient of curvature (C_c)	1.17
Soil classification	SP
Optimum moisture content (%)	10.3
Maximum dry density (KN/m^3)	17.03
Permeability (K) cm/s	1.50×10^{-3}
Angle of friction (ϕ) with omc value	34.95°
Angle of friction (ϕ) with e_{max} value	26.56°
Angle of friction (ϕ) with e_{min} value	33.02°

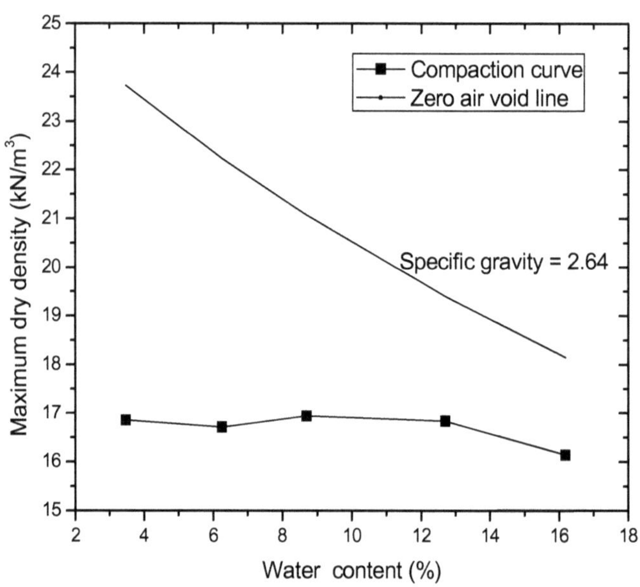

Fig. 3. Compaction curve for Chilika sand

26.56° and 33.02°, respectively. From the results it is clearly evident that sand having maximum void ratio exhibits less angle of friction values than the sand having minimum void ratio. The constant head permeability test for Chilika sand was also conducted and the permeability value was found to be 1.50×10^{-3} cm/s.

3.2 Morphological Characterization

The morphology of sediments plays an important role in understanding the size, shape of the sample. In this present study the morphology i.e. shape of the grains of Chilika sand was studied by the help of optical microscope. Various surface features of the grains was also studied scanning electron micrographs for Chilika sand was obtained.

3.3 Particle Shapc Analysis

The particle shape of Chilika sand sample in terms of roundness index (R_i), elongation (E_R), flatness (F_R), sphericity (ψ), shapefactor (S_F) was determined by optical microscope method using ZEISS Axio optical microscope attached with image analyzer. Generally most of the particles were found to be sub-rounded, sub-angular and angular shaped based on roundness index. According to (Kwan and Mora 2000) the roundness indices ranges from 0 to 1. According to Powers (1953), the particle is said to be angular if it ranges from 0.17 to 0.25, sub-angular 0.25 to 0.35, sub-rounded 0.35 to 0.49, rounded 0.49 to 0.70 and well-rounded 0.70 to 1.00. In this present study the analysis was done for Chilika sand having particle size of 600 μ and 425 μ. A representative number of particles from each particle sizes were considered for this present study. The results for Identifications of different shape, size, flatness of Chilika sand

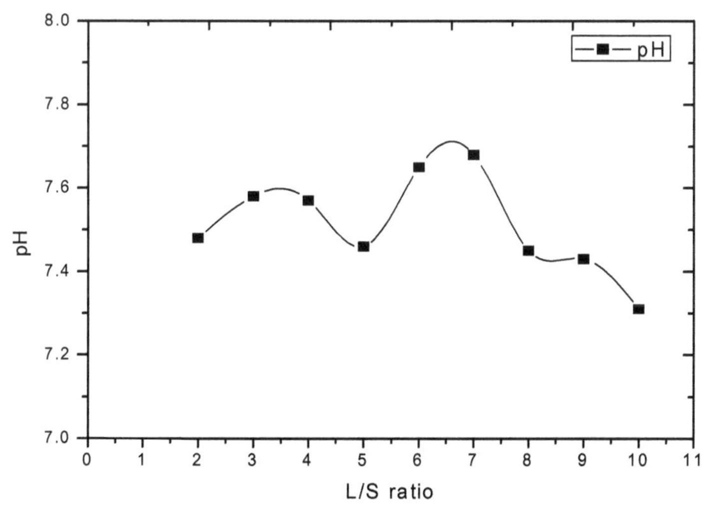

Fig. 4. The variation of pH with liquid to solid ratio for Chilika sand

particles according to shape parameters are given in Table 2 and Identification of the shape of Chilika sand particles based on roundness index are shown in Table 3. Elongation and flatness of Chilika sand was determined and defined as the ratio of intermediate dimension (d_i) of sand grain to largest dimension of sand grain (d_l) and ratio of smallest dimension of sand grain (d_s) to intermediate dimension (d_i) of sand grain respectively. Also the roundness index (R_i) was calculated using Eq. 1 (Wadell 1932). The shape factor (S_F) and sphericity (ψ) was also determined by using Eqs. 2 and 3 respectively (Yudhbir and Abedinzadeh 1991).

Table 2. Identifications of different shape, size, and flatness of Chilika sand particles according to shape parameters

Particle size	No	Elongation (E_R)	Flatness (F_R)	Sphericity (ψ)	Shape factor (S_F)	Roundness index (R_i)
600 μ	a.	0.58	0.36	0.60	0.45	0.43
	b.	0.96	0.16	0.30	0.40	0.47
	c.	0.74	0.19	0.36	0.37	0.29
	d.	0.95	0.59	0.71	0.75	0.22
	e.	0.66	0.52	0.74	0.59	0.36
425 μ	a.	0.75	0.37	0.56	0.52	0.29
	b.	0.73	0.35	0.55	0.51	0.20
	c.	0.74	0.33	0.53	0.49	0.27
	d.	0.98	0.40	0.55	0.63	0.32
	e.	0.64	0.36	0.58	0.48	0.30

Table 3. Identification of the shape of Chilika sand particles based on roundness index

Sl.no.	600 MICRON		425 MICRON	
	Particles	Shape	Particles	Shape
1.		sub rounded		sub-angular
2.	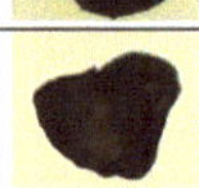	sub-rounded	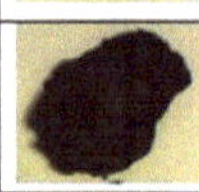	angular
3.		sub-angular	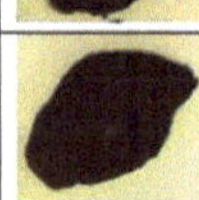	sub-angular
4.		angular	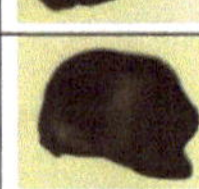	sub-angular
5.	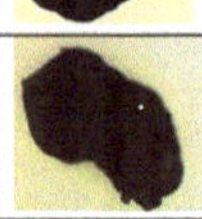	sub-rounded	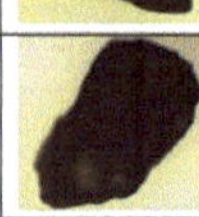	sub-angular

$$R_i = \frac{\sum\left(\frac{r}{R}\right)}{N} \tag{1}$$

Where, r is the radius of corners of sand grain, R is the radius of inscribed circle, N is the number of corner.

$$S_F = \frac{d_S}{\sqrt{d_L \times d_I}} \tag{2}$$

$$\Psi = \sqrt[3]{\frac{d_S \times d_I}{d_L^2}} \tag{3}$$

Where, d_s is the smallest dimension of sand grain, d_i is intermediate dimension of sand grain, d_l is largest dimension of sand grain. The largest and intermediate dimension of sand particle was calculated and the smallest dimension of sand grain was measured using Vernier scale of optical microscope.

3.4 Zingg Shape Classification

According to Zing (1935) particles are classified in to four categories; flat and elongated, elongated, spherical and flat shaped (Yudhbir and Abedinzadeh 1991). The classifications of Chilika sand grains are shown in Figs. 5 and 6, where most of the sand particles are found to be elongated and few particles are found to be both flat and elongated. Zinggs diagram for different particle shapes belongs to 600 μ and 425 μ (Zingg 1935) are shown in Figs. 5 and 6 respectively.

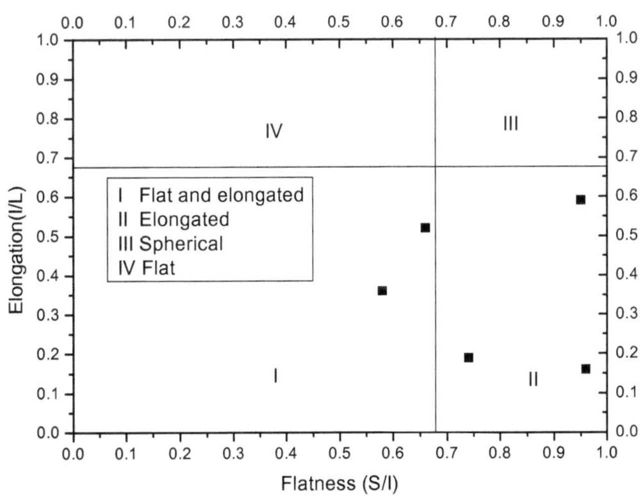

Fig. 5. Zinggs diagram for different particle shapes that belongs to 600 μ (Zingg 1935)

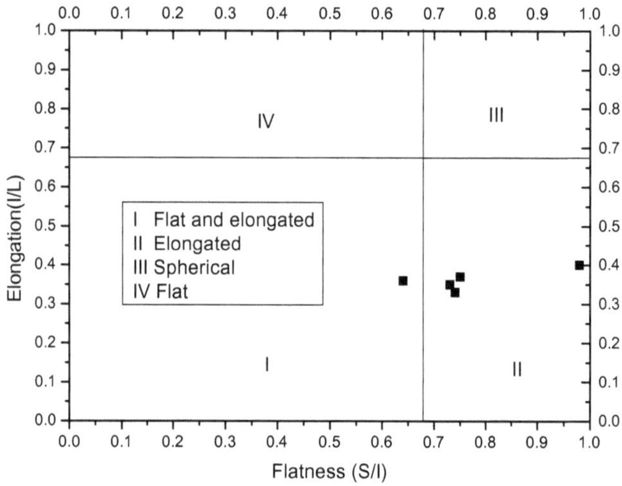

Fig. 6. Zinggs diagram for different particle shapes that belongs to 425 μ (Zingg 1935)

Fig. 7. Scanning electron microscope image of Chilika sand

The SEM analysis was done for Chilika sand sample and shown in Fig. 7. It was observed that different shape, size and angularity of sand grains are present.

3.5 Mineralogical Characterization

The X-Ray Diffraction analysis for Chilika sand is presented in Fig. 8 and it was found that the Chilika sand mostly contains quartz, silica and aluminium oxide.

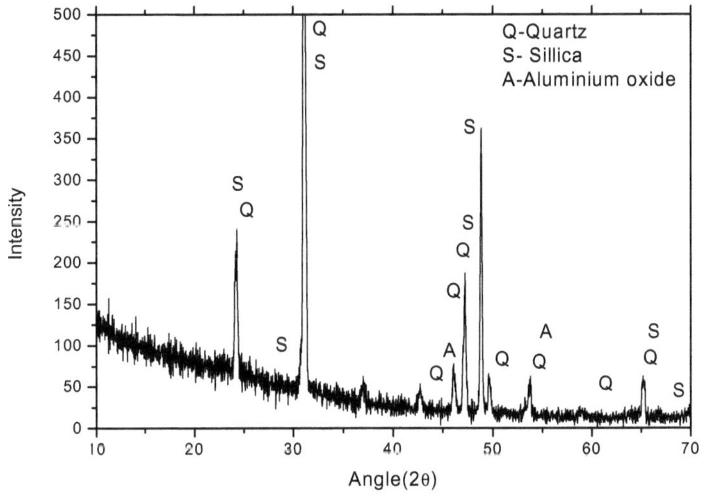

Fig. 8. XRD analysis for Chilika sand

3.6 Chemical Characterization

The pH of Chilika sand sample was determined as per IS:2720 (Part XXVI) - 1987, and found to be 7.74 and also the pH corresponding to different liquid to solid ratio (L/S) varying between 2 and 10 was measured. The sand sample of about 30 g is mixed with 30 g sample is mixed and stirred with distilled water. The solution is kept for 1 h and then the pH value of the solution was determined. The results of pH with different liquid to solid ratios (L/S) are shown in Fig. 4. From the Fig. 4 it is found that the pH value increases and then decrease with increase in liquid to solid ratios of the Chilika sand.

4 Stabilization Using Biopolymer

The effect of biopolymer stabilization on the Compaction characteristics, Shear strength and permeability tests with biopolymer. The effect of biopolymer stabilization on the wind erosion of the dredged sand was studied in terms of penetration test as per Chen et al. (2013), The guar gum solution was added at the rate of litre per meter square to the sand surface and allowed to penetrate up to depth 4 mm. Penetration curve for Chilika sand sample obtained from the test is shown in Fig. 9. The permeability test of Chilika sand was conducted in the flexiblewall permeameter with addition of biopolymer guar gum solution. The value of permeability of Chilika sand was found to be 2.4×10^{-4} cm/s.

The standard Proctor compaction curve with addition of guar gum for Chilika sand is shown in Fig. 11. The optimum moisture content was found to be 14.83% and maximum dry density 16.95 kN/m^3. Results indicate that the value for OMC increases whereas the maximum dry density decreases with addition of biopolymer guar gum.

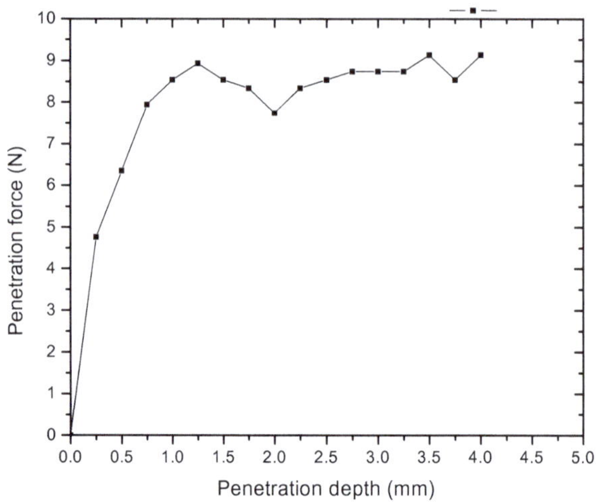

Fig. 9. Penetration curve for Chilika sand with addition of biopolymer

Also direct shear test (DST) test with addition of biopolymer guar gum was conducted at OMC value obtained with addition of biopolymer guar gum. The angle of friction value (ϕ) was found to be 27.47°. From the result it is clearly evident that sand having less angle of friction value as compared to that value obtained from direct shear test at omc without addition of biopolymer guar gum.

4.1 Dispersion Test with Biopolymer

Dispersion test for Chilika sand was conducted by preparing a sand sample with the 10.9 cm height mould and having dia 4.9 cm by adding two percentage of guar gum

Fig. 10. Dispersion test of Chilika sand with addition of biopolymer

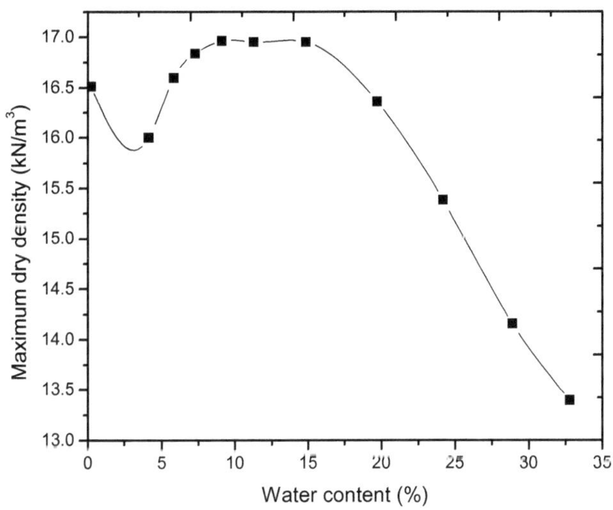

Fig. 11. Compaction curve for Chilika sand with addition of biopolymer guargum

solution with the sand. Then submerge the sand sample that was extracted from the mould in a container having water and allowed the sand sample to disperse for an hour or more till it dispersed. The rate of dispersion of sand sample was less as stabilized by biopolymer guar gum and shown in Fig. 10.

5 Conclusion

Grain size distribution tests for Chilika sand it was found that the sand is with 80.1% of fine sand and 19.7% of medium sand and it is poorly graded sand (SP) as compared to other dredged sands. The specific gravity of Chilika sand is 2.64. The optimum moisture content was found to be 10.3% and maximum dry density 17.03 kN/m^3. The maximum and minimum void ratio values of Chilika sand were found to be 0.91 and 0.66, respectively. Three different direct shear test (DST) tests were conducted. The angles of friction values (ϕ) were found to be 34.95°, 26.56° and 33.02°, respectively at OMC, e_{max} and e_{min} values. From the results it is clearly evident that sand having maximum void ratio exhibits less angle of friction values than the sand having minimum void ratio. From the constant head permeability test for Chilika sand the permeability value was found to be 1.50×10^{-3} cm/s. The morphology of Chilika sand was studied by the help of optical microscope. Various surface features of the grains was also studied scanning electron micrographs for Chilika sand was obtained. The particle shape of Chilika sand was studied for various shape parameters. Generally most of the particles were found to be sub-rounded, sub-angular and angular shaped based on roundness index. From the zings shape classification it was found that the sand particles are elongated and few particles are found to be both flat and elongated. The SEM analysis was done for Chilika sand and it was observed that different shape, size and angularity of sand grains are present. The X-Ray Diffraction analysis for

Chilika sand is conducted and found that Chilika sand mostly contains quartz, silica and aluminium oxide. The pH value was found 7.72 pH value increases and then decrease with increase in liquid to solid ratios of the Chilika sand.

The effect of biopolymer stabilization on the wind erosion of the dredged sand was also studied in terms of penetration test, permeability and dispersion tests. Penetration curve for Chilika sand sample obtained and maximum depth of penetration was found 1.3 mm. The value of permeability of Chilika sand with biopolymer stabilization was found to be 2.4×10^{-4} cm/s. The rate of dispersion of sand sample was less as stabilized by biopolymer guar gum.

Hence the present study indicates that the Chilika sand can be used in various construction works as well as fill materials. From this present study it can be concluded that the dredged sand from Chilika at Satapada site can also be used as a embankment fill material.

Acknowledgments. The authors would like to thank Chilika Development Authority, Bhubaneswar, India for providing useful information about Chilika lake which enable us to complete the present study that is based on my research work.

References

ASTM D 422-63, Standard Test Method for Particle-Size Analysis of Soils. ASTM International, West Conshohocken, PA (2007)

ASTM D2487-11, Standard Practice for Classification of Soils for Engineering Purposes (Unified Soil Classification System), ASTM International, West Conshohocken, PA (2011)

ASTM D1557-12e1, Standard Test Methods for Laboratory Compaction Characteristics of Soil Using Standard Effort ((12 400 ft-lbf/ft3(600 kN-m/m^3)), ASTM International, West Conshohocken, PA (2012)

ASTM D854-14, Standard Test Methods for Specific Gravity of Soil Solids by Water Pycnometer, ASTM International, West Conshohocken, PA (2014)

ASTM D4253 Test Methods for Maximum Index Density and Unit Weight of Soils using a Vibratory Table

ASTM D2216 Test Methods for Laboratory Determination of Water (Moisture) Content of Soil and Rock by Mass

ASTM D4791, Standard test method for flat particles, elongated particles, or flat and elongated particles in coarse aggregate, ASTM, West Conshohocken (2005)

Chen, R., Zhang, L., Budhu, M.: Biopolymer stabilization of mine tailings. J. Geotech. Geoenviron. Eng. **139**, 1802–1807 (2013). doi:10.1061/(ASCE)GT.1943-5606.0000902. ASCE

Chilika Development Authority, Bhubaneswar. A publication on strengthening of wetland research & training centre and bio-diversity conservation on Chilika lagoon. www.chilika.com

Kwan, A.K.H., Mora, C.F., Chan, H.C.: Particle shape analysis of coarse aggregate using digital image processing. Scopus (2000). doi:10.1016/S0008-8846(99)00105-2

Naik, P.K., Pati, G.C., Choudhury, A., Naik, K.C.: Conservation of Chilika Lake, Odisha, India. In: Proceedings of Taal 2007 (2007)

Powers, M.C.: A new roundness scale for sedimentary particles. J. Sediment. Petrol. **23**, 117–119 (1953). doi:10.1306/D4269567-2B26-11

Sahu, B.K., Pati, P., Panigrahy, R.C.: Environmental conditions of Chilika lake during pre and post hydrological intervention. J. Coast Conserv. **18**, 285–297 (2014). doi:10.1007/s11852-014-0318-z. Springer

Shashank, B., Sharma, S., Sowmya, S., Latha, R., Meenu, P., Singh, D.N.: State-of-the-art on geotechnical engineering perspective on bio-mediated processes. Environ. Earth Sci. **75**, 270 (2016). doi:10.1007/s12665-015-5071-6. Springer

Subudhi, D.P., Signum, S., Das, A., Asutosh, A.T., Yadav, D., Das, M.R.: Geotechnical characterization of dredged mud from Chilika lake. Int. J. Adv. Res. Civ. Struct. Environ. Infrastruct. Eng. Developing **2**, 36–41 (2014). ISSN: 2320-723X

Wadell, H.: Volume, shape, and roundness of rock particles. J. Geol. **40**, 443–451 (1932). doi:10.1086/623964

Wang, D., Abriak, N.E., Zentar, R.: One-dimensional consolidation of lime-treated dredged harbour sediments. Eur. J. Environ. Civ. Eng. **19**, 199–218 (2014). doi:10.1080/19648189.2014.939309. Taylor and Francis

Yudhbir, J., Abedinzadeh, R.: Quantification of particle shape and angularity using the image analyzer. ASTM International (1991). doi:10.1520/GTJ10574J

Zingg, Th.: Beitrage zur schotteranalyse. Min. Petrog. Mitt. Schweiz. **15**, 39–140 (1935)

Correlations Between Sediments Characteristics from Tunisian Harbors

Amel Missaoui[1,2(✉)], Imen Said[1], Marwa Zelleg[1], Essaieb Hamdi[1], and Zoubeir Lafhaj[2]

[1] Ecole Nationale d'Ingénieurs de Tunis, LR14ES03-Ingénierie Géotechnique, Université de Tunis El Manar, BP 37 Le Belvédère, 1002 Tunis, Tunisia

[2] Ecole Centrale de Lille, Villeneuve d'Ascq, France

Abstract. Sediments are defined as continental and marine deposits of particles coming from preexisting rocks alteration and suspended solids precipitation. The accumulation of these materials, due to continued intervention of natural as well as anthropogenic activities, presents a serious environmental issue. Consequently, a dredging operation becomes necessary in order to maintain marine commercial activities. Before this operation, a detailed knowledge of sediments characteristics is required. This step of investigation plays an important role in the further management of dredged materials. In order to optimize and facilitate the characterization phase, correlations between properties of sediments must be determined.

This paper aims to study the correlations between characteristics of marine sediment dredged from different Tunisian harbors. In this context, sixteen samples were extracted from studied harbors. The first part of the paper is devoted to present the methods used for the characterization of sediments, including the total organic carbon, pore volume and specific surface area. The second part deals with the Results analysis. The last part is dedicated to the analysis of relationships between properties. Significant correlations between both couples (TOC vs pore volume) and (TOC vs SSA) were determined from Tunisian sediments characteristics.

Keywords: Marine sediments · Tunisian harbors · Geotechnical and chemical characterization · Correlation

1 Introduction

Sediments present a complex and heterogeneous type of soil with specific characteristics, which can change from region to another. The huge quantities of marine sediments, extracted from harbors worldwide, obliged authorities and researchers to find out their characteristics (physical, chemical, geotechnical, mineralogical…). This full comprehension of sediments properties allowed their best management and monitoring. Some of the most important properties to be determined in sediments were the specific surface area, its porosity, and the total organic carbon content. In one hand, the specific surface area presents a fundamental property, which controls the reaction between sediments particles and surrounding environment such as biological process, contaminants

D.N. Singh and A. Galaa (eds.), *Contemporary Issues in Geoenvironmental Engineering*, Sustainable Civil Infrastructures,
DOI 10.1007/978-3-319-61612-4_31

adsorption, and ecosystem level (Rawlins et al. 2010). According to Mayer (1994), sediments with the higher specific surface area have a tendency to gather larger organic matter contents. Burone et al. (2003) showed in his study a strong correlation between total organic content and organic carbon content and recommended the use of this correlation to determine TOC directly from OM content. This technique could be only used for sediments sampled from a similar environment. The TOC analyses are requested especially with contaminants assessment (Schumacher 2002). The TOC content reflected the nature of sampled area and the quality and richness of benthic fauna. The aquatic species richness reached their peaks at low to moderate organic carbon contents (Hyland et al. 2005). Thus, the determination and the analysis of total organic carbon and specific surface area are necessary to assess the sediments properties and their interactions with the surrounding ecosystem. In the other hand, the pore volume in sediments appeared to be an important parameter to assess the water and organic contents. Mayer et al. (2004) studied the correlation between organic matter concentrations and small mesopores in soils and sediments. Results showed that only small part of the total organic matter content was found in small mesopores. The present study aims to establish correlations between Total organic content and both specific surface area and pore volume in sediments dredged from two different Tunisian harbors.

This paper is composed of three parts. The first part is devoted to present the methods used for the geotechnical and chemical characterization of sediments, including organic matter, Total organic carbon, pore volume and specific surface area. The second part deals with the Results analysis. The last part studies the relationship between the determined properties of sediments. Significant correlations between sediments characteristics were presented and commented.

2 Description of Sediments Sampling from Both Harbors

Sediments were collected from two harbors located in the southeast of Tunisia, as presented in Fig. 1. The first one is Gabes harbor, which is located in the middle of the Gulf of Gabes. This harbor is situated next to the most important chemical group in Tunisia. In 2002, it was confirmed that 4 to 4.5 mm^3/year of industrial wastes, mainly phosphogypsum, were discharged into the sea. Eight samples (designed GS1 to GS8) were collected from this harbor. The collected sediments were stored in hermetic barrels and transported to the laboratory for further characterization. The samples were extracted using an appropriate bucket and a diver. Extraction was made in the first 50 cm of the upper layer. The second harbor is located in Zarzis. It is situated approximately 30 km south of the island of Djerba, 50 km from the Gulf of Gabes and 80 km from the border with Libya. Zarzis harbor provides commercial exchanges of the region dealing mainly with the export of marine salt, representing approximately 80% of the harbor total traffic (CJB Environnement and EAM 2006), crude petroleum and white oil products imports. The present harbor has an oil wharf located in the extension of the commercial quay on the west side. Sediment samples were extracted from eight different locations throughout the harbor. Hereinafter, samples from Zarzis harbor are designated ZS1 to ZS8.

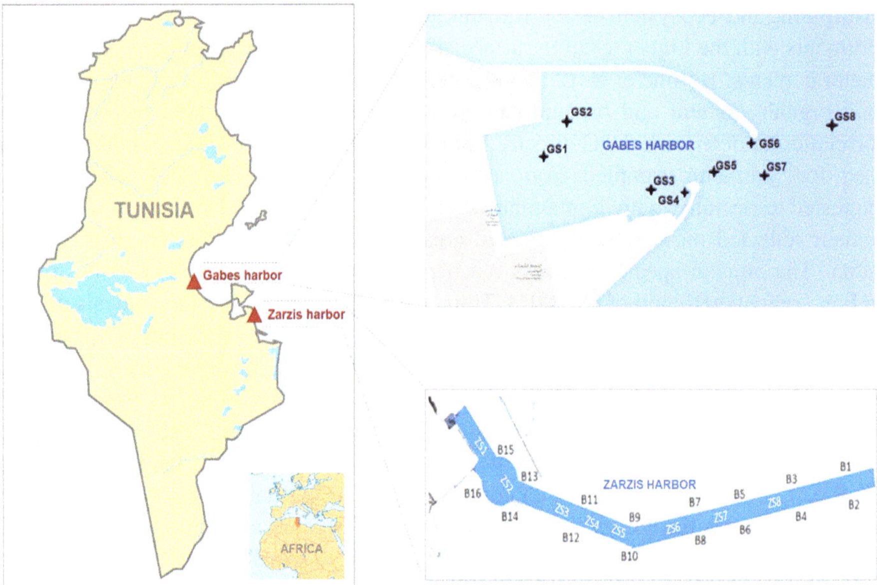

Fig. 1. Location of Gabes and Zarzis harbors and sampling description

3 Methods

3.1 Specific Surface Area and Pore Volume

The specific surface area SSA was determined by N_2 gas adsorption on dried sediments. For this purpose, an apparatus type Micrometrics (Tristar II Plus 2.02) was used. The analysis Bath Temperature was 77.3 K. The test allowed the collection of adsorption-desorption isotherms. Basing on these isotherms the Brunauer, Emmett, and Teller (BET) surface areas and the pore volume in sediments were carried out.

3.2 Total Organic Carbon

The total organic carbon (TOC) was carried out basing on the French standard NF EN 13137. The TOC was calculated from the difference between the concentration of total carbon (TC) and the total inorganic carbon (TIC). The first parameter TC was determined via measurement of carbon dioxide released during combustion (AFNOR 2001).

4 Results and Analysis

4.1 Sediments Characteristics

In order to evaluate relationships between Tunisian sediments from both Gabes and Zarzis harbor, a complete characterization of Gabes and Zarzis harbors were determined (Missaoui et al. 2015). The grain size analysis conducted on sediments revealed that GS samples were silty sands and sandy silts however, ZS samples were silty sands. It has been noticed in previous researches that lesser the grain distribution, the greater the specific surface area was (Horowitz and Elrick 1987). In addition, researchers demonstrated that sediments with fine grain distribution presented more organic matter content (OM) (Nelson 1983). The full characterization of Tunisian sediments revealed that the OM content in GS ranged from 3.49% up to 24.8%, however, in ZS, the organic matter content varied between 2.8% and 20.9%. It was shown in previous studies that the high organic matter content was responsible for greater water content (Soane 1990). This was confirmed by water contents determined in GS (87.2%–533.2%) and ZS (94.2%–364.6%).

In this paper, only specific surface area, pore volumes, Total organic carbon, and Total nitrogen content of sediments were reported. Table 1 displays the characteristics of different samples extracted from Gabes and Zarzis harbors. This section is devoted to the analysis of obtained results.

– *Specific surface area*

Table 1 presents the specific surface areas (SSA) of samples extracted from Gabes and Zarzis harbors. Gabes sediments (GS) reported relatively high specific areas, that ranged from 8.6 m^2/g up to 38.8 m^2/g. These results were related to the fine grain size distribution reported in these sediments (Missaoui et al. 2015). Zarzis sediments (ZS) revealed low values that ranged over the following interval [2.7 m^2/g – 10.4 m^2/g], which reflected their sandy aspect. GS results were comparable to those reported in sediments sampled from the Cantabrian estuaries in Northern Spain (Coz et al. 2007).

Table 1. Characteristics of sediments from Gabes and Zarzis harbors (Specific surface area, Pore volume, and TOC).

Samples		Specific surface area (m^2/g)	Pore volume (cm^3/g)	TOC (%)
Gabes harbor	GS1	9.3	0.025	3.2
	GS2	29.5	0.077	8.1
	GS3	8.6	0.022	1.9
	GS4	21.0	0.054	5.4
	GS5	38.8	0.091	9.9
	GS6	19.8	0.052	5.4
	GS7	25.6	0.067	6.1
	GS8	9.9	0.027	1.6
Zarzis harbor	ZS1	5.9	0.025	3.5
	ZS2	10.4	0.037	6.5
	ZS3	3.5	0.012	1.4
	ZS4	2.8	0.009	1.6
	ZS5	4.1	0.014	2.6
	ZS6	4.5	0.015	1.5
	ZS7	4.5	0.016	1.5
	ZS8	2.7	0.010	0.9

In fact, SSA revealed to be significantly correlated to the capacity of adsorption and desorption between sediments particles and contaminants dissolved ions (Petersen et al. 1996). Thus, sediments with high SSA could be easily contaminated with heavy metals.

– *Pore volume*

Table 1 displays the pore volume determined in Gabes and Zarzis harbors. Results ranged from 0.009 cm^3/g in ZS8 up to 0.091 cm^3/g in GS5. It was noticed that Gabes sediments reported the highest pore volumes. GS results were comparable to those reported before in clay minerals (Illite (0.02 cm^3/g) and Montmorillonites (0.11 cm^3/g)) (Diamond 1970). The pore volumes in sediments could be used to estimate the capacity to absorb and transfer organic and inorganic species (Hay et al. 2011).

– *Total organic carbon*

Table 1 depicts the Total organic carbons in GS and ZS samples. Results varied between 0.9% in ZS8 and 9.9% in GS5. Gabes sediments reported the higher organic contents with a mean value of 5.2%, which were comparable to results reported in the study of Bel Hadj Ali et al. (2014). Zarzis harbor presented lower values with a mean Total organic carbon of 2.44%. ZS organic carbon contents were similar to those determined in the marine sediments from Port of Santos (Brazil) (Buruaem et al. 2013).

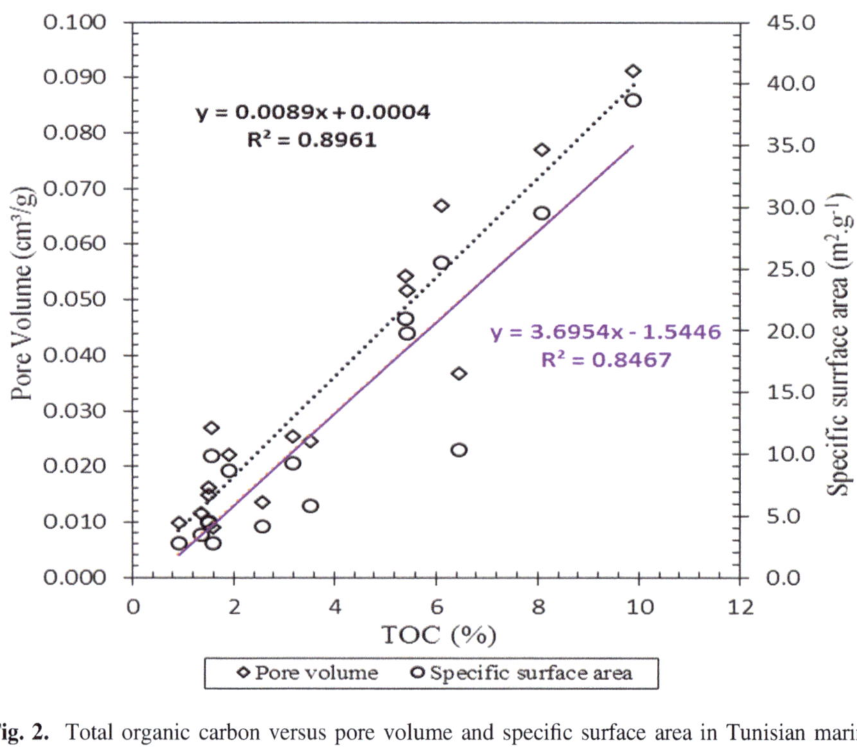

Fig. 2. Total organic carbon versus pore volume and specific surface area in Tunisian marine sediments

4.2 Interrelationships Between TOC Content, Pore Volume and Specific Surface Area in Tunisian Sediments

Figure 2 displays the relationships between TOC and pore volume and between TOC and SSA. A linear regression was performed on both relationships. A significant correlation between TOC and pore volume was reported with a regression coefficient of 0.896. In the study of (Mayer et al. 2004), it was shown that the pore volumes in sediments can be converted to organic matter masses. Thus, the significant relationship between TOC and pore volume could be justified. In Tunisian sediments, A significant correlation between TOC and SSA was also determined (R^2 = 0.846). In previous research, significant correlations were only found between SSA and organic content to fine particles ratio (OC/(clay) (Zeraat et al. 2011). The significant correlations depicted in Fig. 2 could be justified by the important relation between pore volume and SSA, which were determined via the same N_2 adsorption test. The latter relation was presented in the research of Chilingar et al. (1963). In one hand, as reported in the study of Mayer (1994), sediments specific surface area have a relevant relationship with organic matter contents. In the other hand, the significant correlation between total organic content and organic carbon content demonstrated in the study of (Burone et al. 2003). The latter relationships were justified by the characteristics of Tunisian sediments. Figure 3 presents the correlation between Organic matter and both pore volume and

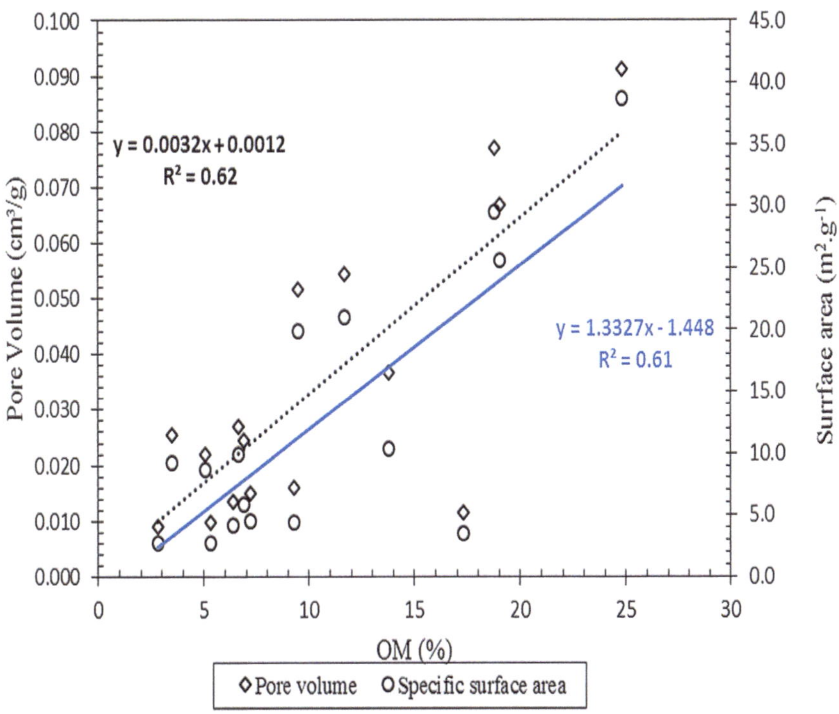

Fig. 3. Organic matter content (OM) versus pore volume and specific surface in Tunisian marine sediments

Table 2. Total organic carbon content in marine sediments worldwide.

Location	TOC (%)	References
Hudson Bay, Canada (America)	0.23–1.27	(Hare et al. 2014)
Gabes harbor, Tunisia (Africa)	7.10–7.56	(Bel Hadj Ali et al. 2014)
Wuljin east sea, Korea (Asia)	0.06–1.75	(Lee et al. 2011)
Port of Santos, Brazil (America)	1.31–3.88	(Buruaem et al. 2013)
Ubatuba bay (America)	0.05–3.57	(Burone et al. 2003)

SSA in Tunisian sediments. Results showed that significant correlations ($R^2 > 0.6$) between couples (TOC, pore volume) and (TOC, SSA) were determined (Table 2).

5 Conclusion

The objective of this study was to establish correlations between total organic content and both specific surface area and pore volume in sediments. For this purpose, sixteen samples were extracted from two different Tunisian harbors. Results revealed that SSA in marine sediments ranged from 2.7 m^2/g up to 38.8 m^2/g. However, pore volume varied between 0.009 cm^3/g and 0.091 cm^3/g. The TOC content ranged from 0.9% up to 9.9%. Analysis of determined characteristics showed significant correlations between both couples (TOC vs pore volume) and (TOC vs SSA). Basing on these significant relationships between different characteristics, the time and cost consuming during sediments characterization could be minimized.

Acknowledgments. This study was funded by the partnership Hubert Cruien "Utique" of the French Ministry of Foreign Affairs and the Tunisian Ministry of Higher Education and Scientific Research in Tunisia (Project N° 14 G 1116). The authors express their acknowledgements to the Tunisian Merchant Marine and Ports Office and the Tunis International Center for Environmental Technologies.

References

AFNOR. NF EN 13137 Characterization of waste - Determination of total organic carbon (TOC) in waste, sludges and sediments (2001)

Bel Hadj Ali, I., et al.: Characterization of Tunisian marine sediments in Rades and Gabes harbors. Int. J. Sedim. Res. **29**(3), 391–401 (2014)

Burone, L., Muniz, P., Pires-Vanin, A., Maria, S.: Spatial distribution of organic matter in the surface sediments of Ubatuba Bay (Southeastern – Brazil). Anais Da Academia Brasileira De Ciencias **75**, 77–90 (2003)

Buruaem, L.M., et al.: Integrated quality assessment of sediments from harbour areas in Santos-São Vicente Estuarine system, Southern Brazil. Estuar. Coast. Shelf Sci. **130**, 179–189 (2013). http://www.sciencedirect.com/science/article/pii/S0272771413002710. Accessed 15 Jan 2016

Chilingar, G.V., Main, R., Sinnokrot, A.: Relationship between porosity, permeability, and surface area of sediments. J. Sedim. Petrol. **33**(3), 759–765 (1963)

CJB Environnement and EAM. Etude d'identification des sites de rejet des matériaux de dragage dans les ports de commerce Tunisiens (2006)

Coz, A., et al.: Physico-chemical and environmental characterisation of sediments from Cantabrian estuaries (Northern Spain). Aquat. Ecosystem Health Manage. **10**(1), 41–46 (2007)

Diamond, S.: Pore size distributions in clays. Clays Clay Miner. **18**, 7–23 (1970)

Hare, A.A., et al.: Characterization of sedimentary organic matter in recent marine sediments from Hudson Bay, Canada, by Rock-Eval pyrolysis. Organic Geochem. **68**, 52–60 (2014). http://www.sciencedirect.com/science/article/pii/S0146638014000084. Accessed 26 Feb 2014

Hay, M.B., et al.: Characterization of the intragranular water regime within subsurface sediments: pore volume, surface area, and mass transfer limitations. Water Resour. Res. **47**(10), 1–19 (2011)

Horowitz, A.J., Elrick, K.A.: The relation of stream sediment surface area, grain size and composition to trace element chemistry. Appl. Geochem. **2**(4), 437–451 (1987)

Hyland, J., et al.: Organic carbon content of sediments as an indicator of stress in the marine benthos. Marine Ecol. Prog. Ser. **295**, 91–103 (2005)

Lee, C., et al.: Geotechnical characterization of marine sediments in the Ulleung Basin, East Sea. Eng. Geol. **117**(1–2), 151–158 (2011) http://linkinghub.elsevier.com/retrieve/pii/S0013795210002188. Accessed 21 Feb 2014

Mayer, L.M., et al.: Organic matter in small mesopores in sediments and soils. Geochim. Cosmochim. Acta **68**(19), 3863–3872 (2004)

Mayer, L.M.: Surface area control of organic carbon accumulation in continental shelf sediments. Geochim. Cosmochim. Acta **58**(4), 1271–1284 (1994)

Missaoui, A., et al.: Chemical and physical characterization of Gabes marine sediments. In: 16th African Regional Conference on Soil Mechanics and Geotechnical Engineering, pp. 529–536 (2015)

Nelson, C.S.: Bottom sediments of Lake Rotoma. New Zealand J. Marine Freshwater Res. **17**, 185–204 (1983). %5C%5Cniwa-ham%5Cshared%5CProjects%5CHRC08201%5CWorking%5CBibliography%5CLiterature database (EKW updated 180908)%5CRegion specific Ngati Pikiao Te Arawa%5CNelson 1983 bottom sediments of Lake Rotoma.pdf

Petersen, L.W., et al.: Relations between specific surface area and soil physical and chemical properties. Soil Sci. **161**, 1–13 (1996)

Rawlins, B.G., et al.: Applied geochemistry estimating specific surface area of fine stream bed sediments from geochemistry. Appl. Geochem. **25**(9), 1291–1300 (2010). http://dx.doi.org/10.1016/j.apgeochem.2010.05.009

Schumacher, B.A.: Methods for the determination of total organic carbon (TOC) in soils and sediments (2002)

Soane, B.D.: The role of organic matter in soil compactibility: a review of some practical aspects. Soil Tillage Res. **16**(1–2), 179–201 (1990)

Zeraat, M., Farhad, P., Ali, K.: The relationship between specific surface area and soil organic carbon in loess-derived soils of Northern Iran, pp. 1–2 (2011)

Some Studies on Mechanical Behaviour of Cement Stabilized Dredged Soil from Flood Spill Channel of Jhelum River Srinagar

B.A. Mir(✉)

Department of Civil Engineering, National Institute of Technology, Srinagar 196006, J&K, India
p7mir@nitsri.net, bashiriisc@yahoo.com, bamiriitb@gmail.com

Abstract. Soft soil deposits usually have a low bearing capacity and undergo excessive settlement over a long period of time. Using such materials either as a foundation medium or construction material is almost impossible without some means of improving their adverse properties. This study deals with an overview of the dredged soil stabilization generated by dredging the flood spill channel of Jhelum River, Srinagar. But concern over environmental affects of dredging, disposal and the increasing unavailability of suitable disposal sites, has put pressure for characterization of this material. Thus, disturbed and undisturbed samples of dredged soil from different locations of flood channel were collected for conduct of various field and laboratory tests for determination of physical and mechanical properties as per relevant standard procedures. On the basis of geotechnical investigations, it has been observed that dredged soil is composed of silt, sand, clay and organic matter, all important components of topsoil, which is not suitable as a construction material in its in-situ condition. Therefore, an attempt has been made to study some physical and mechanical properties of cement stabilized dredged soil for its utilization in bulk in various geotechnical applications. The soil-cement mixing is adopted to improve a soft clay foundation, which provides stability during the construction stage. Test specimens were prepared with dredged soil using a range of cement (by dry weight of the soil) at $0.95\gamma_{dmax}$ and optimum moisture content and subjected to different tests as per relevant standard procedures. On the basis of geotechnical investigations, it is observed that there is significant improvement in the mechanical behaviour of cement stabilized dredged soil. Thus the main objective of present study was a two fold-first, to overcome challenges posed by dredged soil and second, to adopt its economical and environmentally beneficial utilization in various geotechnical applications as a construction material.

1 Introduction

Soft Soil deposits usually have a low bearing capacity and undergo excessive settlement over a long period of time. One of the important problems geotechnical engineers are often to deal with is the reliable assessment of mechanical behaviour of soft soils. Using such materials either as a foundation medium or construction material is almost

D.N. Singh and A. Galaa (eds.), *Contemporary Issues in Geoenvironmental Engineering*, Sustainable Civil Infrastructures,
DOI 10.1007/978-3-319-61612-4_32

impossible without some means of improving their adverse properties. This study deals with the solid waste generated by dredging the flood spill channel of Jhelum River, Srinagar. Jhelum River is essentially the lifeline of Kashmir valley, forming the backbone of all its economic activities. Numerous hydropower projects on it and its tributaries such as Kishanganga, Uri and Lower Jhelum are vital to electricity generation in the state of Jammu and Kashmir. However, River Jhelum is also the cause of large and small scale flooding in the region. Floods have been reported frequently throughout history in Kashmir. One of the most disastrous floods, which occurred in September 2014, in which a large area of South and Central Kashmir, especially the Srinagar city was submerged, which caused intolerable devastating damage to residential buildings and shopping complexes (Fig. 1). In view of this, various flood control measures and methods have been recommended from time to time. One among them is flood spill channel which carries excess water from main river body during flooding from Padshahi Bagh, Srinagar and running into Wular Lake. But unfortunately, the present capacity of the channel is 5000 cusecs, which is significantly less than its design capacity of 17000 cusecs in 1902. This decrease is due to the encroachments and sediment deposition in the flood channel over the years. In such a scenario, the threat of flooding is evident and thus various proposals are being formulated, one among them being dredging out sediment deposited in the bed of the flood channel (Fig. 2). Dredging of the river bed and flood channels generates dredged soil in large quantity posing serious health and environmental problems. Concern over environmental affects of dredging, disposal and the increasing unavailability of suitable disposal sites, has put pressure on Geotechnical Engineers for characterization of this material. This material can be a valuable resource for many practical purposes. Depending on the type of environment, excavated material may comprise of gravel, sand, silt or soft clays. On the basis of its characterization, it may be put to various uses such as fill material, subgrade construction, reclamation, landscaping, agriculture, covers for landfills, constructing wetlands for water quality improvement, bank stabilization, creation of islands, wildlife habitat wetlands and amongst others. A number of techniques are available for dewatering and improving engineering properties of dredged material and other soft soils (Haliburton 1978; Chan et al. 2011). All improvement techniques require construction equipment, and selection of appropriate equipment is important. Detailed investigations (Bartos 1977; Spigolon 1993; Winfield and Lee 1999; Dermatas et al. 1999; Harrington and Smith 2013) have been reported for characterization of dredged material and depending on the sediment type and processing requirements, dredged material may be used as: concrete aggregates (sand or gravel); backfill material; raw material for brick manufacturing (clay with less than 30 percent sand); ceramics, such as tile; pellets for insulation or lightweight backfill or aggregate (clay); and raw material for the production of riprap or blocks for the protection of dikes and slopes against erosion (rock, mixture). Dredged materials may also be used for environmental enhancement of wetlands, fisheries and other habitats for wildlife utilization (Fig. 3). Beneficial use of dredged material is an integral and necessary part of dredge material management process (Aarninkhof and Luijendijk 2010; Mir and Mir 2004). For all above applications, a brief study about geotechnical characterization of dredged material forms an important consideration. Therefore, an attempt has been made to study some physical and mechanical properties of dredged

Fig. 1. Devastating flood in Kashmir-2014

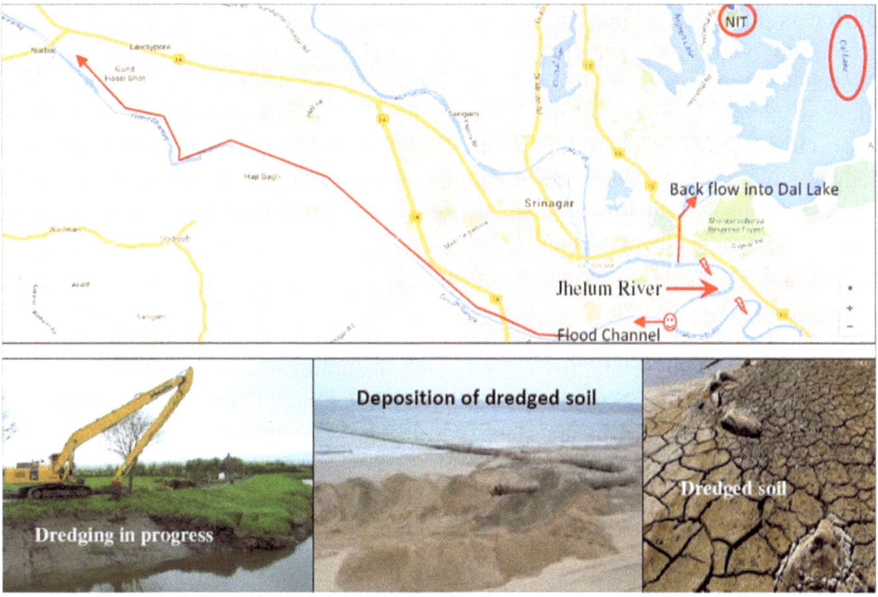

Fig. 2. Dredging and dredged soil deposits along flood channel shore line

Fig. 3. Utilization of dredged soil for land creation and beach nourishment

soil for its utilization in bulk in various geotechnical applications. In this study, disturbed and undisturbed samples of dredged soil from different locations of flood channel were collected for conduct of various field and laboratory tests for determination of physical, index and mechanical properties as per relevant standard procedures. On the basis of geotechnical investigations, it has been observed that in-situ dredged soil is composed of silt, sand, clay and organic matter, all important components of topsoil, which is not suitable as a construction material in its in-situ condition. These soils exhibit high water content and low workability, which pose difficulties for various construction projects. However, treated dredged soil can be utilized as a resource for various engineering applications and a stabilizer for improving behaviour of fine grained soils (Grubb et al. 2008; Grubb et al. 2010-I; Grubb et al. 2010-II; Chrysochoou et al. 2010). There are numerous additives available for stabilization of such problematic soils. However, the choice and effectiveness of an additive depends on the type and in-situ state of soil deposit. High water content and low workability of marginal soils are not suitable for construction of highway construction project. However, addition of cement as an additive proves as an effective stabilizer for better control of workability during compaction of these cement admixed marginal soils (Mitchell 1976; Miller and Azad 2000; Balasingam and Sariosseiri 2008). Therefore, in this study, test specimens were prepared with dredged soil using a range of cement (by dry weight of the soil) at $0.95\gamma_{dmax}$ and Optimum moisture content and subjected to different tests as per relevant standard procedures (ASTM 1632; ASTM D1633; PCA 1995; PCA 2006). Although, many researchers (Hernandez and Al-Tabbaa 2005; Osman and Al-Tabbaa 2005) have also reported mechanical behaviour of cement stabilized soft soils, yet there is need for a systematic investigation for examing the effectiveness of cement treatment and to study mechanical properties of these soils for their bulk utilization as an engineered construction material. On the basis of geotechnical investigations, it is observed that the addition of cement improved the drying rate, workability, compaction and strength characteristics of the soils. Significant improvement in unconfined compressive strength and modulus of elasticity are attained by cement treatment of these soils. The main contributions of this study to practice are on quantifying improvement in mechanical behaviour due to cement treatment and highlighting the fact that higher percentages of cement could turn stabilization from beneficial to an extremely dangerous practice. On the whole, the main objective of this study is to successfully manage dredged soil in a manner that is protective of human health and the environment in and around Jhelum River.

2 Cement Stabilization of Soils

Due to rapid urbanization and increase in population by many folds during past few decades, there is a scarcity of good construction sites for construction of various engineering projects and Geotechnical Engineers are required to investigate the suitability of marginal/weak soil deposits. Since marginal/weak soil deposits are considered to have poor engineering characteristics, which cause severe damage to civil engineering structures and facilities. Hence, these soils must be treated prior to construction operations, so that desired properties can be achieved. Ground improvement by treating soft soil with various types of stabilizers is an attractive alternative and often economical compared to other ground improvement methods. Addition of stabilizer alters the engineering properties of the existing soil to create a new site material which is capable to meet certain material requirements. Among various chemical stabilization methods, cement stabilization can be used to improve properties of marginal/weak soils to provide a workable platform for construction various infrastructures/projects (ACI230.1R-90, Brandon et al. 2009). Since in present study, the dredged soil consists of sandy silt with appreciable clayey content, cement stabilization may prove more beneficial. Also, due to lack of aluminates and silicates in silts and granular material, pozzolanic reactions do not take place, therefore, pozzolanic admixtures such as cement need to be used for stabilization of silt dominated soils. Cement being a mechanical additive, is often used to improve the strength and stiffness of soft clayey soils (Mitchell 1981; Uddin et al. 1997). Many other researchers (Croft 1967; Chew et al. 2004) have documented the beneficial effects of cement on the performance of soils. Cement treatment leads to improvement in the mechanical properties of soils. However, the findings of different researchers on the role of Portland cement on compacted properties as well as strength and modulus have not been entirely consistent. For example, some researchers (Balmer 1958; Clough et al. 1981) reported cement treatment increased cohesion while internal friction angle remained constant, while as others (e.g., Chew et al. 2004) stated that internal friction angle increased significantly. Some studies have indicated that at very low cement contents, improvement in strength is due to an increase in friction angle rather than cohesion (Rocha et al. 1961). Research has also shown that cohesion increases with curing time while friction angle remains constant (Nash et al. 1965; Wissa et al. 1965). Abboud (1973) reported that the Mohr-Coulomb failure envelope for cement treated soils is curved and several failure criteria, including Griffith crack theory and modified Griffith crack theory have been proposed to account for the strength behaviour of soil. It has also been reported by various researchers (Thompson 1966; Muhunthan and Sariosseiri 2008) that fine grained soils yield higher increase in cohesion (particularly in compressive strength under UCS test) compared to improvement in internal friction, which signifies that cement stabilization displays brittle behaviour.

3 Stabilization Mechanism

Cement stabilization is one such technique that has been adopted for the improvement of soft clay dominated soils. The major reactions that result in the increase in shear strength of soft soil when cement is added are (Ho and Chan 2011; Kamruzzaman et al. 2002):

1. Isomorphous substitution of calcium in the clay particles decrease the interlayer spacing and causes coagulation of the clay particles, thus reducing the plasticity of the soil.
2. Calcium hydroxide will be formed when quick lime reacts with water as a result of which the pH in the system increases.
3. Calcium silica hydrate (C-S-H) compounds are formed by the dissolution of silica in the clay particles followed by its reaction with calcium oxides. These compounds

form bonds either between the binder particles and soil particles or between the binder particles to form a stiff matrix. The composition and structure of the C-S-H gel and the type and amount of other hydration products can alter due to the presence of organic matter in soil.

The reactions that take place in the soil–cement stabilization can be represented as below:

$$C_3S + H_2O \rightarrow C_3S_2H_x + Ca(OH)_2 \quad (1)$$

$$Ca(OH)_2 \rightarrow Ca^{++} + 2(OH)^- \quad (2)$$

$$Ca^{++} + 2(OH)^- + SiO_2 \rightarrow CS \quad (3)$$

$$Ca^{++} + 2(OH)^- + Al_2O_3 \rightarrow CAH \quad (4)$$

Thus, soil-cement is a simple, highly-compacted mixture of soil, Portland cement, and water. As the cement reacts, or hydrates, the mixture gains strength and improves the engineering properties of the raw soil. Its improved strength and durability, combined with its low first cost and ease of construction, make soil-cement an outstanding value for use as a base and subbase material (Fig. 4).

Thus, the main objective of this study is to successfully manage dredged soil in a manner that is protective of human health and the environment in and around Capital City Srinagar.

Fig. 4. Soil-cement stabilization for roadway construction

4 Materials and Methods

In the present investigation, samples of dredged soil have been collected from three locations of flood spill channel (Fig. 5). After geotechnical evaluation of physical and mechanical properties (Table 1), the weakest sample out of the three chosen sample

Fig. 5. Sample collection site from flood channel

Table 1. Properties of dredged soil

Sr. No.	Properties	S-III
1	Sand size (%)	22
2	Silt size (%) < 75μ	62
3	Clay size (%) < 2μ	16
4	Coefficient of uniformity	NA
5	Coefficient of curvature	NA
6	Suitability number	>50
7	Specific gravity, G	2.52
8	Moisture content, w_n (%)	36.6
9	Liquid limit (%)	38.6
10	Plastic limit (%)	29.8
11	Shrinkage limit (%)	13.9
12	Plasticity index, PI = LL–PL (%)	9.8
13	Plasticity index A-Line, PI_A = 0.73(LL-20) (%)	13.6
14	Plasticity index U-Line, PI_U = 0.9(LL-8) (%)	27.5
15	Clay mineral type	Kaolinite
16	Classification	MI
17	Liquidity index, LI	0.70
18	Flow index, I_F	15.9
19	Toughness index	0.62
20	Activity	0.61
21	Liquid limit, LL (oven-dried) (%)	32.8

(continued)

Table 1. (*continued*)

Sr. No.	Properties	S-III
22	Free well index, FSI (%)	3.8
23	Optimum moisture content - by light comp. (LC) test (%)	24.6
24	Maximum dry unit weight - by light comp. test (kN/m^3)	14.81
25	Optimum moisture content - by heavy comp. (HC) test (%)	21.4
26	Maximum dry unit weight by heavy comp. test (kN/m^3)	16.1
27	CBR value % (LC)	3.3
28	CBR value % (HC)	4.1
29	Undrained cohesion, c_u - in-situ (kN/m^2)	29.7
30	Direct shear test parameters: in-situ	c_u = 10 kPa
		$\phi = 21^\circ$

was selected for cement stabilization. Though any type of cement can be used to stabilization of soils, the most commonly used is the ordinary Portland cement (Janz and Johansson 2002). Test specimens were prepared with dredged soil using a range of cement from 6% to 24% (with 6% increment by dry weight of the soil) at $0.95\gamma_{dmax}$ and optimum moisture content and subjected to different tests as per relevant standard procedures. Various laboratory tests performed to characterize the stabilized dredged soil include determination of engineering properties such as compaction tests and shear strength tests. All the tests were carried out as per the relevant Standards [ASTM].

5 Results and Discussions

The properties, which are of greatest importance in the characterization of a material to be used as a resource material, are physical and mechanical properties. Thus, the cement stabilized dredged soil samples have been characterized for establishing their suitability as resource material. Cement stabilization improve the engineering properties of marginal soils such as, compaction characteristics, unconfined compressive strength, modulus of elasticity, etc. The test results are described as below:

6 Physical and Index Properties of Dredged Soil

Dredged soils vary in colour. A single sample may exhibit a whole range of colours merging more or less perceptibly into one another in a variety of patterns and forms. The specific gravity of dredged soil (ASTM D854 (14), ASTM D5550 (14)) lies in the narrow range from 2.52 to 2.59 unlike natural soils (2.65 to 2.85). Low specific gravity and a relatively uniform grain size distribution, resulting in low unit weight will result in lower earth pressure. For particles of size finer than 75 micron, hydrometer analysis was also carried out (Fig. 6). Particle size distribution analysis (ASTM D 422–63 (07e2).) data revealed that the dredged soil contained about 62% silt, 22% sand and appreciable fine content. The soil grading is used to know the susceptibility of a soil to

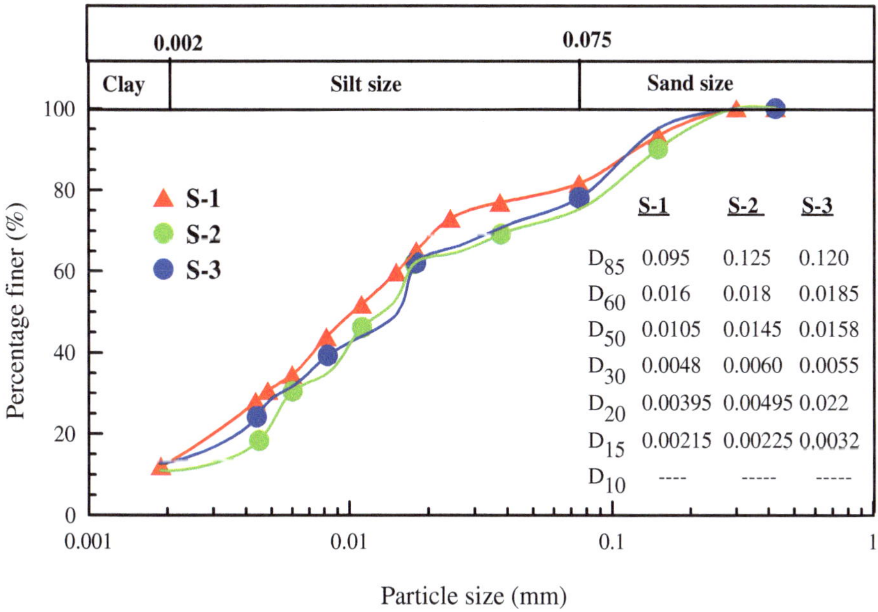

Fig. 6. Particle size distribution curves (Flood Spill Channel Srinagar)

frost action, required for the design of drainage filters, an index to the coefficient of permeability and the shear strength of the soil. The suitability of a backfill material also depends on the soil gradation. Atterberg limits such as liquid limit, plastic limit and shrinkage limit are extensively used in geotechnical engineering, which provide an overall idea for the engineering properties of the soils. Atterberg limit tests were conducted as per relevant standard procedure (ASTM D4318 (0e1), ASTM D427 (04)). Since, the dredged soil is silty in nature, therefore, liquid limit tests were also conducted using cone penetration method for cross-check. However, there was negligible variation between test values. The liquid limit (air dried) and plasticity index of the untreated dredged material varies in the range of 34 - 39% and 8–10 respectively. Based on test results, dredged soil can be classified as silty sand with medium plasticity. However, consistency limits are reduced for cement stabilized dredged sample changing it from medium plasticity to low plasticity soil. The test results are presented in Table 1 and the flow curves of dredged soil are illustrated in Fig. 7(ab). From test results, it is seen that liquid limit (LL) decreases from 38.6% to 24.4% on addition of 6% cement content. This can be attributed due the fact that the pozzalonic behaviour of cement makes the treated soil coarser than original soil samples due to the agglomerations of cement and soil particles and hence water holding capacity of stabilized soil is reduced compared to untreated soil. Further, there is a marginal change in LL values (22–24.4%) with increasing cement content (6–18%), which is due to the cation exchange reaction and flocculation–aggregation for presence of more amount of cement, which reduces plasticity index of soil. It is also seen that liquid limit increases as cement content increases. One possible reason for the increase in the liquid limit of

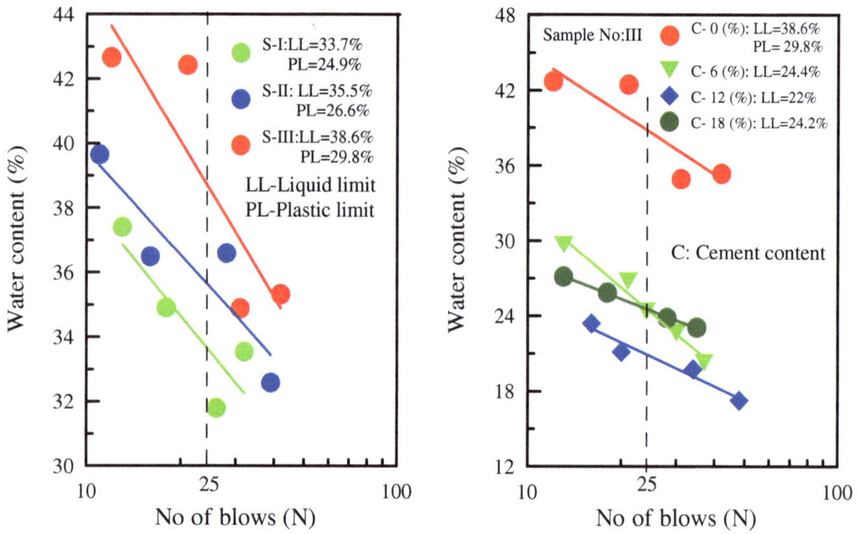

(a). Untreated dredged soil (b). Cement stabilized dredged soil

Fig. 7. Flow curves for dredged soil from flood channel Srinagar

the treated soil may be the microporous nature of the aggregated particles of the material that trap the water within intra-aggregate pores. Similar results have also been reported by other researchers (e.g., Chew et al. 2004).

7 Engineering Properties

The engineering properties of the dredged soil are very much influenced by the factors like source of dredging, disposal point etc. The various engineering properties are described as below:

7.1 Effect of Cement Stabilization on Compaction Characteristics of Dredged Soil

One of the basic and least expensive construction procedures used for soil stabilization is compaction. Compaction improves the engineering properties of foundation material so that the required shear strength is obtained. Compaction is often required when building subgrades or bases for airport pavements, roads, embankments, earthfill dams, or similar structures. Compaction is the process of increasing the density of material by packing the particles close together with reduction in volume of air. The compacted unit weight depends upon the amount and method of energy application and material properties such as grain size, gradation, particle shape, plasticity and the moisture content at compaction. Both standard Proctor and modified compaction tests (ASTM D698 (12e2),

ASTM D1557 (12e1)) were conducted on untreated dredged samples collected from flood channel and weakest sample was chosen for cement stabilization. The test results of the standard and modified compaction test are presented in a plot of dry unit weight versus water content as shown in Fig. 8(ab).

From Fig. 8, it is observed that the maximum dry unit weight of the stabilized samples is higher than that of the unstabilized samples whereas the optimum water content varies in a very narrow range. Since the optimum moisture content (OMC) of low swelling soils is mainly controlled by the liquid limit, therefore, the variation in OMC of the unstabilized and the stabilized samples is negligible. Similar behaviour has also been reported by other researchers e.g. Horpibulsuk (2006). The variation of optimum moisture content (OMC) and maximum dry density (MDD) with cement content for cement stabilized specimens is shown in Fig. 9.

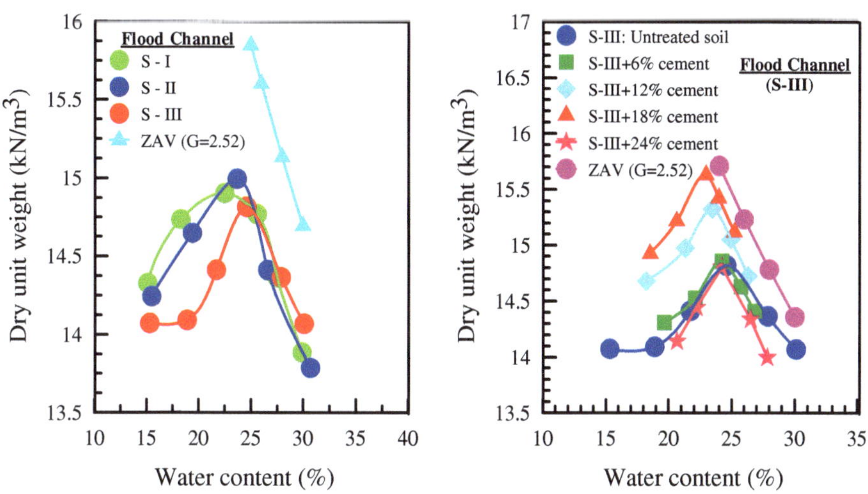

(a). Untreated dredged soil (b). Cement stabilized dredged soil

Fig. 8. Compaction curves for dredged soil from flood channel Srinagar

From Fig. 9, it is seen that OMC decreases and MDD increases with an increase in cement content upto 18% and beyond which it decreases with increase in OMC. This may be attributed due to development of inter-aggregate pores (larger than 0.01 mm) and intra-aggregate pores (smaller than 0.01 mm) during cement stabilization of stabilized dredged soil. During the soil and cement mixing, and with an increase in cement content, the formation of soil–cement clusters occurring due to physicochemical interaction, leads to reduction of the small inter-aggregate pore and to slight increase in the large inter-aggregate pore volume, which results in the decrease of the dry unit weight. However, with time, large inter-aggregate pores are filled due to the growth of cementitious products (Nagaraj et al. 1990; Mitchell 1976). Thus, as a result of that, the total pore volume decreases with time. Therefore, an increase in the dry unit weight at 18% of cement can be observed and beyond which with increase in cement content,

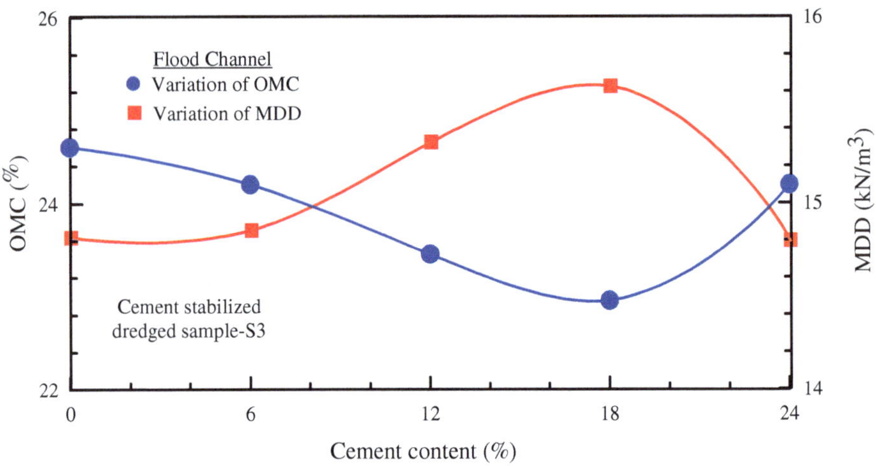

Fig. 9. Variation of optimum moisture content (OMC) and maximum dry density (MDD) with cement content for stabilized dredged soil specimen

MDD decreases and OMC increases. This increase in OMC is probably a consequence of two reasons: the first is, the additional water held with the flocculated soil structure resulting from cement and silica fume interactions, and the other reason is, exceeding water absorption by cement and silica fume as a result of its porous properties (Zhang et al. 1996). Rahman (1987) attributed the increase in dry density due higher specific gravity of cement used as stabilizer.

7.2 Effect of Cement Stabilization on Strength Characteristics of Dredged Soil

For any engineering application of soil, its strength characteristics are essential. In present study, unconfined compression and direct shear tests were conducted on the cement stabilized dredged soil. The test results are described as below:

Unconfined Compression Test

In some special cases, as for checking the short-term stability of foundations and slopes where the rate of loading is fast but drainage is very slow, one of the most common tests is the unconfined compression test. Unconfined compression strength test is the simplest and quickest test for determining the shear strength of cohesive soils. The samples were prepared (ASTM D2166 (13)) and compacted at $0.95\gamma_{dmax}$ and corresponding water content on dry of optimum (for standard Proctor tests) and tested in an unconfined compression testing machine. The variation of unconfined compression shear strength for cement stabilized dredged soil for immediate and 7days curing test series is shown in Fig. 10(a-b). Substantial improvement in strength can be attained with cement stabilization, which is due to the formation of cementitious compounds with increase in cement content. Curing time also increases the strength with cement content and may be due to the formation of secondary cementitious products due to pozzolanic reactions in addition to earlier hydration reaction which imparts further strength to the cement treated soil.

From Fig. 10a, it is observed that stress-strain behaviour of stabilized dredged soil is improved unlike untreated dredged soil behaviour, which was highly non-linear and does not exhibit any peak. The soil consistency is improved and the unconfined compressive strength (UCS) of dredged soil is rapidly improved. From Fig. 10b, it is seen that UCS of dredged soil is further enhanced after curing the cement stabilized dredged soil for 7 days. It is seen that, as cement content increases UCS and effective cohesion of soil also increase. Thus, from test results, it is seen that silty soils with appreciable clay content with low plasticity index are better suited to be stabilized with cement. Cement being a mechanical additive, is often used to improve the strength and stiffness of soft clayey soils. The variation of maximum UCS with cement content is shown in Fig. 11. The maximum shear strength (UCS) for cement stabilized soil has been achieved for an optimum cement content of 19.3%. The strength increase can be classified into different zones. Initially, when cement content is added to the untreated soil, shear strength gradually increases. This may be designated as inert zone. As the cement content increases, the cement per grain contact point increases and, upon hardening, imparts a commensurate amount of bonding at the contact points. This zone is designated as the active zone. Beyond this zone, the strength development slows down while. The incremental gradient becomes nearly zero and does not make any further significant improvement. This zone is referred to as the deterioration zone. In the deterioration zone, the water is not adequate for hydration because of the excess of cement input. Consequently, as cement content increases, the cementitious products and strength decreases. The strength decrease appears when cement content >19.3% (Fig. 11). Similar behaviour has also been reported by other researchers (e.g., Currin

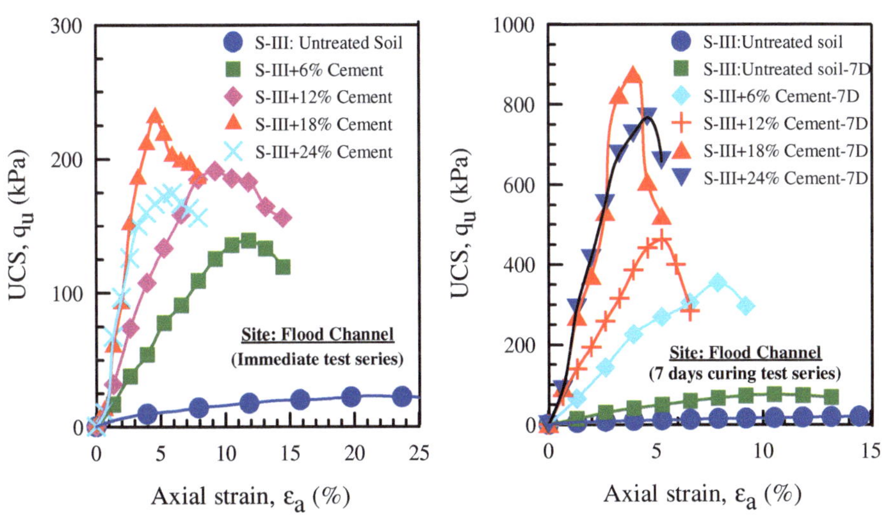

(a). Immediate test series (b). 7 days curing test series

Fig. 10. Stress-strain curve for cement stabilized dredged soil for immediate test series

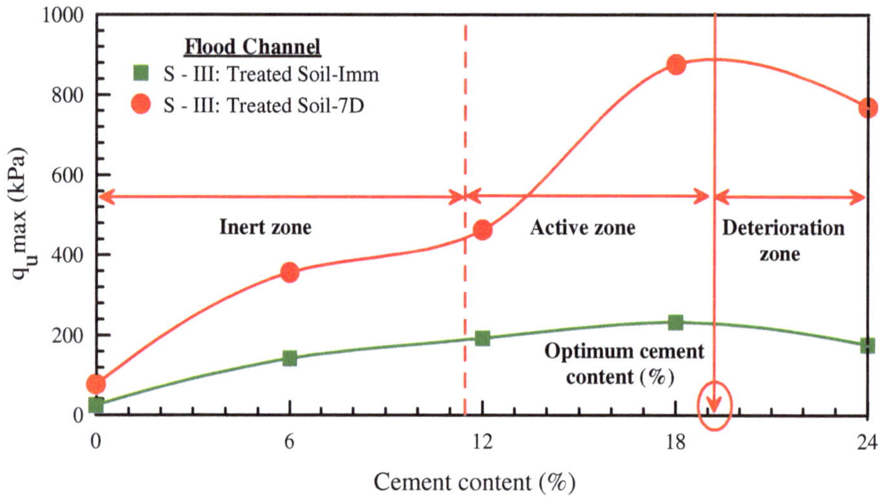

Fig. 11. Variation of maximum unconfined compressive strength (q_u) with cement content for stabilized dredged soil

et al. 1976). Test results also reveal that cement treated materials behave in a more brittle manner than non-treated materials.

The variation of failure strain with cement content is shown in Fig. 12. It has been observed that for 18% cement treated soils, specimens split vertically. Therefore, while increase in strength can be achieved by cement treatment, high percentages of cement should be used with extreme caution in field applications.

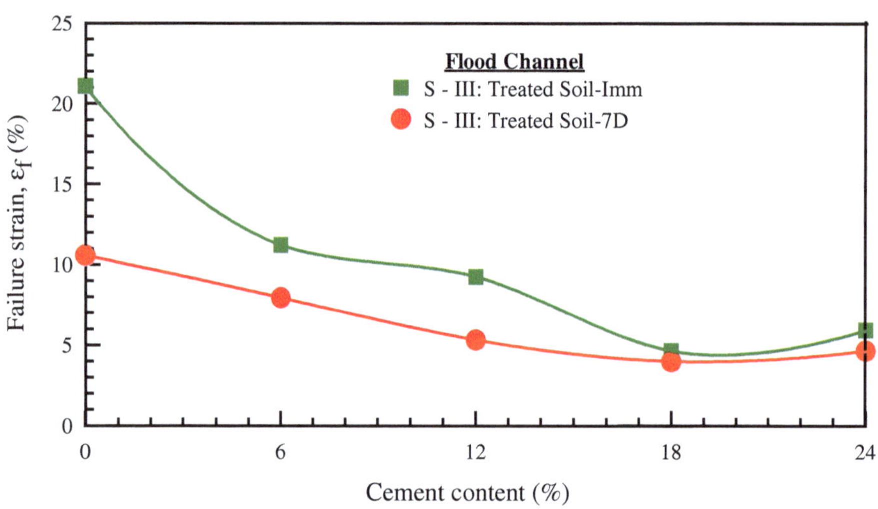

Fig. 12. Variation of failure strain with cement content for stabilized dredged soil

Direct Shear Tests

Direct shear tests were conducted (ASTM D6528 (-07)) on dredged soil compacted at $0.95\gamma_{dmax}$ and corresponding water content on dry side of optimum with and without cement stabilization for immediate and 7 days curing series tests respectively. The Mohr-Coulomb failure envelopes are shown in Figs. 13 and 14. From test results, it is seen that untreated dredged soil exhibits very soft to soft consistency. The angle of internal friction varies in a very narrow range between 21 to 26 degrees indicating that untreated dredged soil exhibits loose denseness. However, after cement treatment, shear strength parameters (c & ϕ) are improved and there is drastic change in the behaviour of untreated and treated dredged soil behaviour. The variation of cohesion (c) and frictional shearing angle (ϕ) is shown in Fig. 15. It is seen that significant improvement in strength and modulus of elasticity are attained by cement treatment of these soils. It is observed that UCS and effective cohesion of soil increase as cement content increases. However, internal friction angle remains relatively constant for cement stabilized dredged soil regardless of cement content and curing time for higher cement content (e.g. > 6%).

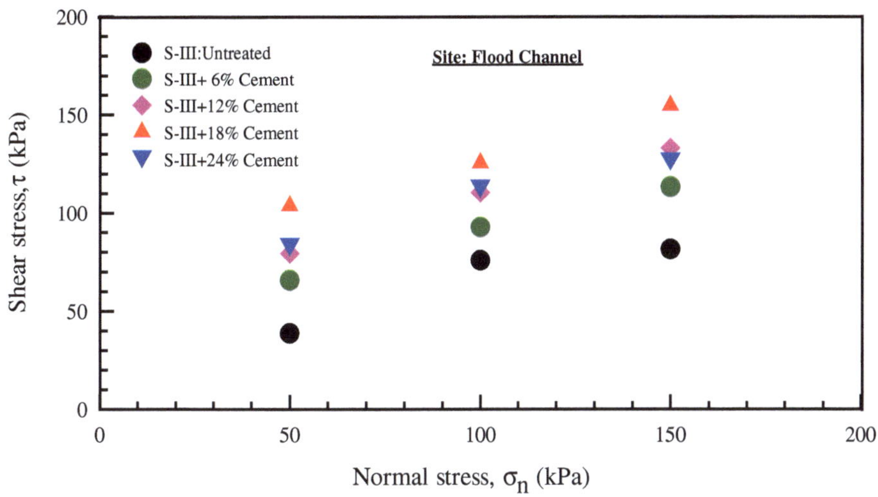

Fig. 13. Strength envelope for cement stabilized dredged soil for immediate test series

California Bearing Ratio Test

California Bearing ratio (CBR) is a measure of the load carrying capacity (resistance to direct penetration) of any soil or granular material, which is expressed as a percentage of the load carrying capacity of a standard crushed rock specimen (which is taken as of 100% value) determined by a penetration test. Resistance to penetration of a rigid plunger is measured and the loads at penetrations of 2.5 mm and 5.0 mm are expressed as percentage of two standard loads (1370 kg and 2055 kg). The higher percentage is taken as the CBR value (usually at 2.5 mm penetration). For highways, standard compaction hammer with a height of fall of 31 cm and weighing 2.6 kg (5.5 lbs) is

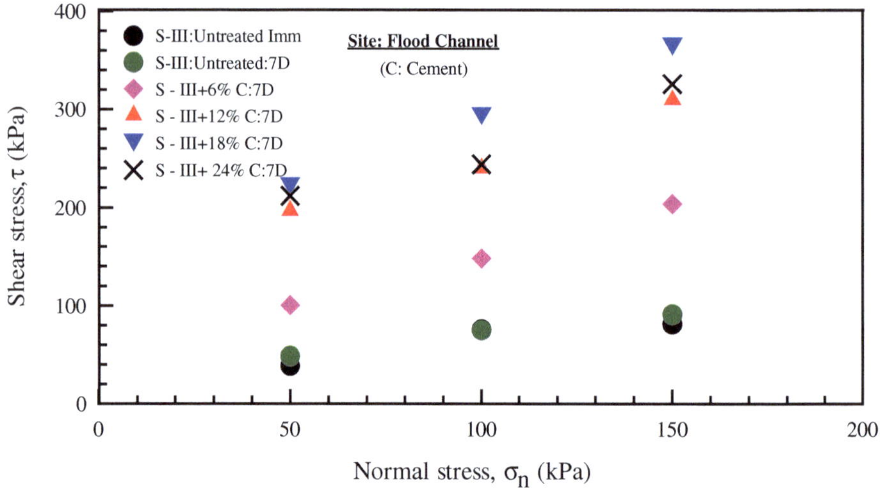

Fig. 14. Strength envelope for cement stabilized dredged soil for 7days test series

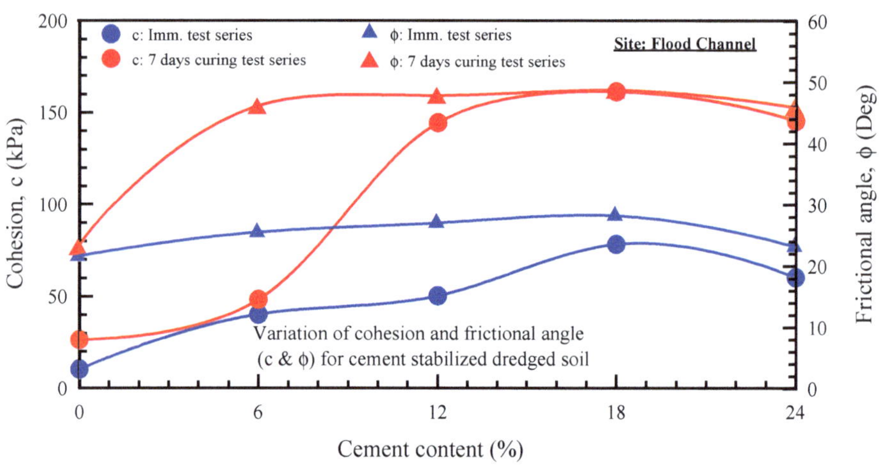

Fig. 15. Variation of shear strength parameters (c & ϕ) for cement stabilized dredged soil for immediate and 7 Days test series

generally used and the sample is compacted in three equal layers with fifty-five blows per layer. For the design or airfield pavements, the modified compaction hammer with a height of fall of 46 cm and weighing 4.89 kg (10 lbs) is generally used and the sample is compacted in five equal layers with fifty-five blows per layer. In present study, soaked CBR tests (4-days) were conducted (ASTM D1883 (16), ASTM D2168 (10)) for all the three samples compacted at $0.95\gamma_{dmax}$ and corresponding water content on dry side of optimum. The test results obtained are given in Table 2.

Table 2. CBR values for dredged material

Properties	S-I	S-II	S-III
CBR % (Standard Proctor compaction test)	4.1	4.7	3.2
CBR % (Modified Proctor compaction test)	4.5	4.9	3.9

8 Conclusions

Based on the test results obtained, the following conclusions can be made:

1. Dredged soil mainly consists of poorly graded silty sand with appreciable fines.
2. Dredged soil exhibits medium compressibility with high rate of loss of shear strength.
3. The compaction characteristics are improved by cement stabilization.
4. Cement content increases unconfined compressive strength and effective cohesion of soil also increase.
5. Internal friction angle remains relatively constant for cement treated soils regardless of cement content and curing time.
6. Cement treatment changes the behaviour of soft dredged soil from normally consolidated to over consolidated state.

Detailed investigation is being carried out for complete characterization of this material collected from flood spill channel Srinagar. Hence, using dredged soil as a resource has a two-fold advantage. First, to avoid increase the water carrying capacity of flood spill channel to avoid flooding by dredging flood channel bed and second, characterize this waste material to use it for various engineering applications as a construction material.

Acknowledgments. The investigation reported in this paper forms a part of the Research Project work on characterization of dredged soils at NIT Srinagar. Thanks are due to Faculty of Geotechnical Engg. Division and supporting staff of the Soil Mechanics laboratory for their timely help during the course of investigation. The Author also thank the IFC Deptt, Govt of J&K for allowing this work on dredged soil from flood spill channel Srinagar for its characterization and bulk utilization

References

Aarninkhof, S., Luijendijk, A.: Safe disposal of dredged material in an environmentally sensitive environment. Port Technol. Int. **47**, 39–45 (2010)

Abboud, M.M.: Mechanical properties of cement-treated soils in relation to their use in embankment construction. Ph. D dissertation, University of California, Berkeley, CA (1973)

ACI 230.1R-90. State-of-the-Art report on soil cement. ACI Mater. J. **87**(4) (1990)

ASTM 1632-96. Standard practice for making and curing soil-cement compression and flexure test specimens in the laboratory. ASTM International, West Conshohocken, PA (2007). www.astm.org

ASTM D1633-00. Standard test methods for compressive strength of molded soil-cement cylinders. ASTM International, West Conshohocken, PA (2007). www.astm.org
ASTM D854 (14). Standard test methods for specific gravity of soil solids by water pycnometer. ASTM International, West Conshohocken, PA (2014). www.astm.org
ASTM D5550 (14). Standard test method for specific gravity of soil solids by gas pycnometer. ASTM International, West Conshohocken, PA (2014). www.astm.org
ASTM D 422–63 (07e2). Standard test method for particle size analysis of soils. Annual Book of ASTM Standards, Philadelphia, Sec. 4, vol. 04-08, pp. 10-16
ASTM D4318 (0e1). Standard test methods for liquid limit, plastic limit, and plasticity index of soils. ASTM International, West Conshohocken, PA (2010). www.astm.org
ASTM D427 (04). Test method for shrinkage factors of soils by the mercury method. ASTM International, West Conshohocken, PA (2004). www.astm.org
ASTM D698 (12e2). Standard test methods for laboratory compaction characteristics of soil using standard effort (12 400 ft-lbf/ft^3 (600 kN-m/m^3)). ASTM International, West Conshohocken, PA (2012). www.astm.org
ASTM D1557 (12e1). Standard test methods for laboratory compaction characteristics of soil using modified effort (56,000 ft-lbf/ft^3 (2,700 kN-m/m^3)). ASTM International, West Conshohocken, PA (2012). www.astm.org
ASTM D2166/D2166 M (13). Standard test method for unconfined compressive strength of cohesive soil. ASTM International, West Conshohocken, PA (2013). www.astm.org
ASTM D6528 (07). Standard test method for consolidated undrained direct simple shear testing of cohesive soils. ASTM International, West Conshohocken, PA (2007). www.astm.org
ASTM D1883 (16). Standard test method for california bearing ratio (CBR) of laboratory-compacted soils. ASTM International, West Conshohocken, PA (2016). www.astm.org
ASTM D2168 (10). Standard test methods for calibration of laboratory mechanical-rammer soil compactors. ASTM International, West Conshohocken, PA (2010). www.astm.org
Balasingam, B., Sariosseiri, F.: Interpretation of geotechnical properties of cement treated soil. Research report-Compaction control of marginal soils in fills, Washington State Univ., Pullman, WA 99164-2910 (2008)
Balmer, G.G.: Shear strength and elastic properties of soil-cement mixture under triaxial loading. Portland Cement Association Research and Development Laboratories (1958)
Bartos, Jr., M.J.: Classification and engineering properties of dredged material. Technical Report D-77-17, U.S. Army Engineer Waterways Experiment Station, Vicksburg, MS (1977)
Brandon, T., Brown, J., Daniels, W., DeFazio, T., Filz, G., Musselman, J., Forshaand, C., Mitchell, J.: Rapid stabilization/polymerization of wet clay soils. Literature Review, Air Force Research Laboratory, Blacksburg (2009)
Chan, C.M., Mizutani, T., Kikuchi, Y., Kawabata, Y.: On the strength characteristics of dredged clay solidified with recycled cement. Eur. J. Sci. Res. **51**(4), 457–466 (2011)
Chew, S.H., Kamruzzaman, A.H.M., Lee, F.H.: Physiochemical and engineering behaviour of cement treated clays. J. Geotech. Geoenviron. Eng. **130**(7), 696–706 (2004)
Chrysochoou, M., Grubb, D.G., Drengler, K.L.: Stabilized dredged material III: a mineralogical perspective. J. Geotech. Geoenviron. Eng. **136**, 1037–1050 (2010)
Clough, G.W., Sitar, N., Bachus, R.C., Rad, N.S.: Cemented sands under static loading. J. Geotech. Eng. Div. ASCE **107**(GT6), 799–817 (1981)
Croft, J.B.: The influence of soil mineralogical composition on cement stabilization. Geotechnique, London, England **17**, 119–135 (1967)
Currin, D.D., Allen, J.J., Little, D.N.: Validation of soil stabilization index system with manual development. Report No. FJSRL-TR-0006, Frank J. Seisler Research Laboratory, United States Air Force Academy, Colorado (1976)

Dermatas, D., Dutko, P., Moon, D.H.: Use of dredged materials as fill in transportation-related projects, p. 71. Center for Environmental Engineering, Stevens Institute of Technology, Hoboken (1999)

Grubb, D.G., Chrysochoou, M., Smith, C.J.: Dredged material stabilization: The role of mellowing on cured properties. In: Khire, M.H., Alshawabkeh, A.N., Reddy, K.N. (eds.) Proceedings of the GeoCongress 2008: Geotechnics of Waste Management and Remediation, pp. 772–780. ASCE, Reston (2008)

Grubb, D.G., Chrysochoou, M., Smith, C.J., Malasavage, N.E.: Stabilized dredged material-I: Parametric study. J. Geotech. Geoenviron. Eng. **136**(8), 1011–1024 (2010a)

Grubb, D.G., Malasavage, N.E., Smith, C.J., Chrysochoou, M.: Stabilized dredged material-II: Geomechanical behaviour. J. Geotech. Geoenviron. Eng. **136**, 1025–1036 (2010b)

Haliburton, T.A.: Guidelines for dewatering/densifying confined dredged material. Synthesis report, Technical Report DS-78-11, US Army Engineer Waterways Experiment Station, Vicksburg, Miss (1978)

Harrington, J., Smith, G.: Guidance on the beneficial use of dredged material in Ireland. Cork Institute of Technology, Ireland (2013)

Hernandez, M.F., Al-Tabbaa, A.: Mechanical properties of laboratory-scale wet soil-mixed organic soils: Laboratory study. Project Rep. Prepared for the National Deep Mixing (NDM) (2005)

Ho, M.H., Chan, C.M.: Some mechanical properties of cement stabilized Malaysian soft clay. World Acad. Sci. Eng. Technol. **74**, 24–31 (2011)

Horpibulsuk, S., Katkan, W., Sirilerdwattana, W., Rachan, R.: Strength development in cement stabilized low plasticity and coarse grained soils. Soils Found. **46**(3), 351–366 (2006)

Janz, M., Johansson, S.E.: The Function of Different Binding Agents in Deep Stabilization. Swedish Deep Stabilization Research Center, p. 9. SGI, Linkoping (2002)

Kamruzzaman, A.H.M., Chew, S.H., Lee, F.H.: Microstructure of cement treated Singapore marine clay. Soils and Foundations, 80–90 (2002)

Miller, G.A., Azad, S.: Influence of soil type on stabilization with cement kiln dust. Constr. Build. Mater. **14**(2), 89–97 (2000)

Mitchell, J.K.: The properties of cement-stabilized soil. In: Proceedings of Residential Workshop on Materials and Methods for Low Cost Road, Rail and Reclamation Works, pp. 365–404. Unisearch Ltd., Leura (1976)

Mitchell, J.K.: Fundamentals of soil behaviour. John Wiley & Sons, Inc., New York (1976)

Mitchell, J.K.: Soil improvement state of the art report. In: 10th International Conference on Soil Mechanics and Foundation Engineering, vol. 4, pp. 509–565 (1981)

Mir, B.A., Mir, F.A.: Applications and geotechnical evaluation of dredged soil obtained from Dal Lake in Srinagar. In: Proceedings of the National Conference on Soils and their applications in Civil Engineering, pp. 26–37 (2004)

Muhunthan, B., Sariosseiri, F.: Interpretation of geotechnical properties of cement treated soils. Research Report WA-RD 715.1. Pullman Washington State University, Washington (2008)

Nash, J.K.T.L., Jardine, F.M., Humphrey, J.D.: The economic and physical feasibility of soil-cement dams. In: Proceedings of the 6th International Conference on Soil Mechanics and Foundation Engineering, vol. 2, pp. 517–521 (1965)

Nagaraj, T.S., Vatasala, A., Srinivasa, M.: Discussion on change in pore size distribution due to consolidation of clays. Geotechnique **40**(2), 303–305 (1990)

Osman, A., Al-Tabbaa, A.: Behaviour of cement stabilized soft clay in aggressive environments-laboratory study. Project Report. Prepared for Federal Highways Administration, Dept. of Eng., Univ. of Cambridge, U.K., p. 203 (2005)

Portland Cement Association (PCA). Soil-cement laboratory handbook. Rep. No. EB052.07S, Skokie, Ill., p. 59 (1995)

Portland Cement Association (PCA). Guide to cement treated base (CTB). Rep. No. EB236.01, Skokie, Ill., p. 26 (2006)
Rahman, M.A.: Effect of cement- rice husk ash mixture on the geotechnical properties of the lateritic soils. J. Soils Found. **27**(2), 61–65 (1987)
Rocha, M., Folque, J., Esteves, V.P.: The application of cement stabilized soils in the construction of earth dams. In: 5th International Conference on Soil Mechanics and Foundation Engineering, Paris, France (1961)
Spigolon, S.J.: Geotechnical site investigation strategy for dredging projects: Report 2, Geotechnical factors in the dredgeability of sediments. Contract Report DRP-93-3, prepared for U.S. Army Corp of Engineers, Washington, DC (1993)
Thompson, M.R.: Shear strength and elastic properties of lime-soil mixtures. Highway Res. Rec. **139**, 1–14 (1966). Washington, D.C.
Uddin, K., Balasubramaniam, A.S., Bergado, D.T.: Engineering behaviour of cement-treated Bangkok soft clay. Geotech. Eng. **2891**, 89–119 (1997)
Winfield, L.E., Lee, C.R.: Dredged material characterization tests for beneficial use suitability. DOER Technical Notes Collection (TN DOER-C2), U.S. Army Engineer Research and Development Center, Vicksburg, MS (1999)
Wissa, A.E.Z., Ladd, C.C., Lambe, T.W.: Effective stress-strength parameters of stabilized soils. In: Proceedings of the 6th International Conference on Soil Mechanics and Foundation Engineering, pp. 412–416 (1965)
Zhang, M.H., Lastra, R., Malhotra, V.M.: Rice husk ash paste and concrete. Cem. Concr. Res. **26** (6), 963–9777 (1996)

Dredged Marine Sediment as Raw Material in Civil Engineering Applications

Marwa Zelleg[1(✉)], Imen Said[1], Amel Missaoui[1], Zoubeir Lafhaj[2], and Essaieb Hamdi[1]

[1] Ecole Nationale d'Ingénieurs de Tunis, LR 14 ES 03 Laboratoire de Recherche d'Ingénierie Géotechnique, Université de Tunis El Manar, Tunis, Tunisie
[2] Ecole Centrale de Lille, Villeneuve d'ASCQ, France

Abstract. Huge amounts of dredged marine sediments are generated periodically in order to maintain waterways and ensure ships navigation. After dredging, marine sediment are either discharged at sea or disposed in lands. Nowadays, dredged sediment reuse in civil engineering applications become more and more a beneficial tool for waste material management. Indeed, various reuse ways have been studied such as road construction, clays bricks, cementitious materials and paving blocks. Sediment characteristics are the primary decision maker in choosing the adequate valorization way. Thus, laboratory identification is an essential step in sediment management. Physical, mechanical, chemical, mineralogical and environmental properties of dredged marine sediments define the possible civil engineering applications according to standards. This works is divided into two main sections. In the first part, a literature review of sediment reuse is presented. The choice of alternative reuse is based on geotechnical, mechanical, chemical, mineralogical and environmental characterization of marine sediment. Possible sediment reuse in civil engineering applications are classified according to material proprieties. The second part, deals with two different applications of marine sediment dredged from two Tunisian harbors. Sediment extracted from the first study area, Zarzis commercial harbor, was reused as a new material for road construction. The mechanical tests of various formulations conducted on different mixtures are compaction and compression. The experimental results showed the feasibility of Zarzis sediment with 3% lime addition in road applications in base-course material. Sediment from Gabes harbor is the second investigated sample. Based on its characteristics, this material could be reused in concrete formulation. Results revealed that the substitution of 20% of sand volume by Gabes channel sediments provided a concrete with higher compression strength than reference one.

Keywords: Marine sediment · Road reuse · Concrete · Tunisian harbors · Paving blocks · Clay bricks

1 Introduction

Sediments are defined as the union of organic or mineral particles transported, precipitated and accumulated at the bottom of lakes, rivers (river sediment) or harbors (marine sediments) (Adams et al. 1992). Sediments in estuaries and harbor areas are

D.N. Singh and A. Galaa (eds.), *Contemporary Issues in Geoenvironmental Engineering*, Sustainable Civil Infrastructures,
DOI 10.1007/978-3-319-61612-4_33

essential components of the marine ecosystem by ensuring the development of various aquatic organisms. These sediments are in constant contact with pollutants coming primarily from industrial waste and boat traffic. The pollutants trapped in sediments can migrate to the fauna, flora and individuals and damage their health (US EPA 1998). It can threaten a major ecological and environmental hazard (Rosenberg 1977; Adams et al. 1992; Hu et al. 2006).

Marine sediments must be periodically dredged in harbors all over the world to ensure sufficient draft for navigation, to maintain harbor waterways and to reduce the risk of pollution of the marine ecosystem. According to statistics (ICES 2011; Le Guyader 2012), Netherland presents the greater dredged sediments quantity with a volume of 122 million m^3 and over the next 10 years, 33.5 million m^3 of river sediments are to be dredged in France.

Despite its importance, dredging does not allow a full resolution of the problem. Dredged sediment management presents an ecological and economic problem because of the large volumes and sediment pollution. Dredging operations managers are required to adapt the least damaging solution for the environment. Thus, conventional and classical solutions are used namely sea discharged and land disposal with or without confinement. Related to its inexpensive cost, the dumping at sea is an option adopted by many dredging managers. Indeed, costs depend only on the distance between the disposal site and the dredging area. This solution is only allowed when heavy metals concentrations and organic pollutants in sediments are below acceptability thresholds fixed by standards. However, hydrodynamic agents in the discharge site may contribute to sediment remobilization towards the harbor. On the other hand, land disposal requires a monitoring and permanent maintenance. Sea dumping and land disposal are constrained by national and international regulations and conventions such as European commission (OSPAR Convention 1992; London Convention (1972, 2013); Bray 2008; European Parliament and Council 2008).

New alternatives are being developed for the treatment and reuse of sediment in order to monetize the costs of dredging and eliminate the migration risk of contaminates to the environment. Researchers have started to study paths to reuse marine sediments into various construction materials such as clay bricks (Hamer et al. 1999; Hamer and Karius 2002; Weng et al. 2003; Lafhaj et al. 2008), road construction (Maher et al. 2006; Kamali et al. 2008; Zentar et al. 2008; Scordia et al. 2008), cementitious materials (Agostini et al. 2007; Wang 2009; Limeira et al. 2011) and even paving blocks (Said et al. 2015).

The choice of the suitable way of reuse depends on sediments proprieties. Thus, physical, chemical and environmental characterization is an essential step for the management of marine sediments. In this work, a methodology of marine sediment reuse based on its characteristics has been established in the first section. Possible sediment reuse in civil engineering applications are classified according to sediment proprieties. The second part deals with two different applications of marine sediment dredged from two Tunisian harbors. First, Sediment extracted from Zarzis (Tunisia) commercial harbor, was reused as a new material for road construction. The experimental results of mechanical tests (compaction test, compression tests) on various formulations have demonstrated the feasibility of the beneficial reuse of Zarzis sediment in road applications. Then, based on its characteristics, sediment from Gabes has been successfully reused in concrete formulation. Laboratory study was conducted to improve this possible valorization path.

2 Literature Review

Beneficial reuse of dredged marine sediments in the construction applications as a fine or coarse aggregate have been studied by several researchers (Said et al. 2015; Kang et al. 2016; Kazi Aoual-Benslafa et al. 2014; Liu et al. 2016; Marmin et al. 2014; Aziz et al. 2016; Cantré et al. 2013). Limeira et al. (2011, 2010) have investigated the suitability of marine sediments in the production of harbor pavement. They have indicated that raw materials used in the production of pavement has achieved compressive strength criteria required by the Spanish Standard. Junakova et al. (2015) focused on reuse of coarse-grained marine sediments (0–4 mm) in place of natural aggregate and fine-grained sediments in place of cement. Results of this study showed that concrete made from coarse-grained marine sediments with 20% substitution is suitable as raw material (Junakova et al. 2015). Liu et al. (2016) have focused on chloride salt content of dredged marine sand for the production of reinforced concrete. Said et al. (2015) aimed to use dredged sediment from Rades Harbor for the fabrication of paving block as a partial substitution of silica sand. They have concluded that studied sediments can be evaluated as fine aggregate in the production of paving blocks and optimum substitution ratio should be 19%.

The choice of reuse methods is directly influenced by sediment's characteristics. For example, clayey sediment can be reused in brick construction. Indeed, many studies showed that river and marine sediments have been used as natural clay replacement when producing bricks that comply with construction standards and legislative environmental requirements (Hamer and Karius 2002; Samara et al. 2009). On the other hand, sandy and silty sediments present an advantageous raw material for road construction. In particular, fine sediments are difficult to reuse in road construction in comparison to sands because of their high water content, the presence of organic matter, their weak mechanical behaviour and the presence of pollutants in some cases (Zentar et al. 2012). In addition, the Atterberg limits of sediments reflect their capability to attract water on their surface. It has been opined by researchers that the fabric structure of sediments, physico-chemistry of clay-water systems and organic carbon content influence the Atterberg limits which in turn reflects the mineralogy of the sediments. Chemical characterization of sediments is important in order to gain knowledge about the presence of minerals and micro-organisms in terms of the metal oxides in the sediments. In order to understand the potential threat that sediments can pose to human health and to ecosystem, environmental characterization is necessary. It has been established that around 43% of heavy metals and organic matter are sorbed on sediments' particle surface due to the high specific surface area (i.e., higher surface adsorption and the ionic attraction).

3 Zarzis Marine Sediment as Raw Material in Road Construction

3.1 Sampling Site Description

The studied sample, designed as ZS, were dredged from Zarzis harbor. The harbor is located in the southeast of Tunisia as shown in Fig. 1. It is situated approximately 30 km south of the Djerba island, 50 km from the Gulf of Gabes and 80 km from the border with Libya. The present harbor provides commercial exchanges of the region dealing mainly on the export of marine salt, representing approximately 80% of the harbor total traffic (CJB and EAM 2006), crude petroleum and white oil products imports. Zarzis harbor has an oil wharf located in the extension of the commercial quay on the west side.

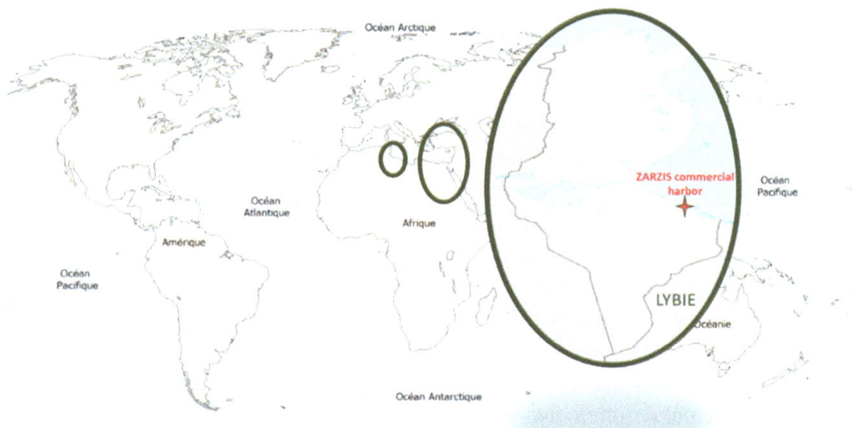

Fig. 1. Zarzis harbor location

3.2 Sediment Characterization

Table 1 resumes ZS characterization results. The studied sediments are mainly silty sands with an average of 75% sand, 18% silt and 7% clay fraction. Results show the presence of 15% organic matter that indicates that ZS are moderately organic. The methylene blue value, which represents the adsorption capacity of methylene blue on the particle surfaces, is low and coincides with the grain size distribution. However, the plasticity of studied dredged sediment is high.

For future reuse, it is necessary to investigate heavy metals concentration in sediment leaches. Indeed, the solid fraction test assess the total concentration of various trace elements considered as contaminants. However, sediments do not present environmental issues if heavy metals are trapped and not mobilized under runoff or rainfall infiltration effect. For this purpose, the pollution degree of ZS was evaluated by environmental analyzes using ICP-OES to sediment leachates collected by the leaching test according to English standard EN 12457-3.

Table 1. Zarzis sediment characterization

Water content (%)	209.6
Sand percentage (%)	75
Silt percentage (%)	18
Clay percentage (%)	7
Methylene blue	1.2
WL (%)	83.5
WP (%)	66.13
IP	17.37
Organic matter (%)	15
Particle density (g/cm^3)	1.7

Results indicate that heavy metals concentrations in the leachate of ZS were under ICP apparatus detection limit. Therefore, these concentrations were below the threshold limit defining the acceptability of sea discharge. Thus, ZS may be incorporated into the formulation of a new material without the need to go through a treatment process.

3.3 Mix Design

An experimental study was performed on raw mixed sediments with percentages of cement or lime to evaluate the potential of their reuse in road construction. The percentages of lime or cement mixed with fine sediments were fixed to 3% of the dry mass of raw sediments. Three formulations were tested in this work, as depicted in Table 2.

Table 2. Formulation of Zarzis sediment

Mixture		% Lime	% Cement
ZS0	Raw sediment	0	0
ZS1	Sediment + 3% lime	3	0
ZS2	Sediment + 3% cement	0	3

3.4 Testing Methods

Proctor test, I-CBR (Immediate California Bearing Ratio) test and Unconfined compression test are the most recommended to determine the feasibility of a specific material to be used in road engineering. The ICBR index defines the capacity of a material to support the circulation of building machines, and measures the ratio of force required for a circular piston to penetrate into a granular medium in a CBR mould at the speed of 1.27 $\pm$ 0.1 mm/min. The prescribed I-CBR values for different road layers are specified in French standard NF P 98-115: for a sub-base material – prescribed value of 35 with a minimum value of 25; for a base-course material – prescribed value of 45 with a minimum value of 35.

Unconfined compression tests were performed on monolithic samples, which were prepared at the optimum moisture content and at the optimum dry density as defined by the modified Proctor tests. Before the mechanical tests, the samples with a diameter of 50 mm and a height of 100 mm were cured under sealed conditions in a specified room with a constant temperature of 20 °C.

3.5 Results and Analyses

ICBR values were evaluated at the optimum moisture content and at the optimum dry density as defined by the modified Proctor. Proctor results are synthetized in Table 3.

Figure 2 depicts ICBR results for tested formulations. Mixture ZS0 and ZS2 show a low bearing capacity comparing to ZS1. Bearing capacity (ICBR) for ZS0 and ZS2 is acceptable for its use in sub-base material. On the other hand, mixture ZS1 present a high CBR index. Thus, mixture ZS1 using 3% of lime could be employed as base-course material.

Table 3. Proctor results

Mixture	Optimum water content (%)	Optimum dry density (T/m^3)
ZS0	22	0.9
ZS1	27	1
ZS2	24	0.94

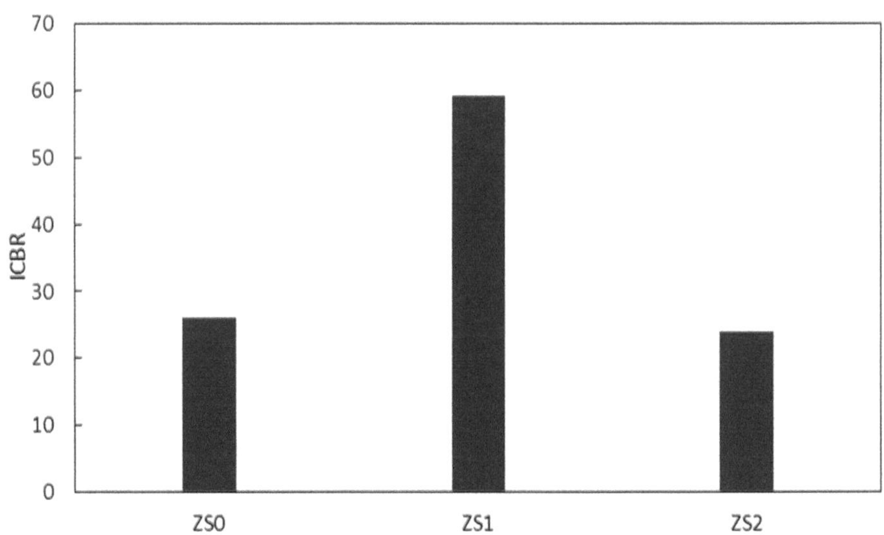

Fig. 2. ICBR results

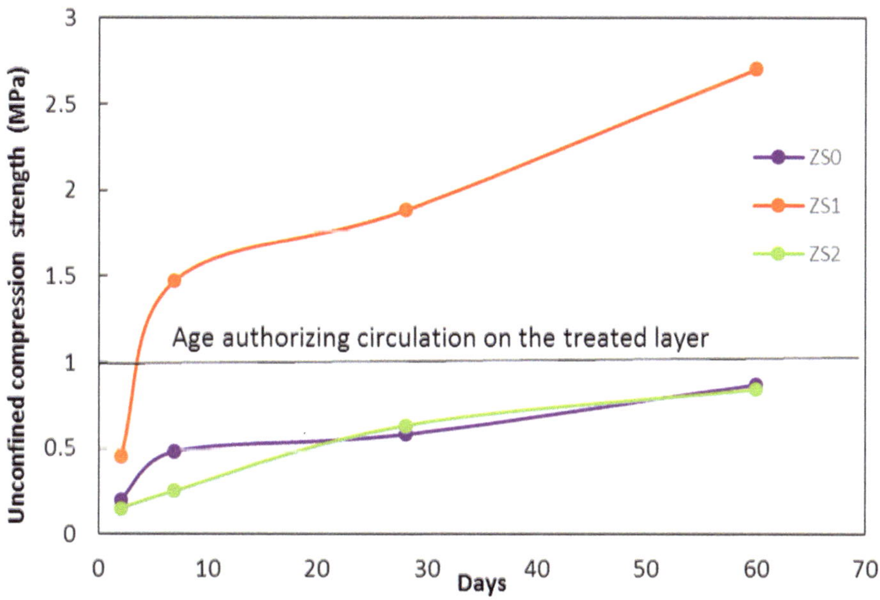

Fig. 3. Effect of binder content and curing time on unconfined

Figure 3 shows the development of unconfined compressive strength of cured samples at different binder contents in relation to curing time. It is clear that depending on time, the assessment of achievement mechanical performance differs from one formulation to another. The test for determining the age authorizing the circulation is related to the mechanical performance in compression of the material studied. A road layer is useful when its unconfined compressive strength is greater than 1 MPa. From Fig. 3, this value was not reached for mixture ZS0 and ZS2 and was reached on the third day for the mixture ZS1.

Results of ZS formulations are compared to Wang's studies (Wang et al. 2012). Based on the mineralogical composition and physical characteristics of Dunkirk sediments, Wang assessed the effects of cement and lime through, modified Proctor compaction, and unconfined compressive strength tests. The potential of sediments solidified with cement or lime for road construction is evaluated through a proposed methodology from I-CBR value. In his research, Wang characterized Dunkirk sediment as sandy soils. He concluded that, from the point of view of mechanics and applicability in road construction, 6% cement is an economic and reasonable amount to improve the engineering properties of sediments.

In other study, Tran (Trant 2009) tried to reuse silty clay sediment on road application. This study necessitated the use of corrective sands in formulations in order to obtain better results in terms of compacting and compression.

4 Gabes Marine Sediment as Raw Material in Concrete

4.1 Sampling Site Description and Sediment Characterization

Studied sediment were collected from Gabes commercial harbor. The harbor is located in the middle of the gulf of Gabes, in the south east of Tunisia as showed in Fig. 4. This harbor is located next to the most important chemical industry in Tunisia. Since 1974, it serves as a commercial and industrial purpose. Its activity is characterized by an industrial vocation which mainly assures transit of chemicals for neighborhoods industries. Commercial traffic of Gabes harbor consists primarily on the importation of sulfur and ammonia and the exportation of phosphoric acid and phosphate fertilizer.

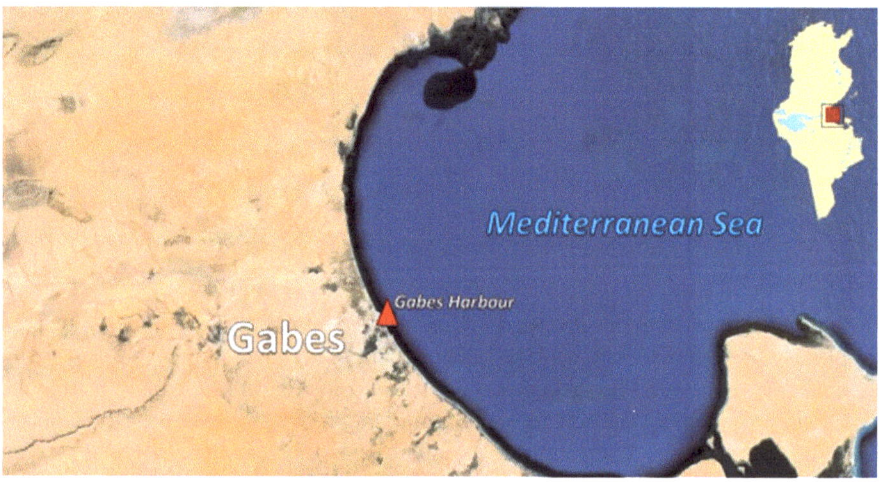

Fig. 4. Map of the study area in Gabes, Tunisia (Google earth 2015)

4.2 Sediment Characterization

Table 4 illustrates the physical characteristics of Gabes channel sediment. Gabes sediments presented high water content equal to 108%. According to the grain size distribution, Gabes channel sediments were classified as silty sands. The negligible fraction of clay particles may be beneficial for the sand substitution in concrete. According to plasticity index, Gabes channel sediments were considered as plastic silty soils. Gabes channel sediments displayed a specific gravity between 2.6 g/cm^3 and 2.8 g/cm^3, which is a property of most inorganic soils. The latter result was confirmed by the organic matter content in Gabes Channel sediments that classified them as weakly organic soils. Gabes harbor sediments displayed high calcium carbonate content that categorized them as marl soils. Thus, according to physical properties, Gabes channel sediments present interesting results as sand substitution in concrete.

Table 4. Physical properties of Gabes Channel

Water content (%)	108.9	High water content
Gravel fraction (%)	1.6	Silty sand
Sand fraction (%)	58	
Silt fraction (%)	36.8	
Clay fraction (%)	3.6	
Methylene blue	1.48	Silty soil
WL (%)	67.6	Plastic silty soil
WP (%)	42.15	
Organic matter (%)	5.2	Weakly organic soil
Calcium carbonate content (%)	40.29	Marl soil
Particle density (g/cm^3)	2.61	–

4.3 Mix Design

This work investigates the feasibility of Gabes marine sediments reuse in concrete. The aim of the study was to compare fresh and hardened properties of reference concrete to seven designed mixtures of Gabes sediment concrete (CGS). CGS samples were carried out by sand volume substitution with pretreated Gabes channel sediments. Reference concrete formulation was taken from a Tunisian concrete plant.

4.4 Results and Analyses

In order to evaluate mixture strength, the compressive strength of concrete specimens was investigated in accordance with the French standard NF EN 12390-3. Tests were performed in triplicate. At 28 days, CGS20, with 20% of sediment, had the higher compressive strength (32 MPa).

An optimal replacement of 20% can be deduced and above this substitution rati compressive strength of CGS decreases. Comparable tendency was observed in the study of Agostini et al. (2007), where treated sediments were reused in cementitious materials. In his study, the author tested three mortars for which a given sand volume was replaced by the same sediment volume (33%, 66% and 100%). An optimal substitution ratio was observed around 33%.

5 Conclusion

This study investigated the reuse of marine sediment as raw material in civil engineering applications. It showed the importance of sediment characteristics in choosing the adequate reuse way. This research also treated two different reuse ways: Road construction and concrete. Sediment extracted from the first study area, Zarzis commercial harbor, was reused as a new material for road construction. Based on characterization results, Zarzis sediment was tested as raw material for roads. An experimental study was performed on raw mixed sediments with percentages of cement (3%) or lime (3%) to evaluate their potential reuse in this field. The mechanical tests conducted on these mixtures (CBR value equal to 59) showed the feasibility of Zarzis sediment with 3% lime addition application in road applications in base-course material.

Sediment from Gabes harbor was the second investigated sample. According to physical characterization, Gabes channel sediments were classified as silty sands. The particle density and organic matter content classified it as weakly organic soils with only 5.2% organic content. In addition, Gabes harbor sediments displayed high calcium carbonate content (40%) that categorized them as marl soils. This could explain the recorded compressive strength of designed concrete. Results revealed that the substitution of 20% of sand volume by Gabes channel sediments provided a concrete with higher compression strength than reference one.

Acknowledgments. This study was funded by the partnership Hubert Cruien "Utique" of the French Ministry of Foreign Affairs and the Tunisian Ministry of Higher Education and Scientific Research in Tunisia (Project No. 14 G 1116). The authors express their acknowledgements to the Tunisian Merchant Marine and Ports Office and the Tunis International Center for Environmental Technologies.

References

Adams, W.J., Kimerle, R.A., Barnett Jr., J.W.: Sediment quality and aquatic life assessment. Environ. Sci. Technol. **26**(10), 1865–1875 (1992)

Agostini, F., Skoczylas, F., Lafhaj, Z.: About a possible valorisation in cementitious materials of polluted sediments after treatment. Cement Concr. Compos. **29**(4), 270–278 (2007)

Aziz, M., Ghorbel, E., Bibi, M.: The possibility of using dredging sludge in manufacturing cements: optimization of heat treatment cycle and ratio replacement. Constr. Build. Mater. **106**, 330–341 (2016)

Cantré, S., Saathoff, F.: Investigation of dredged materials in combination with geosynthetics used in dike construction. Procedia Eng. **57**, 213–221 (2013)

CJB Environnement INC, EAM Environmental Assessment and Management: Etude d'identification des sites de rejet des matériaux de dragage dans les ports de commerce Tunisiens, OMMP Office de la Marine Marchande et des ports (2006)

European Parliament and Council: Marine Strategy Framework Directive 2008/56/EC (2008)

Junakova, N., Junak, J., Balintova, M.: Reservoir sediment as a secondary raw material in concrete production. Clean Technol. Environ. Policy **17**, 1161–1169 (2015)

Hamer, K., Waschkowitz, C., Isenbeck-Schröter, M., Schulz, H.D.: Harbour sediments for brick production. In: Ressourcen-Umwelt-Management, Schriftenreihe der Gesellschaft für Umwelt Geowissenschaften (GUG), Köln, pp. 223–240 (1999)

Hamer, K., Karius, V.: Brick production with dredged harbour sediments. An industrial-scale experiment. Waste Manage. **22**, 521–530 (2002)

Hu, Z., Saman, W.R.G., Ronald, R.N., Wu, D., Zhang, D., Masatoshi, M., Kong, H.: Removal of PCDD/Fs and PCBs from sediment by oxygen free Pyrolysis. J. Environ. Sci. **18**(5), 989–994 (2006)

Helcom Helsinki Commission: Baltic Marine Environment Protection Commission. Convention on the Protection of the Marine Environment of the Baltic Sea Area (2008)

ICES: Report of the working group of the effects of extraction of marine sediments on the marine ecosystem. In: International Council for the Explotation of the Sea. - Delft, the Netherlands, 93 pages (2011)

Kazi Aoual-Benslafa, F., Kerdal, D., Mekerta, B., Semcha, A.: The use of dredged sediments as sand in the mortars for tunnel lining and for environmental protection, p. 39. Arab. J. Sci, Eng (2014)

Kamali, S., Bernard, F., Abriak, N.E., Degrugilliers, P.: Marine dredged sediments as new materials resource for road construction. Waste Manage. **28**, 919–928 (2008)

Lafhaj, Z., Samara, M., Agostini, F., Boucard, L., Skoczylas, F., Depelsenaire, G.: Polluted river sediments from the north region of France: treatment with Novosol® process and valorization in clay bricks. Constr. Build. Mater. **22**(5), 755–762 (2008)

Le Guyader, C.: Enquête «Dragage 2009» Synthèse des données. Ministère de l'écologie, du développement durable, des transports et du logement. Centre d'Etudes Techniques Maritimes et fluviales, pp. 1–39 (2012)

London Convention: Specific guidelines for the assessment of dredged material, Convention on the Prevention of Marine Pollution by Dumping of Wastes and Other (2013)

London Convention: Convention on the prevention of marine pollution by dumping of wastes and other matter (1972)

Limeira, J., Etxeberria, M., Agulló, L., Molina, D.: Mechanical and durability properties of concrete made with dredged marine sand. Constr. Build. Mater. **25**, 4165–4174 (2011)

Limeira, J., Agullo, L., Etxeberria, M.: Dredged marine sand in concrete: an experimental section of a harbor pavement. Constr. Build. Mater. **24**, 863–870 (2010)

Liu, W., Cui, H., Dong, Z., Xing, F., Zhang, H., Lo, T.Y.: Carbonation of concrete made with dredged marine sand and its effect on chloride binding. Constr. Build. Mater. **120**, 1–9 (2016)

Maher, A., Douglas, W.S., Jafari, F.: Field placement and evaluation of stabilized dredged material (SDM) from the New York/New Jersey Harbor. Mar. Georesour. Geotechnol. **24**, 251–263 (2006)

Marmin, S., Dauvin, J., Lesueur, P.: Collaborative approach for the management of harbour-dredged sediment in the Bay of Seine (France). Ocean Coast. Manage. **102**, 328–339 (2014)

OSPAR Convention: Convention for the protection of the marine environment of the North-East Atlantic (1992)

US EPA, ACOE: Evaluation of material proposed for discharge to waters of the US – Testing Manual (Inland Testing Manual). EPA-823-B-98-004, Office of Water and the Department of the Army US Army Corps of Engineers, Washington DC (1998)

Rosenberg, R.: Effects of dredging operations on estuarine benthic macrofauna. Mar. Pollut. Bull. **8**(5), 102–104 (1977)

Samara, M., Zoubeir, L., Christophe, C.: Valorization of stabilized river sediments in fired clay bricks: factory scale experiment. J. Hazard. Mater. **163**, 701–710 (2009)

Trant, N.T.: Valorisation de sédiments marins et fluviaux en technique routière. Thèse de Doctorat, Ecole Nationale Supérieure des Mines de Douai, 187 p. (2009)

Kang, G., Tsuchida, T., Athapaththu, A.M.R.G.: Engineering behavior of cement treated marine dredged clay during early and later stages of curing. Eng. Geol. **209**, 163–174 (2016)

Said, I., Missaoui, A., Lafhaj, Z.: Reuse of Tunisian marine sediments in paving blocks: factory scale experiment. J. Clean. Prod. **102**, 66–77 (2015)

Scordia, P.Y., Lafhaj, Z., Skoczylas, F., Mongouin, T.: Characterization and valorization of polluted and treated river sediments in road. Eur. J. Environ. Civil Eng. **7**(4), 451–469 (2008)

Wang, D., Abriak, N., Zentar, R., Xu, W.: Solidification/stabilization of dredged marine sediments for road construction. Environ. Technol. **33**(1), 95–101 (2012)

Wang, H.Y.: Durability of self-consolidating lightweight aggregate concrete using dredged silt. Constr. Build. Mater. **23**, 2332–2337 (2009)

Weng, C.H., Lin, D.F., Chiang, P.C.: Utilization of sludge as brick materials. Adv. Environ. Res. **7**, 679–685 (2003)

Zentar, R., Wang, D., Abriak, N.E., Benzerzour, M., Chen, W.: Utilization of siliceous–aluminous fly ash and cement for solidification of marine sediments. Const. Buil. Mater. **35**, 856–863 (2012)

Zentar, R., Dubois, V., Abriak, N.E.: Mechanical behaviour and environmental impacts of a test road built with marine dredged sediment. Resour. Conserv. Recycl. **52**, 947–954 (2008)

Author Index

D.N. Singh and A. Galaa (eds.), *Contemporary Issues in Geoenvironmental Engineering*, Sustainable Civil Infrastructures,
DOI 10.1007/978-3-319-61612-4

MIX
Papier aus verantwortungsvollen Quellen
Paper from responsible sources
FSC® C105338

If you have any concerns about our products,
you can contact us on
ProductSafety@springernature.com

In case Publisher is established outside the EU,
the EU authorized representative is:
Springer Nature Customer Service Center GmbH
Europaplatz 3, 69115 Heidelberg, Germany

Printed by Libri Plureos GmbH
in Hamburg, Germany